Quantum
Computing and Information
A Scaffolding Approach

Peter Y. Lee
Huiwen Ji
Ran Cheng

Copyright © 2024 Polaris QCI Publishing

This work is subject to copyright. All rights are reserved. No part of this publication may be reproduced, reprinted, translated, uploaded to electronic storage systems, or transmitted, in any form or by any means, electronic, mechanical, or otherwise, without the prior written permission of the publisher. Circumventing digital rights management, unauthorized distribution, sharing, or sale of this publication is a violation of law and is subject to criminal prosecution.

ISBN 978-1-961-88000-9 (ebook)

ISBN 978-1-961-88001-6 (paperback, b/w)

ISBN 978-1-961-88002-3 (hardcover, b/w)

ISBN 978-1-961-88003-0 (paperback, color)

ISBN 978-1-961-88004-7 (hardcover, color)

Library of Congress Control Number (LCCN) 2024901045

First edition, March 2024

This document is typeset using LaTeX.

Quantum circuit drawings are created using the yquant package from https://github.com/projekter/yquant.

Publisher website: https://polarisqci.com

About This Book

In a world where quantum computing stands at the crossroads of computation and quantum mechanics, *Quantum Computing and Information: A Scaffolding Approach* offers a meticulously designed pathway for mastering this transformative technology. As part of an educational series, this book serves as a comprehensive resource for beginning graduate students, senior undergraduates, and anyone invested in understanding the quantum computational landscape.

The book follows a "scaffolding approach," inspired by pedagogical theories from Lev Vygotsky and Jerome Bruner, guiding readers through complex subject matter without overwhelming them. Through the gradual introduction of concepts, layered reinforcement, and practical exercises, the book facilitates deep learning. Employing ample illustrations, tables, and special boxes for highlights and key concepts, the text makes quantum computing accessible without diluting its intricacies.

Four major sections unfold a comprehensive learning journey: from understanding the basics of quantum systems, through the manipulation of these systems with quantum gates, to the fascinating phenomenon of entanglement, and finally, to essential quantum algorithms, error correction techniques, and quantum information theory.

Whether you are a novice to quantum computing or have some experience in the field, this book offers a structured and incremental approach to gaining a robust understanding. Get ready to embark on an enlightening voyage through the captivating realm of quantum computing.

About the Authors

Dr. Peter Y. Lee holds a Ph.D. in Electrical Engineering from Princeton University. His research at Princeton focused on quantum nanostructures, the fractional quantum Hall effect, and Wigner crystals. Following his academic tenure, he joined Bell Labs, making significant contributions to the fields of photonics and optical communications and securing over 20 patents. Dr. Lee's multifaceted expertise extends to educational settings; he has a rich history of teaching, academic program oversight, and computer programming. Dr. Lee is currently on the faculty of Fei Tian College, New York.

Dr. Huiwen Ji earned her Ph.D. in Chemistry at Princeton University, where she specialized in the solid-state chemistry of binary and ternary chalcogenides, a field intricately tied to quantum properties and topological surface states. This rigorous academic background laid the foundation for her subsequent research endeavors, blending quantum physics, materials chemistry, and structure-property relationships in solid-state functional materials. In her roles as a Postdoctoral Scholar at the University of California, Berkeley, and a Research Scientist at Lawrence Berkeley National Lab, she further delved into the nuances of advanced material science. Recognized for her significant contributions, Dr. Ji has received accolades such as the ACS PRF Doctoral New Investigator Award and the NSF CAREER Award. She currently serves as a faculty member at the University of Utah.

Dr. Ran Cheng earned his Ph.D. in Physics from the University of Texas at Austin with a focus on theoretical condensed matter physics. After receiving his doctorate, he became a postdoctoral researcher at Carnegie Mellon University to further his inquiry into magnetic materials. He is now a faculty member at the University of California, Riverside, where he explores three core research domains: spintronics, topological materials, and quantum magnets. A recognized pioneer in the burgeoning field of antiferromagnetic spintronics, Dr. Cheng was honored with the NSF CAREER award and the DoD MURI award alongside a cadre of distinguished physicists.

Contents

Level Indicators

Unmarked content: Foundational, appropriate for all readers, including newcomers to quantum computing.

✳: Typically an initial conceptual overview accessible to general readers, followed by math-intensive segments for a senior undergraduate or early graduate-level audience with a background in linear algebra. Assumes familiarity with unmarked content.

✶: Advanced exploration, extending beyond the ✳-level topics.

Preface ix

Reviews xiii

About Quantum Computing and Information xv

I Qubits & Qudits: Foundations

1 Quantum Mechanics Through Photons 3
1.1 Introducing Quantum Mechanics 4
1.2 Understanding Photons 7
1.3 The Quantum State Postulate 10
1.4 The Quantum Observable Postulate 17
1.5 The Quantum Measurement Postulate 24
1.6 The Uncertainty Principle 33
1.7 ✶ Further Readings on Quantum Mechanics 36
1.8 Topic Reviews 37
1.9 Summary and Conclusions 40
    Problem Set 1 41

2 Fundamentals of Spin Systems 43
2.1 Spin, Angular Momentum, and Magnetic Moment 44
2.2 Spin-1/2 States and Pauli Matrices 45

2.3	∗General Spin State Representation		47
2.4	The Bloch Sphere		50
2.5	Spin Measurement		51
2.6	Summary and Conclusions		55
	Problem Set 2		56

3 A Framework for Qubits and Qudits — 59

3.1	Physical Qubit Systems	60
3.2	Qubit and Qudit States	60
3.3	Change of Basis	65
3.4	∗General Formulation of Quantum Measurement	71
3.5	✸Application to Quantum State Tomography	80
3.6	Summary and Conclusions	81
	Problem Set 3	83

4 Dynamics of Quantum Systems — 85

4.1	The Evolution Postulate of Quantum Mechanics	86
4.2	∗The Schrödinger Equation	87
4.3	Stationary Nature of Energy Eigenstates	91
4.4	Universal Quantum Computing and Annealing	94
4.5	∗Larmor Precession and Rabi Oscillations	96
4.6	✸Further Exploration	102
4.7	✸Deferred Proofs	104
4.8	Summary and Conclusions	106
	Problem Set 4	107

II Quantum Gates & Elementary Circuits

5 Single-Qubit Quantum Gates — 111

5.1	Quantum Versus Classical Logic Gates	112
5.2	Common Single-Qubit Gates	115
5.3	From Gate Sequences to Quantum Circuits	123
5.4	Quantum Random Number Generator	126
5.5	The BB84 Quantum Key Distribution (QKD) Protocol	128
5.6	The Quantum Coin Game	135
5.7	∗The No-Cloning Theorem: Proof Outline	138
5.8	Summary and Conclusions	139
	Problem Set 5	140

6 Multi-Qubit Systems — 143

6.1	Systems of Two Qubits	144

6.2	∗Measurements of Two-Qubit Systems	148
6.3	Multi-Qubit System States	156
6.4	∗Measurements of Multi-Qubit Systems	162
6.5	∗Time Evolution of Multi-Qubit States	167
6.6	Summary and Conclusions	171
	Problem Set 6	172

7 Multi-Qubit Quantum Gates ... 175

7.1	Common Multi-Qubit Gates	176
7.2	Universal Sets of Qubit Gates	185
7.3	Boolean Representation of Quantum Gates	187
7.4	Equivalent Gate Sequences	191
7.5	∗Exploratory Topics	200
7.6	Summary and Conclusions	202
	Problem Set 7	203

III Quantum Entanglement

8 Bell States ... 207

8.1	Maximally Entangled States	208
8.2	∗Bell Basis	209
8.3	Bell State Creation	211
8.4	Bell Measurement	214
8.5	Bell State Conversion	215
8.6	∗Deferred Proofs	217
8.7	∗Generalization: GHZ States	218
8.8	Summary and Conclusions	219
	Problem Set 8	220

9 Entanglement and Bell Inequalities ... 223

9.1	Classical Correlation vs. Quantum Entanglement	224
9.2	The EPR Paradox	227
9.3	Bell Inequalities	229
9.4	Bell-CHSH Inequality with Classical Correlation	231
9.5	Bell-CHSH Inequality with Quantum Entanglement	233
9.6	Experimental Verification	236
9.7	The No-Communication Theorem	240
9.8	∗Derivation of Bell-CHSH Quantity for Bell States	241
9.9	∗Further Exploration	246
9.10	Summary and Conclusions	247

Problem Set 9 248

10 Key Applications of Entanglement 251
10.1 Review of Preliminaries 252
10.2 Superdense Coding 254
10.3 Quantum Teleportation 257
10.4 Entanglement Swapping 266
10.5 Quantum Gate Teleportation 270
10.6 E91 Quantum Key Distribution Protocol 283
10.7 Summary and Conclusions 288
      Problem Set 10 289

IV Quantum Computation & Information

11 Quantum Algorithms: A Sampler 295
11.1 The Deutsch-Jozsa Algorithm 297
11.2 QUBO, VQE, QAOA, and AQC 305
11.3 Quantum Bomb and Quantum Money 322
11.4 Summary and Conclusions 335
      Problem Set 11 336

12 Quantum Error Correction: A Primer 341
12.1 Preliminary Concepts 342
12.2 *Mixed States, Density Operators, and CPTP Maps 346
12.3 Error Mechanisms in Quantum Computing 369
12.4 Introduction to Error Correction Codes 384
12.5 Summary and Conclusions 399
      Problem Set 12 400

13 Fundamentals of Quantum Information 405
13.1 Quantum Probability Essentials 408
13.2 Quantum Entropy and Information 426
13.3 *Core Theorems in Quantum Information 437
13.4 Further Exploration 444
13.5 Summary and Conclusions 444
      Problem Set 13 445

V Supporting Materials

Essential Mathematics: Quick References **453**
A   Complex Numbers 453
B   Trigonometry 454
C   Linear Algebra for QCI 456
D   Pauli Matrices 463

Bibliography **467**

List of Figures **477**

List of Tables **479**

Index **481**

Journey Forward **490**

Preface

This book serves as a part of a series of textbooks initially crafted for the Master of Science in Quantum Computing Program at Fei Tian College, Middletown, New York. The series aspires to offer a pedagogically sound, systematic approach to teaching and learning quantum computing. It includes the following titles:

- Mathematical Foundations of Quantum Computing
- Quantum Computing and Information: A Scaffolding Approach (current book)
- Quantum Algorithms and Applications: A Scaffolding Approach

While each book functions as a standalone guide to its respective topic, collectively they furnish a comprehensive understanding of quantum computing.

Designed for beginning graduate students and senior undergraduates, this book also includes markers to aid both entry-level and more advanced readers.

Quantum Computing and Information (QCI) is a complex discipline, comprising an intricate web of knowledge that spans advanced mathematics, quantum mechanics, and sophisticated algorithms. Navigating this multidimensional landscape requires an approach to learning and teaching that acknowledges the inherently linear nature of reading and lectures, while also addressing the multi-faceted structure of the subject matter.

Effective teaching—and, by extension, effective learning—is not merely the transmission of information but a dynamic process of constructing understanding within a cognitive framework. This involves connecting new knowledge with existing cognitive structures, akin to weaving new threads into an ever-expanding web of understanding in the brain.

The scaffolding approach, central to this text, is inspired by educational theories such as Lev Vygotsky's Zone of Proximal Development and Jerome Bruner's instructional strategies, refined through the authors' extensive experience in academia. It recognizes that learning in such a complex field as QCI involves navigating a path that gradually ascends from foundational concepts to advanced applications.

This book operationalizes the scaffolding approach through several key strategies:

- Progressive Introduction of Concepts: The content is structured to ensure learners are adequately prepared for each new concept, mirroring the idea of assembling a multidimensional puzzle piece by piece.
- Grounding in Simple Examples: Initially, straightforward examples and intuitive explanations are used to anchor new topics, which gradually evolve into more abstract and generalized concepts.

- Spiral Learning: Key ideas are revisited from multiple perspectives and in varying contexts, embodying the spiral nature of learning where concepts are deepened and expanded upon with each iteration.
- Active Engagement: Exercises and problems are integrated to promote deep learning through practical application, allowing learners to apply and test their understanding in meaningful ways.
- Cognitive Load Management: The book is designed with clarity in mind, featuring abundant illustrations and tables to complement the text, thereby reducing the cognitive load and demystifying complex mathematical and theoretical constructs.

By incorporating these elements, the scaffolding approach in this text not only imparts knowledge but also cultivates the skills necessary to navigate the multifaceted and interlinked domains of QCI. It reflects a sophisticated blend of the art and science of teaching, tailored to the unique challenges of this cutting-edge field.

Parts of the Book

The book is divided into four primary sections:

- Part I lays the groundwork for understanding quantum systems, commencing with simpler entities like photons and advancing to more complex systems such as qubits and qudits.
- Part II transitions from understanding individual quantum systems to manipulating them using quantum gates, acting as a bridge to more advanced quantum operations.
- Part III delves into the fascinating and essential quantum phenomenon of entanglement, starting with simple Bell States and expanding into its broader theory and applications.
- Part IV explores critical components of quantum computing and information: it offers a sampler of quantum algorithms, introduces the fundamentals of error correction, and delves into the principles of quantum information, forming a comprehensive overview of quantum computation and information.

Navigation Aids

To assist readers in navigating the book, the layout incorporates the following features:

> **Concept Box**
>
> This box is reserved for important concepts, postulates, and theorems.

> **Highlight Box**
>
> This box highlights corollaries, summaries, implications, and other key points.

Exercise 0.1 The exercises interspersed throughout the text serve to hone your skills and reinforce your understanding of the material discussed.

Each chapter concludes with a comprehensive Problem Set designed for thorough mastery. These problems are generally more challenging than the in-text exercises, providing an opportunity for deeper engagement with the content.

This indicator is used for tips, alerts, connections between concepts, and pieces of advice.

Info Box

This box provides supplementary context, associated concepts, and additional information.

Level Indicators

In order to accommodate learners with different academic backgrounds, some sections are marked by level indicators.

Unmarked content serves as the foundational basis for this subject and is appropriate for all readers, including those new to quantum computing.

Items marked with ∗ typically begin with an introductory conceptual overview suitable for general readers, followed by math-intensive content designed for a senior undergraduate or early graduate-level audience with a background in linear algebra. Both components of ∗ presuppose familiarity with unmarked foundational content.

Items marked with ✳ target further exploration and delve into topics that extend beyond the material covered at the ∗ level.

Acknowledgements

To Ms. Elsie He and Ms. Jing Hunter, we express our deep gratitude for your artistic talents in the design of our book cover. Your perceptive interpretation of the subject matter and steadfast commitment to crafting a cover that is both visually arresting and thematically profound are truly valued.

Our heartfelt gratitude goes to Ms. Nathalie Chiao and Ms. Jenny Chang. Your editorial expertise has significantly enhanced the quality, readability, and coherence of our manuscript. Your meticulous attention to detail and commitment to clear communication have been invaluable.

To our esteemed colleagues and students, Mr. Yiyong Huang, Dr. Jason Wang, Dr. Joseph Zhao, and Dr. James Yu, we are immensely grateful. Your intellectual curiosity, unwavering support, and insightful questions have been pivotal in motivating us and shaping the approach of this book. Your feedback has been crucial in ensuring we provide practical understanding alongside a clear educational progression.

We also wish to express our sincerest thanks to our knowledgeable reviewers. Your expert reviews and constructive suggestions have played a key role in enhancing the rigor and accessibility of this text, enabling us to present complex concepts in an engaging and informative manner.

- Robert J. Cava, Professor of Chemistry, Princeton Quantum Initiative, Princeton University
- Andrew Kent, Professor of Physics, The Center for Quantum Phenomena, New York University
- Shuwang Li, Professor of Applied Mathematics, Illinois Institute of Technology
- Stephen A. Lyon, Professor of Electrical and Computer Engineering, Princeton Quantum Initiative, Princeton University
- Leonid Pryadko, Professor of Physics, University of California at Riverside

Finally, we are indebted to the wider quantum computing community, upon whose advancements we build. Our goal is to distill and share this knowledge, making it accessible to students and inspiring further exploration into this fascinating field. Special thanks to the following influential experts:

- Scott Aaronson, Professor of Theoretical Computer Science, University of Texas at Austin; director of its Quantum Information Center
- John Preskill, Feynman Professor of Theoretical Physics, California Institute of Technology
- Ryan O'Donnell, Professor of Computer Science, Carnegie Mellon University
- John Watrous, Technical Director for Education, IBM Quantum

Dedication

This book is dedicated to all the inquisitive minds and relentless spirits who believe in the power of learning and the unbounded possibilities of human intellect.

Reviews

“ Quantum Computing is definitely going to impact our future lives. This book adheres to a pedagogical methodology that balances theoretical rigor with accessibility. The scaffolding approach that the authors use guides the reader through the learning journey. This makes the book not only academically rigorous but also effective as a teaching tool.”

— Robert J. Cava, Professor of Chemistry
Princeton Quantum Initiative, Princeton University

“ This impressive book covers the burgeoning field of quantum information, bridging the fundamentals of quantum mechanics and its present and future applications in secure communication and quantum computing. The author's approach is rigorous – including all the necessary linear algebra – while the book is highly readable and accessible. It will benefit a wide range of audiences with different backgrounds, from undergraduate students learning quantum mechanics to experts who want a deep understanding of quantum information protocols.”

— Andrew Kent, Professor of Physics
The Center for Quantum Phenomena, New York University

“ Quantum computing is poised to be one of the first major technological developments of the 21st century. This book assumes a student has a solid background in quantum mechanics, which allows it to introduce the broad field of quantum information and computing in depth. At the same time it covers important topics from multiple angles, which is invaluable in guiding students who are first learning the material. It will serve well both for teaching and as a reference.”

— Stephen Lyon, Professor of Electrical and Computer Engineering
Princeton Quantum Initiative, Princeton University

" This textbook is elegantly crafted, utilizing a unique 'scaffolding approach' to render complex topics in quantum computing easily comprehensible for newcomers to the field. It is invaluable for both educators and students of quantum computing.

Quantum Computing and Information: A Scaffolding Approach offers a comprehensive and insightful introduction to quantum computing. Targeted at upper-division undergraduates with a foundational grasp of linear algebra or first-year graduates, it serves as an excellent resource for a one-semester course. The authors employ a lucid and engaging style, ensuring that complex topics are accessible. Their original illustrations and tables, meticulously designed to complement the text, enhance comprehension. Additionally, the textbook provides both concise and detailed examples, aiding entry-level students in grasping fundamental concepts. A well-considered balance between straightforward exercises (to consolidate specific knowledge) and problems (to integrate a broader understanding) is maintained."

— *Shuwang Li, Professor of Applied Mathematics*
Illinois Institute of Technology

About Quantum Computing and Information

Introduction

Quantum Computing and Information (QCI) herald a paradigm shift in computational and information sciences, leveraging quantum mechanical principles to tackle complex problems across diverse fields such as cryptography, finance, and material science. Quantum computing, a cornerstone of QCI, focuses on utilizing quantum mechanics for computation, encompassing the development of algorithms, processors, and software. Extending beyond computing, quantum information pertains to the processing, storage, and transmission of information, while quantum sensing utilizes quantum properties for precise measurements. As QCI transitions from theoretical exploration to practical applications, its potential to revolutionize industry and academia becomes increasingly apparent.

Potential and Challenges

Quantum computing promises unprecedented speed in performing calculations for certain problem types, far exceeding classical computing capabilities. Algorithms like Shor's and Grover's exemplify quantum computing's superiority in tasks such as integer factorization and database searching. Yet, this potential is tempered by challenges in scalability, quantum noise, and operational costs. As we venture into systems with 100 qubits and beyond, the inadequacy of classical computers to simulate quantum systems becomes evident, highlighting the quantum advantage.

The NISQ Era and Beyond

Currently, we find ourselves in the Noisy Intermediate-Scale Quantum (NISQ) era, characterized by quantum processors that hold the promise of computations beyond classical simulation yet lack comprehensive error correction capabilities. This era has spurred the development of hybrid quantum-classical algorithms to optimize the computational power of NISQ devices amid noise and limited qubit coherence.

Transitioning from the NISQ era, we are beginning to witness the emergence of Quantum Utility. This new phase is not merely quantified by the number of qubits but by the capacity for reliable, significant computations that address real-world problems, marking a critical step from academic inquiry to industrial application. Quantum utility signifies the commencement of quantum computing's tangible benefits, driven by technological advancements that address enterprise-level challenges.

Workforce and Education

The burgeoning interest in QCI has precipitated a growing demand for professionals versed in quantum technologies. In response, educational institutions are integrating quantum computing into their curricula, anticipating a rise in the quest for quantum talent. Accessible software platforms will likely play a key role in democratizing quantum computing education, akin to the role of machine learning platforms today.

Required Expertise

The interdisciplinary nature of QCI necessitates expertise in linear algebra, quantum mechanics, computer science, and aspects of electrical engineering, among others. Proficiency with quantum programming languages and platforms such as Qiskit, Cirq, Q#, or Braket becomes increasingly important. As the field evolves, so too will the requisite skill set.

Learning Pathway

For those new to the field, a structured learning approach is recommended. Starting with foundational concepts like linear algebra and quantum mechanics before progressing to practical applications in quantum algorithms and programming ensures a solid base. Keeping abreast of the latest developments through journals, conferences, and academic courses is essential in a rapidly advancing field.

Are You Ready?

QCI represents not just a technological leap but a fundamental reimagining of computation and information processing. At this pivotal moment, the future of computing is poised for transformation, laden with challenges but rich in opportunity. The invitation is clear: embrace this cutting-edge technology, arm yourself with the necessary knowledge, and contribute to a future powered by the vast potential of QCI.

A Pioneering Vision for Quantum Computing

> " Nature isn't classical, and if you want to make a simulation of nature, you'd better make it quantum mechanical, and it's a wonderful problem, because it doesn't look so easy."
>
> — *Richard Feynman, 1981*
>
> First Conference on the Physics of Computation
> Photo credit: Smithsonian Institution

Qubits & Qudits: Foundations

1	Quantum Mechanics Through Photons	3
2	Fundamentals of Spin Systems	43
3	A Framework for Qubits and Qudits	59
4	Dynamics of Quantum Systems	85

1. Quantum Mechanics Through Photons

Contents

1.1	**Introducing Quantum Mechanics**	**4**
1.1.1	A Brief History	4
1.1.2	Principles of Quantum Mechanics	5
1.1.3	Our Approach	6
1.2	**Understanding Photons**	**7**
1.2.1	What Are Photons?	7
1.2.2	Polarization of Photons	7
1.2.3	A Curious Light Polarization Experiment	9
1.3	**The Quantum State Postulate**	**10**
1.3.1	Classical Variables to Quantum Vectors	10
1.3.2	Rectilinear Polarization States	11
1.3.3	Basis States	12
1.3.4	Diagonal Polarization States	13
1.3.5	Circular Polarization States	15
1.3.6	General Polarization States	16
1.3.7	The Quantum State Postulate	17
1.4	**The Quantum Observable Postulate**	**17**
1.4.1	Macroscopic Quantities to Quantum Observables	18
1.4.2	Observables for Rectilinear Polarizations	20
1.4.3	Observables for Diagonal Polarizations	21
1.4.4	Observables for Circular Polarizations	21
1.4.5	Basis Equivalence and Dirac Notation	22
1.5	**The Quantum Measurement Postulate**	**24**
1.5.1	Measurement Probability	24
1.5.2	State Collapse	26
1.5.3	∗ Statistical Average	29
1.5.4	Quantum Interference	31
1.6	**The Uncertainty Principle**	**33**
1.6.1	Compatible versus Incompatible Observables	34
1.6.2	∗ The Uncertainty Inequality	35

1.7	✷ Further Readings on Quantum Mechanics	36
1.8	Topic Reviews	37
1.8.1	Review of Postulates 1-3	37
1.8.2	Review of Photons as Quantum Systems	38
1.9	Summary and Conclusions	40
	Problem Set 1	41

Quantum mechanics (QM) serves as the cornerstone for understanding the fundamental nature of our universe. It is the theoretical framework that governs the behavior of matter and energy on microscopic scales. Its principles have profound implications for emerging fields such as quantum computing, quantum communication, and quantum information. However, the abstract and counterintuitive nature of QM can present significant challenges for those unfamiliar with traditional physics and mathematics.

Photons offer a gateway into the world of quantum mechanics that is both accessible and illustrative. As the simplest system that embodies the core principles of QM, photons provide an opportunity to bypass the mathematical complexities often associated with this field. Photon phenomena can be easily visualized, enabling a more intuitive understanding. Moreover, photons play a crucial role as a key physical platform for quantum computing and quantum communication. This makes them not only a pedagogical tool but also a practical one, bridging theory and application.

The purpose of this chapter is to introduce the first three principles of Quantum Mechanics through the lens of photon behavior. The aim is to emphasize the basic concepts and principles essential to quantum computing, providing an easier learning path for readers who have not come from a physics background. By grounding the discussion in tangible photon phenomena, the chapter will seek to demystify QM, laying a solid foundation for further exploration and study in the interconnected fields of quantum computing, communication, and information.

1.1 Introducing Quantum Mechanics

1.1.1 A Brief History

At the turn of the 20th century, classical physics had successfully explained many macroscopic phenomena, including Newtonian mechanics, electromagnetism, thermodynamics, and more. Yet, certain physical observations, such as the photoelectric effect and the black body radiation problem, could not be explained within this classical framework.

Quantum mechanics emerged as a revolutionary theory that could account for these discrepancies. It became the fundamental theory describing matter, energy, and motion at atomic and subatomic scales. Technologies such as semiconductor devices, lasers, and nuclear energy, and emerging fields like quantum computing, are rooted in quantum mechanics.

Key figures in the development of quantum mechanics include Planck, Einstein, Heisenberg, Bohr, Born, Schrödinger, and Dirac. Theories and notations named after these pioneers continue to be vital in current research and applications.

Quantum physics supersedes its classical counterparts, rendering classical physics as a special case of quantum mechanics. However, a full unification between quantum mechanics and general relativity remains an open question, with theories like superstring theory attempting to bridge this gap.

1.1.2 Principles of Quantum Mechanics

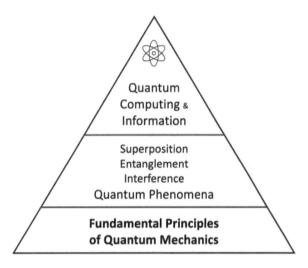

Figure 1.1: Role of Quantum Mechanics in Quantum Computing

As depicted in Fig. 1.1, the core components of quantum computing and information are grounded in specific quantum mechanical phenomena. Phenomena such as quantum superposition, entanglement, and interference are essential, but these complex behaviors are themselves derived from more fundamental principles of quantum mechanics.

Often referred to as postulates or axioms, these principles are the foundational pillars of quantum mechanics, analogous to how Newton's laws form the foundation of classical mechanics. Below is a brief introduction to the five principles, contextualized within the realm of quantum computing:

1. Quantum States: A quantum state is a complex vector that contains all the information about a quantum system. Quantum bits, or qubits, replace classical bits in quantum computing, and the state of a qubit is described as a superposition of the 0 and 1 states. This means a qubit can represent both states simultaneously, a feature that is unique to quantum systems. This principle enables parallel computation and has the potential to vastly increase computational efficiency for specific problems.

2. Quantum Observables: In quantum mechanics, observables are measurable physical quantities represented by Hermitian operators. In quantum computing, these operators extract specific outcomes from a system's quantum state during measurement.

3. Quantum Measurements: Quantum measurements facilitate information extraction from a quantum system, collapsing the state into one of the possible outcomes. This principle is essential in quantum algorithms, where measurements obtain the final result, and it is fundamental in quantum error correction.

4. Quantum Evolution: Quantum evolution governs the change in quantum states over time according to the Schrödinger equation. In quantum computing, quantum gates control this evolution, and algorithms are constructed using sequences of these gates.

5. Quantum Composition: Quantum composition involves building complex quantum systems from simpler ones. In quantum computing, this principle enables the creation of intricate quantum circuits from basic quantum gates, fostering the design of complex algorithms and enhancing the understanding of quantum computational processes.

These principles underpin the theoretical foundation of quantum computing, shaping the design, function, and comprehension of quantum algorithms, circuits, and systems. For those in the fields of quantum computing and quantum information, understanding these principles is vital, as they directly inform the practical aspects of computation and algorithm design.

In this chapter, we will focus on the first three principles, with the remaining two explored in subsequent chapters.

1.1.3 Our Approach

We strive to make Quantum Mechanics accessible to non-physics majors, tailored specifically for quantum computing and information (QCI), without sacrificing the rigor of core concepts. Our scaffolding approach unfolds the myriad of QM concepts gradually but systematically across three chapters (1-3) focused on photons, $\frac{1}{2}$-spins, and qubits/qudits.

Traditional courses in quantum mechanics often commence with complex mathematical formulations of differential equations, applicable to a wide range of quantum systems. While this approach is mathematically rigorous, it does not align well with the specialized needs for understanding QCI.

In contrast, our approach uses linear algebra and emphasizes two-level quantum systems, particularly qubits, focusing on principles and techniques that are essential to quantum computing. By using tangible examples involving photons and electron spins, we render these abstract concepts more accessible. This methodology ensures an intuitive grasp of the subject matter without compromising academic rigor.

Furthermore, we acknowledge that QCI is an interdisciplinary field, bridging the domains of quantum physics, computer science, and mathematics. By situating the subject matter within this broader interdisciplinary context, we offer a more comprehensive perspective that extends beyond the scope of a conventional course

1.2 Understanding Photons

in quantum physics, thereby preparing the reader for the multifaceted landscape of QCI.

1.2 Understanding Photons

1.2.1 What Are Photons?

Photons are the particles of light. In fact, they constitute the particles of all electromagnetic waves, including radio waves, microwaves, visible light, X-rays, gamma rays, and more. You may wonder, "We have never heard that radio waves have photons; do they?" The answer is yes. In the case of radio waves, since their frequency is low and the wavelength is large, ranging from meters to 100 meters, we find ourselves in the classical regime where the quantum effects of the photons are not apparent.

However, quantum effects become crucial in the behavior of electromagnetic radiation under certain conditions, such as interactions at small scales (atomic and subatomic), high energy regimes, and strongly confined systems. Under these circumstances, classical electromagnetic theories fall short in describing phenomena like energy quantization and wave-particle duality. In fields like quantum computing and quantum communication, it's not just that quantum effects are prominent; they are intentionally harnessed and manipulated, requiring a quantum treatment of light as well.

In the quantum regime, the energy of the electromagnetic waves is quantized, and the energy of each packet or photon depends on the frequency ($\nu = \omega/2\pi$), not on the amplitude, of the electromagnetic radiation:

$$E = h\nu = \hbar\omega \tag{1.1}$$

where h is the Planck constant, and $\hbar = h/2\pi$ is the reduced Planck constant.

In our everyday classical regime, an electromagnetic wave can be viewed as a flux of photons. For example, when looking at a computer screen, our eyes receive somewhere between 10^{14} to 10^{16} photons per second. However, in quantum computing and communication, we can transmit, receive, and measure a single photon at a time.

1.2.2 Polarization of Photons

In classical electromagnetism, polarization refers to the orientation of the electric field vector's oscillations in an electromagnetic wave. A polarized light wave consists of electric and magnetic fields oscillating in specific directions as the wave propagates. Linear polarization, where the electric field vector oscillates along a single axis, is common, but circular and elliptical polarization also exist. Various optical devices such as polarizers can manipulate or analyze classical light wave polarization.

In the quantum realm, polarization takes on a more abstract representation. The polarization of a photon corresponds to an intrinsic property related to its angular momentum. A photon is treated as a quantum particle, and its polarization is a two-dimensional complex vector, corresponding to a quantum state. Unlike the continuous oscillations of the electric field in the classical description, the quantum description deals with discrete outcomes. Quantum polarization can still be linear,

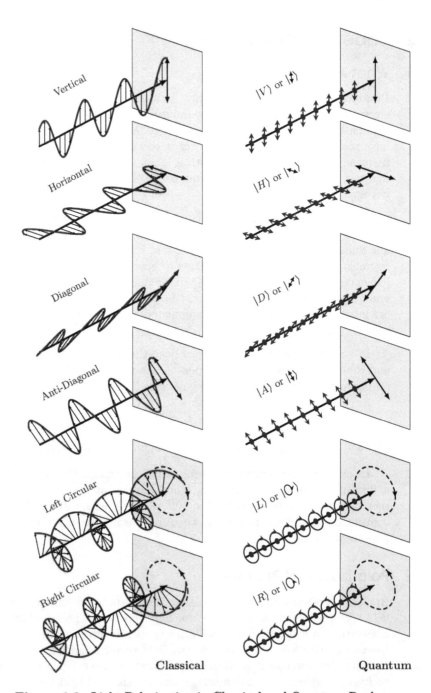

Figure 1.2: Light Polarization in Classical and Quantum Realms

Note: R and L in circular polarization refer to the rotation direction of the electric field vector as seen in the direction of propagation.

1.2 Understanding Photons

circular, or elliptical, and is vital in various quantum technologies such as quantum cryptography and quantum computing. Quantum polarization states can also be superposed and entangled, giving rise to distinctly quantum behaviors with no classical analogs.

Figure 1.2 illustrates light polarizations in the classical realm as waves of electric field vibration and as flux of photons in the quantum realm. The relationship between classical and quantum polarization descriptions lies in their conceptual connection and experimental consistency. While the classical view describes continuous wave properties, and the quantum view deals with particle-like attributes, polarization measurements in both realms yield consistent results. The quantum polarization state of a photon can be related to the corresponding classical wave's electric field orientation. This connection enables us to use photons as a model system to introduce the fundamental principles of quantum mechanics conveniently.

Table 1.1 lists key polarization states examined in this chapter, corresponding to Fig. 1.2. Additionally, elliptically polarized light is the most general state and can be considered as a linear combination of various orthogonal polarizations. Unpolarized light is a uniform statistical combination of these orthogonal states.

Category	Orthogonal States	Symbols
Rectilinear	Vertical	$|V\rangle$ or $|\updownarrow\rangle$
	Horizontal	$|H\rangle$ or $|\leftrightarrow\rangle$
Diagonal	Diagonal (45°)	$|D\rangle$ or $|\nearrow\rangle$
	Anti-Diagonal (135°)	$|A\rangle$ or $|\nwarrow\rangle$
Circular	Right(handed)	$|R\rangle$ or $|\circlearrowright\rangle$
	Left(handed)	$|L\rangle$ or $|\circlearrowleft\rangle$

Table 1.1: Key Photon Polarization States

1.2.3 A Curious Light Polarization Experiment

Figure 1.3 depicts an experiment involving light polarization. In the background, there are some unpolarized light sources. Two polarizers are placed in front of the light sources, with one (H) oriented horizontally and the other (V) vertically. When observing either of these polarizers alone, we notice that about half of the light is transmitted. However, in the region where the two polarizers overlap, the light is entirely blocked.

A third polarizer (D), oriented at 45°, is then inserted between polarizers H and V. Intuitively, one might expect this third polarizer to block even more light, and that all the light would still be completely blocked where the three polarizers overlap. This intuition holds in the regions where H and D or V and D overlap, with about 25% of light being transmitted. Surprisingly, however, in the region where all three polarizers overlap, approximately 12.5% of the light is now transmitted. This situation does not occur if the third polarizer is placed on top or beneath H and V or if it is also oriented horizontally or vertically.

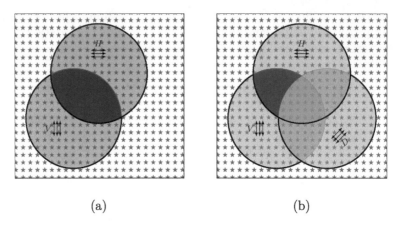

Figure 1.3: A Light Polarization Experiment

This intriguing behavior can be understood within the classical regime using traditional electromagnetic theory. In fact, you can demonstrate the experiment using Polaroid films as polarizers.

In the quantum regime, where it is possible to experiment with one photon at a time, the result is exactly the same. This raises a puzzling question: if H and V polarizations alone can block all photons, why do some photons pass after inserting another polarizer in between?

Quantum mechanics provides a comprehensive explanation for this experiment, even quantitatively. We will proceed to explain this experiment in terms of quantum mechanics, but first, we need to introduce some fundamental principles of quantum mechanics.

1.3 The Quantum State Postulate

1.3.1 Classical Variables to Quantum Vectors

The foundational principle of quantum mechanics is the principle of the quantum state. To understand this, let's first contrast it with classical physics, where the state of systems is described by recognizable physical quantities.

In a classical mechanical system, such as a ball moving through space, the state is characterized by position, velocity, and acceleration. In the case of a classical electromagnetic wave, the state is described by the **E** and **B** vectors, representing the electric and magnetic field strengths.

Quantum systems deviate substantially from their classical counterparts. Instead of conventional quantities like position or velocity, the state of a quantum system is represented by a special vector known as the state vector or wave function.

For a two-level quantum system like photon polarization, the state vector is two-dimensional with components α and β, both of which are generally complex numbers. Unlike classical vectors, where components are real numbers, the squared magnitudes of α and β relate to the probability of specific measurements. This

1.3 The Quantum State Postulate

probabilistic nature requires the state vector to be normalized, meaning the sum of the squares of these components equals one.

> **Postulate 1: Quantum State**
>
> (Simplified version) The state of a quantum system is completely described by a normalized complex vector.

For instance, in a two-level system such as photon polarization, the quantum state is a two-dimensional vector, typically expressed as

$$|\psi\rangle = \begin{bmatrix} \alpha \\ \beta \end{bmatrix}. \tag{1.2}$$

To declare $|\psi\rangle$ as normalized means:

$$\langle \psi | \psi \rangle = |\alpha|^2 + |\beta|^2 = 1. \tag{1.3}$$

While photon polarization involves a two-dimensional state vector, the dimensionality of state vectors in other quantum systems can be much larger, even extending to infinity.

(i) The vector representing a quantum state is often referred to as the *state vector*, or simply, the *state*. In some contexts, especially in infinite-dimensional space, it is also called the *wavefunction*.

(i) Throughout this text, we use the *Dirac notation* for vectors and operators, offering a concise and powerful way to represent quantum states and operations. In this notation, $|\psi\rangle$ represents a column vector (where ψ is a label), while $\langle\psi|$ is its adjoint (the complex conjugate transpose, $\langle\psi| \equiv |\psi\rangle^\dagger$). The inner product between $|\phi\rangle$ and $|\psi\rangle$ is expressed as $\langle\phi|\psi\rangle$. For further resources on Dirac notation and linear algebra, please refer to Appendix C.

Exercise 1.1 Warm-up exercise on state vectors using Dirac notation.
Given $|a\rangle = \begin{bmatrix} \sqrt{2} + i\sqrt{3} \\ \sqrt{2} - i\sqrt{3} \end{bmatrix}$,

(a) find $\langle a|$,

(b) normalize $|a\rangle$, and

(c) find a normalized vector $|b\rangle$ such that $\langle a|b\rangle = 0$.

1.3.2 Rectilinear Polarization States

Let's first take vertical and horizontal Rectilinear polarizations as examples to elucidate the Quantum State Postulate. We will begin by selecting vertical polarization as our reference state and assigning it the simplest normalized two-dimensional vector:

$$|V\rangle = \begin{bmatrix} 1 \\ 0 \end{bmatrix}. \tag{1.4}$$

 The content inside $|\cdot\rangle$ serves merely as a label. For instance, vertical polarization can be expressed in various ways such as $|\text{vertical}\rangle$, $|\updownarrow\rangle$, or $|V\rangle$.

What is the state vector for horizontal polarization, denoted as $|H\rangle$? Experimental evidence (refer to § 1.2.3) shows that vertically polarized photons will not pass through a horizontal polarizer, indicating that vertical and horizontal polarizations are mutually exclusive.

In quantum terms, mutually exclusive states correspond to disjoint measurement outcomes. Such states cannot coexist; if a system is in one of these states, it cannot simultaneously be in another. This exclusivity leads to the orthogonality of the states, meaning their inner product is zero: $\langle V|H\rangle = 0$.

> **Implication 1.1: Orthogonal States**
>
> Mutually exclusive quantum states are orthogonal; that is, their inner products are 0. Since these states are both normalized and orthogonal, they are referred to as *orthonormal*.

Given this orthogonality, the simplest normalized vector representing $|H\rangle$ and orthogonal to $|V\rangle$ is

$$|H\rangle = \begin{bmatrix} 0 \\ 1 \end{bmatrix}. \tag{1.5}$$

1.3.3 Basis States

We recognize in linear algebra that a complete set of orthonormal vectors can serve as a basis. In our context, the vectors $|V\rangle$ and $|H\rangle$ are orthonormal. Additionally, since these vectors are linearly independent, they span the entire two-dimensional space, thus forming a complete set. The completeness ensures that any possible vector in the space can be represented as a linear combination of these vectors, allowing them to function as basis states. This leads to the expression of any polarization state, $|P\rangle$, as

$$|P\rangle = \alpha|V\rangle + \beta|H\rangle = \begin{bmatrix} \alpha \\ \beta \end{bmatrix}. \tag{1.6}$$

where α and β are complex numbers satisfying the normalization condition $|\alpha|^2 + |\beta|^2 = 1$.

> **Implication 1.2: Superposition States**
>
> A general quantum state exists as a linear superposition of basis states.

1.3 The Quantum State Postulate

In forthcoming sections, we will elaborate on this idea using examples of diagonal and circular polarizations. However, before delving into those specifics, let's discuss the nature of our chosen basis states further.

We have designated the set $\{|V\rangle, |H\rangle\}$ as our *standard basis* for the system. In the parlance of quantum computing, this standard basis is also commonly referred to as the *computational basis*, represented by $\{|0\rangle, |1\rangle\}$. For our photon polarization system, the equivalence between these sets is formalized as

$$|0\rangle \equiv |V\rangle, \quad |1\rangle \equiv |H\rangle. \tag{1.7}$$

Therefore, Eq. 1.6 can be reformulated in terms of the computational basis as

$$|\psi\rangle = \alpha |0\rangle + \beta |1\rangle. \tag{1.8}$$

Equations 1.6 and 1.8 can be likened to the representation of a two-dimensional real vector $\mathbf{v} = a\hat{x} + b\hat{y}$ for easier visualization.

1.3.4 Diagonal Polarization States

Let's look at diagonal polarization. We said earlier that photon polarization is a system that's very easy to visualize in quantum mechanics. Let's illustrate why we can, in fact, use regular vectors to represent the polarization states. As shown in Fig. 1.4, the vertical polarization $|V\rangle$ is along the y axis, the horizontal polarization $|H\rangle$ is along the x axis, and the diagonal polarization $|D\rangle$ would be along the $y = x$ line. Thus, you would expect the diagonal polarization state to be proportional to a combination, or in quantum mechanics terms, a superposition of the vertical state and the horizontal state, represented by $|D\rangle \sim (|V\rangle + |H\rangle)$.

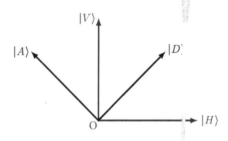

Figure 1.4: Relationship between H/V and D/A Polarizations

We have already assigned vectors to the vertical and horizontal polarizations, so the diagonal polarization should be proportional to $\begin{bmatrix} 1 \\ 1 \end{bmatrix}$. Only we have a problem; this vector is not normalized, meaning if you take the inner product of this vector with itself, which gives you $1^2 + 1^2 = 2$, it is not 1. So, in order to make this vector normalized, we need to divide it by $\sqrt{2}$. This correction leads us to the actual representation of the diagonal state:

$$|D\rangle = \frac{1}{\sqrt{2}} \begin{bmatrix} 1 \\ 1 \end{bmatrix}, \tag{1.9}$$

or in Dirac notation,

$$|D\rangle = \frac{1}{\sqrt{2}}\left(|V\rangle + |H\rangle\right). \tag{1.10}$$

Now, let's look at the anti-diagonal polarization state $|A\rangle$; it is 135°, along the $y = -x$ line. This polarization state, you can imagine it should be $|A\rangle \sim (|V\rangle - |H\rangle)$, which is $\begin{bmatrix} 1 \\ -1 \end{bmatrix}$. Once again, this vector is not normalized. So, to normalize it, we need to divide it by $\sqrt{2}$, therefore

$$|A\rangle = \frac{1}{\sqrt{2}}\left(|V\rangle - |H\rangle\right) = \frac{1}{\sqrt{2}}\begin{bmatrix} 1 \\ -1 \end{bmatrix}. \tag{1.11}$$

The D and A polarizations are also mutually exclusive, reflected as $|D\rangle$ and $|A\rangle$ being orthogonal, i.e., $\langle D|A\rangle = 0$.

■ **Example 1.1** Let's verify $\langle D|A\rangle = 0$. In matrix representation,

$$\langle D|A\rangle = \frac{1}{\sqrt{2}}\begin{bmatrix} 1 & 1 \end{bmatrix}\frac{1}{\sqrt{2}}\begin{bmatrix} 1 \\ -1 \end{bmatrix} = 0.$$

In Dirac notation,

$$\langle D|A\rangle = \frac{1}{\sqrt{2}}\left(\langle V| + \langle H|\right)\frac{1}{\sqrt{2}}\left(|V\rangle - |H\rangle\right)$$
$$= \frac{1}{2}\left(\langle V|V\rangle - \langle V|H\rangle + \langle H|V\rangle - \langle H|H\rangle\right)$$
$$= \frac{1}{2}(1 - 0 + 0 - 1) = 0.$$

■

Since $|D\rangle$ and $|A\rangle$ are orthonormal, they also form a basis for photon polarization states. A general polarization state can be represented in this basis as

$$|P\rangle = \gamma|D\rangle + \delta|A\rangle, \tag{1.12}$$

where γ and δ are complex numbers satisfying $|\gamma|^2 + |\delta|^2 = 1$.

In particular,

$$|V\rangle = \frac{1}{\sqrt{2}}\left(|D\rangle + |A\rangle\right), \tag{1.13a}$$

$$|H\rangle = \frac{1}{\sqrt{2}}\left(|D\rangle - |A\rangle\right). \tag{1.13b}$$

In the context of quantum computing, $|D\rangle$ and $|A\rangle$ correspond to the $|+\rangle$ and $|-\rangle$ states, which we will study later.

Exercise 1.2 Compare Equation 1.12 and Equation 1.6, and derive formulas for

(a) γ and δ in terms of α and β

1.3 The Quantum State Postulate

(b) α and β in terms of γ and δ

using both matrix representation and Dirac notation.

1.3.5 Circular Polarization States

So far, we have only examined linear polarizations. In linear polarization, the polarization vector stays fixed as the photon propagates. But in circular polarization, as depicted in Fig. 1.2, the polarization vector rotates as the photon propagates. With right circular polarization $|R\rangle$, the turning of the polarization vector is right-handed as we look in the direction of propagation. When the photon travels one wavelength, the polarization vector will turn by 360° clockwise.

Let's look at how we can construct a state vector for this quantum state. We'll utilize the xy coordinate system, as for photon polarizations, we have the luxury to do so. The difference between linear polarization $|D\rangle$ and circular polarization $|R\rangle$ is that for $|D\rangle$, the horizontal vector and the vertical vector are always in phase, but for $|R\rangle$, the state vector goes through the y axis first then turns to the x axis; or equivalently, the x ($|H\rangle$) lags the y ($|V\rangle$) component by a 90° phase.

So we need to introduce a phase factor. A 90° lagging phase corresponds to a phase factor $e^{-i\pi/2} = -i$. Therefore, we expect the state vector $|R\rangle$ to be

$$|R\rangle = \frac{1}{\sqrt{2}}\left(|V\rangle - i|H\rangle\right) = \frac{1}{\sqrt{2}}\begin{bmatrix}1\\-i\end{bmatrix}. \tag{1.14}$$

Again, the $\sqrt{2}$ factor comes from the normalization requirement.

Of course, we could also argue that the y component *leads* the x component by a 90° phase, and correspondingly $|R\rangle = \frac{1}{\sqrt{2}}(i|V\rangle + |H\rangle)$. This expression and Eq. 1.14 should be physically equivalent. Note that the two differ just by a global phase factor i:

$$(i|V\rangle + |H\rangle) = i(|V\rangle - i|H\rangle).$$

This leads to the following observation:

Implication 1.3: Global Phase

Global phase of a quantum state vector does not have physical significance. That is, $e^{i\phi}|\psi\rangle$ represents the same quantum state as $|\psi\rangle$, where ϕ is a real number representing a phase angle.

What about left-handed circular polarization? It is the opposite to right polarization, with the H component leading the V component by a 90° phase. So the state vector (up to a global phase) is

$$|L\rangle = \frac{1}{\sqrt{2}}\left(|V\rangle + i|H\rangle\right) = \frac{1}{\sqrt{2}}\begin{bmatrix}1\\i\end{bmatrix}. \tag{1.15}$$

Exercise 1.3 Verify that $|L\rangle$ and $|R\rangle$ are normalized and orthogonal with each other, using both matrix representation and Dirac notation.

Similar to $|H\rangle$ and $|V\rangle$, and $|D\rangle$ and $|A\rangle$, $|L\rangle$ and $|R\rangle$ also form a basis for photon polarization states, because they are orthonormal. A general polarization state can be represented in this basis as

$$|P\rangle = \mu |L\rangle + \nu |R\rangle. \tag{1.16}$$

Exercise 1.4 Represent $|V\rangle$ and $|H\rangle$ in terms of $|L\rangle$ and $|R\rangle$.

1.3.6 General Polarization States

We have established that any polarization state, denoted by $|P\rangle$, can be represented in the form (see Eq. 1.6):

$$|P\rangle = \alpha |V\rangle + \beta |H\rangle. \tag{1.17}$$

Here, α and β are complex numbers subject to the normalization condition $|\alpha|^2 + |\beta|^2 = 1$. These coefficients can be parameterized by two real angles θ and ϕ as:

$$\alpha = \cos\theta, \quad \beta = \sin\theta e^{i\phi}. \tag{1.18}$$

The normalization condition is naturally satisfied due to $\cos^2\theta + \sin^2\theta = 1$, allowing for an arbitrary phase $e^{i\phi}$ between $|V\rangle$ and $|H\rangle$. Thus, $|P\rangle$ becomes:

$$|P\rangle = \cos\theta |V\rangle + \sin\theta e^{i\phi} |H\rangle. \tag{1.19}$$

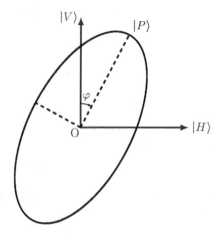

Figure 1.5: Elliptical Polarization

This general form accounts for elliptical polarization, as illustrated in Fig. 1.5. Rectilinear, diagonal, and circular polarizations emerge as special cases, as summarized in the table below.

Polarization State	θ	ϕ
$\lvert V\rangle$	0	0
$\lvert H\rangle$	$\pi/2$	0
$\lvert D\rangle$	$\pi/4$	0
$\lvert A\rangle$	$-\pi/4$	0
$\lvert L\rangle$	$\pi/4$	$\pi/2$
$\lvert R\rangle$	$\pi/4$	$-\pi/2$

Exercise 1.5 Validate that Eq. 1.19 reproduces the special cases listed above.

Exercise 1.6 Show that when $\phi = 0$ Eq. 1.19 represents a linear polarization state. What is its angle relative to $\lvert V\rangle$?

We have previously seen that $\lvert P\rangle$ can be alternatively expressed in different bases (Eqs. 1.12 and 1.16).

1.3.7 The Quantum State Postulate

In § 1.3.1 we have introduced the Quantum State Postulate in complex vector space, which is adequate for quantum computing in most cases. More generally, quantum states reside in a Hilbert space.

> **Postulate 1: Quantum State**
>
> The state of a quantum system is completely described by a normalized complex vector in a Hilbert space.

A Hilbert space ($\mathcal{H}$) is a vector space equipped with an inner product that is complete, meaning that it contains all its limit points. Its dimension is $N = 2^n$ for a system of n qubits, and can even be infinite. But for our purpose in quantum computing, we can consider a Hilbert space $\mathcal{H}$ as a complex vector space, denoted as $\mathbb{C}^N$.

For a two-level system, the quantum state can be described by:

$$|\psi\rangle = \begin{bmatrix} \alpha \\ \beta \end{bmatrix}. \tag{1.20}$$

Here $\alpha, \beta \in \mathbb{C}$ satisfying $\langle\psi|\psi\rangle = |||\psi\rangle|^2 = 1$, and the state space is $\mathbb{C}^2$.

1.4 The Quantum Observable Postulate

As we delve into the intricate landscape of quantum mechanics, we encounter a key juncture where the State Postulate seamlessly gives way to the Measurement Postulate. Central to navigating this landscape are observables, fundamental quantities that lie at the core of understanding measurable properties within quantum systems. The Quantum Observable Postulate, our starting point, stands as the foundation of this exploration.

In the context of Quantum Computing, where harnessing quantum phenomena holds immense potential, measurement assumes paramount importance. Measurement serves as the gateway through which we extract information from quantum systems, shaping the very essence of Quantum Computing.

Our journey begins with the Observable Postulate, followed by the subsequent unveiling of the Measurement Postulate. Each step reveals distinct facets of the quantum realm. Through a series of subsections, we unravel the transition from classical macroscopic quantities to the enigmatic world of quantum observables. Along this path, we delve into illustrative examples, unravel the intricacies of mathematical representation, and unveil the profound interplay between observables and quantum states. This section, serving as both a guide and an exploration, establishes the groundwork for a deeper voyage into the boundless potential of Quantum Computing.

1.4.1 Macroscopic Quantities to Quantum Observables

The principle of quantum states has been introduced through illustrative examples involving photon polarization. According to this principle, the state of a quantum system finds complete description within a normalized state vector. It is worth noting the term 'completely,' which we've hitherto left unexplored. Additionally, we've established that the components of these state vectors are generally complex numbers.

One might inquire: how do physical attributes like light intensity, energy, position, and velocity fit into this framework? These attributes are categorized as quantum observables (also referred to as measurables), and they constitute the focus of our forthcoming exploration.

In quantum mechanics, observables are represented by Hermitian operators, symbolized as M in the subsequent discussion. In this context, M can signify properties such as position or light intensity. These operators take the form of matrices, possessing eigenvalues and eigenvectors that relate to the measurable outcomes attainable from the quantum state system.

Postulate 2: Quantum Observables

Any physically measurable quantity is represented by a Hermitian operator (matrix) M.

The eigenvalues of M, denoted as $\{\lambda_i\}$, represent the possible outcomes when measuring M.

For each eigenvalue λ_i, the corresponding eigenvector $|\lambda_i\rangle$ is the state that yields that particular measurement outcome with certainty.

The Observable Postulate is illustrated in Fig. 1.6. This postulate states that any physically measurable quantity in quantum mechanics can be associated with a Hermitian operator known as an observable, represented here by M. When this operator acts on its eigenstate $|\lambda_i\rangle$, the outcome is the state itself scaled by a corresponding eigenvalue λ_i. For a d-dimensional system (i.e., the operator M is a $d \times d$ matrix), there are d such pairs of eigenvalues and eigenvectors, and the index i runs from 1 to d, accounting for each dimension of the system. Eigenvalues

1.4 The Quantum Observable Postulate

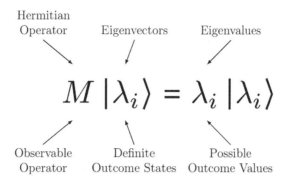

Figure 1.6: Illustration of the Quantum Observable Postulate

may be degenerate, meaning different eigenvectors can correspond to the same eigenvalue. For further details on eigenvectors and eigenvalues in the context of quantum observables, see Appendix C and Ref [9].

This concept may appear quite mathematical, but it will become clearer with examples that we will explore next. In quantum computing, this postulate informs how we can extract information from qubits. The readout is essentially a measurement, and thus the postulate helps in understanding how algorithms will behave when translated into physical operations.

Since quantum observables are Hermitian operators (matrices), let's first review the properties of those in order to understand this postulate:

- Definition of Hermitian operator: $M^\dagger = M$.

- Hermitian operators have real eigenvalues, ensuring that measurements yield real-numbered outcomes.

- The eigenstates associated with distinct eigenvalues of a Hermitian operator are guaranteed to be orthogonal. (A mathematical nuauce: while the eigenvalues are unique, the eigenvectors corresponding to them are not. Specifically, non-degenerate eigenvectors can differ by an overall phase factor, which has no observable physical consequence.)

- Hermitian operators possess a complete set of eigenstates, allowing any quantum state to be expanded as a linear combination of these eigenstates.

Exercise 1.7 For any complex matrix A, prove that $AA^\dagger$ is a Hermitian matrix.

Exercise 1.8 Find the eigenvalues and eigenvectors for

$$\begin{bmatrix} a - \alpha & b + i\beta \\ b - i\beta & a + \alpha \end{bmatrix},$$

where $a, b, \alpha, \beta \in \mathbb{R}$.

1.4.2 Observables for Rectilinear Polarizations

Let's examine linear polarizations again. According to the observable principle, the reason we can measure these states with definite answers (pass or blocked) means that there is an observable for each case. That observable is a particular Hermitian matrix, and the eigenvectors of this matrix should be the vectors corresponding to those definite states. We will demonstrate how this works.

Let's start with the vertical polarization $|V\rangle$, and we use the vertical observable M_V to measure it:

$$M_V = \begin{bmatrix} 1 & 0 \\ 0 & 0 \end{bmatrix}. \tag{1.21}$$

We know when we use a vertical polarizer to measure a photon with vertical polarization, we get a definite yes answer. If we assign a value 1 to yes (pass) and 0 to no (blocked) to the measurement outcome, we expect $|V\rangle$ to be an eigenvector of M_V with an eigenvalue 1:

$$M_V |V\rangle = 1 |V\rangle = |V\rangle, \tag{1.22}$$

which you can easily verify. Similarly, when we use a vertical polarizer to measure a photon with horizontal polarization, we get a definite no answer. There is no maybe answer. We expect $|H\rangle$ to be an eigenvector of M_V with an eigenvalue 0:

$$M_V |H\rangle = 0 |H\rangle = 0. \tag{1.23}$$

This demonstrates an important concept derived from the Quantum Observable principle:

> **Implication 2.1: Quantization of Measurement Outcomes**
>
> Unlike classical systems where observables can take continuous ranges of values, quantum observables often have discrete eigenvalues.

The same concept also underlies why atoms have discrete energy levels, and that qubits exist as two-level systems. It is explained by the observable principle because only possible outcome of a quantum measurement is one of the eigenvalues of the corresponding observable operator. These eigenvalues are often discrete numbers.

Similarly, if we use a horizontal polarizer to measure vertical and horizontal polarizations, we are using an observable M_H (which is complimentary to M_V):

$$M_H = \begin{bmatrix} 0 & 0 \\ 0 & 1 \end{bmatrix}, \tag{1.24}$$

$$M_H |V\rangle = 0, \tag{1.25a}$$
$$M_H |H\rangle = |H\rangle. \tag{1.25b}$$

1.4.3 Observables for Diagonal Polarizations

Now, let's look at the diagonal polarizations $|D\rangle$ and $|A\rangle$. They are also mutually exclusive quantum states, and the two vectors are orthonormal. We will measure them using a diagonal polarizer, represented by an observable operator M_D. This round, instead of verifying the observable matrix, let's derive it, which involves essential skills for anyone working on quantum computing. Let

$$M_D = \begin{bmatrix} a & b \\ c & d \end{bmatrix}, \quad (1.26)$$

and we require

$$M_D |D\rangle = |D\rangle, \text{ i.e., } \begin{bmatrix} a & b \\ c & d \end{bmatrix} \frac{1}{\sqrt{2}} \begin{bmatrix} 1 \\ 1 \end{bmatrix} = \frac{1}{\sqrt{2}} \begin{bmatrix} 1 \\ 1 \end{bmatrix}, \quad (1.27a)$$

$$M_D |A\rangle = 0, \quad \text{i.e., } \begin{bmatrix} a & b \\ c & d \end{bmatrix} \frac{1}{\sqrt{2}} \begin{bmatrix} 1 \\ -1 \end{bmatrix} = 0. \quad (1.27b)$$

Through some algebra, we can determine that $a = b = c = d = \frac{1}{2}$, so:

$$M_D = \frac{1}{2} \begin{bmatrix} 1 & 1 \\ 1 & 1 \end{bmatrix}. \quad (1.28)$$

The preceding demonstration of M_V and M_D leads to another implication of the Observable Postulate:

Implication 2.2: Measurement Basis

The choice of observable determines the basis in which the measurement outcome is to be expressed.

Exercise 1.9 Derive the observable operator M_A for measuring $|D\rangle$ with outcome 0 and $|A\rangle$ with outcome 1, similar to how M_D is derived in Eq. 1.27.

1.4.4 Observables for Circular Polarizations

The circular polarizations $|L\rangle$ and $|R\rangle$ are also mutually exclusive quantum states. There should be an observable operator M_L which has $|L\rangle$ as eigenvector with eigenvalue 1, and $|R\rangle$ with eigenvalue 0; and vice versa for the complimentary observable M_R.

This round, we will demonstrate a more general method to derive the matrices of M_L and M_R. According to the spectral decomposition theorem, a Hermitian matrix can be expressed as the sum of its eigenvalues times the outer products of the eigenvectors:

$$M = \sum_i \lambda_i |\lambda_i\rangle\langle\lambda_i|. \quad (1.29)$$

Therefore,

$$M_L = 1\,|L\rangle\langle L| + 0\,|R\rangle\langle R| \qquad (1.30\text{a})$$

$$= \frac{1}{\sqrt{2}}\left(|V\rangle + i\,|H\rangle\right)\frac{1}{\sqrt{2}}\left(\langle V| - i\,\langle H|\right) \qquad (1.30\text{b})$$

$$= \frac{1}{2}\left(|V\rangle\langle V| - i\,|V\rangle\langle H| + i\,|H\rangle\langle V| + |H\rangle\langle H|\right), \qquad (1.30\text{c})$$

or, in matrix form,

$$M_L = \frac{1}{2}\begin{array}{c}\\ |V\rangle \\ |H\rangle\end{array}\begin{array}{cc}\langle V| & \langle H| \\ \begin{bmatrix} 1 & -i \\ i & 1 \end{bmatrix}\end{array}. \qquad (1.31)$$

The above derivation of M_L is a key skill for anyone learning quantum computing. Through M_L we have also demonstrated that even though an observable operator may contain complex numbers, its associated eigenvalues are real numbers, owing to a property of Hermitian matrices.

> **Exercise 1.10** Using both the matrix representation and Dirac notation, verify that
> $$M_L\,|L\rangle = |L\rangle,$$
> $$M_L\,|R\rangle = 0.$$

> **Exercise 1.11** Derive the observable operator M_R for measuring $|L\rangle$ with outcome 0 and $|R\rangle$ with outcome 1, similar to how M_L is derived in Eq. 1.30.

1.4.5 Basis Equivalence and Dirac Notation

Quantum states and observables are not tied to a specific basis; rather, their vector or matrix representation changes when expressed in different bases. According to the Observable Postulate, each observable's eigenvectors serve as a basis for quantum states. These bases, formed by the eigenvectors of Hermitian operators, are mathematically interchangeable for describing quantum systems.

Dirac's bra-ket notation offers a consistent, basis-agnostic framework for representing these quantum entities. For a succinct review of Dirac notation in the context of linear algebra, refer to Appendix C.

> **Basis-Independence in Dirac Bra-Ket Notation**
>
> The Dirac bra-ket notation abstractly and universally represents quantum states and operators, highlighting their inherent basis-independence.

We previously examined photon polarization states and their associated observables. Table 1.2 consolidates their vector and matrix representations in both rectilinear and diagonal bases.

As a demonstration of the power of the Dirac notation, let's derive the matrix of M_L in the $\{|D\rangle, |A\rangle\}$ basis. We treat the equations in Table 1.2 as basis-independent relations, and work through the algebra to obtain M_L in terms of $|D\rangle$ and $|A\rangle$:

1.4 The Quantum Observable Postulate

	Relations	
$\lvert D\rangle = \frac{1}{\sqrt{2}}(\lvert V\rangle + \lvert H\rangle)$		$\lvert A\rangle = \frac{1}{\sqrt{2}}(\lvert V\rangle - \lvert H\rangle)$
$\lvert L\rangle = \frac{1}{\sqrt{2}}(\lvert V\rangle + i\lvert H\rangle)$		$\lvert R\rangle = \frac{1}{\sqrt{2}}(\lvert V\rangle - i\lvert H\rangle)$
$M_V = \lvert V\rangle\langle V\rvert$	$M_D = \lvert D\rangle\langle D\rvert$	$M_L = \lvert L\rangle\langle L\rvert$

Quantity	$\{\lvert V\rangle, \lvert H\rangle\}$ Basis	$\{\lvert D\rangle, \lvert A\rangle\}$ Basis
$\lvert V\rangle$	$\begin{bmatrix}1\\0\end{bmatrix}$	$\frac{1}{\sqrt{2}}\begin{bmatrix}1\\1\end{bmatrix}$
$\lvert H\rangle$	$\begin{bmatrix}0\\1\end{bmatrix}$	$\frac{1}{\sqrt{2}}\begin{bmatrix}1\\-1\end{bmatrix}$
$\lvert D\rangle$	$\frac{1}{\sqrt{2}}\begin{bmatrix}1\\1\end{bmatrix}$	$\begin{bmatrix}1\\0\end{bmatrix}$
$\lvert A\rangle$	$\frac{1}{\sqrt{2}}\begin{bmatrix}1\\-1\end{bmatrix}$	$\begin{bmatrix}0\\1\end{bmatrix}$
$\lvert L\rangle$	$\frac{1}{\sqrt{2}}\begin{bmatrix}1\\i\end{bmatrix}$	$\frac{1}{2}\begin{bmatrix}1+i\\1-i\end{bmatrix}$
$\lvert R\rangle$	$\frac{1}{\sqrt{2}}\begin{bmatrix}1\\-i\end{bmatrix}$	$\frac{1}{2}\begin{bmatrix}1-i\\1+i\end{bmatrix}$
M_V	$\begin{bmatrix}1 & 0\\0 & 0\end{bmatrix}$	$\frac{1}{2}\begin{bmatrix}1 & 1\\1 & 1\end{bmatrix}$
M_D	$\frac{1}{2}\begin{bmatrix}1 & 1\\1 & 1\end{bmatrix}$	$\begin{bmatrix}1 & 0\\0 & 0\end{bmatrix}$
M_L	$\frac{1}{2}\begin{bmatrix}1 & -i\\i & 1\end{bmatrix}$	$\frac{1}{2}\begin{bmatrix}1 & i\\-i & 1\end{bmatrix}$

Table 1.2: Summary of Photon Polarization States and Observables

$$M_L = |L\rangle\langle L| \tag{1.32a}$$

$$= \frac{1}{\sqrt{2}}(|V\rangle + i|H\rangle)\frac{1}{\sqrt{2}}(\langle V| - i\langle H|) \tag{1.32b}$$

$$= \frac{1}{2}((|D\rangle + |A\rangle) + i(|D\rangle - |A\rangle))\frac{1}{2}((\langle D| + \langle A|) - i(\langle D| - \langle A|)) \tag{1.32c}$$

$$= \frac{1}{2}(|D\rangle\langle D| + i|D\rangle\langle A| - i|A\rangle\langle D| + |A\rangle\langle A|) \tag{1.32d}$$

$$= \frac{1}{2}\begin{bmatrix} 1 & i \\ -i & 1 \end{bmatrix}. \tag{1.32e}$$

Exercise 1.12 In our standard formulation, matrices M_V, M_D, and M_L have eigenvalues of 1 (pass) for the first eigenvector and 0 (blocked) for the second. Consider an alternative assignment where the second eigenvector has an eigenvalue of -1 instead of 0. Derive the matrices M_V, M_D, and M_L under this new scheme, and represent them in the $|V\rangle, |H\rangle$ basis. You will discover the Pauli matrices.

While quantum computing commonly employs orthonormal basis states tied to specific observables, it's worth noting that mathematically, basis states need only to be linearly independent and complete; orthonormality is not a requirement.

1.5 The Quantum Measurement Postulate

The Observable Postulate and the Measurement Postulate constitute fundamental tenets within the realm of quantum mechanics. These postulates delineate the behavior of quantum systems and the intricate process of measurement. Specifically, the Observable Postulate stipulates that every observable quantity, such as position and light intensity, is symbolized by a Hermitian operator in quantum mechanics. On the other hand, the Measurement Postulate details the methodology of measuring an observable quantity within a quantum system.

In the ensuing discussion, we will delve into the intricacies of the Measurement Postulate, dissecting it into two key components: measurement probability and state collapse. These principles form the bedrock of quantum mechanics, providing insight into the interdependencies between measurements, states, and probabilities. We will elucidate these concepts using the context of photon polarization, employing it as a visual aid to comprehend the embodiment of these principles within quantum systems.

Quantum measurement is a complex subject that continues to evolve with advancements in quantum computing and quantum information science. The standard quantum measurement postulates presented in this section serve as foundational concepts. These will be extended in § 3.4 to encompass a general measurement framework, including more complex scenarios such as Positive Operator-Valued Measures (POVMs), thereby offering increased flexibility and broader applicability.

1.5.1 Measurement Probability

The Observable Postulate informs us that if a quantum state takes the form of an

1.5 The Quantum Measurement Postulate

eigenstate $|\lambda_i\rangle$ of an observable M, the measurement outcome will unequivocally yield λ_i. In essence, distinguishing eigenstates of an observable remains unambiguous.

However, when we endeavor to measure a general quantum state—such as a superposition state $|\psi\rangle$ as presented in Eq. 1.17—that does not conform to an eigenstate of M, multiple outcomes come into play, inherently characterizing quantum measurements as probabilistic phenomena.

> **Postulate 3a: Measurement Probability**
>
> The outcome of measuring an observable M on a quantum state $|\psi\rangle$ corresponds to one of its eigenvalues. The probability associated with each eigenvalue λ_i is given by $|\langle\lambda_i|\psi\rangle|^2$ (the Born rule).

This postulate offers a multi-faceted understanding of quantum measurement, leading to several significant implications, as detailed below:

> **Implication 3.1: Non-deterministic Outcomes**
>
> Quantum measurements inherently entail probabilistic outcomes—a distinct contrast to classical physics where measurements are deterministic provided knowledge of both initial conditions and governing equations.

> **Implication 3.2: Quantum State and Measurement Probabilities**
>
> The Born rule establishes the mathematical framework that links quantum states to the probabilities of measurement outcomes.

> **Implication 3.3: Total Probability**
>
> Quantum state vectors maintain normalization across all bases, aligning with the fundamental requirement that probabilities of mutually exclusive events must collectively sum to 1.

Photon Polarization Experiments

The experiments shown in Fig. 1.7 exemplify the Born rule. Photons are prepared in the vertical polarization $|\psi\rangle = |V\rangle$. In part (a), they are measured with a vertical polarizer. The passing probability is $|\langle V|\psi\rangle|^2 = |\langle V|V\rangle|^2 = 1$, thus 100% of the photons pass through.

In part (b), where the photons are measured with a horizontal polarizer, the passing probability becomes $|\langle H|\psi\rangle|^2 = |\langle H|V\rangle|^2 = 0$, and consequently, no photons pass through.

In part (c), when the photons are measured with a diagonal polarizer, the passing probability is $|\langle D|\psi\rangle|^2 = |\langle D|V\rangle|^2 = 0.5$. As a result, 50% of photons pass through while the rest are blocked. This is reflected in the figure by the doubled spacing between the photons.

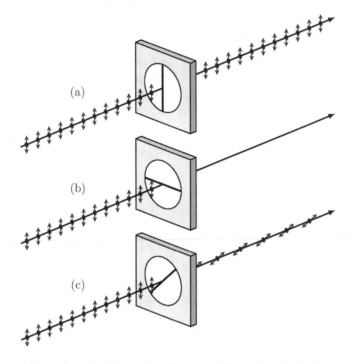

Figure 1.7: Measuring Photons Through a Polarizer

Exercise 1.13 What is the passing probability of vertically polarized photons $|\psi\rangle = |V\rangle$ when measured with a device that passes left circular polarization M_L?

1.5.2 State Collapse

In classical physics, measurements can often be conducted in such a way that the system remains undisturbed. The deterministic framework of classical physics enables these non-invasive measurements. However, quantum mechanics is fundamentally different. Owing to the quantization of energy and other properties, along with the limitations imposed by the Heisenberg Uncertainty Principle, any measurement in a quantum system inevitably involves a disturbance to that system. For example, measuring an electron's position necessarily affects its momentum. This departure from classical intuition is one of the defining traits of quantum mechanics.

This divergence between classical and quantum paradigms leads us to an essential feature of quantum theory: the postulate of measurement state collapse. In quantum systems, a measurement forces a transition in the state of the system. Specifically, according to Postulate 3b, when an observable M is measured on a quantum state $|\psi\rangle$, the state collapses to one of the eigenstates $|\lambda_i\rangle$ corresponding to the measured eigenvalue λ_i. This state collapse is immediate and irreversible.

> **Postulate 3b: Measurement State Collapse**
>
> After the measurement, the quantum state collapses to the eigenstate associated with the measured eigenvalue.

1.5 The Quantum Measurement Postulate

The immediate implication of this postulate is that measurements in quantum mechanics are fundamentally non-reversible processes.

Implication 3.4: Irreversible Effect

Measurements permanently alter the quantum state, forcing it into one of the eigenstates of the measured observable. This non-reversibility is non-classical and affects the system's subsequent dynamics.

In other words, the act of measurement irrevocably changes the quantum state. Once the system is in an eigenstate of the observable M, subsequent measurements with the same observable will consistently yield the same result.

Implication 3.5: Subsequent Measurements

Once a measurement has been made, the system's state collapses to a definite eigenstate. Therefore, any repeated measurements with the same observable will yield identical results.

The Copenhagen Interpretation and von Neumann Framework

The measurement state collapse postulate aligns with the Copenhagen interpretation of quantum mechanics, primarily attributed to Niels Bohr and Werner Heisenberg. John von Neumann later provided a mathematically rigorous version of this perspective, delineating deterministic unitary evolution from non-deterministic measurement-induced collapse. While several interpretations of quantum mechanics exist, in this text, we adopt the von Neumann framework. Further discussions on various interpretations can be found in § 9.2.

Examples

Repeated Measurements with Photon Polarization

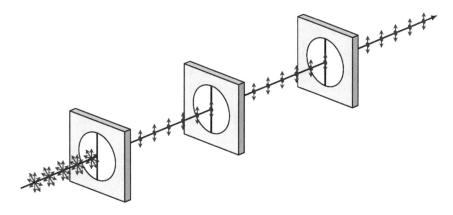

Figure 1.8: Repeated Identical Measurements

Figure 1.8 serves as a concrete example of this principle. Initially, the photons are unpolarized, implying a mixture of different polarization states. (We will discuss such mixed quantum states and their representation by density matrices in a later section.) When the photons pass through the first vertical polarizer, their states collapse to vertical polarization. Subsequent measurements with vertical polarizers will then result in a 100% transmission rate, without blocking or altering the photon states.

Cross-Polarizer Experiments

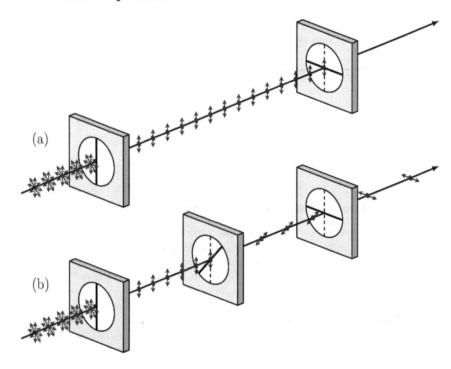

Figure 1.9: Measuring Photons Through Crossed Polarizers

Armed with the Measurement Postulate, we can elucidate the perplexing cross-polarizer experiment discussed in § 1.2.3. Fig. 1.9 shows the quantum rendition of this experiment, where light is treated as a photon stream.

In part (a), we introduce unpolarized photons to a vertical polarizer. Approximately 50% of the photons pass, consistent with a representation of unpolarized photons as being 50% in $|V\rangle$ and 50% in $|H\rangle$. Photons that do pass are forced into the vertical polarization state $|V\rangle$. Subsequent measurement using a horizontal polarizer yields a zero probability of passage, $|\langle H|V\rangle|^2 = 0$, aligning with experimental findings.

In part (b), a diagonal polarizer is inserted between the vertical and horizontal polarizers. The probability of photons in state $|V\rangle$ passing this diagonal polarizer is $|\langle V|D\rangle|^2 = 50\%$. The photons that pass are then in state $|D\rangle$. Measuring these photons with a horizontal polarizer gives a passage probability of 50%, resulting in an overall probability of $50\% \times 50\% \times 50\% = 12.5\%$, which matches the experimental observations.

1.5 The Quantum Measurement Postulate

The crux of this experiment lies in the Measurement Postulate-induced state collapse to $|D\rangle$ upon passing the diagonal polarizer. Absent this quantum mechanical feature, the photons would not navigate through all three polarizers in part (b), especially considering they are entirely blocked by just two polarizers in part (a).

1.5.3 *Statistical Average

In quantum mechanics, physical observables are represented by Hermitian operators M. The macroscopic measurements we make yield the statistical average of these observables, calculated in accordance with the postulates discussed earlier.

Consider the eigenvectors $|\lambda_i\rangle$ of M. They form an orthonormal basis, allowing us to express a general quantum state as

$$|\psi\rangle = \sum_i c_i |\lambda_i\rangle, \tag{1.33}$$

where the coefficients c_i are complex numbers satisfying $\sum_i |c_i|^2 = 1$.

Upon measuring a superposition state using the operator M, the result is probabilistic. The possible outcomes are the eigenvalues λ_i, each occurring with a probability $P_i = |\langle \lambda_i | \psi \rangle|^2$. The statistical average of these outcomes can be calculated as:

$$\langle M \rangle = \sum_i P_i \lambda_i = \sum_i |\langle \lambda_i | \psi \rangle|^2 \lambda_i. \tag{1.34}$$

Interestingly, this can be simplified to a basis-independent form:

> **Implication 3.6: Statistical Average**
>
> The statistical average (expected value, or expectation value) for measuring an observable M on a quantum state $|\psi\rangle$ is given by:
>
> $$\langle M \rangle = \langle \psi | M | \psi \rangle. \tag{1.35}$$

To build intuition for this expression, consider the case where $|\lambda_0\rangle = |0\rangle$, $|\lambda_1\rangle = |1\rangle$, and $|\psi\rangle = \alpha |0\rangle + \beta |1\rangle$.

$$\begin{aligned}
\langle M \rangle &= |\alpha|^2 \lambda_0 + |\beta|^2 \lambda_1 \\
&= \langle \psi | 0 \rangle \langle 0 | \psi \rangle \lambda_0 + \langle \psi | 1 \rangle \langle 1 | \psi \rangle \lambda_1 \\
&= \langle \psi | (\lambda_0 |0\rangle\langle 0| + \lambda_1 |1\rangle\langle 1|) |\psi\rangle \\
&= \langle \psi | M | \psi \rangle,
\end{aligned}$$

where we have used the spectral decomposition property of Hermitian matrix M:

$$M = \sum_i \lambda_i |\lambda_i\rangle \langle \lambda_i|. \tag{1.36}$$

The general proof is similar to our simple example:

Proof.

$$\langle M \rangle = \sum_i |\langle \lambda_i | \psi \rangle|^2 \lambda_i$$

$$= \sum_i \langle \psi | \lambda_i \rangle \langle \lambda_i | \psi \rangle \lambda_i$$

$$= \sum_i \langle \psi | (\lambda_i |\lambda_i\rangle \langle \lambda_i |) |\psi \rangle$$

$$= \langle \psi | \left(\sum_i \lambda_i |\lambda_i\rangle \langle \lambda_i | \right) |\psi \rangle$$

$$= \langle \psi | M | \psi \rangle .$$

□

Example with Photon Polarization: Malus' Law

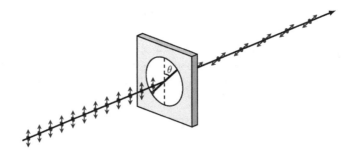

Figure 1.10: Measuring Photons: Malus' Law

To illustrate Eq. 1.35, we consider a scenario involving vertically polarized photons measured with a polarizer oriented at an angle θ, as depicted in Fig. 1.10. The state of the photon after passing through the polarizer can be described by

$$|\theta\rangle = \cos\theta |V\rangle + \sin\theta |H\rangle = \begin{bmatrix} \cos\theta \\ \sin\theta \end{bmatrix}. \tag{1.37}$$

The associated observable operator for the polarizer is given by

$$M_\theta = |\theta\rangle \langle \theta| = \begin{bmatrix} \cos^2\theta & \cos\theta \sin\theta \\ \cos\theta \sin\theta & \sin^2\theta \end{bmatrix}. \tag{1.38}$$

To find the statistical average of M_θ, we apply Eq. 1.34:

$$\langle M \rangle = |\langle \theta|V\rangle|^2 \cdot 1 = \cos^2\theta. \tag{1.39}$$

Alternatively, using Eq. 1.35, we obtain

$$\langle M \rangle = \langle V|M_\theta|V\rangle = \cos^2\theta. \tag{1.40}$$

Both methods yield the statistical average as $\cos^2\theta$, which confirms Malus' Law: the output-to-input intensity ratio is $\cos^2\theta$. Since light intensity is proportional to photon flux, this result provides a quantum-mechanical rationale for Malus' law.

1.5 The Quantum Measurement Postulate

One may ask why we should prove Malus' law using quantum mechanics when it can be explained using classical electromagnetic theory. There are several compelling reasons. Firstly, using photons as an example helps develop a foundational understanding of quantum mechanics. Secondly, the quantum theory of photons is essential for exploring advanced quantum phenomena such as photon entanglement. Lastly, this example serves to illustrate how classical physics emerges as a statistical limit of quantum physics.

> **Exercise 1.14** Suppose we measure circularly polarized photons $|L\rangle$ with the polarizer M_θ. Calculate the statistical average for this case.

> **Exercise 1.15** Suppose we measure $|V\rangle$ photons using a specialized device D rather than a polarizer. This device yields a $+1$ for photons polarized at an angle θ and -1 for the orthogonal polarization at $\theta + \pi/2$. Calculate $\langle D \rangle$.

1.5.4 Quantum Interference

Quantum mechanics fundamentally challenges our classical understanding of waves and particles with its concept of wave-particle duality. This is beautifully encapsulated by the phenomenon of interference. Unlike classical interference, which involves the superposition of wave amplitudes, quantum interference incorporates the quantum State Postulate, which permits particles like photons to exist in superposition states, and the Observables and Measurements Postulates, which govern the measurement of these superpositions.

Quantum Interference

Quantum interference is the phenomenon where particles like photons exist in superposition states and interfere with themselves upon measurement.

The simplest manifestation of quantum interference is the two-path interference. This phenomenon occurs when particles like photons or electrons have the potential to traverse two or more paths simultaneously and interfere with themselves. This mechanism serves as a cornerstone for studying quantum interference, entanglement, and the implementation of qubits.

1 Mach-Zehnder Interferometer

The Mach-Zehnder interferometer, depicted in Fig. 1.11, is an exemplary experimental setup to observe photon two-path interference. Upon entering the interferometer, the photon encounters a beam splitter (labeled BS1), which is essentially a half mirror – with 50% transmission and 50% reflection. This device effectively splits the photon into two paths: the transmitted path A and the reflected path B.

 It is important to note a single photon can travel along two paths simultaneously, a phenomenon related to particle-wave duality. It is not that some photons travel in path A while others travel in path B. This has been extensively verified using "quantum eraser experiments" [61].

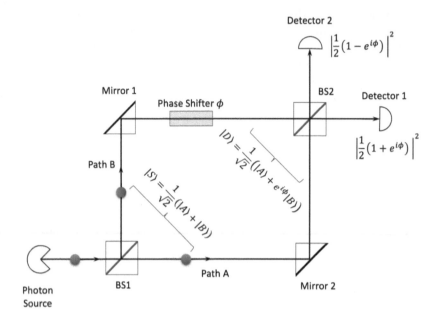

Figure 1.11: Mach-Zehnder Interferometer

The post-splitting superposition state is given by

$$|S\rangle = \frac{1}{\sqrt{2}}(|A\rangle + |B\rangle), \tag{1.41}$$

where $|A\rangle$ and $|B\rangle$ represent the photon states along paths A and B, respectively.

Following the first beam splitter, the photon is in a superposition state and may accrue a phase shift ϕ along path B relative to path A. This phase shift modifies the state to

$$|D\rangle = \frac{1}{\sqrt{2}}(|A\rangle + e^{i\phi}|B\rangle). \tag{1.42}$$

After traversing paths A and B, the photon encounters the second beam splitter (BS2). BS2 mixes the photon state from paths A and B and sends it to Detector 1 and Detector 2. The photon detectors, coupled with BS2, effectively measure the photon state in the basis $\frac{1}{\sqrt{2}}(|A\rangle + |B\rangle) \equiv |S\rangle$ and $\frac{1}{\sqrt{2}}(|A\rangle - |B\rangle)$, commonly denoted as $|+\rangle$ and $|-\rangle$ in quantum computing. (The beam splitter itself is functionally equivalent to a Hadamard gate, a topic to be elaborated on later.) Specifically, the observable operator associated with Detector 1 in the figure is $M_S = |S\rangle\langle S|$, which leads to the following probability for detecting the photon in state $|S\rangle$:

$$P = |\langle S|D\rangle|^2 = \left|\frac{1}{2}(1 + e^{i\phi})\right|^2 = \cos^2\frac{\phi}{2}. \tag{1.43}$$

Notice that P shows a sinusoidal dependence on ϕ. If $\phi = \pi$, the detection probability drops to zero, illustrating the principle of destructive interference.

Exercise 1.16 Using the quantum formulation of two-path interference, calculate the interference pattern in a Mach-Zehnder interferometer if (a) only the first beam splitter, and (b) both beam splitters, have a split ratio of 40-60 instead of 50-50.

2 Double-Slit Experiment

A related demonstration of photon two-path interference is the double-slit experiment, as depicted in Figure 1.12, where a single photon is fired at a barrier with two slits. Even though it's a single photon, it acts as if it has gone through both slits simultaneously, creating an interference pattern on a detector screen beyond the barrier.

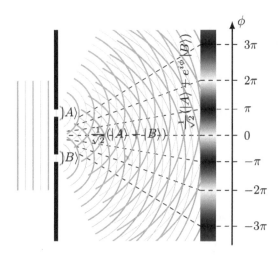

Figure 1.12: Double Slit Interference Experiment

Exercise 1.17 ✱ Investigate the delayed-choice quantum eraser experiments, which are related to the double-slit experiment but delve into more captivating quantum effects associated with photon interference. Recommended sources include, but are not limited to, https://en.wikipedia.org/wiki/Quantum_eraser_experiment and https://en.wikipedia.org/wiki/Delayed-choice_quantum_eraser.

1.6 The Uncertainty Principle

The Measurement Postulate in quantum mechanics leads to several profound implications, one of which is the Uncertainty Principle. Though historically known as the Heisenberg Uncertainty Principle (HUP), it can be interpreted as a direct corollary of the Measurement Postulate.

The Uncertainty Principle has two forms. The first form distinguishes between compatible and incompatible observables. The second form presents the principle as a mathematical inequality. We will focus on the first version here, given its relevance to quantum computing.

1.6.1 Compatible versus Incompatible Observables

When performing a quantum measurement, the measured value is always an eigenvalue of the associated observable operator. Subsequently, the quantum state collapses to the corresponding eigenstate. Eigenstates of an operator represent states where measurement outcomes are certain, occurring with probability 1.

Two observables can either be compatible or incompatible, indicating whether both can be measured with 100% accuracy. This is often referred to as simultaneous measurements, although not necessarily at the same moment in time. Incompatibility between observables implies that measurements of one observable disturb those of the other. The Uncertainty Principle encapsulates this condition as follows:

> **The Uncertainty Principle**
>
> If operators A and B, representing two observables, do not commute, i.e., $[A, B] \neq 0$, then these observables cannot be measured simultaneously with perfect accuracy (i.e., with probability 1). In other words, they are incompatible.

Proof. Assume, for the sake of contradiction, that A and B can be measured simultaneously with perfect accuracy. Then, there must exist a complete set of simultaneous eigenstates $\{|\phi_i\rangle\}$, satisfying:

$$A|\phi_i\rangle = a_i|\phi_i\rangle, \quad B|\phi_i\rangle = b_i|\phi_i\rangle, \tag{1.44}$$

where a_i and b_i are the eigenvalues of A and B respectively.

If A and B possess simultaneous eigenstates $|\phi_i\rangle$, we have:

$$AB|\phi_i\rangle - BA|\phi_i\rangle = a_i b_i |\phi_i\rangle - b_i a_i |\phi_i\rangle = 0. \tag{1.45}$$

This implies that the commutator $[A, B] = AB - BA = 0$, contradicting our initial assumption $[A, B] \neq 0$.

Therefore, if A and B do not commute, they cannot have simultaneous eigenstates and thus cannot be measured simultaneously with perfect accuracy. □

> **Implications of the Uncertainty Principle**
>
> Oberservables A and B can be measured simultaneously
>
> $\Leftrightarrow [A, B] = 0$
>
> $\Leftrightarrow A$ and B share eigenvectors $\{|\phi_i\rangle\}$
>
> $\Leftrightarrow A = \sum_i a_i |\phi_i\rangle\langle\phi_i|, \quad B = \sum_i b_i |\phi_i\rangle\langle\phi_i|.$

Examples with Photon Polarization

To better understand the Uncertainty Principle in action, we can consider photon polarization as an illustrative example.

1.6 The Uncertainty Principle

Vertical and Horizontal Polarization

Suppose we aim to measure the vertical polarization $|V\rangle$ and the horizontal polarization $|H\rangle$. Do these observables commute? Indeed, they do:

$$A \equiv M_V = \begin{bmatrix} 1 & 0 \\ 0 & 0 \end{bmatrix},$$

$$B \equiv M_H = \begin{bmatrix} 0 & 0 \\ 0 & 1 \end{bmatrix},$$

$$[A, B] = AB - BA = 0.$$

This means we can measure both $|V\rangle$ and $|H\rangle$ with 100% certainty.

Vertical and Diagonal Polarization

Now, consider measuring the vertical polarization $|V\rangle$ and the diagonal polarization $|D\rangle$:

$$A \equiv M_V = \begin{bmatrix} 1 & 0 \\ 0 & 0 \end{bmatrix},$$

$$B \equiv M_D = \frac{1}{2}\begin{bmatrix} 1 & 1 \\ 1 & 1 \end{bmatrix},$$

$$[A, B] = AB - BA \neq 0.$$

Since $[A, B] \neq 0$, the observables for $|V\rangle$ and $|D\rangle$ are incompatible. In practice, if we measure $|V\rangle$ with 100% certainty, subsequent measurements for $|D\rangle$ will show both pass and block with 50% probability.

Applications in Quantum Communication: BB84 Protocol

In the context of the BB84 Quantum Key Distribution (QKD) protocol, consider a string of photons with varying polarizations.

- Case 1: Vertical and Horizontal - A sequence like {H, V, V, V, H, V, H, ...} can be measured with either a vertical or horizontal polarizer, confirming the polarization of each photon with 100% certainty.

- Case 2: Diagonal - For a sequence {D, A, D, A, A, A, D, ...}, a 45° or 135° diagonal polarizer can measure the polarization states, since the states are orthogonal.

- Case 3: Mixture - If the sequence mixes vertical, horizontal, and diagonal polarizations, like {H, V, D, V, H, A, D, H, ...}, perfect measurement is not possible. This stems from the incompatibility of the observable operators for $|V\rangle$ and $|D\rangle$, as shown above.

> **Exercise 1.18** Show that the measurements of left circular polarization $|L\rangle$ and the diagonal polarization $|D\rangle$ are incompatible by demonstrating that their observable operators do not commute. Explain what happens when photons in state $|D\rangle$ are measured for circular polarization and diagonal polarization.

1.6.2 ∗The Uncertainty Inequality

The Uncertainty Principle can also be formally expressed in terms of standard deviations of measurements for the operators A and B. This formulation is particularly historical, often presented in the context of position and momentum or energy and time.

$$\Delta A \cdot \Delta B \geq \frac{1}{2} |\langle [A, B] \rangle|, \tag{1.46}$$

where $\Delta A = \sqrt{\langle A^2 \rangle - \langle A \rangle^2}$ and similarly for ΔB. Here, $[A, B] = AB - BA$ represents the commutator of A and B.

This inequality form of the Uncertainty Principle is not an arbitrary construct; it stems from the mathematical foundation provided by the Cauchy-Schwarz inequality: $|\langle u|v \rangle|^2 \leq \langle u|u \rangle \langle v|v \rangle$.

Although the underlying mathematics are intriguing, we will not delve into those details in this text.

1.7 *Further Readings on Quantum Mechanics

For students and researchers aiming to deepen their understanding of quantum mechanics, we recommend the following textbooks.

- Junichiro Kono. *Quantum Mechanics for Tomorrow's Engineers: New Edition.* English. New edition. Cambridge University Press, Sept. 29, 2022, page 350. ISBN: 978-1108842587.

  This text presents a pioneering approach to quantum mechanics, tailored for students in quantum computing and engineering. This comprehensive undergraduate textbook bridges the gap between abstract quantum principles and their tangible applications in the field of quantum engineering.

  Kono's work is unique in its focus on current technologies, steering clear of historical methodologies to prepare students for contemporary engineering challenges. The book offers a robust introduction to quantum information fundamentals and demonstrates their practicality through a plethora of real-world examples. These include quantum well infrared photodetectors, solar cells, and quantum teleportation, as well as cutting-edge topics like quantum computing, and band gap engineering.

- Claude Cohen-Tannoudji, Bernard Diu, and Franck Laloë. *Quantum Mechanics, Vol 1: Basic Concepts, Tools, and Applications.* Wiley, 2019. ISBN: 978-3-527-34553-3.

  The series of three volumes, authored by Nobel Prize laureate Claude Cohen-Tannoudji and colleagues, stands as a seminal contribution to the field of quantum mechanics. This comprehensive series not only introduces the fundamentals but also delves into more intricate aspects of the subject.

  The first volume is particularly noteworthy for its excellent introductory content. It provides a clear and detailed exposition of the fundamental postulates of quantum mechanics. The text excels in going beyond simple definitions, integrating rich discussions and intuitive explanations that shed light on complex concepts. A remarkable aspect of this volume is its focus on two-level systems. These systems form the cornerstone of qubit systems in quantum computing and quantum information science, and the dedicated sections offer valuable insights directly relevant to current research and technological advancements in these areas.

The series continues to offer in-depth knowledge in its subsequent volumes. Volume 2, titled "Angular Momentum, Spin, and Approximation Methods," and Volume 3, "Fermions, Bosons, Photons, Correlations, and Entanglement," extend the discussion to more advanced topics. These volumes provide a comprehensive understanding of the subject, making the series an indispensable resource for anyone keen to explore the depths of quantum mechanics.

- Richard P. Feynman, Robert B. Leighton, and Matthew Sands. *Quantum Mechanics (Feynman Lectures on Physics, Volume 3)*. Basic Books, Oct. 4, 2011. ISBN: 978-0465025015.

This volume, part of the renowned "Feynman Lectures on Physics," provides an inspirational and insightful exploration of quantum mechanics. Authored by the legendary physicist Richard P. Feynman, this book is celebrated for its ability to convey complex ideas in an accessible manner. Feynman's unique teaching style combines rigorous scientific understanding with a deep appreciation of the beauty underlying quantum phenomena. This text is an excellent resource for anyone seeking a deeper conceptual understanding of quantum mechanics, presented in a way that is both engaging and thought-provoking.

- J. J. Sakurai and Jim Napolitano. *Modern Quantum Mechanics*. Cambridge University Press, 2017. ISBN: 978-1-108-47322-4.

This text offers a comprehensive and detailed examination of advanced quantum mechanics concepts and techniques, particularly useful for those involved in quantum simulation. The book adeptly covers a range of advanced topics including Symmetry in Quantum Mechanics, Scattering Theory, Approximation Methods, Identical Particles, and Relativistic Quantum Mechanics. Its clear and structured approach makes complex ideas more accessible, while providing practical insights into how these principles are applied in modern quantum research and technology.

1.8 Topic Reviews

In this chapter, we navigated through the intricate landscape of quantum mechanics using photons as our guide. We delved into a myriad of core concepts and mathematical formulations, serving both as an introduction and a focused exploration into the quantum realm. To aid in assimilating this rich tapestry of knowledge, this section provides a concise review of the essential concepts and equations introduced.

1.8.1 Review of Postulates 1-3

- **Postulate 1: Quantum State**
  - States of quantum systems are described by normalized complex vectors in a Hilbert space.

- **Implications from Postulate 1**
  - Orthogonal states are mutually exclusive, rendering them orthonormal.
  - A general quantum state can exist as a superposition of basis states.

- The global phase of a state vector has no physical significance.
- **Postulate 2: Quantum Observables**
  - Physically measurable quantities correspond to Hermitian operators.
  - Measurement outcomes and states are defined by the eigenvalues and eigenvectors of these operators.
- **Implications from Postulate 2**
  - Quantum observables often have discrete eigenvalues, leading to quantization of measurement outcomes.
  - The choice of observable specifies the basis in which the measurement will be expressed.
  - Dirac bra-ket notation universally represents quantum states and operators in a basis-independent manner.
- **Postulate 3: Quantum Measurements**
  - Measurement outcomes and probabilities are governed by the Born rule.
  - Measurements induce a collapse of the quantum state.
- **Implications from Postulate 3**
  - Quantum measurements are inherently probabilistic.
  - A measurement irreversibly alters the state.
  - Subsequent measurements yield identical results.
  - Statistical average for measuring an observable can be calculated.
  - The Born rule connects quantum states to measurement probabilities.
  - Probabilities of all possible outcomes must sum to 1.
- **Cross-Postulate Implications**
  - Quantum Interference: A phenomenon that arises due to the superposition of states and subsequent measurements.
  - The Uncertainty Principle: Limitations exist for the simultaneous measurement of non-commuting observables.

See **Table 1.2**: Summary of Photon Polarization States and Observables

1.8.2 Review of Photons as Quantum Systems

- **Photon as Quantum Systems**
  - Photons are elementary particles and the quanta of electromagnetic fields, including light.
  - Photons are vital in various quantum technologies such as lasers, quantum cryptography, and quantum computing.
  - The polarization state of a photon can be represented as a normalized complex vector in a Hilbert space.

1.8 Topic Reviews

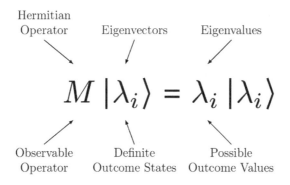

General superposition state: $|\psi\rangle = \sum_i c_i |\lambda_i\rangle$ with $\langle\psi|\psi\rangle = \sum_i |c_i|^2 = 1$.
Probability of outcome λ_i: $|\langle\lambda_i|\psi\rangle|^2 = |c_i|^2$. Post-measurement state: $|\lambda_i\rangle$.
Expected value (statistical average) of M: $\langle M \rangle = \sum_i |c_i|^2 \lambda_i = \langle\psi|M|\psi\rangle$.

Figure 1.13: Summary of the Observable and Measurement Postulates

  - Three essential sets of orthogonal photon polarization states: rectilinear, diagonal, and circular.

- **Characteristics of Photons**

  - Polarization states that are orthogonal are mutually exclusive, thus aligning with the orthonormal property in quantum mechanics.

  - A photon's polarization can exist as a superposition of basis polarization states.

  - Quantum polarization states can also be superposed and entangled, giving rise to distinctly quantum behaviors with no classical analogs.

  - Quantum interference is observed when photons are sent through two-path systems, supporting the superposition principle.

- **Photon Observables**

  - Polarization of a photon serves as an example of a quantum observable.

  - Measurements related to polarization can be defined through appropriate Hermitian operators.

- **Photon Measurements**

  - Measurement of a photon's polarization yields results consistent with the Born rule.

  - After measurement, the photon state collapses to the eigenstate corresponding to the measured polarization angle.

- **Characteristics of Photon Measurements**

  - Similar to other quantum systems, photon measurements are probabilistic.

- The act of measuring the polarization collapses the photon's state, altering it irreversibly.

- Repeated measurements on the same photon would yield the same polarization angle, provided the photon is not altered between measurements.

See **Table 1.2**: Summary of Photon Polarization States and Observables

1.9 Summary and Conclusions

Foundational Principles

In this chapter, we have delved into the foundational principles that govern quantum computing and quantum information science (QCI). Our discussion emphasized the importance of understanding these principles, not just for theoretical clarity but also for practical applications.

We began by examining quantum states, described by normalized complex vectors in a vector space, setting the stage for the introduction of quantum bits or qubits. We explored the critical concept of superposition, which underlies the parallel computational capabilities unique to quantum systems.

We then transitioned to the discussion of quantum observables, represented by Hermitian operators. These operators define physically measurable quantities, and their eigenvalues and eigenvectors are intimately related to the outcomes and probabilities of quantum measurements.

Next, we addressed the essential rules governing quantum measurements, focusing on the Born rule for calculating probabilities and the concept of state collapse post-measurement. We underscored the inherent probabilistic nature of quantum mechanics and the irreversible effect of measurements on quantum states.

Further, we discussed ancillary but important concepts such as quantum interference, the uncertainty principle, and the lack of physical significance for the global phase of a quantum state. These discussions enriched our understanding of the subtleties and complexities involved in quantum mechanics and quantum computation.

General Framework

While our approach has been tailored towards two-level quantum systems like qubits, the principles discussed are general and applicable to higher-dimensional quantum systems as well. Throughout the chapter, we have presented examples using photons to make the concepts more relatable, without compromising on academic rigor.

Our framework positions QCI as an interdisciplinary field, combining insights from quantum physics, computer science, and mathematics. This multidisciplinary lens offers a more comprehensive understanding, preparing the reader for the rich and diverse landscape of quantum computing and information.

Upcoming Topics

In the upcoming chapters, we will initially focus on half-integer spins, another physical realization of a qubit system. Following this, we will introduce a general theoretical framework for qubits. Subsequently, we will explore the remaining principles of Quantum Evolution and Quantum Composition, further deepening our understanding of the theory and practice of Quantum Computing and Information.

Problem Set 1

1.1 Compare Eqs. 1.12 and 1.16, and derive formulas for

    (a) γ and δ in terms of μ and ν

    (b) μ and ν in terms of γ and δ

using both matrix representation and Dirac notation.

1.2 Show that when $\phi = \frac{\pi}{2}$, Eq. 1.19 represents an elliptical polarization state with major radius aligned with $|V\rangle$ or $|H\rangle$. What are its major and minor radii?

1.3 Investigate the relationship between the tilt angle φ, and the major and minor radii of the ellipse in Fig. 1.5, with respect to θ and ϕ.

1.4 Elaborate the conditions for a, c, and d, such that

$$\begin{bmatrix} a & \sin\theta e^{i\phi} + i\cos\theta e^{-i\phi} \\ c & d \end{bmatrix}$$

is a Hermitian matrix, where $\phi, \theta \in \mathbb{R}$ and $a, c, d \in \mathbb{C}$.

1.5 A general linear polarization state $|\theta\rangle$ at angle θ relative to $|V\rangle$, and its orthogonal state $|\theta_\perp\rangle$, are given by

$$|\theta\rangle = \begin{bmatrix} \cos\theta \\ \sin\theta \end{bmatrix}, \quad |\theta_\perp\rangle = \begin{bmatrix} \sin\theta \\ -\cos\theta \end{bmatrix}.$$

Derive the corresponding observable operator M_θ.

1.6 Derive the observable operator M_P for the general (elliptical) polarization state given in Eq. 1.19.

1.7 Consider photons polarized at an angle α relative to $|V\rangle$, and measure them using a polarizer oriented at an angle β. Show that the statistical average of the measurement is $\cos^2(\alpha - \beta)$, which depends only on the relative angle $\alpha - \beta$.

1.8 Elaborate the conditions of α, β, δ and γ under which measuring photons against the general polarization state $\alpha|V\rangle + \beta|H\rangle$ is compatible with measuring against another general polarization state $\delta|V\rangle + \gamma|H\rangle$.

2. Fundamentals of Spin Systems

Contents

2.1	**Spin, Angular Momentum, and Magnetic Moment**	**44**
2.1.1	Fermions, Bosons, and Anyons	44
2.1.2	1/2 Spin	44
2.1.3	Magnetic Moment	45
2.2	**Spin-1/2 States and Pauli Matrices**	**45**
2.2.1	Spin-1/2 States and Observables	45
2.2.2	Properties of Pauli Matrices	46
2.3	✶ **General Spin State Representation**	**47**
2.3.1	✶ Defining the Spin Observable for Any Direction	47
2.3.2	✶ Spin State in Any Direction	48
2.3.3	General Spin State	49
2.3.4	Spin State and Polarization State Correspondence	49
2.4	**The Bloch Sphere**	**50**
2.4.1	Key States on the Bloch Sphere	51
2.4.2	Representation of Pauli X, Y, and Z on the Bloch Sphere	51
2.5	**Spin Measurement**	**51**
2.5.1	The Stern-Gerlach Experiment	52
2.5.2	Cascaded Stern-Gerlach Experiments	54
2.5.3	✶ Modern Spin Experiments	54
2.6	**Summary and Conclusions**	**55**
	Problem Set 2	**56**

In this chapter, we explore the fundamental principles of quantum mechanics as they pertain to spin systems. Understanding the intricacies of spin is vital, particularly for fields that heavily rely on quantum mechanics, such as quantum computing. Along the way, we will introduce two essential tools: the Pauli matrices and the Bloch sphere.

2.1 Spin, Angular Momentum, and Magnetic Moment

In classical physics, objects like a spinning tennis ball possess angular momentum. This angular momentum is a vector quantity directed along the axis of rotation, and its magnitude can take continuous values. However, for quantum particles such as electrons, protons, and neutrons, as well as composite particles like atoms, the scenario is fundamentally different. The angular momentum of these particles is quantized and can only take discrete values. Moreover, these particles can have a permanent, non-zero angular momentum, as if they are perpetually spinning. This intrinsic form of angular momentum is known as *spin*.

Spin is a pervasive phenomenon in physics with a plethora of applications, including but not limited to electron microscopy, magnetic resonance, and quantum computation.

2.1.1 Fermions, Bosons, and Anyons

Particles can be classified into two broad categories based on their spin: fermions and bosons. Fermions have half-integer spins ($\frac{1}{2}$, $\frac{3}{2}$, ...) while bosons possess integer spins (0, 1, 2, ...). Essentially all particles of conventional matter in our observable universe, including electrons, protons, and neutrons, are fermions. Photons, which are particles of light, have an integer spin of 1, classifying them as bosons. This distinction between integer and half-integer spins arises from the principles of special relativity, which dictate how particles' wave functions transform under the Lorentz group.

In addition to fermions and bosons, there is another intriguing class of quasiparticles known as anyons. Unlike fermions and bosons, which adhere strictly to integer or half-integer spin values, anyons exhibit more exotic behaviors in two-dimensional systems. Their spins can take on a range of values between integers and half-integers, leading to unconventional quantum statistics. Anyons are currently of great interest, especially in the burgeoning field of topological quantum computing, where they are considered as potential carriers of quantum information.

2.1.2 1/2 Spin

Particles with a spin of $\frac{1}{2}$, such as electrons and some atomic nuclei, are of particular interest in the domain of quantum computing. As illustrated in Fig. 2.1, these particles exhibit two discrete quantum states, commonly termed as up and down states. These are mathematically represented by $|\uparrow\rangle$ and $|\downarrow\rangle$, or equivalently $|0\rangle$ and $|1\rangle$. The angular momentum corresponding to these states is $S_u = \frac{\hbar}{2}$ and $S_d = -\frac{\hbar}{2}$, where $\hbar$ is the reduced Planck constant.

In addition to serving as two-level quantum systems that can act as qubits, the mathematical framework that describes spin-$\frac{1}{2}$ systems can easily be generalized to model qubits, forming a cornerstone for quantum computing algorithms.

One fascinating aspect of spin-$\frac{1}{2}$ particles is their unusual rotational behavior. Unlike classical objects that return to their original state after a 360° rotation, spin-$\frac{1}{2}$ particles require a 720° rotation for their wavefunction to return to its original state. This intriguing property, fundamentally linked to the topology of quantum states, significantly departs from classical expectations.

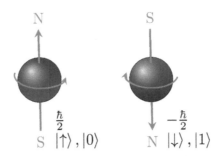

Figure 2.1: 1/2 Spin

Given its significance in quantum computing, our subsequent discussions will concentrate primarily on spin-$\frac{1}{2}$ systems.

2.1.3 Magnetic Moment

For charged quantum particles like electrons and protons, the intrinsic spin is associated with a magnetic moment. Mathematically, this magnetic moment is proportional to the spin, denoted as $\vec{\mu} \propto \vec{S}$. This relationship is fundamental to technologies that exploit magnetic interactions, such as Magnetic Resonance Imaging (MRI), quantum computing, and quantum sensing.

2.2 Spin-1/2 States and Pauli Matrices

2.2.1 Spin-1/2 States and Observables

Spin, as an intrinsic form of angular momentum, is represented by a vector in three-dimensional space. For particles with spin-$\frac{1}{2}$, this vector assumes two quantized values, $\frac{\hbar}{2}$ and $-\frac{\hbar}{2}$, in any chosen direction, where $\hbar$ is the reduced Planck constant. In the z-direction, these states are typically referred to as up and down, denoted as $|0\rangle \equiv |\uparrow\rangle$ and $|1\rangle \equiv |\downarrow\rangle$, which constitute the computational basis. Mathematically, these states are defined as:

$$|0\rangle = \begin{bmatrix} 1 \\ 0 \end{bmatrix}, \quad |1\rangle = \begin{bmatrix} 0 \\ 1 \end{bmatrix}. \tag{2.1}$$

The observable S_z, associated with measuring spin in the z-direction, has these states as its eigenstates, and $\frac{\hbar}{2}$ and $-\frac{\hbar}{2}$ as the corresponding eigenvalues. S_z can be represented through its spectral expansion as:

$$S_z = \frac{\hbar}{2}|0\rangle\langle 0| - \frac{\hbar}{2}|1\rangle\langle 1| = \frac{\hbar}{2}Z, \tag{2.2}$$

where Z is the Pauli Z matrix, represented as

$$Z = \begin{bmatrix} 1 & 0 \\ 0 & -1 \end{bmatrix}. \tag{2.3}$$

In quantum computing contexts, we often work directly with Z instead of S_z, given that they differ only by a factor of $\frac{\hbar}{2}$.

Similarly, in the x-axis, the state vectors are expressed as:

$$|+\rangle = \frac{1}{\sqrt{2}}(|0\rangle + |1\rangle) = \frac{1}{\sqrt{2}}\begin{bmatrix} 1 \\ 1 \end{bmatrix}, \quad (2.4a)$$

$$|-\rangle = \frac{1}{\sqrt{2}}(|0\rangle - |1\rangle) = \frac{1}{\sqrt{2}}\begin{bmatrix} 1 \\ -1 \end{bmatrix}. \quad (2.4b)$$

The associated observable is the Pauli X matrix:

$$X = |+\rangle\langle+| - |-\rangle\langle-| = \begin{bmatrix} 0 & 1 \\ 1 & 0 \end{bmatrix}. \quad (2.5)$$

In the y-axis, the state vectors are:

$$|+_i\rangle = \frac{1}{\sqrt{2}}(|0\rangle + i|1\rangle) = \frac{1}{\sqrt{2}}\begin{bmatrix} 1 \\ i \end{bmatrix}, \quad (2.6a)$$

$$|-_i\rangle = \frac{1}{\sqrt{2}}(|0\rangle - i|1\rangle) = \frac{1}{\sqrt{2}}\begin{bmatrix} 1 \\ -i \end{bmatrix}. \quad (2.6b)$$

$|+_i\rangle$ and $|-_i\rangle$ are often written as $|i_+\rangle$ and $|i_-\rangle$, or $|i\rangle$ and $|-i\rangle$ in other texts. The associated observable is the Pauli Y matrix:

$$Y = |+_i\rangle\langle+_i| - |-_i\rangle\langle-_i| = \begin{bmatrix} 0 & -i \\ i & 0 \end{bmatrix}. \quad (2.7)$$

Exercise 2.1 Derive the matrix forms of Pauli matrices X and Y based on their bra-ket representations.

The specific forms of these state vectors are not arbitrary but are the result of insightful contributions from physicists, notably Pauli. These forms will be further justified in the subsequent discussion about general spin directions.

2.2.2 Properties of Pauli Matrices

Pauli matrices serve functions beyond acting as spin observables; they play a pivotal role in various domains of quantum computing and quantum physics. As such, a thorough understanding of their properties is crucial. These properties are highlighted below, with more details in Appendix D. (For succinct presentation of certain relations, Pauli matrices are often denoted as $\sigma_1 \equiv X$, $\sigma_2 \equiv Y$, $\sigma_3 \equiv Z$. In the following, we use σ to represent any one of them.)

1. Hermitian Property: $\sigma^\dagger = \sigma$.

2. Unitary: $\sigma^\dagger \sigma = I$.

3. Involutory Property: $\sigma^2 = I$.

4. Commutation Relation: $[\sigma_j, \sigma_k] = 2i\varepsilon_{jkl}\sigma_l$, where ε_{jkl} is the Levi-Civita permutation symbol.

5. **Anti-Commutation Relations:** $\{\sigma_i, \sigma_j\} = 2\delta_{ij}I$, where δ_{ij} is the Kronecker delta function.

6. **Algebraic Completeness:** Pauli matrices (together with I) provides a complete basis for general matrices in $\mathbb{C}^{2\times 2}$.

7. **Spin Angular Momentum:** Pauli matrices represent the spin angular momentum operator in the x, y, and z directions with the correct rotation properties.

> **Why Complex Numbers?**
>
> One may wonder why one of the Pauli matrices contains i. Is it fundamentally necessary or just a matter of convenience? It turns out the presence of i is necessary for the Pauli operators:
>
> 1. to satisfy anti-commutation relation.
> 2. to generate the correct rotation properties of angular momentum for spin-$\frac{1}{2}$ particles.
> 3. to provide a complete, orthogonal basis for 2×2 Hermitian matrices.

Exercise 2.2 Given the eigenvectors $|0\rangle$ and $|1\rangle$ of Pauli matrix Z, $|+\rangle$ and $|-\rangle$ of X, and $|+_i\rangle$ and $|-_i\rangle$ of Y, prove that

$$\begin{aligned}
\langle 0|Z|0\rangle &= 1, & \langle 1|Z|1\rangle &= -1, \\
\langle +|X|+\rangle &= 1, & \langle -|X|-\rangle &= -1, \\
\langle +_i|Y|+_i\rangle &= 1, & \langle -_i|Y|-_i\rangle &= -1,
\end{aligned}$$

and

$$\begin{aligned}
\langle 0|X|0\rangle &= \langle 1|X|1\rangle &= \langle 0|Y|0\rangle &= \langle 1|Y|1\rangle &= 0, \\
\langle +|Y|+\rangle &= \langle -|Y|-\rangle &= \langle +|Z|+\rangle &= \langle -|Z|-\rangle &= 0, \\
\langle +_i|Z|+_i\rangle &= \langle -_i|Z|-_i\rangle &= \langle +_i|X|+_i\rangle &= \langle -_i|X|-_i\rangle &= 0.
\end{aligned}$$

Provide a physical interpretation of these results in terms of measuring the X, Y, and Z components of spins oriented in various directions.

2.3 * General Spin State Representation

In this subsection, we explore the representation of spin states and observables along a general direction in three-dimensional space. This direction is specified by $\hat{u}$, a real unit vector with Cartesian components u_x, u_y, and u_z, where $u_x^2 + u_y^2 + u_z^2 = 1$. In spherical coordinates (see Fig. 2.2), $\hat{u}$ is described by the polar angle θ and azimuthal angle ϕ:

$$\hat{u} = \begin{bmatrix} u_x \\ u_y \\ u_z \end{bmatrix} = \begin{bmatrix} \sin\theta\cos\phi \\ \sin\theta\sin\phi \\ \cos\theta \end{bmatrix}. \tag{2.8}$$

2.3.1 * Defining the Spin Observable for Any Direction

The Pauli matrices X, Y, and Z represent the Cartesian components of the observable of an angular momentum, a three-dimensional vector. Consequently, the spin

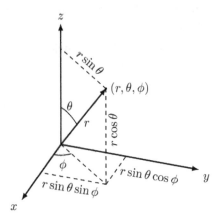

Figure 2.2: Spherical Coordinate System

observable along the direction of $\hat{u}$ can be expressed as:

$$S = u_x X + u_y Y + u_z Z = \sin\theta\cos\phi X + \sin\theta\sin\phi Y + \cos\theta Z. \qquad (2.9)$$

Substituting X, Y, and Z with their respective matrices, we find:

$$S \equiv \sigma_u = \begin{bmatrix} \cos\theta & e^{-i\phi}\sin\theta \\ e^{i\phi}\sin\theta & -\cos\theta \end{bmatrix}. \qquad (2.10)$$

To validate the definition of S in Eq. 2.9, the expected value $\langle S \rangle$ (see § 1.5.3) for a spin state should align with the Cartesian components of $\hat{u}$:

$$\langle +|S|+\rangle = u_x = \sin\theta\cos\phi, \qquad \langle -|S|-\rangle = -u_x, \qquad (2.11a)$$
$$\langle +_i|S|+_i\rangle = u_y = \sin\theta\sin\phi, \qquad \langle -_i|S|-_i\rangle = -u_y, \qquad (2.11b)$$
$$\langle 0|S|0\rangle = u_z = \cos\theta, \qquad \langle 1|S|1\rangle = -u_z. \qquad (2.11c)$$

These relations can be verified through the bra-ket notation using the results from Exercise 2.2 or by directly computing them in matrix form.

Exercise 2.3 Verify Eq. 2.11.

2.3.2 ∗ Spin State in Any Direction

The eigenvectors of S in Eq. 2.10 are the spin states in the $\hat{u}$ and $-\hat{u}$ directions:

$$|+_u\rangle = \cos\frac{\theta}{2}|0\rangle + \sin\frac{\theta}{2}e^{i\phi}|1\rangle = \begin{bmatrix} \cos\frac{\theta}{2} \\ \sin\frac{\theta}{2}e^{i\phi} \end{bmatrix}, \qquad (2.12a)$$

$$|-_u\rangle = -\sin\frac{\theta}{2}|0\rangle + \cos\frac{\theta}{2}e^{i\phi}|1\rangle = \begin{bmatrix} -\sin\frac{\theta}{2} \\ \cos\frac{\theta}{2}e^{i\phi} \end{bmatrix}. \qquad (2.12b)$$

2.3 ✻ General Spin State Representation

Exercise 2.4 Verify that $S|+_u\rangle = |+_u\rangle$ and $S|-_u\rangle = -|-_u\rangle$.

Exercise 2.5 You may find $|+_u\rangle$ and $|-_u\rangle$ expressed differently in other texts:

$$|+_u\rangle = \cos\frac{\theta}{2}e^{-i\frac{\phi}{2}}|0\rangle + \sin\frac{\theta}{2}e^{i\frac{\phi}{2}}|1\rangle = \begin{bmatrix} \cos\frac{\theta}{2}e^{-i\frac{\phi}{2}} \\ \sin\frac{\theta}{2}e^{i\frac{\phi}{2}} \end{bmatrix}, \quad (2.13a)$$

$$|-_u\rangle = -\sin\frac{\theta}{2}e^{-i\frac{\phi}{2}}|0\rangle + \cos\frac{\theta}{2}e^{i\frac{\phi}{2}}|1\rangle = \begin{bmatrix} -\sin\frac{\theta}{2}e^{-i\frac{\phi}{2}} \\ \cos\frac{\theta}{2}e^{i\frac{\phi}{2}} \end{bmatrix}. \quad (2.13b)$$

Explain why this definition of $|+_u\rangle$ and $|-_u\rangle$ is equivalent to the one in Eq. 2.12.

2.3.3 General Spin State

The state $|+_u\rangle$ serves as a general representation of a spin state. In fact, all the specialized states we discussed earlier, such as $|0\rangle$ and $|+\rangle$, including $|-_u\rangle$, can be viewed as special cases of $|+_u\rangle$. This is summarized in Table 2.1.

Spin State	θ	ϕ
$|0\rangle$	0	0
$|1\rangle$	π	0
$|+\rangle$	$\pi/2$	0
$|-\rangle$	$-\pi/2$	0
$|+_i\rangle$	$\pi/2$	$\pi/2$
$|-_i\rangle$	$\pi/2$	$-\pi/2$
$|-_u\rangle$	$\theta + \pi$	ϕ

Table 2.1: Special Spin States

Exercise 2.6 Validate that the expression for $|+_u\rangle$ in Eq. 2.12 reproduces the special cases listed in Table 2.1.

2.3.4 Spin State and Polarization State Correspondence

Even though spin and photon polarization are two distinct physical phenomena, they share many mathematical similarities because they both describe two-level quantum systems. In fact, various spin states can be mapped to photon polarization states, as listed in Table 2.2. The corresponding observables manifest as different matrices, because we assigned $1, 0$ for photon polarization, while $1, -1$ for spin states.

In this section, we extended the representation of spin states and spin observables to an arbitrary direction in three-dimensional space. Understanding these general representations sets the foundation for studying more advanced topics in quantum mechanics.

Spin State	Spin Observable	Polarization State	Polarization Observable
$|0\rangle$	Z	$|V\rangle$	M_V
$|1\rangle$	Z	$|H\rangle$	M_H
$|+\rangle$	X	$|D\rangle$	M_D
$|-\rangle$	X	$|A\rangle$	M_A
$|+_i\rangle$	Y	$|L\rangle$	M_L
$|-_i\rangle$	Y	$|R\rangle$	M_R
$|+_u\rangle$	S	$|P\rangle$	M_P

(See Table 1.2 for definitions of polarization states and observables, and Eq. 1.19 for $|P\rangle$)

Table 2.2: Spin State and Polarization State Correspondence

2.4 The Bloch Sphere

Recalling the expression for the generic spin state $|+_u\rangle$ from Eq. 2.12, we find that each state is specified by the polar angle θ and azimuthal angle ϕ. It is then natural to map each point on the surface of a unit sphere to a corresponding spin state $|+_u\rangle$. Conversely, each possible spin state can also be represented as a point on this sphere. This sphere, known as the Bloch Sphere, serves as a geometrical representation for any qubit state $|\psi\rangle$, as shown in Fig. 2.3.

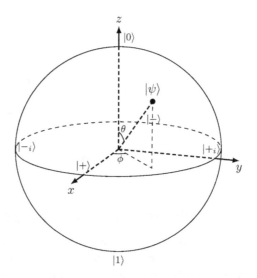

Figure 2.3: Key Points on the Bloch Sphere

> **Bloch Sphere**
>
> The Bloch sphere is an intuitive geometrical representation that maps any qubit

2.5 Spin Measurement

state $|\psi\rangle$ to a point on a unit sphere. The state $|\psi\rangle$ is parameterized by the polar angle θ and azimuthal angle ϕ:

$$|\psi\rangle = \cos\frac{\theta}{2}|0\rangle + \sin\frac{\theta}{2}e^{i\phi}|1\rangle = \begin{bmatrix} \cos\frac{\theta}{2} \\ \sin\frac{\theta}{2}e^{i\phi} \end{bmatrix}. \qquad (2.14)$$

The expression for $|\psi\rangle$ in Eq. 2.14 is the same as for $|+_u\rangle$ but applies universally to any qubit state.

Note that we typically restrict $\theta \in [0, \pi]$ and $\phi \in [0, 2\pi)$, recognizing that states $(-\theta, \phi + \pi)$ and $(-\theta, \phi - \pi)$ are equivalent to (θ, ϕ).

2.4.1 Key States on the Bloch Sphere

As depicted in Fig. 2.3, certain key states correspond to specific points on the Bloch sphere. Specifically, the north and south poles along the z-axis are associated with the $|0\rangle$ and $|1\rangle$ states, respectively. Eigenstates of the Pauli X operator, namely $|+\rangle$ and $|-\rangle$, map to points along the x-axis, while eigenstates of the Pauli Y operator, $|+_i\rangle$ and $|-_i\rangle$, align along the y-axis.

It is worth noting that orthogonal states are positioned at antipodal points on the sphere, separated by an angle of 180°, rather than the 90° one might intuitively expect. For instance, the orthogonal complement of a state $|\psi\rangle$ represented by (θ, ϕ) is found at $(\pi - \theta, \pi + \phi)$ or $(\theta + \pi, \phi)$, up to a global phase of -1. This arrangement reflects the intriguing property of quantum spin, where a full rotation requires a 720° turn, not the classical 360°.

2.4.2 Representation of Pauli X, Y, and Z on the Bloch Sphere

The Pauli operators X, Y, and Z induce specific rotations on the Bloch sphere. For instance, the action of X on a state $|\psi\rangle$, transforming $|\psi\rangle$ to $X|\psi\rangle$, corresponds to a 180° rotation about the x-axis. Similarly, applying Y and Z corresponds to 180° rotations about the y-axis and z-axis, respectively. These rotations are illustrated in Fig. 2.4.

Note: A 180° rotation on the Bloch sphere is associated with the action of the Pauli operators X, Y, or Z. Technically, this rotation is equivalent to applying a Pauli operator up to a global phase factor. However, such a global phase factor does not change the observable properties of a quantum state.

On the Bloch sphere, the effects of these rotations manifest as flips between certain basis states. Specifically, X swaps the states $|0\rangle$ and $|1\rangle$, and $|+_i\rangle$ and $|-_i\rangle$; Z swaps $|+\rangle$ and $|-\rangle$, and $|+_i\rangle$ and $|-_i\rangle$; while Y swaps $|0\rangle$ and $|1\rangle$, and $|+\rangle$ and $|-\rangle$.

Exercise 2.7 Verify the above transformations using bra-ket algebra, for example, confirm that $X|+_i\rangle = |-_i\rangle$, $Y|+\rangle = |-\rangle$, and $Z|-\rangle = |+\rangle$.

2.5 Spin Measurement

In this section, our focus shifts to the measurement of spins. While modern methods

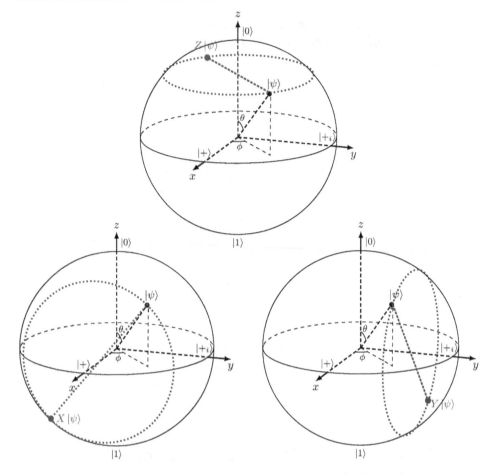

Figure 2.4: Pauli X, Y, and Z as Rotations on the Bloch Sphere

are designed to suit various quantum computing platforms, the Stern-Gerlach (SG) experiment serves as a canonical approach for understanding spin measurement. This experiment provides vital insights into the quantization of angular momentum and serves as a foundational framework for more advanced measurement techniques. We begin by examining the key aspects of the SG experiment.

2.5.1 The Stern-Gerlach Experiment

The Stern-Gerlach experiment, a cornerstone in the study of particle spin, is predicated on directing a beam of particles, such as electrons or atoms, through an inhomogeneous magnetic field (Fig. 2.5). Given the magnetic moments of these particles (§ 2.1.3), their different spin orientations cause them to deviate along the magnetic field gradient and register on a detection screen.

As depicted, the experiment primarily measures the Z observable, due to the magnetic field gradient being mainly along the z-axis. The observable Z has two eigenvalues, which correspond to the spin-up and spin-down states. While the initial spin state of the particles is random, the measurement collapses it to either spin-up or spin-down. Therefore, two distinct spots appear on the screen, in stark contrast to the continuous distribution one would expect for classical angular momentum.

2.5 Spin Measurement

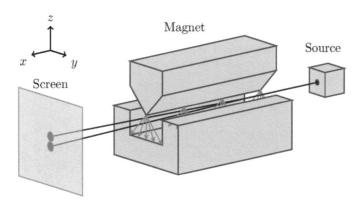

Figure 2.5: Stern-Gerlach Experiment

The original experiment, carried out by Otto Stern and Walther Gerlach in 1922 using silver atoms, conclusively revealed two separate spots, thus providing unequivocal evidence for the quantization of angular momentum and magnetic moment. This groundbreaking discovery laid the essential groundwork for the development of quantum theories.

 The necessity of a magnetic field gradient in the SG experiment can be attributed to the dipole nature of the magnetic moment, which has the same orientation as the spin. In a uniform field, the potential energy depends on the orientation but is position-independent. A gradient in the magnetic field thus gives rise to a force that causes the observed deflection.

A Fascinating Story

Stern described an early episode (Source: Physics Today):

"After venting to release the vacuum, Gerlach removed the detector flange. But he could see no trace of the silver atom beam and handed the flange to me. With Gerlach looking over my shoulder as I peered closely at the plate, we were surprised to see gradually emerge the trace of the beam. ... Finally, we realized what [had happened]. I was then the equivalent of an assistant professor. My salary was too low to afford good cigars, so I smoked bad cigars. These had a lot of sulfur in them, so my breath on the plate turned the silver into silver sulfide, which is jet black, so easily visible. It was like developing a photographic film."

The anecdote underscores a significant point about scientific research: sometimes, the path to discovery is not straightforward and may even be filled with seemingly trivial or random events that turn out to be pivotal. It serves as a testament to the resilience and ingenuity that are often required in the scientific journey. It is not just high-tech equipment or sophisticated theories that make for great science, but also an open mind, attention to detail, and the ability to adapt and learn from unexpected situations. In some instances, as comically illustrated by Stern's account, the constraints or limitations we face might inadvertently lead us to meaningful insights, reaffirming the adage that "necessity is the mother of invention."

2.5.2 Cascaded Stern-Gerlach Experiments

The Measurement Postulate can be well demonstrated using cascaded Stern-Gerlach experiments, as illustrated in Fig. 2.6. These cascaded setups are conceptually analogous to the cross-polarizer experiments detailed in § 1.5.2.

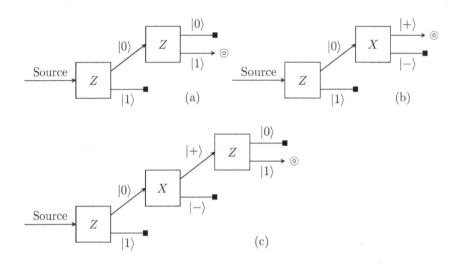

Figure 2.6: Cascaded Stern-Gerlach Experiments

In part (a), a beam of particles with random spin orientations is directed through a Stern-Gerlach apparatus (SGA) oriented along the z-axis. This first SGA measures the spin with respect to the Z observable and bifurcates the beam into two separate paths, each corresponding to one of Z's eigenstates, $|0\rangle$ and $|1\rangle$, in a 50 : 50 ratio. The $|0\rangle$ beam is then directed to a second SGA, also oriented along the z-axis. At this stage, no particles will be detected in the $|1\rangle$ output, confirming that the first measurement collapsed the spin state to $|0\rangle$, which is orthogonal to $|1\rangle$.

Part (b) varies from part (a) by having the second SGA oriented along the x-axis. The measurement outcomes are $|+\rangle$ and $|-\rangle$, the eigenstates of the X observable. Given that $|\langle 0|+\rangle|^2 = |\langle 0|-\rangle|^2 = 0.5$, approximately 25% of the particles will be found in the $|+\rangle$ state.

In part (c), an additional SGA oriented in the x-axis is inserted between the two Z-oriented SGAs. Now, about 12.5% of particles are observed in the $|1\rangle$ output of the final SGA. This is because the middle X-oriented SGA forces the spins into $|+\rangle$ and $|-\rangle$ states, which then feed into the final Z-oriented SGA. This SGA subsequently splits the $|+\rangle$ state into $|0\rangle$ and $|1\rangle$ with a 50 : 50 ratio. The net probability of finding a particle in the $|1\rangle$ state is $0.5 \times 0.5 \times 0.5 = 0.125$ or 12.5%, mirroring the behavior seen in the cross-polarizer experiments.

2.5.3 ✻ Modern Spin Experiments

In quantum mechanics, measurement refers to the extraction of classical information from a quantum system. While early experiments such as Stern-Gerlach provided initial insights into quantum spin, contemporary approaches have significantly evolved, particularly with the advent of quantum computing. One of the leading candidates for building qubits in a quantum computer is the use of spin-based qubits, which often rely on more advanced techniques for precise spin manipulation and measurement [58].

Spin-Based Qubits

Spin-based qubits are typically built using semiconductor quantum dots or trapped ions. The benefit of using these systems is that they offer a higher level of control and scalability. The challenge lies in accurately manipulating and measuring the spin states without undesirable interactions, which would disrupt the quantum information stored in the qubits.

Spin Precession and Rabi Oscillations

Two fundamental phenomena often employed in modern spin experiments are spin precession and Rabi oscillations. Spin precession refers to the motion of a quantum spin vector around an external magnetic field, much like the way a gyroscope precesses around an axis. Rabi oscillations involve driving transitions between different quantum states using an oscillating external field, commonly a microwave field. We will study these two phenomena in §§ 4.5.1 and 4.5.2.

Both phenomena are invaluable for spin manipulation in quantum computing. Spin precession allows for the precise control of qubit states by modulating the external magnetic field. On the other hand, Rabi oscillations are essential for implementing quantum gates, which are the basic operations in quantum computing algorithms.

Measurement Techniques

When a spin-1/2 particle, such as an electron, is measured along a specific axis, the orientation of its spin relative to that axis determines the outcome of the measurement. If the spin state is aligned with the measurement axis, the result is deterministic, yielding a definitive spin value (up or down along that axis). This principle underlies many modern spin experiments and is essential for confirming the theoretical predictions of quantum mechanics.

Advanced measurement techniques such as spin resonance and quantum nondemolition measurements are also integral to the robustness of modern spin experiments. These methods allow for the highly sensitive detection of spin states while minimizing the disturbance to the quantum system, thus enabling the extraction of reliable classical information.

Modern spin experiments have come a long way from their early counterparts, employing sophisticated methods for spin manipulation and measurement. These developments not only deepen our understanding of quantum mechanics but also pave the way for practical applications like quantum computing.

2.6 Summary and Conclusions

In this chapter, we laid the groundwork by introducing the essential elements of quantum spins, their states, and associated observables. A natural segue led us to explore the Pauli matrices, which serve as key observables for spin-1/2 systems. The significance of this subject matter is multifold; not only do spins constitute the basis for spin-based qubits, a prominent category of qubit implementations, but their study also fortifies our comprehension of the first three Postulates of Quantum Mechanics—namely, the state postulate, the observable postulate, and the measurement postulate.

Summary of Key Points

Our investigation started with an exploration of angular momentum, drawing distinctions between its classical and quantum characterizations. Through the resolution of eigenvalue equations for the Pauli matrices, X, Y, and Z, we were able to identify the eigenstates and eigenvalues, thereby building a robust mathematical framework for spin-1/2 systems.

Further, we engaged in a thorough discussion on general spin states and observables that can be aligned in any spatial direction. The geometric representation of these states was made transparent through the Bloch sphere, a construct vital for understanding the geometry of quantum states. The chapter closed with an examination of spin measurement techniques, from the pioneering Stern-Gerlach experiment to its modern-day variants, emphasizing their crucial role in the domain of quantum computing.

Implications and Applications

This chapter's insights are not purely academic; they have wide-ranging practical implications, particularly in the burgeoning field of quantum computing. Techniques like Larmor precession and Rabi oscillations serve as the backbone of stable and scalable quantum circuits. Moreover, the principles elucidated herein have applicability that extends into additional sectors such as quantum cryptography and quantum communication.

Closing Remarks

Mastering the quantum behavior of spins provides us with a refined vocabulary for describing microscopic phenomena. This enriched understanding not only challenges traditional classical viewpoints but is also instrumental in the rapidly advancing realm of quantum technologies. In upcoming chapters, we aim to construct a more general theoretical framework that encompasses the descriptions and dynamics of qubits, laying the groundwork for deeper inquiries into the subject.

Problem Set 2

2.1 Given the spin observable defined in Eq. 2.10:

$$\sigma_u = \begin{bmatrix} \cos\theta & \sin\theta e^{-i\phi} \\ \sin\theta e^{i\phi} & -\cos\theta \end{bmatrix},$$

calculate the following expected values:

(a) $\langle 0|\sigma_u|0\rangle$,

(b) $\langle +|\sigma_u|+\rangle$ where $|+\rangle = \frac{1}{\sqrt{2}}(|0\rangle + |1\rangle)$, and

(c) $\langle +_i | \sigma_u | +_i \rangle$ where $|+_i\rangle = \frac{1}{\sqrt{2}}(|0\rangle + i|1\rangle)$.

2.2 Given that σ_u is the spin observable defined in Eq. 2.10, and $|+_u\rangle$ and $|-_u\rangle$ (Eq. 2.12) are its eigenvectors, using $|+_u\rangle$ and $|-_u\rangle$ in as columns we can construct the following matrix:

$$U = \begin{bmatrix} \cos\frac{\theta}{2} & -\sin\frac{\theta}{2} \\ \sin\frac{\theta}{2}e^{i\phi} & \cos\frac{\theta}{2}e^{i\phi} \end{bmatrix}.$$

(a) Verify that U is unitary, i.e., $U^\dagger U = UU^\dagger = I$.

(b) Verify that $U^\dagger \sigma_u U = Z \equiv \begin{bmatrix} 1 & 0 \\ 0 & -1 \end{bmatrix}$.

The exercise illuminates the construction of a unitary transformation U that rotates a spin (or qubit) in the direction (θ, ϕ) to the z-direction.

2.3 A generic two-level quantum state can be expressed as

$$|\psi\rangle = \alpha|0\rangle + \beta|1\rangle,$$

where α and β are complex numbers satisfying $|\alpha|^2 + |\beta|^2 = 1$.

Find θ and ϕ such that the above $|\psi\rangle$ is equivalent to $|+_u\rangle$ in Eq. 2.12.

2.4 For the spin state $|+_u\rangle$ (see Eq. 2.12) in the direction (θ, ϕ), find the state vector of $|+_u\rangle$ after the following rotations:

(a) around the x-axis by π,

(b) around the y-axis by π,

(c) around the z-axis by π,

(d) around the x-axis by an angle γ,

(e) around the y-axis by an angle γ,

(f) around the z-axis by an angle γ,

(g) by π around the axis specified by polar angle α and azimuthal angle β, and

(h) by γ around the axis specified by polar angle α and azimuthal angle β.

All rotations are assumed to be counterclockwise. You may use the Bloch sphere as a visualization tool.

2.5 Find a general formula for the square root of the 2×2 identity matrix I.

3. A Framework for Qubits and Qudits

Contents

3.1	**Physical Qubit Systems**	60
3.2	**Qubit and Qudit States**	60
3.2.1	General Qubit States	62
3.2.2	The Orthogonal Partner	62
3.2.3	The Bloch Sphere	63
3.2.4	✶ Qudit States	64
3.3	**Change of Basis**	65
3.3.1	Changing Between $|0\rangle$ and $|1\rangle$ to $|+\rangle$ and $|-\rangle$ Bases	66
3.3.2	✶ Change of Basis: Qudits	68
3.3.3	✶ Change from Computational Basis to Eigenvector Basis	70
3.4	**✶ General Formulation of Quantum Measurement**	71
3.4.1	Review of Basic Principles on Quantum Measurements	72
3.4.2	✶ A General Measurement Framework	72
3.4.3	✶ Projective Measurements for Observables	74
3.4.4	✶ Basis-Dependent Measurements	75
3.4.5	✶ Measuring in Alternative Bases	76
3.4.6	✶ Sampling Errors	77
3.4.7	✶ Advanced Topics in Quantum Measurements	79
3.5	**✶ Application to Quantum State Tomography**	80
3.5.1	✶ Extracting Parameters from Alice's Measurements	80
3.5.2	✶ Extracting Parameters from Bob's Measurements	80
3.5.3	✶ Reconstructing the Quantum State	81
3.6	**Summary and Conclusions**	81
	Problem Set 3	83

The chapter aims to develop a comprehensive framework for qubits, which serve as the cornerstone of quantum computing and quantum information systems. While the preceding chapters introduced key concepts using specific examples such as photons and electron spins, this chapter generalizes these ideas to accommodate all forms of qubits. We also extend our discussion to include qudits, or general d-level quantum systems, where relevant.

Two focal topics—change of basis and quantum measurement theory—will be explored in depth. Quantum state tomography will be highlighted as a real-world application to demonstrate these principles.

This framework serves as an essential link, facilitating a smooth transition from fundamental concepts to more advanced topics in the quantum realm.

3.1 Physical Qubit Systems

In the preceding chapters, we discussed photon polarization and $\frac{1}{2}$ spins as forms of physical qubits, which are two-level quantum systems. However, there are several other platforms being actively explored for qubit implementation in quantum computing. Some of these are intrinsically multilevel systems but offer two states that can be effectively isolated, thereby enabling their use as qubits.

The following table (Table 3.1) provides a list of qubit platforms, along with the method of information encoding and the basis states relevant to each system. This table aims to give an overview of the breadth of qubit platforms currently under investigation, thus showcasing the versatility and adaptability of qubit systems.

These qubit implementations are currently subjects of intense research and development, each with its own set of advantages and limitations [52]. Factors such as noise rates, controllability, scalability, and cost vary significantly among different qubit platforms. While some may offer excellent controllability, others might excel in terms of low noise or cost-effectiveness. As the field of quantum computing continues to mature, it remains an open question which of these qubit implementations will emerge as the most practical and widely adopted for large-scale quantum information processing.

> **Exercise 3.1** Conduct an investigation into the qubit systems enumerated in Table 3.1 as well as others you come across. Your investigation should focus on the following aspects for each qubit system:
> - Current state-of-the-art developments
> - Advantages and disadvantages, including noise rates, controllability, scalability, and cost
> - Leading companies and research groups actively working on these platforms
>
> Compile your findings into concise summaries for each qubit system to gain a comprehensive understanding of the current landscape of quantum computing.

3.2 Qubit and Qudit States

Despite the various physical manifestations of qubits, they can all be characterized by a shared mathematical model grounded in quantum mechanics.

3.2 Qubit and Qudit States

Qubit System	Information Encoding	Basis States
Photon polarization	Polarization states	Vertical/horizontal polarizations, or left/right circular polarizations
Photon path	Dual-rail encoding	Path A and path B
Squeezed light	Quadrature states	Amplitude- and phase-squeezed states
Superconductor charge qubit	Charge	Uncharged and charged islands
Superconductor flux qubit	Current	Clockwise and counterclockwise current
Superconductor transmon qubit	Energy levels, with large capacitance	Ground and excited states
Superconductor fluxonium qubit	Energy levels, with large inductance	Ground and excited states
Trapped ion	Electron internal states, hyperfine levels	Ground and excited states
Neutral atoms, Rydberg atoms	Hyperfine levels, Rydberg states	Ground and excited states
Molecular qubit	Molecular states	Rotational or vibrational states
Quantum dot - spin	Electron spin	Up/down spin states
Quantum dot pair	Electron localization	Left dot and right dot
Nitrogen-vacancy (N-V) center	Electron spin	Electron spin projection ($m_s = 0$ or -1)
Cat qubit	Superpositions of coherent states	Coherent state superpositions
Topological system (anyons)	Braiding of excitations	Fusion basis states

Table 3.1: Examples of Qubit Implementations

3.2.1 General Qubit States

The State Postulate posits that the state of a quantum system is encapsulated by a unit vector within a Hilbert space. For qubits, this space constitutes a two-dimensional complex vector space, denoted as $\mathbb{C}^2$.

In the realm of quantum computing, the basis states for an individual qubit are typically represented as $|0\rangle$ and $|1\rangle$, reminiscent of the classical binary states 0 and 1. A general qubit state can be depicted as a linear combination of these basis states:

$$|\psi\rangle = \alpha |0\rangle + \beta |1\rangle. \tag{3.1}$$

In this expression, α and β are complex coefficients. Due to the necessity for the state vector to be normalized, we have the constraint $|\alpha|^2 + |\beta|^2 = 1$. This arises from $|\alpha|^2$ and $|\beta|^2$ representing the probabilities of measuring the qubit in states $|0\rangle$ and $|1\rangle$, respectively, and their cumulative sum equating to 1.

1 Global Phase and Relative Phase

The phase of a complex number, represented as $re^{i\phi}$, is indicated by the ϕ value. Introducing a global phase to a quantum state applies an identical phase factor to both basis states, which leaves the measurement probabilities of each state unaltered. Therefore, a state such as:

$$|\psi'\rangle = \alpha e^{i\phi} |0\rangle + \beta e^{i\phi} |1\rangle,$$

is congruent to $|\psi\rangle$, since both $|\alpha e^{i\phi}|^2$ and $|\beta e^{i\phi}|^2$ equate to their non-phased counterparts. Consequently, global phase lacks physical significance. However, relative phase is of importance as it can influence measurement outcomes across varied bases.

2 Expressing Qubit States with Real Numbers

Acknowledging that global phase is non-consequential and that the state vector is normalized, it is feasible to convey a general qubit state using a mere two real parameters. Adopting the notations $\alpha = \cos\frac{\theta}{2}$ and $\beta = \sin\frac{\theta}{2} e^{i\phi}$, we obtain:

$$|\psi\rangle = \cos\frac{\theta}{2} |0\rangle + \sin\frac{\theta}{2} e^{i\phi} |1\rangle. \tag{3.2}$$

With this parameterization, both the normalization condition and the relative phase factor are inherently satisfied.

Equation 3.2 mirrors $|+_u\rangle$ from Eq. 2.12, illustrating a $\frac{1}{2}$-spin oriented in the direction (θ, ϕ). Now this has been derived devoid of any explicit reliance on the intrinsic nature of the qubits.

> **Exercise 3.2** Show that the state $|\psi\rangle = \alpha |0\rangle + \beta |1\rangle$ represents a $\frac{1}{2}$-spin oriented in the direction:
> $$\hat{u} = (2\operatorname{Re}(\alpha\beta^*), 2\operatorname{Im}(\alpha\beta^*), |\alpha|^2 - |\beta|^2). \tag{3.3}$$

3.2.2 The Orthogonal Partner

The state which stands orthonormal to $|\psi\rangle$ is:

3.2 Qubit and Qudit States

$$|\psi_\perp\rangle = \beta^* |0\rangle - \alpha^* |1\rangle, \quad (3.4)$$

ensuring both $\langle\psi_\perp|\psi_\perp\rangle = 1$ and $\langle\psi|\psi_\perp\rangle = 0$. Here β^* denotes the complex conjugate of β. It is worth noting that the analogous state $e^{i\phi}|\psi_\perp\rangle$, especially $-|\psi_\perp\rangle$, is also orthonormal to $|\psi\rangle$. In its (θ, ϕ) representation, we have:

$$|\psi_\perp\rangle = -\sin\frac{\theta}{2}|0\rangle + \cos\frac{\theta}{2}e^{i\phi}|1\rangle, \quad (3.5)$$

which corresponds to $|-_u\rangle$ in Eq. 2.12.

Exercise 3.3 Given the states:

$$|\psi_1\rangle = |0\rangle,$$
$$|\psi_2\rangle = -\frac{1}{2}|0\rangle + \frac{\sqrt{3}}{2}|1\rangle,$$
$$|\psi_3\rangle = -\frac{1}{2}|0\rangle - \frac{\sqrt{3}}{2}|1\rangle,$$

prove that:

$$|\langle\psi_1|\psi_2\rangle|^2 = |\langle\psi_2|\psi_3\rangle|^2 = |\langle\psi_3|\psi_1\rangle|^2 = \frac{1}{4}.$$

3.2.3 The Bloch Sphere

In § 2.4, we introduced the Bloch sphere in the context of electron spin. Beyond spin, the Bloch sphere provides a versatile visualization for general qubit states. Each point on this sphere represents a distinct qubit state, denoted as $|\psi\rangle$. The vector emanating from the origin to this point, symbolized as $\hat{u}$, is the Bloch vector. This vector is defined by two angles: the polar angle θ and the azimuthal angle ϕ. Likewise, the state vector $|\psi\rangle$ is parameterized using these same angles. Fig. 3.1 elucidates these definitions alongside highlighting key points on the sphere.

The expectation values of the Pauli spin operators X, Y, and Z in the state $|\psi\rangle$ (refer to Eq. 3.2) are given by:

$$\langle X \rangle = \sin\theta\cos\phi, \quad (3.6a)$$
$$\langle Y \rangle = \sin\theta\sin\phi, \quad (3.6b)$$
$$\langle Z \rangle = \cos\theta. \quad (3.6c)$$

These coincide with the components of $\hat{u}$, allowing $|\psi\rangle$ to be interpreted as representing the angular momentum of the spin in three dimemsions.

Exercise 3.4 Prove Eq. 3.6.

An important detail to note is the presence of the term $\frac{\theta}{2}$ within the formulation of $|\psi\rangle$. Moreover, it's intriguing to observe that orthogonal states, such as $|+\rangle$ and $|-\rangle$, occupy antipodal positions on the sphere. This implies they are separated by an angle of 180°, contrary to the 90° separation that might be more intuitively

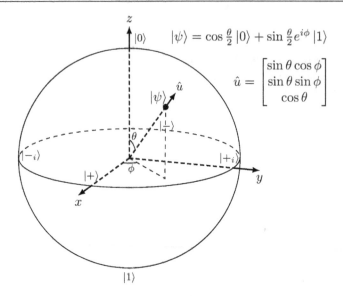

Figure 3.1: Visualization of qubit states on the Bloch Sphere

anticipated. This particular aspect is related to the notion of $\frac{1}{2}$-spin representing an angular momentum.

3.2.4 ∗ Qudit States

Qutrits and qudits denote quantum systems with more than two levels. While qubits are ubiquitous in quantum computing, qutrits (three-level systems) and qudits (general d-level systems) offer more intricate and potentially versatile quantum states. Various quantum systems, including superconducting qubits, Rydberg atoms, and NV centers, possess more than two inherent states. However, control, calibration, and maintaining coherence of these states become substantially more complex and demanding.

Despite these challenges, there's growing enthusiasm in the research community to explore qudits as they present potential pathways for advanced quantum computational techniques. As the study progresses and whenever it benefits pedagogical discussions, the standard qubit framework will be broadened to include discussions on qudits.

In addition, an n-qubit system is mathematically equivalent to a $d = 2^n$ qudit. This equivalence means that formulations developed for qudits can also be applied to multiple qubits that are combined and treated as a single quantum system, broadening the applicability and relevance of qudit-based theoretical frameworks.

The general state of a qudit can be expressed as:

$$|\psi\rangle = \sum_{j=0}^{d-1} c_j |j\rangle, \qquad (3.7)$$

where $c_j \in \mathbb{C}$. This state's normalization constraint is given by:

$$\langle\psi|\psi\rangle = \sum_{j=0}^{d-1} |c_j|^2 = 1. \tag{3.8}$$

The basis states' orthogonality ensures:

$$\langle j|k\rangle = \delta_{j,k}, \tag{3.9}$$

with $\delta_{j,k}$ being the Kronecker delta function.

The completeness condition for the basis states is expressed as:

$$\sum_{j=0}^{d-1} |j\rangle\langle j| = I. \tag{3.10}$$

A quantum state in d dimensions can be parameterized by $2(d-1)$ real parameters due to the global phase factor and the normalization condition. Consequently, the state representation becomes:

$$|\psi\rangle = \begin{bmatrix} \cos\theta_1 \\ \cos\theta_2 \sin\theta_1 e^{i\phi_1} \\ \cos\theta_3 \sin\theta_1 \sin\theta_2 e^{i\phi_2} \\ \vdots \\ \cos\theta_{d-1} \sin\theta_1 \cdots \sin\theta_{d-2} e^{i\phi_{d-2}} \\ \sin\theta_1 \cdots \sin\theta_{d-2} \sin\theta_{d-1} e^{i\phi_{d-1}} \end{bmatrix}. \tag{3.11}$$

Exercise 3.5 ✷ For the state $|\psi\rangle$ in Eq. 3.11, identify at least one vector from the set of $d-1$ vectors that are orthonormal to it.

3.3 Change of Basis

Understanding qubit states and operators across different bases is fundamental to quantum computing. The importance of bases can be distilled into two primary considerations:

1. Perspectives for Quantum States: Alternate bases often illuminate novel perspectives on quantum states, offering insights pivotal for specific quantum operations or algorithms. In this context, while the underlying physical quantum states and observables remain constant, their vector or matrix representations undergo transformation. This idea can be likened to the classical concept of reference frames in Newtonian physics. Inspecting quantum states from varied "viewpoints" mirrors the act of changing the basis for the state vectors.

2. Quantum State Evolution as Basis Transformation: The operations of quantum gates or other unitary transformations on quantum states can be mathematically interpreted as a change of basis in the respective vector spaces. In quantum computing, we often work in the computational basis, but the quantum state itself exists in a Hilbert space that can be represented in any orthonormal basis. Unlike the previous scenario, here the physical quantum states undergo genuine changes.

The rudiments of basis change were introduced in § 1.4.5, utilizing photons as exemplars. This section will expand upon those foundational concepts, aiming to furnish a rigorous framework for changes in basis. In § 4.1, we will investigate the unitary time-evolution that can be viewed as a change of basis.

3.3.1 Changing Between $|0\rangle$ and $|1\rangle$ to $|+\rangle$ and $|-\rangle$ Bases

To build intuition, let's examine a straightforward scenario.

1 Quantum States in Two Bases

A single qubit's quantum states are two-dimensional, necessitating a basis of two orthogonal, normalized states. The standard computational basis utilizes states $|0\rangle$ and $|1\rangle$. Meanwhile, the $\{|+\rangle, |-\rangle\}$ basis (often referred to as the Hadamard basis) defines them as:

$$|+\rangle = \frac{1}{\sqrt{2}}(|0\rangle + |1\rangle), \quad |-\rangle = \frac{1}{\sqrt{2}}(|0\rangle - |1\rangle). \tag{3.12}$$

A general qubit state in the computational basis, $|\psi\rangle = \alpha|0\rangle + \beta|1\rangle$, can be expressed in the $\{|+\rangle, |-\rangle\}$ basis as:

$$|\psi\rangle = \frac{\alpha+\beta}{\sqrt{2}}|+\rangle + \frac{\alpha-\beta}{\sqrt{2}}|-\rangle. \tag{3.13}$$

This provides new coefficients for the states in the revised basis.

Exercise 3.6 Express the θ, ϕ parametrized $|\psi\rangle$ given in Eq. 3.2 in the $|+\rangle, |-\rangle$ basis.

2 Basis Transformation Operator

Transitioning from one orthonormal basis to another in quantum mechanics corresponds with a unitary operator, as unitary transformations preserve the inner product and hence the orthonormality of the basis. (See Appendix C for the definition and properties of unitary operators.) The basis transformation operator U for transitioning from $\{|0\rangle, |1\rangle\}$ to $\{|+\rangle, |-\rangle\}$ corresponds to a quantum evolution where $|0\rangle$ transforms to $|+\rangle$ and $|1\rangle$ to $|-\rangle$. This operator takes a linear combination of states $|0\rangle$ and $|1\rangle$ to the corresponding combination of $|+\rangle$ and $|-\rangle$. Hence,

$$|+\rangle = U|0\rangle, \quad |-\rangle = U|1\rangle. \tag{3.14}$$

Consequently, the operator U is:

$$U = |+\rangle\langle 0| + |-\rangle\langle 1|. \tag{3.15}$$

Proof. The definition of U provides:

$$|+\rangle = U|0\rangle \Rightarrow |+\rangle\langle 0| = U|0\rangle\langle 0|, \tag{3.16a}$$
$$|-\rangle = U|1\rangle \Rightarrow |-\rangle\langle 1| = U|1\rangle\langle 1|. \tag{3.16b}$$

3.3 Change of Basis

Combining the two equations gives:

$$|+\rangle\langle 0| + |-\rangle\langle 1| = U(|0\rangle\langle 0| + |1\rangle\langle 1|). \tag{3.17}$$

Utilizing the completeness property $|0\rangle\langle 0| + |1\rangle\langle 1| = I$, we deduce Eq. 3.15. □

Expanding Eq. 3.15, we obtain another expression for U that directly mirrors its matrix form:

$$U = \frac{1}{\sqrt{2}}(|0\rangle\langle 0| + |0\rangle\langle 1| + |1\rangle\langle 0| - |1\rangle\langle 1|). \tag{3.18}$$

Exercise 3.7 Validate that $U = \frac{1}{\sqrt{2}}\begin{bmatrix} 1 & 1 \\ 1 & -1 \end{bmatrix}$ performs the transformation in Eq. 3.14.

3 Inverse Transformation

Applying U^{-1} to $|+\rangle, |-\rangle$ brings them back to $|0\rangle, |1\rangle$. Given that U is unitary (meaning $U^{-1} = U^\dagger$):

$$|0\rangle = U^\dagger |+\rangle, \quad |1\rangle = U^\dagger |-\rangle. \tag{3.19}$$

Exercise 3.8 Verify that U in Eq. 3.15 is unitary using the bra-ket notation.
Hint: $U^\dagger = |0\rangle\langle +| + |1\rangle\langle -|$.

4 Basis Rotation

As depicted in Fig. 3.2, a basis change can be visualized as a rotation within the Hilbert space's coordinate system. While the vector $|\psi\rangle$ stays physically unchanged, its components with respect to the new basis vary. This variation effectively presents the same quantum state but in a different mathematical form.

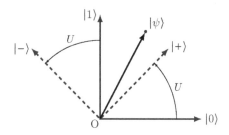

Figure 3.2: Basis Rotation $\{|0\rangle, |1\rangle\} \to \{|+\rangle, |-\rangle\}$

Similar to how a geometric rotation preserves vector length and relative angles in the coordinate system, a basis change via a unitary transformation keeps the inner product intact. This preservation is the reason a change of basis is often conceptualized as a basis rotation.

5 Transformation of Vector Components

The vector components can be transformed directly using $U^\dagger$:

$$\begin{bmatrix} \frac{\alpha+\beta}{\sqrt{2}} \\ \frac{\alpha-\beta}{\sqrt{2}} \end{bmatrix} = U^\dagger \begin{bmatrix} \alpha \\ \beta \end{bmatrix}. \tag{3.20}$$

Exercise 3.9 For a given quantum state $|\psi\rangle = \frac{1}{2}|+\rangle + i\frac{\sqrt{3}}{2}|-\rangle$, compute its representation in $\{|0\rangle, |1\rangle\}$ basis.

6 Computing Vector Components through Projection

In addition to the matrix representation of the transformation operator U and its application to state vectors, there is an intrinsic method for calculating the amplitudes of a quantum state in a specific basis. This approach involves decomposing (or projecting) the state $|\psi\rangle$ into an orthonormal basis. For the standard basis $\{|0\rangle, |1\rangle\}$, the state can be expressed as

$$|\psi\rangle = \langle 0|\psi\rangle |0\rangle + \langle 1|\psi\rangle |1\rangle, \tag{3.21}$$

where the components of $|\psi\rangle$ are the projections $\langle 0|\psi\rangle$ and $\langle 1|\psi\rangle$. With $|\psi\rangle = \alpha|0\rangle + \beta|1\rangle$, these yield α and β as expected.

Similarly, for the basis $\{|+\rangle, |-\rangle\}$, the state can be expressed as

$$|\psi\rangle = \langle +|\psi\rangle |+\rangle + \langle -|\psi\rangle |-\rangle, \tag{3.22}$$

$$\langle +|\psi\rangle = \frac{\alpha+\beta}{\sqrt{2}}, \quad \langle -|\psi\rangle = \frac{\alpha-\beta}{\sqrt{2}}. \tag{3.23}$$

3.3.2 ∗ Change of Basis: Qudits

After acquainting ourselves with the fundamental principles and operations of basis change using the example $\{|0\rangle, |1\rangle\} \to \{|+\rangle, |-\rangle\}$, we now extend them to general qudit systems, where the index (e.g., i or j below) extends to $\{0, 1, \ldots, d-1\}$.

1 Definition and Operator

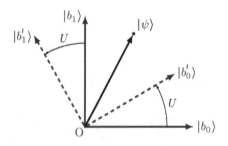

Figure 3.3: Illustration of Basis Rotation $\{|b_i\rangle\} \to \{|b_i'\rangle\}$

3.3 Change of Basis

Given two complete and orthonormal bases $\{|b_i\rangle\}$ and $\{|b'_i\rangle\}$, their orthogonality and completeness relations are:

$$\langle b_i|b_j\rangle = \langle b'_i|b'_j\rangle = \delta_{ij}, \quad \sum_i |b_i\rangle\langle b_i| = \sum_i |b'_i\rangle\langle b'_i| = I. \tag{3.24}$$

To change from the basis $\{|b_i\rangle\}$ to $\{|b'_i\rangle\}$, a unitary operator U is defined such that $|b'_i\rangle = U|b_i\rangle$. Specifically, U is:

$$U = \sum_i |b'_i\rangle\langle b_i| = \sum_{i,j} \langle b_j|b'_i\rangle |b_j\rangle\langle b_i|. \tag{3.25}$$

Exercise 3.10 Derive the formulas of U in Eq. 3.25 from its definition $|b'_i\rangle = U|b_i\rangle$. Hint: see the proof for the $\{|0\rangle, |1\rangle\} \to \{|+\rangle, |-\rangle\}$ case.

2 Vectors and Matrices under Change of Basis

While physical properties, such as quantum states and observables, remain invariant across distinct bases, their vector or matrix representations are transformed by U upon a basis change.

A general qudit state $|\psi\rangle$ can be expressed in either of the bases:

$$|\psi\rangle = \sum_i c_i |b_i\rangle = \sum_i c'_i |b'_i\rangle. \tag{3.26}$$

The coefficients or vector components of $|\psi\rangle$ are obtained as:

$$c_i = \langle b_i|\psi\rangle, \quad c'_i = \langle b'_i|\psi\rangle. \tag{3.27}$$

Considering $\{c_i\}$ as a column vector $|c\rangle$ and $\{c'_i\}$ as $|c'\rangle$, we have:

$$|c\rangle = U|c'\rangle, \quad |c'\rangle = U^\dagger |c\rangle. \tag{3.28}$$

Inner products are scalars and, therefore, remain constant across different bases:

$$\langle u'|v'\rangle = \langle u|v\rangle, \tag{3.29}$$

given that $|u'\rangle = U^\dagger |u\rangle$ and $|v'\rangle = U^\dagger |v\rangle$ according to Eq. 3.28.

Exercise 3.11 Prove the invariance of the inner product under a unitary transformation (see Eq. 3.29).

For an observable A, $\langle u|A|u\rangle$ represents the expected value (statistical average) of A in the state $|u\rangle$ (see § 1.5.3). It is also a scalar and, therefore, should remain constant across different bases. This requires the matrix form of A to transform according to:

$$A' = U^\dagger A U. \tag{3.30}$$

With this transformation, $\langle u|A|u\rangle$ is indeed invariant under a unitary transformation:

$$\langle u'|A'|u'\rangle = \langle U^\dagger u|U^\dagger A U|U^\dagger u\rangle \qquad (3.31a)$$
$$= \langle u|UU^\dagger A UU^\dagger|u\rangle \qquad (3.31b)$$
$$= \langle u|A|u\rangle. \qquad (3.31c)$$

In § 3.4.4, we will demonstrate the basis independence of $\langle u|A|u\rangle$ through an example.

Exercise 3.12 Demonstrate that, if two observables commute in the original basis ($[A, B] = 0$), they will also commute in the transformed basis ($[UAU^\dagger, UBU^\dagger] = 0$).

3.3.3 ∗ Change from Computational Basis to Eigenvector Basis

The action of transitioning to the eigenvector basis of a Hermitian operator is termed as "diagonalizing the operator". This nomenclature stems from the fact that, in the eigenvector basis, the operator is represented as a diagonal matrix with its eigenvalues on the diagonal. Such a transformation finds significant applications in quantum computing. Prominent among these are the quantum phase estimation and Hamiltonian simulation algorithms. Furthermore, as discussed in § 3.4.5, measurements in quantum computing often employ such basis transformations.

Given H is a Hermitian operator with eigenvalues $\{\lambda_i\}$ and normalized eigenvectors $\{\phi_i\}$:

$$H|\phi_i\rangle = \lambda_i|\phi_i\rangle, \qquad (3.32)$$

$\{\phi_i\}$ form a complete, orthonormal basis.

To change basis from the computational basis $\{|i\rangle\}$ to the H eigenbasis $\{|\phi_i\rangle\}$, we can use the unitary basis transformation operator U given by:

$$|\phi_i\rangle = U|i\rangle, \quad U = \sum_i |\phi_i\rangle\langle i|. \qquad (3.33)$$

As a consequence of the spectral decomposition theorem, the matrix form of H is diagonal in the $\{|\phi_i\rangle\}$ basis:

$$H = \sum_i \lambda_i |\phi_i\rangle\langle\phi_i|, \qquad (3.34)$$

while the matrix form of $U^\dagger H U$ is diagonal in the computational basis:

$$U^\dagger H U = \sum_i \lambda_i |i\rangle\langle i|. \qquad (3.35)$$

Eigenbases for Pauli Operators

The eigenbases of the Pauli operators play an important role in quantum computations and measurements.

- The eigenbasis for Pauli Z is the computational basis itself, consisting of the eigenvectors $|0\rangle$ and $|1\rangle$.

- The eigenbasis for Pauli X consists of the states $|+\rangle$ and $|-\rangle$. The unitary transformation relating this basis to the computational basis is the Hadamard operator H, defined as

$$H = \frac{1}{\sqrt{2}} \begin{bmatrix} 1 & 1 \\ 1 & -1 \end{bmatrix}. \tag{3.36}$$

It can be verified that $|+\rangle = H|0\rangle$ and $|-\rangle = H|1\rangle$. In addition, the diagonalization relationship $Z = HXH$ holds. This basis is sometimes referred to as the X basis or the Hadamard basis.

Note that here, H corresponds to U from Eq. 3.33, and X corresponds to H from Eq. 3.32. We have repurposed H as it is the conventional symbol for the Hadamard operator.

- The eigenvectors of the Pauli Y operator are $|+_i\rangle$ and $|-_i\rangle$. For the eigenbasis, we often use the equivalent vectors $|+_i\rangle$ and $i|-_i\rangle$ for the convenience of calculation. The unitary operator that transforms from this eigenbasis to the computational basis is $R_x(-\frac{\pi}{2})$, also denoted simply as R_x. This operator is defined as (see § 5.2.5):

$$R_x = \frac{1}{\sqrt{2}} \begin{bmatrix} 1 & i \\ i & 1 \end{bmatrix}. \tag{3.37}$$

One can verify that $|+_i\rangle = R_x|0\rangle$ and $i|-_i\rangle = R_x|1\rangle$. Furthermore, the diagonalization relationship $Z = R_x^\dagger Y R_x$ holds.

Exercise 3.13 Given the Hermitian operator:

$$A = \begin{bmatrix} 1 & 1 \\ 1 & 0 \end{bmatrix},$$

(a) Find its eigenvalues and the corresponding eigenvectors.

(b) If a system is in a state $|\psi\rangle = |0\rangle$, express this state in the eigenbasis of A.

(c) Consider measuring the state $|\psi\rangle = |0\rangle$ in the eigenbasis of A which is equivalent to measurement using A as observable. What are the probabilities associated with each eigenvalue?

3.4 ∗General Formulation of Quantum Measurement

Quantum measurement plays an essential role in quantum computing. It serves as the bridge that transfers information from the quantum realm to the classical world. Moreover, quantum measurement is an indispensable tool for investigating decoherence, a major source of error in contemporary quantum computing systems. Although intricate, mastering this topic is essential for understanding quantum system operations. In this section, we will formulate a general quantum measurement framework based on basic principles of observable and measurement in quantum mechanics discussed in Chapters 1 and 2.

3.4.1 Review of Basic Principles on Quantum Measurements

The Observable Postulate and the Measurement Postulate are two fundamental postulates in quantum mechanics that describe the behavior of quantum systems and the act of measuring those systems.

The Observable Postulate states that every observable quantity in quantum mechanics is represented by a Hermitian operator. An observable quantity is a physical property of a quantum system that can be measured, such as position, momentum, energy, or spin. The Hermitian operator associated with an observable quantity is called an observable operator, and its eigenvalues represent the possible outcomes of a measurement of that observable quantity.

The Measurement Postulate, on the other hand, describes the process of measuring an observable quantity of a quantum system. According to the Composition Postulate, the act of measuring an observable quantity of a quantum system will cause the system to collapse into one of the eigenstates of the observable operator associated with that observable quantity, and the corresponding eigenvalue will be the result of the measurement. This collapse is also known as the wavefunction collapse, and it is a fundamental feature of quantum mechanics.

These postulates lay the groundwork for understanding quantum measurement in its most straightforward form. However, modern quantum computing and quantum information often employ more advanced measurement schemes such as POVM (Positive Operator-Valued Measure), weak measurements, and nondestructive measurements. In order to accommodate such complex scenarios, it is necessary to extend these standard postulates into a general quantum measurement framework.

3.4.2 ∗ A General Measurement Framework

Quantum measurements are typically characterized by a set of measurement operators $\{M_i\}$, representing potential measurement outcomes.

1 Measurement Probability

Given a quantum system in state $|\psi\rangle$, the probability of realizing outcome i upon measurement is:

$$P_i = \langle \psi | M_i^\dagger M_i | \psi \rangle. \tag{3.38}$$

This formula mirrors the squared magnitude of the vector $M_i |\psi\rangle$, denoted as $\|M_i |\psi\rangle\|^2$.

The product $M_i^\dagger M_i$ is positive semidefinite, that is, a Hermitian matrix with non-negative eigenvalues, ensuring all probabilities $P_i \geq 0$.

Exercise 3.14 Demonstrate that a matrix defined as $M_i^\dagger M_i$ is positive semidefinite.

3.4 ∗ General Formulation of Quantum Measurement

For the total probability to sum up to 1, i.e., $\sum P_i = 1$, the operators $\{M_i\}$ must adhere to the completeness condition:

$$\sum_i M_i^\dagger M_i = I. \tag{3.39}$$

Exercise 3.15 Show that when M_i meets the completeness requirement Eq. 3.39, the sum $\sum P_i$ equals 1.

In this broad portrayal of quantum measurement, the operators $\{M_i\}$ need not be orthogonal.

2 Measurement Average

If a numerical value λ_i corresponds to each M_i, the expected value of λ is:

$$\langle \lambda \rangle = \sum_i P_i \lambda_i = \sum_i \langle \psi | M_i^\dagger M_i | \psi \rangle \lambda_i. \tag{3.40}$$

3 Post Measurement State

A salient feature of quantum measurement is the alteration of the quantum state upon measurement. After the measurement, the system transitions to:

$$|\psi_i\rangle = \frac{M_i |\psi\rangle}{\sqrt{P_i}}. \tag{3.41}$$

The factor $\sqrt{P_i}$ in the denominator guarantees the normalization of $|\psi_i\rangle$, ensuring it remains a legitimate quantum state.

4 Measurement Probability via State Overlap

The likelihood of observing a specific measurement result can be deduced from the overlap (inner product) of the pre- and post-measurement states:

$$P_i = |\langle \psi_i | \psi \rangle|^2. \tag{3.42}$$

Exercise 3.16 Using Eq. 3.41, confirm that the probability of measuring $|\psi\rangle$ and subsequently obtaining $|\psi_i\rangle$ is given by Eq. 3.42.

5 Non-Unitary Transformation

It is pertinent to point out that Eq. 3.41 does not generally represent a unitary transformation, underscoring a distinctive trait of quantum measurement. Should we perform back-to-back measurements with the identical M_i on the post-measurement state (assuming no intermediate state alterations, excluding general POVM measurements), outcomes will be consistent, and the state remains unperturbed.

6 POVM Measurements

The measurement framework outlined here can be extended to the Positive Operator-Valued Measure (POVM) formalism. In the POVM context, the focus is on the measurement outcomes rather than the state of the system after the measurement has been performed. Thus, $\{M_i^\dagger M_i\}$ is replaced with a set of POVM operators $\{E_i\}$, each being a positive semidefinite Hermitian operator that corresponds to a potential measurement outcome, with $\sum_i E_i = I$. Note that given E_i, its decomposition $E_i = M_i^\dagger M_i$ is non-unique. All the preceding equations in this section are still valid except Eq. 3.41.

3.4.3 *Projective Measurements for Observables

In light of our general formulation, measurements pertaining to a quantum observable emerge as a special case known as projective measurement.

Consider an observable H, represented by a Hermitian operator. This operator possesses a collection of real eigenvalues λ_i with corresponding eigenvectors $|\phi_i\rangle$. These eigenvectors construct a complete orthonormal basis. Accordingly, the measurement operators M_i become the projection operators associated with the basis states $|\phi_i\rangle$:

$$M_i = \Pi_i = |\phi_i\rangle\langle\phi_i|. \tag{3.43}$$

Being projection operators, beyond the completeness relationship in Eq. 3.39, the set $\{M_i\}$ adheres to the orthogonal and idempotent properties:

$$M_i M_j = \delta_{ij} M_i. \tag{3.44}$$

Exercise 3.17 Validate Eq. 3.44.

Exercise 3.18 Employing the relationship $\sum_i |\phi_i\rangle\langle\phi_i| = I$, verify that Eq. 3.39 is valid for projective measurements.

Subsequently, the probability formula in Eq. 3.38 for acquiring outcome λ_i is recast as:

$$P_i = \langle\psi|\Pi_i|\psi\rangle = |\langle\psi|\phi_i\rangle|^2. \tag{3.45}$$

Likewise, the formula for the expected value of M in Eq. 3.40 transforms to:

$$\langle H \rangle = \sum_i P_i \lambda_i = \langle\psi|H|\psi\rangle, \tag{3.46}$$

making use of the spectral decomposition property $H = \sum_i \lambda_i |\phi_i\rangle\langle\phi_i|$.

3.4 * General Formulation of Quantum Measurement 75

Exercise 3.19 Confirm that Eq. 3.40 yields Eqs. 3.45 and 3.46.

Finally, the post-measurement state from Eq. 3.41 simplifies to $|\phi_i\rangle$. In this sense, the projection operator essentially selects the respective basis state, i.e., projects the original state onto the basis state.

Our discussions highlight that when the measurement operators $\{M_i\}$ are projection operators linked to the eigenstates of an observable, the overarching measurement framework naturally aligns with observable-centric measurement.

3.4.4 * Basis-Dependent Measurements

Basis dependence is an essential facet of quantum measurement. When quantum systems are measured in different bases, their outcomes differ based on the chosen basis.

To elucidate, a quantum state measured in a specific basis is essentially being measured against that basis's states. For instance, when measuring the state $|\psi\rangle = \alpha|0\rangle + \beta|1\rangle$ in the computational basis, it splits into post-measurement states $|0\rangle$ and $|1\rangle$, with associated probabilities of $|\alpha|^2$ and $|\beta|^2$. This phenomenon is depicted in Fig. 3.4, part (a).

In a similar vein, if $|\psi\rangle$ is measured in the $\{|+\rangle, |-\rangle\}$ basis, as illustrated in part (b), it results in post-measurement states $|+\rangle$ and $|-\rangle$ with probabilities $\frac{|\alpha+\beta|^2}{2}$ and $\frac{|\alpha-\beta|^2}{2}$.

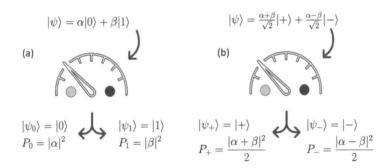

Figure 3.4: Measurements in $\{|0\rangle, |1\rangle\}$ and $\{|+\rangle, |-\rangle\}$ Bases

Traditionally, distinct bases measurements correspond to different observables. For the purpose of quantum computing, let's assign $+1$ and -1 as standard measurement values. Thus, the pertinent observables can be expressed as:

$$M_{0,1} = |0\rangle\langle 0| - |1\rangle\langle 1| = Z, \text{ corresponding to the } \{|0\rangle, |1\rangle\} \text{ basis}; \quad (3.47a)$$
$$M_{+,-} = |+\rangle\langle +| - |-\rangle\langle -| = X, \text{ associated with the } \{|+\rangle, |-\rangle\} \text{ basis}. \quad (3.47b)$$

Given this convention, the measurements in the provided example can also be dubbed as the Z and X measurements.

When an observable M undergoes measurement in different bases, its expected value, $\langle M \rangle$, remains constant (see § 3.3.2). In the above example, the expected value of X calculated in the $\{|+\rangle, |-\rangle\}$ basis through probability argument is

$$\langle X \rangle = (+1)P_+ + (-1)P_- = \frac{|\alpha+\beta|^2}{2} - \frac{|\alpha-\beta|^2}{2} = \alpha\beta^* + \alpha^*\beta. \quad (3.48)$$

But as demonstrated in § 1.5.3, $\langle X \rangle$ is given by $\langle \psi|X|\psi \rangle$, which is independent of the measurement basis. Indeed, one can verify that $\langle \psi|X|\psi \rangle = \alpha\beta^* + \alpha^*\beta$ by computing the inner product directly in the computational basis.

Exercise 3.20 The expected value of Z calculated through a probability argument is:

$$\langle Z \rangle = (+1)P_0 + (-1)P_1 = |\alpha|^2 - |\beta|^2. \quad (3.49)$$

Verify $\langle \psi|Z|\psi \rangle$ yields the same value in both the computational and $\{|+\rangle, |-\rangle\}$ bases.

3.4.5 ∗ Measuring in Alternative Bases

In quantum computing, the computational basis is the de facto standard for measurements, primarily due to the native support from most quantum hardware. Yet, certain scenarios call for measurements in non-computational bases. In such instances, unitary transformations, or basis rotations, are employed to shift to the desired basis before carrying out the measurement.

Suppose one wishes to measure a qudit state, denoted as $|\psi\rangle$, within an orthonormal basis $\{|\phi_i\rangle\}$. The initial task is to identify a suitable unitary transformation, represented by U, that facilitates the transition from the targeted measurement basis $\{|\phi_i\rangle\}$ to the computational basis $\{|i\rangle\}$, up to a global phase factor:

$$U = \sum_i |i\rangle\langle\phi_i|. \quad (3.50)$$

Upon establishing U, the subsequent step involves measuring the state $U|\psi\rangle$ within the computational basis. Interestingly, the outcome probabilities yielded from this process mirror those obtained by measuring $|\psi\rangle$ directly in the $\{|\phi_i\rangle\}$ basis. The congruence of the results can be verified as follows:

$$P_i = |\langle i|U|\psi\rangle|^2 \quad (3.51a)$$

$$= \left|\sum_j \langle i|j\rangle\langle\phi_j|\psi\rangle\right|^2 \quad (3.51b)$$

$$= |\langle\phi_i|\psi\rangle|^2. \quad (3.51c)$$

An essential distinction to highlight is that, post-measurement, the state achieved in the computational basis, $|i\rangle$, does not coincide with the anticipated state, $|\phi_i\rangle$, within the chosen measurement basis. Nevertheless, within the quantum computing context, this incongruence is typically inconsequential since the primary emphasis

3.4 ∗ General Formulation of Quantum Measurement

is on obtaining the measurement readout. If a scenario necessitates the post-measurement state, employing $U^\dagger$ effectively transforms $|i\rangle$ into the desired state:

$$U^\dagger |i\rangle = \sum_j |\phi_j\rangle \langle j|i\rangle = |\phi_i\rangle. \tag{3.52}$$

This strategy for conducting measurements in non-standard bases finds a notable application in the Bell measurement, a topic we shall delve into in § 8.4.

3.4.6 ∗ Sampling Errors

In quantum computing, we often execute a quantum circuit repeatedly to obtain the empirical expectation values of certain observables (see § 1.5.3). Due to the finite number of runs for the circuit, empirical values may deviate from theoretical expectation values. This deviation is known as the sampling error.

In this subsection, we will use the Pauli operators (X, Y, Z) to demonstrate this important aspect of quantum measurement. The Pauli operators are unique in that they are both unitary and Hermitian, enabling them to serve as both quantum gates and observables.

1 Z Measurement

A measurement in the computational basis is equivalent to measuring the Z observable. In this scenario, the readout can be assigned a value of $+1$ for $|0\rangle$ and -1 for $|1\rangle$, as illustrated in the subsequent diagram. Through a collection of such readouts, we can calculate the empirical expectation value and variance of Z, and estimate the sampling error, as will be discussed next.

$$|\psi\rangle \longrightarrow \boxed{\text{\reflectbox{$\measuredangle$}}} \overset{\pm 1}{=\!=\!=} \Rightarrow \langle \psi | Z | \psi \rangle$$

2 Expected Value

Assume a quantum state $|\psi\rangle = \alpha|0\rangle + \beta|1\rangle$. The theoretical expectation value for the Z observable is

$$\langle Z \rangle \equiv \langle \psi | Z | \psi \rangle = |\alpha|^2 - |\beta|^2 = 2|\alpha|^2 - 1, \tag{3.53}$$

since $|\alpha|^2 + |\beta|^2 = 1$.

To compute this empirically, the circuit is run N times with each measurement readout $z_i \in \{\pm 1\}$. The empirical expectation value is then:

$$\langle Z \rangle \approx \frac{1}{N} \sum_{i=1}^{N} z_i. \tag{3.54}$$

3 Standard Deviation

Sampling error pertains to the deviation of the empirical $\langle Z \rangle$ from the theoretical value. The standard deviation serves as an estimate for this error.

The theoretical variance of the observable Z in the state $|\psi\rangle$ is given by

$$(\Delta Z)^2 = \langle Z^2 \rangle - \langle Z \rangle^2. \qquad (3.55)$$

Given that $\langle Z^2 \rangle = |\alpha|^2 + |\beta|^2 = 1$, the theoretical variance for the Z observable in this state becomes

$$(\Delta Z)^2 = 1 - \langle Z \rangle^2 = 1 - (|\alpha|^2 - |\beta|^2)^2 = 4|\alpha\beta|^2. \qquad (3.56)$$

And consequently, the theoretical standard deviation is:

$$\Delta Z = 2|\alpha\beta|. \qquad (3.57)$$

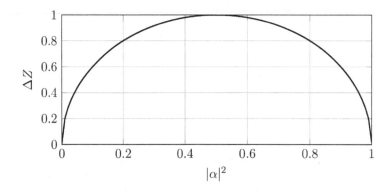

Figure 3.5: Plot of ΔZ as a function of $|\alpha|^2$

The relationship between ΔZ and $|\alpha|^2$ is depicted in Fig. 3.5. This curve elucidates the statistical nature of quantum measurement. When $|\alpha|^2 = 0.5$, the standard deviation ΔZ reaches its maximum value of 1, while the mean $\langle Z \rangle$ is zero. In addition, when $|\psi\rangle$ is a basis state—corresponding to cases where either $\alpha = 0$ or $\beta = 0$—the standard deviation ΔZ is zero. This behavior is in agreement with the established principle that basis states can be measured with zero uncertainty.

4 Sampling Error of the Mean

The sampling error for the expected value $\langle Z \rangle$ is usually characterized by the standard error of the mean (SEM), expressed as:

$$\text{SEM}_{\langle Z \rangle} = \frac{\Delta Z}{\sqrt{N}}, \qquad (3.58)$$

where ΔZ is the standard deviation of the individual measurements and N is the total number of measurements. Thus,

$$\text{SEM}_{\langle Z \rangle} = \frac{2|\alpha\beta|}{\sqrt{N}}. \qquad (3.59)$$

As the number of measurements N becomes large, the standard error of the mean (SEM) approaches zero. This indicates that the empirical average $\langle Z \rangle$ converges to the true expected value.

3.4 ∗ General Formulation of Quantum Measurement

Exercise 3.21 Consider a collection of N qubits, each in the state $|\psi\rangle = \alpha|0\rangle + \beta|1\rangle$. If the goal is to measure the expected value $\langle Z \rangle$ with a 99% accuracy, what is the minimum value of N required to achieve this level of precision?

5 X Measurement

Measuring the X observable corresponds to making measurements in the $\{|+\rangle, |-\rangle\}$ basis. If the quantum state is expressed as $|\psi\rangle = a|+\rangle + b|-\rangle$, the equations developed for Z measurement remain applicable. However, the coefficients a and b are related to α and β via:

$$a = \frac{1}{\sqrt{2}}(\alpha + \beta), \quad b = \frac{1}{\sqrt{2}}(\alpha - \beta). \tag{3.60}$$

The relationship between a and b with α and β allows us to translate the expected values and standard deviations calculated in the $\{|+\rangle, |-\rangle\}$ basis to the computational basis, and vice versa.

Exercise 3.22 Derive formulas for the theoretical standard deviation (ΔX) for the X observable in terms of α and β.

As elaborated in § 3.4.4, meauring in the $\{|+\rangle, |-\rangle\}$ basis can be done by transforming the quantum state with the basis-change unitary operator (here the Hadamard operator H) before measuring in the computational basis. Through repeated readouts, we can calculate the empirical expectation value X and estimate its sampling error, similar to our preceding discussion for Z.

$$|\psi\rangle \;—\boxed{H}\boxed{\overset{\pm 1}{\measuredangle}}\Rightarrow \langle\psi|X|\psi\rangle$$

3.4.7 ∗ Advanced Topics in Quantum Measurements

The preceding quantum measurement framework can be extended to include more complex scenarios such as POVMs, weak measurements, and nondestructive measurements, which allow for greater flexibility and applicability in quantum information science.

POVM: In quantum information and computation, POVMs are often used to model more complex measurement scenarios, including noisy or partial measurements. They are particularly useful in the context of open quantum systems, where they describe the various possible outcomes of measurements on systems interacting with their environment.

Weak Measurement: Traditional measurement collapses the state vector into one of the eigenstates of the observable. Weak measurements, on the other hand, only slightly perturb the quantum state, yielding only partial information about the observable. Weak measurements can be understood as a statistical limit of a series of less invasive measurements.

Nondestructive Measurements: The Measurement Postulate implies that a measurement fundamentally changes the state of the system. Nondestructive measurements try to sidestep this by obtaining information about the quantum state without significantly altering it, which is particularly useful in quantum error correction and certain quantum computation protocols.

The following texts are recommended for readers intesterested in these topics:

- Kurt Jacobs. *Quantum Measurement Theory and its Applications*. English. 1st. Cambridge University Press, 2014, page 554. ISBN: 978-1-107-02548-6.

- Mark M. Wilde. *Quantum Information Theory*. Cambridge University Press, 2017. ISBN: 978-1-107-17616-4.

3.5 *Application to Quantum State Tomography

The topic of this section is quantum state tomography, a powerful technique used to reconstruct the state of a quantum system. We apply the frameworks developed in this chapter on quantum states, basis transformation, and quantum measurements to this process, providing a cohesive case study.

Quantum state tomography involves reconstructing unknown quantum states based on measurement outcomes. As an illustrative example, let's assume we have $N = 2000$ qubits, all in an identical, yet unknown state, $|\psi\rangle$. These qubits are divided equally between Alice and Bob, with each receiving $N_a = 1000$ qubits. Alice measures her qubits in the computational basis, $\{|0\rangle, |1\rangle\}$, and observes $N_{a0} = 241$ qubits in the $|0\rangle$ state. Conversely, Bob uses the $\{|+\rangle, |-\rangle\}$ basis and detects $N_{b+} = 852$ qubits in the $|+\rangle$ state.

3.5.1 *Extracting Parameters from Alice's Measurements

From § 3.2, we know that a general qubit state expressed in the $\{|0\rangle, |1\rangle\}$ basis can be represented as:

$$|\psi\rangle = \cos\frac{\theta}{2}|0\rangle + \sin\frac{\theta}{2}e^{i\phi}|1\rangle. \tag{3.61}$$

Given this, the measurement probabilities for the states $|0\rangle$ and $|1\rangle$ are, respectively, $P_0 = \cos^2\frac{\theta}{2}$ and $P_1 = \sin^2\frac{\theta}{2}$. Using these probabilities, θ can be deduced from the difference $P_0 - P_1$:

$$\cos\theta = P_0 - P_1. \tag{3.62}$$

From Alice's data, we deduce:

$$P_0 \approx \frac{241}{1000}, \quad P_1 \approx \frac{1000 - 241}{1000}. \tag{3.63}$$

This results in an approximate value of $\theta \approx 2.12$ radians.

3.5.2 *Extracting Parameters from Bob's Measurements

Given the state in the $\{|+\rangle, |-\rangle\}$ basis, the expression becomes:

$$|\psi\rangle = \frac{\cos\frac{\theta}{2} + e^{i\phi}\sin\frac{\theta}{2}}{\sqrt{2}}|+\rangle + \frac{\cos\frac{\theta}{2} - e^{i\phi}\sin\frac{\theta}{2}}{\sqrt{2}}|-\rangle. \quad (3.64)$$

From this expression, using the difference in probabilities $P_+ - P_-$, we can derive $\cos\phi\sin\theta$:

$$\sin\theta\cos\phi = P_+ - P_-. \quad (3.65)$$

Exercise 3.23 Derive Eq. 3.65 starting from Eq. 3.64.

From Bob's observations, we gather:

$$P_+ \approx \frac{852}{1000}, \quad P_- \approx \frac{1000 - 852}{1000}. \quad (3.66)$$

With θ previously derived from Alice's measurements, we can calculate ϕ using Eq. 3.65, arriving at $\phi \approx 0.604$.

3.5.3 ✻Reconstructing the Quantum State

With the deduced values of θ and ϕ, the original quantum state can be reconstructed using Eq. 3.61:

$$|\psi\rangle \approx 0.491|0\rangle + 0.871 e^{0.604i}|1\rangle. \quad (3.67)$$

Exercise 3.24 ✻ **Sampling Errors** (See § 3.4.6)

Quantum measurements inherently exhibit stochastic behavior, leading to only approximate acquisition of P_0, P_1, P_+, and P_- through repeated measurements. This phenomenon is termed sampling error. The precision of these values is related to the sample size N, i.e., the number of measurements. Such sampling intricacies are endemic to quantum computing.

(a) Deduce the average sampling errors for P_0, P_1, P_+, and P_- in terms of N, N_a, N_b, etc.

(b) Calculate the average sampling errors for θ and ϕ.

(c) For the illustrated scenario, how large should N be to ensure the errors in θ and ϕ remain within a 0.1% threshold?

3.6 Summary and Conclusions

Comprehensive Framework for Qubits and Qudits

In the current chapter, we embarked on a journey to build a comprehensive framework for qubits and qudits. Serving as the backbone of quantum computing and quantum information systems, qubits and their generalized counterparts, qudits, are integral to our understanding of quantum phenomena. The chapter's aim was to transcend the limitations of specific examples like photons and electron spins, and to present a universal model that caters to the broad spectrum of qubit manifestations.

The topics of change of basis and quantum measurement theory were given special emphasis, not only for their theoretical significance but also for their practical applications in quantum computing. This chapter's examination of quantum state tomography illustrated the tangible applications of the aforementioned principles, highlighting their relevance in real-world scenarios.

Physical Manifestations and Mathematical Models

The diverse platforms that are actively being researched for qubit implementation underscore the versatility and adaptability inherent to quantum systems. We presented an overview of these platforms, drawing attention to their distinct methods of information encoding and their specific basis states.

Despite these various physical representations, a unified mathematical model for qubits emerged, anchored firmly in the tenets of quantum mechanics. This mathematical abstraction enables us to draw parallels across different qubit systems, underscoring the universal principles that underpin them.

Basis Rotations and Unitary Transformations

A considerable portion of this chapter was allocated to understanding quantum states and operators across diverse bases. The importance of bases in quantum mechanics was likened to reference frames in classical physics, offering readers a familiar touchstone. At the heart of our discourse is the concept that changing a basis aligns with a unitary transformation. We explored the transformation of vectors, operators, inner products, and operator averages in-depth.

This detailed inspection of basis changes proves invaluable in two key areas: firstly, alternate bases often shed light on nuanced insights into quantum states, proving essential for understanding quantum operations or algorithms. Secondly, operations triggered by quantum gates or other unitary transformations can be construed as a change of basis within the associated vector spaces. While both scenarios revolve around a change of basis, the nuances distinguishing them are subtle.

A General Formulation for Quantum Measurement

Our dialogue transitioned towards a generalized formulation of quantum measurements, rooted in the observable and measurement postulates of quantum mechanics. This discussion sets the stage for future explorations into decoherence, as well as advanced measurement strategies like POVM and weak measurements prevalent in contemporary quantum computing and quantum information contexts.

A salient insight was recognizing that the observable-based measurements, detailed in previous chapters, represent a specific case within the expansive framework of quantum measurement: the projective measurement.

Additionally, we embarked on a comprehensive examination of measurements across varying bases, augmenting our understanding of the quantum observation process through diverse reference frames.

Upcoming Topics

Having established a robust foundation on qubits and qudits, their physical embodiments, and the mathematical structures characterizing them, we are poised to explore the dynamics of quantum systems. The forthcoming chapter will address the temporal evolution of quantum states. Governed by the Schrödinger equation, the dynamics of qubit systems holds the key to understanding the mechanisms by which quantum information is processed, manipulated, and observed, especially through the lens of quantum gates and circuits.

Our journey will initially concentrate on the dynamics of a single qubit. This will set the stage to navigate the complexities of multi-qubit systems, entanglements, and interactions. These foundational discussions will be instrumental as we transition to more advanced realms, such as quantum algorithms, quantum error correction, and the architectural design of extensive quantum systems.

Problem Set 3

3.1 Consider a unitary operator U with an eigenvector $|\psi\rangle$ such that $U|\psi\rangle = e^{2\pi i \theta}|\psi\rangle$.

The phase estimation algorithm applies U^{2^j} on $|\psi\rangle$ for different values of $j = 0, 1, 2, \ldots$. You task here is to compute $U^{2^j}|\psi\rangle$.

3.2 (Computational basis to general basis transformation.) Derive the matrix U for basis change $\{|0\rangle, |1\rangle\} \to \{|\psi\rangle, |\psi_\perp\rangle\}$ where $|\psi\rangle$ is defined by Eq. 3.2 and $|\psi_\perp\rangle$ by Eq. 3.5. Then, using U and using inner product expressions, determine the representation of $|\varphi\rangle = \alpha|0\rangle + \beta|1\rangle$ in the new basis.

Hint: $U = |\psi\rangle\langle 0| + |\psi_\perp\rangle\langle 1|$.

3.3 Consider a four-level quantum system (qudit with $d = 4$). The following orthonormal basis states span the state space $\mathbb{C}^4$:

$$|\phi_0\rangle = \frac{1}{\sqrt{2}}(|0\rangle + |3\rangle), \qquad (3.68a)$$

$$|\phi_1\rangle = \frac{1}{\sqrt{2}}(|1\rangle + |2\rangle), \qquad (3.68b)$$

$$|\phi_2\rangle = \frac{1}{\sqrt{2}}(|0\rangle - |3\rangle), \qquad (3.68c)$$

$$|\phi_3\rangle = \frac{1}{\sqrt{2}}(|1\rangle - |2\rangle). \qquad (3.68d)$$

(a) Verify these basis states are orthonormal given that $\{|0\rangle, \ldots, |3\rangle\}$ are.

(b) Determine the transformation operator between the computational basis and the new basis in both bra-ket and matrix notation.

3.4 The transformation from the computational basis to the four-dimensional Hadamard basis is given by

$$H_4 = \frac{1}{2}\begin{bmatrix} 1 & 1 & 1 & 1 \\ 1 & -1 & 1 & -1 \\ 1 & 1 & -1 & -1 \\ 1 & -1 & -1 & 1 \end{bmatrix}. \qquad (3.69)$$

(a) Find the basis states of the Hadamard basis $\{|h_i\rangle\}$.

(b) Calculate the representation of $|\psi\rangle = \sum_i c_i |i\rangle$ in the Hadamard basis, where $c_i \in \mathbb{C}$.

(c) Calculate the representation of $|\psi\rangle = \sum_i c_i |h_i\rangle$ in the computational basis, where $c_i \in \mathbb{C}$.

3.5 Design a setup for measuring the Y observable, analogous to the setup provided for the X measurement detailed in § 3.4.6.5.

Hint: Consider the unitary transformation needed to map the eigenstates of Y to the computational basis.

3.6 Consider a collection of independent qubits all in the identical state $|\psi\rangle = \alpha |0\rangle + \beta |1\rangle$.

(a) Suppose $|\alpha|^2 = 0.99$. What is the probability of measuring 10 qubits and find all of them in $|0\rangle$? What is the probability of measuring 10 qubits and find 9 of them in $|0\rangle$ and 1 in $|1\rangle$?

(b) Suppose we measure 10 qubits and find each in the state $|0\rangle$. Estimate the probability for $|\alpha|^2 > 0.99$, assuming $|\alpha|^2$ has a uniform distribution.

3.7 Derive formulas for the theoretical expectation value ($\langle Y \rangle$) and standard deviation (ΔY) for the Pauli-Y observable in terms of α and β for the qubit state $|\psi\rangle = \alpha |0\rangle + \beta |1\rangle$.

3.8 Two orthonormal bases in $\mathbb{C}^d$, denoted by $\{u_j\}$ and $\{v_j\}$, are termed mutually unbiased if all vector pairs, one from each basis, possess inner products with magnitude $\frac{1}{\sqrt{d}}$. That is, $|\langle u_j | v_k \rangle| = \frac{1}{\sqrt{d}}$. Determine a basis mutually unbiased to the computational basis for $d = 2$ and $d = 4$.

4. Dynamics of Quantum Systems

Contents

4.1	The Evolution Postulate of Quantum Mechanics	86
4.1.1	Properties of Unitary Transformation	87
4.2	✻ The Schrödinger Equation	87
4.2.1	✻ The Time-independent Schrödinger Equation	88
4.2.2	✻ The Time-dependent Schrödinger Equation	89
4.2.3	✻ Trotterization	90
4.3	Stationary Nature of Energy Eigenstates	91
4.3.1	Time Evolution Operator in the Basis of Energy Eigenstates	91
4.3.2	Stationary Energy Eigenstates	91
4.3.3	The Ground State and Excited States	92
4.3.4	Qubit as a Two-Level System	93
4.4	Universal Quantum Computing and Annealing	94
4.4.1	Gate-based Universal Quantum Computing	94
4.4.2	Adiabatic Quantum Computation and Quantum Annealing	95
4.5	✻ Larmor Precession and Rabi Oscillations	96
4.5.1	✻ Larmor Precession	96
4.5.2	✻ Rabi Oscillations	99
4.6	✻ Further Exploration	102
4.6.1	✻ Time Evolution Equations for Relativistic Systems	103
4.6.2	✻ Open Quantum Systems	103
4.7	✻ Deferred Proofs	104
4.8	Summary and Conclusions	106
	Problem Set 4	107

In the preceding chapters, we have explored how a qubit's state is represented by its state vector. As we extend our discussion to include the dynamics of qubits, we encounter the question of how a qubit's state evolves over time. Such evolution can result from various interactions, like the application of a quantum gate or quantum annealing processes. These interactions drive the changes we observe over time, which are described as the dynamics of quantum systems.

In the realm of quantum mechanics, dynamics pertains to the evolution of quantum states as a result of physical interactions. The equations and principles of quantum mechanics provide the framework for understanding this evolution. They allow us to predict and analyze how quantum states change, interact, and develop over time, which is fundamental to both the theoretical understanding and practical applications of quantum physics.

The dynamics of a quantum system is principally governed by the Evolution Postulate of quantum mechanics, a fundamental theorem dictating how quantum states evolve temporally. For qubit systems, this evolution is predominantly governed by the Schrödinger equation. Mastery of this fundamental equation is paramount to comprehending the intricate mechanisms by which quantum information is processed, manipulated, and observed in a wide array of quantum systems.

In this chapter, the exploration will introduce the general concepts of the dynamics of quantum system, using examples of a single qubit. Subsequent chapters will broaden this examination to encompass systems of multiple qubits, probing their interactions, entanglements, and the means by which their dynamics can be both controlled and exploited for potent computational tasks. This foundational comprehension of the dynamics of qubits will act as a nexus to more advanced subjects such as quantum algorithms, quantum error correction, and the architecture of large-scale quantum systems.

4.1 The Evolution Postulate of Quantum Mechanics

The Time Evolution Postulate is a fundamental principle in quantum mechanics, describing how quantum states change over time.

> **Postulate 4: Time Evolution**
>
> The time evolution of the state of a closed quantum system is governed by a unitary transformation.

Mathematically, we can express this Postulate as:

$$|\psi(t)\rangle = U(t) |\psi(0)\rangle, \qquad (4.1)$$

where $|\psi(t)\rangle$ is the quantum state of the system at time t, $U(t)$ is the time-evolution operator that describes how the state at time $t = 0$ evolves to the state at time t, and $|\psi(0)\rangle$ is the initial state of the system.

The necessity of the time-evolution operator being unitary in a closed quantum system stems from the conservation of total probability, which must remain equal to one. Essentially, a quantum state vector must always be normalized since its components represent probability amplitudes, and their square sum must equal one. As the quantum state naturally evolves over time, this normalization must be preserved. Unitary transformations fulfill this requirement by maintaining the inner product, thus ensuring the coherence and consistency of the behavior of the system.

4.1.1 Properties of Unitary Transformation

Unitary transformations are fundamental to quantum mechanics as they govern the evolution of the states of closed quantum systems. They are also essential in quantum computing, possessing several key properties that enable quantum state manipulation. For simplicity, we will denote $U(t)$ as U in the following discussion.

1. Preservation of Inner Products: The overlap of two quantum states, $|\phi\rangle$ and $|\psi\rangle$, remains invariant under the same unitary evolution. Explicitly,

$$\langle U\phi|U\psi\rangle = \langle \phi| U^\dagger U |\psi\rangle = \langle \phi|\psi\rangle. \tag{4.2}$$

A consequence of this property is that an orthonormal basis remains orthonormal under a unitary evolution.

Moreover, as $|\langle\phi|\psi\rangle|^2$ represents the probability of measuring state $|\phi\rangle$ in state $|\psi\rangle$, or vice versa, such probabilities remain unchanged under unitary evolution, preserving the probabilistic interpretation of quantum mechanics.

2. Reversibility: Unitary transformations are reversible, meaning that they can be undone by applying the inverse transformation using U^{-1}.

3. Linearity: Superposition principle holds during the unitary evolution. A unitary transformation applied to a superposition of quantum states results in a superposition of transformed states.

4. Unit Eigenvalues: The eigenvalues of a unitary transformation are complex numbers with an absolute value of 1.

5. Orthonormal Row and Column Vectors: The column vectors of a unitary matrix form an orthonormal set, and the same applies to the row vectors when transposed.

Unitary transformations can be *interpreted* as a change of basis in Hilbert space, effectively rotating the quantum state vector while preserving the structure of the space. In particular, quantum gates apply unitary transformations to quantum states. While computations are commonly performed in the computational basis, quantum states exist in a Hilbert space representable in any orthonormal basis. See § 3.3.2 for an in-depth discussion on the change of basis concept.

4.2 *The Schrödinger Equation

The Schrödinger equation is a cornerstone of non-relativistic quantum mechanics, governing the time evolution of quantum states in systems like electrons in atoms, quantum dots, and superconducting qubits. This equation links the time-dependent evolution of a quantum state to the system's Hamiltonian operator, often simply referred to as the Hamiltonian, H. The Hamiltonian represents the total energy observable of the system and thus acts as a crucial connection between the abstract quantum state and measurable physical quantities, like energy levels.

The Schrödinger equation is expressed as:

$$i\hbar \frac{\partial}{\partial t} |\psi(t)\rangle = H(t) |\psi(t)\rangle, \tag{4.3}$$

where $|\psi(t)\rangle$ is the quantum state vector at time t, and $\hbar$ is the reduced Planck constant.

While we will not delve into solving this equation here, we will focus on key concepts and results essential to understanding quantum gates, quantum hardware, and providing a foundation for further study.

Hermitian Property of the Hamiltonian

As a consequence that the solution of the Schrödinger equation must represent a unitary evolution given by Eq. 4.1, the Hamiltonian H must be a Hermitian operator. We will provide a proof of this property at the end of this section (see § 4.7).

From a physical point of view, the Hermitian condition for the Hamiltonian is necessarily true, since H represents the total system energy, and as such, its eigenvalues must be real numbers.

4.2.1 ∗ The Time-independent Schrödinger Equation

In the simplest case where the Hamiltonian H does not change with time, the Schrödinger equation is expressed as:

$$i\hbar \frac{\partial}{\partial t} |\psi(t)\rangle = H |\psi(t)\rangle. \tag{4.4}$$

This equation has a general solution of the form:

$$|\psi(t)\rangle = e^{-\frac{i}{\hbar}Ht} |\psi(0)\rangle, \tag{4.5}$$

which can be verified by differentiating both sides over t (see § 4.7).

The exponential term

$$U(t) = e^{-\frac{i}{\hbar}Ht} \tag{4.6}$$

is the unitary time-evolution operator, and it determines how the quantum state $|\psi(0)\rangle$ at the initial time evolves to the state $|\psi(t)\rangle$ at time t.

Here the exponential function of operator A is defined as

$$e^A \equiv \exp(A) \equiv \sum_{n=0}^{\infty} \frac{1}{n!} A^n, \tag{4.7}$$

where A is a normal operator, meaning $AA^\dagger = A^\dagger A$. Normal operators include Hermitian and unitary operators.

Properties of the Exponential Operator

The exponential operator e^A has several important properties:

Product Property

In general, $e^{A+B} \neq e^A e^B$. However,

$$e^{A+B} = e^A e^B, \quad \text{if } [A, B] = 0. \tag{4.8}$$

We will provide a proof of this property at the end of this section (see § 4.7).

4.2 ∗ The Schrödinger Equation

Unitary Property

If A is Hermitian, then e^{iA} is unitary. A proof is outlined below:

$$\left(e^{iA}\right)\left(e^{iA}\right)^{\dagger} = e^{iA} e^{-iA^{\dagger}} \quad (4.9a)$$
$$= e^{iA} e^{-iA} \quad (4.9b)$$
$$= e^{iA-iA} = e^{0} = I. \quad (4.9c)$$

This property confirms that $U(t)$ in Eq. 4.6 is unitary, consistent with the Time Evolution Postulate that the state of a closed quantum system is governed by a unitary transformation.

Generalized Euler Formula

Consider the case where $A^2 = I$ (i.e., A is involutory) and θ is a real number. In this scenario, the following relationship holds:

$$e^{i\theta A} = \cos\theta I + i\sin\theta A. \quad (4.10)$$

This equation can be analogized to the Euler formula, $e^{i\theta} = \cos\theta + i\sin\theta$. A detailed proof of this property is provided at the end of this section (see § 4.7). Readers are encouraged to study the mathematics behind this proof, as the techniques and concepts involved are highly relevant to the field of quantum computing.

> **Exercise 4.1** Express e^{iX} as a single 2×2 matrix, where X is the Pauli-X matrix.

> **Exercise 4.2** ∗ Express $e^{i(I+Z)/2}$ to a single 2×2 matrix, where Z is the Pauli-Z matrix and I is the identity matrix.

4.2.2 ∗ The Time-dependent Schrödinger Equation

When the Hamiltonian varies with time, denoted as $H(t)$, the formulation of the time-evolution operator becomes more complex. For cases where $H(t_1)$ commutes with $H(t_2)$ for any t_1 and t_2 within the time domain of interest, the solution to the time-dependent Schrödinger equation is given by

$$U(t) = \exp\left(-\frac{i}{\hbar}\int_0^t H(t')dt'\right). \quad (4.11)$$

In scenarios where $H(t_1)$ and $H(t_2)$ do not commute, the solution is expressed using a time-ordered exponential, known as the Dyson series:

$$U(t) = \mathcal{T}\exp\left(-\frac{i}{\hbar}\int_0^t H(t')dt'\right). \quad (4.12)$$

Here, $\mathcal{T}$ represents the time-ordering operator, which organizes the terms in the exponential based on their time arguments, arranging factors corresponding to later times to the left of those associated with earlier times.

It can be shown ([8]) that the time-evolution operator $U(t)$ is unitary even when $H(t)$ is time-dependent, as it preserves the inner product and the norm of state vectors. This ensures that the probabilities remain well-defined and conserved throughout the evolution.

While the Dyson series is fundamental to a comprehensive understanding of time-dependent quantum mechanics, it is not crucial for a general quantum computing audience. For practical applications in quantum computing and simulation, the Trotterization approximation is often more relevant.

4.2.3 * Trotterization

Trotterization, also known as the Trotter-Suzuki decomposition, is a method used in quantum mechanics to approximate the evolution of quantum systems with non-commuting terms in their Hamiltonians. This technique is particularly useful in quantum computing for simulating complex quantum systems, where exact solutions are often computationally infeasible.

When dealing with a time-varying Hamiltonian $H(t)$, the Trotterization formula can be used to approximate the time-evolution operator $U(t)$. Consider dividing the total evolution time T into N small intervals, $\Delta t = T/N$. The Trotterization formula for a time-dependent Hamiltonian is:

$$U(T) \approx \prod_{n=1}^{N} \exp\left(-\frac{i}{\hbar} H(t_n) \Delta t\right), \tag{4.13}$$

where $t_n = n\Delta t$ represents the time at the nth interval. This product is ordered such that factors corresponding to later times are placed to the left of those associated with earlier times.

This formula approximates the time-ordered exponential (Eq. 4.12) by assuming the Hamiltonian remains approximately constant over each interval Δt. Each term in the product is an exponential of the Hamiltonian evaluated at a specific time t_n, scaled by Δt. The accuracy of this approximation increases as $N \to \infty$ (i.e., as $\Delta t \to 0$).

Trotterization is also useful for Hamiltonians containing non-commuting terms, common in many real-world quantum systems. For example, consider a Hamiltonian of the form $H = H_1 + H_2$, even when time-independent. In such cases, the exact solution to the Schrödinger equation can be challenging to compute, as $e^{i(H_1+H_2)} \neq e^{iH_1} e^{iH_2}$. Trotterization allows us to approximate the time-evolution operator over a small interval Δt:

$$U(T) \approx \left(\exp\left(-\frac{i}{\hbar} H_1 \Delta t\right) \exp\left(-\frac{i}{\hbar} H_2 \Delta t\right)\right)^N. \tag{4.14}$$

The effectiveness of Trotterization depends on the size of Δt and the nature of the Hamiltonian. For rapidly changing or highly non-commutative Hamiltonians, more sophisticated versions, such as higher-order Suzuki-Trotter expansions, might be necessary.

> **Exercise 4.3** Consider a Hamiltonian $H = \alpha X + \beta Z$, where X and Z are the Pauli matrices, and $\alpha, \beta \in \mathbb{R}$. Write an approximate solution to the Schrödinger equation using Trotterization.

4.3 Stationary Nature of Energy Eigenstates

As a consequence of the time-independent Schrödinger equation, energy eigenstates are stationary, as explained below. This property plays an important role in many areas of quantum physics and quantum computing.

Let $|E_k\rangle$ be the eigenvectors of the Hamiltonian operator H, with the corresponding energy eigenvalues E_k. Since H is Hermitian, E_k has real values. When these eigenvalues are discrete, they are commonly referred to as the energy levels of the system.

4.3.1 Time Evolution Operator in the Basis of Energy Eigenstates

According to the spectral decomposition theorem in linear algebra, H can be expressed as:

$$H = \sum_k E_k |E_k\rangle\langle E_k|. \tag{4.15}$$

Each term in the sum projects the state vector onto the k-th energy eigenstate $|E_k\rangle$ and scales it by the corresponding energy eigenvalue E_k.

The time-evolution operator $U(t)$ can be expressed as:

$$U(t) = e^{-\frac{i}{\hbar}Ht} = \sum_k e^{-\frac{iE_k t}{\hbar}} |E_k\rangle\langle E_k|. \tag{4.16}$$

This equation describes the time-evolution operator $U(t)$ as a sum over all energy eigenstates with the corresponding energy eigenvalues. Each term in the sum evolves the k-th energy eigenstate $|E_k\rangle$ by a phase factor $e^{-\frac{iE_k t}{\hbar}}$. This equation can be viewed as the solution to the time-independent Schrödinger equation in the basis of the energy eigenstates.

4.3.2 Stationary Energy Eigenstates

Equation 4.16 shows that if a system is initially in an energy eigenstate $|\psi(0)\rangle = |E_k\rangle$, then its time evolution is simply a phase factor multiplying the initial state:

$$|\psi(t)\rangle = U(t)|\psi(0)\rangle = e^{-\frac{iE_k t}{\hbar}} |E_k\rangle. \tag{4.17}$$

This means that if a system is prepared in an energy eigenstate, it will remain in that state with a time-dependent phase factor. Such a state is called a stationary state. Stationary states are quantum states that do not change with time, except for an overall phase factor.

If the system is not in an energy eigenstate initially, then its time evolution will involve a superposition of energy eigenstates, and the amplitudes of these energy eigenstates will evolve in time according to the corresponding phase factors. For example, say the system is initially in a superposition of $|E_0\rangle$ and $|E_1\rangle$:

$$|\psi(0)\rangle = \alpha |E_0\rangle + \beta |E_1\rangle. \tag{4.18}$$

At time t, the system state $|\psi(t)\rangle$ is given by

$$|\psi(t)\rangle = U(t)|\psi(0)\rangle = e^{-\frac{iE_0 t}{\hbar}} \alpha |E_0\rangle + e^{-\frac{iE_1 t}{\hbar}} \beta |E_1\rangle. \tag{4.19}$$

It's evident that the probabilities of finding the system in either energy state remain constant, as $|e^{-\frac{iE_0 t}{\hbar}} \alpha|^2 = |\alpha|^2$ and $|e^{-\frac{iE_1 t}{\hbar}} \beta|^2 = |\beta|^2$.

The above discussion highlights the stationary nature of energy eigenstates in an ideal quantum system, where the probability distributions of these states remain constant over time, assuming the system is perfectly isolated and not subjected to any external forces or interactions.

However, in practice, real quantum systems are never completely closed or isolated. They invariably interact with their surroundings to some extent, leading to deviations from the idealized stationary behavior. These interactions can cause transitions between different energy states and introduce phenomena such as decoherence and dissipation, significantly influencing the system's behavior.

Exercise 4.4 Consider a Hamiltonian H:

$$H = X = \begin{bmatrix} 0 & 1 \\ 1 & 0 \end{bmatrix}.$$

(a) Diagonalize this Hamiltonian to find its eigenvalues and eigenvectors.

(b) Given the time-evolution operator $U(t) = e^{-iHt}$, compute $U(t)$ for a time t, using the eigendecomposition of H.

(c) If we start with a state $|0\rangle$, and we evolve it using $U(t)$, the resulting state will be $U(t)|0\rangle$. Compute the probabilities of measuring $|0\rangle$ and $|1\rangle$ after this evolution.

4.3.3 The Ground State and Excited States

The ground state, $|E_0\rangle$, represents the lowest energy state the system can occupy. In non-ideal, real-world quantum systems, this state is typically the most stable configuration, often realized as the system's natural state in the absence of external energy inputs. It serves as the foundational reference point for defining other possible states of the system, known as the excited states. These excited states, having higher energy levels than the ground state, can be populated due to environmental interactions or external stimuli, highlighting the dynamic nature of quantum systems in realistic settings.

An excited state, $|E_1\rangle$ for example, is any state of a system that has a higher energy than the ground state. A qubit or particle has to absorb energy of the amount $E_1 - E_0$ to be "excited" from the ground state to this less stable state. This energy can be supplied through various means, such as by applying an external electric or magnetic field, or by exposing the system to electromagnetic radiation (e.g., microwave or laser) of angular frequency

$$\omega = \frac{E_1 - E_0}{\hbar}. \tag{4.20}$$

The excitation energy can also be provided, usually unintendedly, through thermal energy, which is why some quantum hardware operate at low temperatures.

4.3 Stationary Nature of Energy Eigenstates

A superposition state, which is a combination of several possible states, can be achieved through careful control of the energy input and the system's interactions.

An excited state or superposition state is inherently unstable because it represents a higher energy configuration of the system. When a system is in an excited state, it has multiple pathways to release the excess energy and return to the ground state. This process is known as relaxation. The mechanisms for relaxation can vary, including the emission of photons or phonons, collisions with other particles, or other interactions that allow the system to shed its excess energy. The tendency to return to the ground state also give rise to the short lifetime of qubits - referred to decoherence in quantum computing.

These concepts lay the foundation for understanding a wide range of quantum behaviors and are central to fields like quantum computing hardware, where control of ground and excited states enables the manipulation of qubits for computational purposes.

4.3.4 Qubit as a Two-Level System

A qubit is a two-level system, meaning it involves only two energy eigenstates: the ground state $|E_0\rangle$ and the first excited state $|E_1\rangle$. These states form the computational basis:

$$|0\rangle \equiv |E_0\rangle, \quad |1\rangle \equiv |E_1\rangle. \tag{4.21}$$

In quantum computing, qubits are typically initialized in the ground state $|0\rangle$. Quantum gates are designed to manipulate the qubits by inducing transitions between energy eigenstates or creating superpositions. The stationary nature of energy eigenstates ensures that, in an idealized and isolated quantum system, the qubits remain in their initial states until a quantum gate or external force acts upon them.

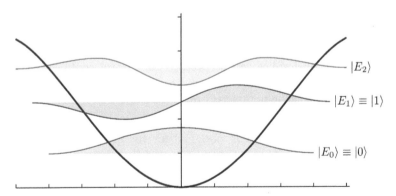

Figure 4.1: Example of a Practical Multi-Level System

In a practical quantum system, such as a superconducting qubit, energy levels are often more complex than a simple two-level system. These levels might include $|E_0\rangle$, $|E_1\rangle$, and $|E_2\rangle$, for example, where only $|E_0\rangle$ and $|E_1\rangle$ are actively used for the qubit operation, and the existence of $|E_2\rangle$ is considered undesirable.

To mitigate potential problems, the system is meticulously engineered so that the energy levels $|E_0\rangle$, $|E_1\rangle$, and $|E_2\rangle$ are unequally spaced, i.e., $E_1 - E_0 \neq E_2 - E_1$. This careful design ensures that energy specifically targeting the transition between $|E_0\rangle$ and $|E_1\rangle$ will not inadvertently affect the transition between $|E_1\rangle$ and $|E_2\rangle$. By minimizing the interaction with the undesirable level $|E_2\rangle$, the system gains greater stability and control, critical factors for the effective functioning of quantum computation. This is illustrated in Fig. 4.1.

Exercise 4.5 Given the Hamiltonian $H = \hbar\omega \begin{bmatrix} -1 & 0 \\ 0 & 1 \end{bmatrix}$, where $\omega \in \mathbb{R}$, find the energy levels E_0 and E_1, and corresponding eigenstates $|E_0\rangle$ and $|E_1\rangle$.

4.4 Universal Quantum Computing and Annealing

In the rapidly advancing field of quantum computing, two predominant paradigms have emerged as foundational approaches to quantum computation: gate-based universal quantum computing and quantum annealing. While gate-based universal quantum computing is designed for a broad range of algorithms and can theoretically solve any computable problem, quantum annealing is particularly tailored for solving optimization problems by finding the minimum of a given objective function. The principle of unitary evolution of quantum states forms the foundation for these quantum computing paradigms, which we elucidate in the following subsections.

4.4.1 Gate-based Universal Quantum Computing

Gate-based quantum computers modulate qubit states using a series of predesigned unitary operations, known as quantum gates. Similar to classical digital computers where data processing occurs through logic gates, gate-based quantum computers process information by sequentially applying quantum gates to an initial state, iteratively evolving it into the final state that represents the computation result.

Quantum gates can be conceptualized as discrete counterparts to the continuous time-evolution operators in the Schrödinger equation. Implementing a quantum gate mandates a controlled alteration of the qubit's Hamiltonian such that the intended transformation applies to its state. This control could entail utilizing external elements like electromagnetic radiation (e.g., microwave pulses) or modulating other system-specific parameters (e.g., for superconducting qubits, trapped ions, or quantum dots).

Given a Hamiltonian H designed for the gate's function and an active time interval Δt, the unitary evolution operator $U(\Delta t)$, as informed by the Schrödinger equation, is represented as:

$$|\psi(\Delta t)\rangle = U(\Delta t)|\psi(0)\rangle. \tag{4.22}$$

Assuming the gate Hamiltonian H remains consistent over Δt, $U(\Delta t)$ can be articulated as:

$$U(\Delta t) = e^{-\frac{iH\Delta t}{\hbar}}. \tag{4.23}$$

4.4 Universal Quantum Computing and Annealing

Exercise 4.6 Consider a one-qubit quantum gate defined by the Hamiltonian $H = \frac{\hbar}{2}Z$. Derive the unitary evolution operator $U(\Delta t)$ over an interval Δt and elucidate the influence of this gate on an arbitrary qubit state.

4.4.2 Adiabatic Quantum Computation and Quantum Annealing

Adiabatic Quantum Computation (AQC) is based on the adiabatic theorem of quantum mechanics, wherein the gradually varied time evolution of a quantum system's Hamiltonian transitions the system from its initial to a desired final state.

1 Adiabatic Quantum Computation (AQC)

AQC leverages the inherent unitary evolution of quantum states governed by the time-dependent Schrödinger equation. According to the adiabatic principle, if a quantum system is initially in an eigenstate (usually the ground state) of a Hamiltonian and the Hamiltonian changes sufficiently slowly, and without encountering level crossings, the system will remain predominantly in the instantaneous eigenstate of the evolving Hamiltonian, avoiding transitions to other energy states. The probability of the system deviating from the desired eigenstate and ending up in other states decreases exponentially with the slowness of the change in the Hamiltonian.

The approach involves encoding a computational problem into a quantum system, where the ground state of this system corresponds to the problem's solution. The process starts with an initial, simple Hamiltonian, H_{initial}, with an easily identifiable ground state. This Hamiltonian then undergoes a slow (adiabatic) evolution to a final Hamiltonian, H_{target}, encoding the solution to the problem.

The Hamiltonian evolution is described by:

$$H(t) = (1 - s(t))H_{\text{initial}} + s(t)H_{\text{target}}, \qquad (4.24)$$

where $s(t)$ is a scheduling parameter varying from 0 to 1 during the process, smoothly transitioning from H_{initial} to H_{target}.

Consider, for example, a spin qubit in a time-dependent magnetic field. The initial Hamiltonian $H_{\text{initial}} = Z$ has a ground state $|0\rangle$, corresponding to the spin aligned with the magnetic field in the z direction. As the field slowly rotates to the x direction, represented by $H_{\text{target}} = X$, the qubit's ground state evolves to $|+\rangle$, provided the change is adiabatic. In more complex systems with many interacting qubits, the final Hamiltonian H_{target} is not as straightforward and represents the complex problem to be solved.

2 Quantum Annealing: An Implementation of AQC

Quantum annealing is an implementation within the AQC framework. Unlike methods relying on quantum gates, it allows the system to naturally evolve under its quantum-mechanical dynamics. The term "annealing" refers to the process in which the Hamiltonian is adiabatically changed, akin to the metallurgical technique of slowly cooling materials to reach a stable state.

Implementing quantum annealing requires precise control over physical parameters characterizing the system's Hamiltonian. For instance, in superconducting quantum annealing systems, the Hamiltonian is implemented by adjusting magnetic fields and couplings between qubits. By carefully tuning these parameters, the system is guided through the desired Hamiltonian evolution.

Quantum annealing computers are particularly effective for problems modeled by the Ising model, which can represent many optimization challenges. We will delve deeper into the Ising model in § 6.5.3 and explore its application in quantum annealing and other quantum algorithms in § 11.2.

4.5 * Larmor Precession and Rabi Oscillations

This section focuses on Larmor precession and Rabi oscillations, two cornerstone phenomena that shed light on the behavior of quantum systems subject to external fields. The dynamics of qubit systems is guided by the Schrödinger equation, which outlines their unitary evolution. Larmor precession and Rabi oscillations serve as archetypal cases to better grasp this evolution. By exploring these phenomena, we aim to demonstrate effective strategies for solving the Schrödinger equation. Their significance is further underscored by their relevance to quantum computing, especially in the practical aspects such as hardware design and quantum gate implementation.

4.5.1 * Larmor Precession

Larmor precession, or spin precession, occurs when a quantum particle with intrinsic angular momentum, or spin, is placed in an external magnetic field. The interaction between the particle's magnetic moment and the external field causes the particle's spin to precess, or rotate, around the direction of the magnetic field.

In the context of quantum computing, Larmor precession is often used as a mechanism to manipulate and control the state of a qubit. By applying the external magnetic field for a designed time interval, the qubit's spin can be precisely controlled, allowing for the implementation of specific quantum gates and operations. Larmor precession is also used in magnetic resonance imaging (MRI) to create images of the body's internal structures, and in nuclear magnetic resonance (NMR) spectroscopy to analyze the properties of molecules and materials.

1 Solution to the Schrödinger Equation

The Hamiltonian, representing the energy of interaction, of a spin with an external magnetic field B in the z-direction can be expressed as:

$$H = -\frac{\hbar\omega}{2}Z, \qquad (4.25)$$

where ω is the Larmor frequency (see details explained at the end of this subsection).

The dynamics of the system is governed by the Schrödinger equation (Eq. 4.4), which has a solution in the form of the time-evolution operator $U(t)$ given by Eq. 4.6. Due to the property $Z^2 = I$ for Pauli matrix Z, we can simplify the matrix exponential in $U(t)$ into a regular matrix (refer to Eq. 4.10):

4.5 ∗ Larmor Precession and Rabi Oscillations

$$U(t) = e^{-iHt/\hbar} = e^{i\omega t Z/2} \tag{4.26a}$$

$$= \cos\frac{\omega t}{2} I + i\sin\frac{\omega t}{2} Z \tag{4.26b}$$

$$= \begin{bmatrix} e^{\frac{i\omega t}{2}} & 0 \\ 0 & e^{-\frac{i\omega t}{2}} \end{bmatrix}. \tag{4.26c}$$

If the spin is initially in the direction of (θ, ϕ), its state vector is:

$$|\psi(0)\rangle = \begin{bmatrix} \cos\frac{\theta}{2} \\ \sin\frac{\theta}{2} e^{i\phi} \end{bmatrix}. \tag{4.27}$$

The time evolution of the system is given by:

$$|\psi(t)\rangle = U(t)|\psi(0)\rangle \tag{4.28a}$$

$$= \begin{bmatrix} e^{\frac{i\omega t}{2}} & 0 \\ 0 & e^{-\frac{i\omega t}{2}} \end{bmatrix} \begin{bmatrix} \cos\frac{\theta}{2} \\ \sin\frac{\theta}{2} e^{i\phi} \end{bmatrix} \tag{4.28b}$$

$$= e^{\frac{i\omega t}{2}} \begin{bmatrix} \cos\frac{\theta}{2} \\ \sin\frac{\theta}{2} e^{i(\phi - \omega t)} \end{bmatrix}, \tag{4.28c}$$

which is the analytic solution to the Schrödinger equation for Larmor precession.

2 Interpretation

From Eq. 4.28 we see that the azimuthal angle of the spin direction at time t is given by

$$\phi(t) = \phi - \omega t, \tag{4.29}$$

while the polar angle θ remains constant.

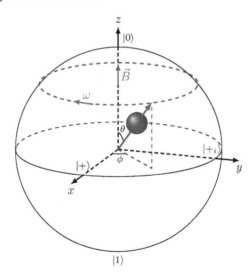

Figure 4.2: Larmor Precession on the Bloch Sphere

This can be visualized as the state vector of the spin rotating around the z-axis (the direction of the magnetic field), tracing out a circle on the surface of the Bloch sphere, as illustrated in Fig. 4.2. The angular frequency of the rotation is the Larmor frequency ω. This type of motion is known as precession, hence the name Larmor precession.

Larmor precession can be likened to the behavior of a spinning top. Just as a top precesses around the vertical direction when not perfectly upright due the influence of gravity, a spin in a magnetic field precesses around the field direction if it's not aligned with the field. This precession is caused by the interaction between the spin's magnetic moment and the magnetic field, similar to the gravitational torque on the top. In the case of an open system, various interactions with the environment, often undesirable, can cause the spin to gradually align more closely with the magnetic field direction over time.

3 Implementation of the $R_z(\gamma)$ Gate

Larmor precession can be utilized to precisely control a spin-based qubit by aligning the magnetic field at an angle θ relative to the spin direction, and then activating the magnetic field for a time interval Δt. The spin direction will precess by an angle $\gamma = \omega \Delta t$. This is exactly the Pauli-Z rotation gate $R_z(\gamma)$, discussed in § 5.2.5.

> **Exercise 4.7** With the $|\psi(t)\rangle$ given in Eq. 4.28, calculate the expected values $\langle\psi(t)|X|\psi(t)\rangle$, $\langle\psi(t)|Y|\psi(t)\rangle$, and $\langle\psi(t)|Z|\psi(t)\rangle$.

4 ✷ The Hamiltonian

The Hamiltonian for a spin in an external magnetic field $\vec{B}$, which is the interaction energy between the magnetic moment of the spin $\vec{\mu}$ and the magnetic field, is given by:

$$H = -\vec{\mu} \cdot \vec{B}. \tag{4.30}$$

The magnetic moment $\vec{\mu}$ of a spin is proportional to its spin angular momentum $\vec{S}$:

$$\vec{\mu} = \gamma \vec{S}, \tag{4.31}$$

In the case of a magnetic field aligned with the z-axis, the interaction energy becomes:

$$H = -\mu_z B = -\gamma B S_z. \tag{4.32}$$

The z-component of spin angular momentum, S_z, can be expressed in terms of the Pauli matrix Z as:

$$S_z = \frac{\hbar}{2} Z. \tag{4.33}$$

Combining these relations, we obtain:

$$H = -\frac{\hbar \gamma B}{2} Z = -\frac{\hbar \omega}{2} Z, \tag{4.34}$$

where $\omega = \gamma B$ is referred to as the Larmor frequency. This is the same as Eq. 4.25.

4.5 ∗ Larmor Precession and Rabi Oscillations

> ### ∗ A Clarification of the Sign of the Hamiltonian
>
> In our discussion so far, we have focused on a generic exploration of Larmor precession involving a spin in a magnetic field oriented along the z-direction, assuming the spin particle is possively charged. For an electron, the Hamiltonian in Eq. 4.34 has an opposite sign: $H = \frac{\hbar\omega}{2} Z$. An astute reader might note a seeming inconsistency: the ground state for an electron in this scenario would be $|1\rangle$, not $|0\rangle$. To dispel any confusion, we delve into the nuances of different conventions in traditional quantum mechanics and quantum computing, as well as the distinctions between various physical systems.
>
> - The lowest energy state, or ground state, corresponds to the magnetic moment being aligned with the magnetic field. For electrons, which possess a negative charge, this alignment translates to their spin being antiparallel to the magnetic field, traditionally referred to as spin-down in quantum mechanics. Conversely, for nuclei containing positively charged protons, the ground state corresponds to the spin being parallel to the magnetic field, or spin-up.
>
> - In quantum computing, the state $|0\rangle$ is commonly associated with spin-up and is conventionally designated as the ground state. This choice is more a matter of computational convenience and is not necessarily consistent with all physical qubit systems.
>
> - Furthermore, Larmor precession can be viewed as a perturbation to a base Hamiltonian that defines the ground state. This perspective is particularly relevant when Larmor precession serves as a model for the manipulation and control of qubit states. The application of a magnetic field (or other fields) introduces this perturbation, resulting in the characteristic precessional motion of the spin.
>
> This clarification aims to bridge the gap between theoretical quantum mechanics and practical quantum computing conventions, enhancing the understanding of these fundamental concepts.

4.5.2 ∗ Rabi Oscillations

Rabi oscillations are a phenomenon that occurs when a two-level quantum system (e.g., a qubit or spin-1/2 particle) is subjected to an oscillatory driving field. This phenomenon is named after the physicist Isidor Isaac Rabi, who first observed it in the context of atomic transitions. The oscillation occurs when the frequency of the external driving force is resonant with the energy difference between the two energy eigenstates. In this case, the probability of finding the electron in one of the energy levels oscillates periodically with time, completing one full oscillation within a Rabi cycle.

Rabi oscillations play a crucial role in many areas of quantum physics, such as atomic physics, quantum optics, and quantum computing. In the context of quantum computing, Rabi oscillations can be used to manipulate qubits by driving transitions between their computational basis states (e.g., $|0\rangle$ and $|1\rangle$). In this case, carefully controlled electromagnetic pulses can be applied to induce Rabi oscillations, which allow for precise control of the superposition state of the qubit.

1 Solution to the Schrödinger Equation

The Hamiltonian of the two-level quantum system (under the rotating wave approximation, see below) is given by

$$H = \frac{\hbar\Omega}{2}Y, \quad (4.35)$$

where Y is the Pauli-Y matrix and Ω is known as the Rabi frequency, which characterizes the strength of the coupling between the system and the external field.

Larmor precession and Rabi oscillations, while physically distinct phenomena, share mathematical similarities, particularly in their involvement of a single Pauli operator in the Hamiltonian.

In quantum computing, Larmor precession is often modeled to represent z-axis rotations of the qubit state, while Rabi oscillations are typically associated with x- and y-axis rotations. In our main discussion, we choose the Y Pauli operator in the Hamiltonian to demonstrate the creation of the $|+\rangle$ superposition state.

As a heads-up, it's noteworthy that the implementation of the most essential two-qubit gates, such as the CNOT and the ZZ gates, involves ZX and ZZ interactions. These interactions and their roles in multi-qubit systems will be explored in Chapters 6 and 7. The physical principles underlying these interactions are similar to those of Larmor precession and Rabi oscillations discussed here.

The dynamics of the system is governed by the Schrödinger equation (Eq. 4.4), which has a solution in the form of the time-evolution operator $U(t)$ given by Eq. 4.6. Due to the property $Y^2 = I$ of the Pauli matrix Y, we can simplify the matrix exponential in $U(t)$ into a regular matrix:

$$U(t) = e^{-iHt/\hbar} = e^{-i\Omega t Y/2} \quad (4.36a)$$

$$= \cos\frac{\Omega t}{2} I - i\sin\frac{\Omega t}{2} Y \quad (4.36b)$$

$$= \begin{bmatrix} \cos\frac{\Omega t}{2} & -\sin\frac{\Omega t}{2} \\ \sin\frac{\Omega t}{2} & \cos\frac{\Omega t}{2} \end{bmatrix}. \quad (4.36c)$$

In the case where the initial state is $|\psi(0)\rangle = |0\rangle$, we have:

$$|\psi(t)\rangle = U(t)|\psi(0)\rangle \quad (4.37a)$$

$$= \begin{bmatrix} \cos\frac{\Omega t}{2} & -\sin\frac{\Omega t}{2} \\ \sin\frac{\Omega t}{2} & \cos\frac{\Omega t}{2} \end{bmatrix}\begin{bmatrix}1\\0\end{bmatrix} \quad (4.37b)$$

$$= \begin{bmatrix} \cos\frac{\Omega t}{2} \\ \sin\frac{\Omega t}{2} \end{bmatrix}. \quad (4.37c)$$

2 Interpretation

From Eq. 4.37 we see that the probabilities of finding the system in $|0\rangle$ and $|1\rangle$ at time t are given by

$$P_0(t) = \cos^2\frac{\Omega t}{2}, \quad (4.38a)$$

$$P_1(t) = \sin^2\frac{\Omega t}{2}. \quad (4.38b)$$

As depicted in Fig. 4.3, the system oscillates between $|0\rangle$ and $|1\rangle$ with a period of $2\pi/\Omega$. The key transitions are:

4.5 ∗ Larmor Precession and Rabi Oscillations

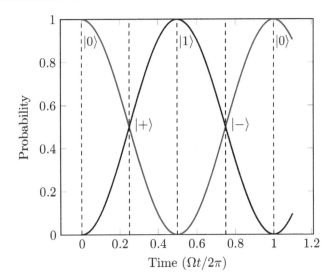

Figure 4.3: Plot of the probabilities $P_0(t)$ and $P_1(t)$ as functions of time.

- $t = 0$: system is in $|0\rangle$.
- $t = 0.25 \cdot 2\pi/\Omega$: system is in $|+\rangle \equiv \frac{1}{\sqrt{2}}(|0\rangle + |1\rangle)$.
- $t = 0.5 \cdot 2\pi/\Omega$: transitioned to $|1\rangle$.
- $t = 0.75 \cdot 2\pi/\Omega$: system is in $|-\rangle \equiv \frac{1}{\sqrt{2}}(|0\rangle - |1\rangle)$.
- $t = 1.0 \cdot 2\pi/\Omega$: back to $|0\rangle$.

3 Implementation of the Rotation Gates

Rabi oscillations, which involve the periodic exchange of population between quantum states, serve as a cornerstone for precise qubit control in quantum computing. They are essential for implementing various quantum gates and facilitating quantum information processing.

By applying a driving field for a time interval Δt, the qubit state undergoes a rotation by an angle $\gamma = \Omega \Delta t$ about the y-axis. This operation corresponds to the $R_y(\gamma)$ gate. In fact, the unitary operator $U(\Delta t)$ in Eq. 4.36 aligns exactly with the matrix representation of $R_y(\gamma)$ (refer to § 5.2.5 for details).

Through Rabi oscillations driven by Hamiltonians involving the X operator, the R_x gate can also be implemented. Other fundamental gates such as X, Y, H, S, and T can be realized by employing combinations of the R_x, R_y, and R_z gates, a topic to be discussed in Chapter 5.

Intuitively, Rabi oscillations can facilitate transformations of a qubit from the state $|0\rangle$ to $|1\rangle$ (analogous to an X gate operation) or to $|+\rangle$ (analogous to a Hadamard gate operation). Furthermore, this mechanism can be extended to coupled states between qubits, thereby enabling entanglement operations.

4 The Hamiltonian

The Hamiltonian for Rabi oscillations has a general form

$$H = \frac{\hbar\Omega}{2}(\cos\phi X + \sin\phi Y), \tag{4.39}$$

where ϕ is a phase factor that determines the specific combination of the Pauli matrices X and Y involved in the Hamiltonian. In the special case where $\phi = 0$, we have the X term; if $\phi = \pi/2$, we have the Y term. In the above discussion, for simplicity, we used only the Y term to illustrate the fundamental concepts.

The specific Hamiltonian for a real physical system depends on the direction of the external driving field, and its phase of oscillation.

Exercise 4.8 Given the Hamiltonian $H = \frac{\hbar\Omega}{2}X$, find a solution to the Schrödinger equation with initial condition $|\psi(0)\rangle = |1\rangle$. Calculate the probability $P_0(t) = |\langle 0|\psi(t)\rangle|^2$.

5 *The Rotating Wave Approximation

The Hamiltonian for Rabi oscillations, representing the energy of a two-level system in a high-frequency oscillatory field, can become quite complex due to the field's time-varying nature. The Rotating Wave Approximation (RWA) is a technique used in quantum mechanics to simplify this Hamiltonian, as described in [8]. It is particularly useful for analyzing interactions between a quantum system and an oscillating external field, such as laser or microwave radiation. The RWA is most valid near resonance, where the frequency of the external field closely matches the system's natural transition frequency, denoted by $\hbar\omega \approx E_1 - E_0$.

In the RWA, the Hamiltonian is transformed into an interaction representation, separating the free evolution of the system from its interaction with the driving field. The key step in RWA is to neglect the rapidly oscillating terms in the interaction Hamiltonian, which effectively average out to zero over time, focusing only on terms that oscillate at the difference frequency between the driving field and the natural frequency of the system.

This simplification yields an effective two-level Hamiltonian, as exemplified in Eq. 4.39, focusing primarily on resonant transitions. Within this framework, the effective Hamiltonian for Rabi oscillations can be viewed as a perturbation to the base Hamiltonian of the quantum system. This approximation greatly facilitates the analysis of quantum dynamics and phenomena, making the mathematical treatment more manageable, as demonstrated in our example. However, it's important to note that the RWA may not always be applicable, especially when far from resonance or when the terms omitted by the approximation become significant. As with any approximation in physics, its applicability must be carefully assessed within the specific physical context.

4.6 *Further Exploration

These topics are seldom covered in the context of quantum computing, except in specialized quantum simulations. They are included here for background information only. For further details, interested readers are directed to Ref. [8].

4.6.1 ✳ Time Evolution Equations for Relativistic Systems

A common misconception is that quantum mechanics is synonymous with the Schrödinger equation. However, this equation is not universally applicable to all quantum systems. In the realm of relativistic systems, which involve high energies or particles moving at speeds close to the speed of light, different equations govern their time evolution.

For fermions, particles with half-integer spin, the Dirac equation is used to describe their evolution over time. The Dirac equation is given by:

$$i\hbar \frac{\partial}{\partial t} |\psi(t)\rangle = H_D |\psi(t)\rangle, \qquad (4.40)$$

where H_D is the Dirac Hamiltonian operator.

For bosons, particles with integer spin, the Klein-Gordon equation is used to describe their evolution over time. The Klein-Gordon equation is given by:

$$\Box \phi + \frac{m^2 c^2}{\hbar^2} \phi = 0, \qquad (4.41)$$

where $\Box = \frac{1}{c^2} \frac{\partial^2}{\partial t^2} - \nabla^2$ is the d'Alembertian operator, a second-order differential operator defined in spacetime, m is the mass of the particle, c is the speed of light, and $\hbar$ is the reduced Planck constant.

4.6.2 ✳ Open Quantum Systems

In quantum mechanics, the Evolution Postulate and the Schrödinger equation are traditionally used to describe the dynamics of closed quantum systems, which are isolated from external influences or interactions. However, in the realm of quantum computing, qubits often undergo controlled interactions with external fields, such as microwave, lasers, or magnetic fields. This situation might appear to conflict with the closed-system framework, but it can be reconciled using perturbation theory.

In such scenarios, the external controls are typically treated as perturbations to the system's Hamiltonian. This approach allows for an analysis of the system's evolution under the influence of these controls. Indeed, Larmor precession and Rabi oscillations, as discussed in this chapter, are examples of how quantum computing often employs such controlled interactions. By modeling these interactions as perturbations, the foundational principles of quantum mechanics are extended to encompass the interaction of qubits with external control mechanisms. This extension is crucial for the precise control and manipulation of qubit states, forming a cornerstone of quantum information processing. It is this nuanced application of the Schrödinger equation and perturbation theory that bridges the gap between the theoretical model of closed-system dynamics and the interactive environment essential to quantum computation.

In quantum mechanics, open quantum systems that interact with their environment typically undergo non-unitary evolution. Such systems are comprehensively described by frameworks like quantum master equations or quantum trajectories, which account for time-dependent system-environment interactions. A notable instance of non-unitary evolution is the measurement process, where the state vector collapses to an eigenstate of the observable being measured. Additionally, decoherence, a phenomenon where qubit states lose coherence due to uncontrolled environmental influences and noise, represents another form of non-unitary evolution. These non-unitary processes can be effectively modeled using Completely Positive Trace-Preserving (CPTP) maps, providing a mathematical framework essential in quantum information theory. We will delve into CPTP maps and quantum channels in § 12.2.7.

4.7 *Deferred Proofs

1 Hermitian Property of the Hamiltonian

As a result of the unitary evolution postulate (Eq. 4.1), the Hamiltonian operator H in the Schrödinger equation (Eq. 4.3) must be Hermitian. This can be seen as follows:

Consider an infinitesimal time interval Δt. Eq. 4.1 can be expressed as:

$$|\psi(\Delta t)\rangle - |\psi(0)\rangle = (U(\Delta t) - I)|\psi(0)\rangle. \tag{4.42}$$

At the same time, the Schrödinger equation Eq. 4.3 gives us

$$|\psi(\Delta t)\rangle - |\psi(0)\rangle = -i\frac{\Delta t}{\hbar}H|\psi(0)\rangle. \tag{4.43}$$

Comparing the two equations above, we obtain:

$$U(\Delta t) = I - i\frac{\Delta t}{\hbar}H. \tag{4.44}$$

Since $U(\Delta t)$ is unitary,

$$I = U(\Delta t)U(\Delta t)^\dagger = \left(I - i\frac{\Delta t}{\hbar}H\right)\left(I + i\frac{\Delta t}{\hbar}H^\dagger\right) = I - i\frac{\Delta t}{\hbar}H + i\frac{\Delta t}{\hbar}H^\dagger. \tag{4.45}$$

Here since Δt is infinitesimally small, we have ignored the $(\Delta t)^2$ term.

Apparently, for the above equation to hold, we must have $H = H^\dagger$, which indicates H is a Hermitian operator. Conversely, we can also show if H is Hermitian, the time evolution operator U as a solution to the Schrödinger equation is unitary.

2 Derivation of the General Solution to SE (Eq. 4.5)

Given the definition of e^A in Eq. 4.7, we can show that

$$\frac{\partial}{\partial t}e^{-\frac{i}{\hbar}Ht} = -\frac{i}{\hbar}He^{-\frac{i}{\hbar}Ht}. \tag{4.46}$$

Differentiate Eq. 4.5 over t:

$$\frac{\partial}{\partial t}|\psi(t)\rangle = \frac{\partial}{\partial t}\left(e^{-\frac{i}{\hbar}Ht}|\psi(0)\rangle\right) \tag{4.47a}$$

$$= -\frac{i}{\hbar}He^{-\frac{i}{\hbar}Ht}|\psi(0)\rangle \tag{4.47b}$$

$$= -\frac{i}{\hbar}H|\psi(t)\rangle. \tag{4.47c}$$

This is the same as the time-independent Schrödinger equation Eq. 4.4.

3 Exponential Product Formula (Eq. 4.8)

We will prove that

$$e^{A+B} = e^A e^B, \quad \text{if } [A, B] = 0. \tag{4.8}$$

If $[A, B] = 0$, meaning A and B commute, we can freely exchange the order of A and B in their product, $AB = BA$.

Given the definition of e^A in Eq. 4.7, we first consider the expansion of e^{A+B}:

$$e^{A+B} = \sum_{n=0}^{\infty} \frac{1}{n!}(A+B)^n. \tag{4.48}$$

Using the binomial theorem, $(A+B)^n$ can be expanded as:

$$(A+B)^n = \sum_{k=0}^{n} \binom{n}{k} A^k B^{n-k}. \tag{4.49}$$

Then

$$e^{A+B} = \sum_{n=0}^{\infty} \frac{1}{n!} \sum_{k=0}^{n} \binom{n}{k} A^k B^{n-k} \tag{4.50a}$$

$$= \sum_{n=0}^{\infty} \sum_{k=0}^{n} \frac{1}{k!(n-k)!} A^k B^{n-k}. \tag{4.50b}$$

Rearranging the sums, this becomes:

$$e^{A+B} = \sum_{k=0}^{\infty} \sum_{n=k}^{\infty} \frac{1}{k!(n-k)!} A^k B^{n-k} \tag{4.51a}$$

$$= \left(\sum_{k=0}^{\infty} \frac{1}{k!} A^k \right) \left(\sum_{l=0}^{\infty} \frac{1}{l!} B^l \right) \tag{4.51b}$$

$$= e^A e^B. \tag{4.51c}$$

Thus, we have shown that $e^{A+B} = e^A e^B$ when A and B commute.

4 Derivation of Eq. 4.10

We want to prove that, if $A^2 = I$ and θ is a real number, then

$$e^{i\theta A} = \cos\theta I + i\sin\theta A. \tag{4.10}$$

Given the definition of the exponential of an operator,

$$e^{i\theta A} = \sum_{n=0}^{\infty} \frac{1}{n!}(i\theta A)^n, \tag{4.52}$$

we separate the series into even and odd terms:

$$e^{i\theta A} = \sum_{n=0}^{\infty} \frac{1}{(2n)!}(i\theta A)^{2n} + \sum_{n=0}^{\infty} \frac{1}{(2n+1)!}(i\theta A)^{2n+1}. \qquad (4.53)$$

Given $A^2 = I$, we can simplify the terms.

For even powers, $(i\theta A)^{2n} = (i^2)^n \theta^{2n} A^{2n} = (-1)^n \theta^{2n} I$.

For odd powers, $(i\theta A)^{2n+1} = (i\theta A)(-1)^n \theta^{2n} A^{2n} = i(-1)^n \theta^{2n+1} A$.

Thus, the series can be rewritten as:

$$\begin{align}
e^{i\theta A} &= \sum_{n=0}^{\infty} \frac{(-1)^n \theta^{2n}}{(2n)!} I + \sum_{n=0}^{\infty} \frac{i(-1)^n \theta^{2n+1}}{(2n+1)!} A & (4.54a) \\
&= \left(\sum_{n=0}^{\infty} \frac{(-1)^n \theta^{2n}}{(2n)!} \right) I + i \left(\sum_{n=0}^{\infty} \frac{(-1)^n \theta^{2n+1}}{(2n+1)!} \right) A & (4.54b) \\
&= \cos\theta I + i \sin\theta A. & (4.54c)
\end{align}$$

In the final step, we recognize the Taylor series expansions of $\cos\theta$ and $\sin\theta$. Therefore, we have proved that $e^{i\theta A} = \cos\theta I + i\sin\theta A$ when $A^2 = I$.

4.8 Summary and Conclusions

Temporal Evolution of Quantum States

In this chapter, we delved into the complex dynamics of quantum systems, centering on the key inquiry: How do qubit states evolve over time? The foundational principles of quantum evolution govern both the operation of quantum gates and the process of quantum annealing. To elucidate these principles further, we employed the case study of a single qubit as a representative example.

Core Principles in Quantum Mechanics

Anchoring our discussions was the Time Evolution Postulate, a cornerstone in quantum mechanics, which posits the deterministic unitary evolution of closed quantum systems. This provided the scaffolding to introduce the Schrödinger equation, the vital equation connecting the temporal evolution of quantum states with their energy via the Hamiltonian. The stationary nature of energy eigenstates, a direct implication of the time-independent Schrödinger equation, was also delineated, elucidating the distinction between ground states and excited states and their pertinence in quantum computing.

Quantum Computing Paradigms

The realm of quantum computing has witnessed the emergence of two cardinal paradigms: Gate-based universal quantum computing and adiabatic quantum computing, including quantum annealing. Their distinctions and foundational underpinnings were examined, underscoring their unique methodologies and applications, all rooted in the principle of unitary evolution of quantum states.

Insights into Quantum Systems

Larmor Precession and Rabi Oscillations, two quintessential phenomena in quantum mechanics, were explored to furnish insights into the behavior of quantum systems under the influence of external fields. These served as pedagogical tools, illuminating essential techniques for tackling the Schrödinger equation, particularly relevant to the domain of quantum hardware.

Perspective on Quantum Systems

The chapter culminated with an exploration into topics not directly applied in quantum computing, yet important for a holistic understanding of the subject. These encompassed relativistic time evolution equations and the domain of open quantum systems.

Upcoming Topics

In the next chapter, we will explore quantum gates, which are central to quantum computing. Building on our foundational understanding of quantum mechanics, we will study the basic single-qubit gates before advancing to topics like multi-qubit gates, entanglement, quantum teleportation, and the E91 quantum key distribution protocol. These discussions will connect our theoretical understanding of quantum mechanics with practical quantum computing applications.

Problem Set 4

4.1 Consider a one-qubit quantum gate defined by the Hamiltonian $H = \hbar\omega X$, where X is the Pauli-X operator. Derive the unitary evolution operator $U(\Delta t)$ over an interval Δt and elucidate the influence of this gate on an arbitrary qubit state.

4.2 Consider $U = u_x X + u_y Y + u_z Z$, where X, Y, Z are the Pauli operators, and u_x, u_y, u_z are real numbers with $u_x^2 + u_y^2 + u_z^2 = 1$. Express $e^{i\gamma U}$ as a single 2×2 matrix, where γ is a real number.

Hint: Show that $U^2 = I$ first.

4.3 Express e^X as a single 2×2 matrix, where X is the Pauli-X operator. (Note this is not the same as e^{iX}.)

4.4 Find a 2×2 Hamiltonian H such that e^{-iH} equals the Hadamard matrix $\frac{1}{\sqrt{2}}\begin{bmatrix} 1 & 1 \\ 1 & -1 \end{bmatrix}$.

4.5 Given the Hamiltonian $H = \hbar\omega \begin{bmatrix} 0 & \alpha \\ \alpha & 1 \end{bmatrix}$, where $\alpha, \omega \in \mathbb{R}$, find the energy levels E_0 and E_1, and corresponding eigenstates $|E_0\rangle$ and $|E_1\rangle$.

4.6 Consider a Hamiltonian $H = \begin{bmatrix} 0 & -i \\ i & 0 \end{bmatrix}$.

- (a) Diagonalize this Hamiltonian to find its eigenvalues and eigenvectors.

- (b) Given the time-evolution operator $U(t) = e^{-iHt}$, compute $U(t)$ for a time t, using the eigendecomposition of H.

(c) If we start with a state $|0\rangle$, and we evolve it using $U(t)$, the resulting state will be $U(t)|0\rangle$. Compute the probabilities of measuring $|0\rangle$ and $|1\rangle$ after this evolution.

4.7 (Rabi oscillation with general Hamiltonian.) Consider the Hamiltonian $H = \frac{\hbar\Omega}{2}(\cos\phi X + \sin\phi Y)$, where $\phi \in \mathbb{R}$.

(a) Find the matrix form of $U(t) = e^{-iHt/\hbar}$.

(b) Given the initial condition $|\psi(0)\rangle = |0\rangle$, calculate the probability evolution $P_0(t) = |\langle 0|\psi(t)\rangle|^2$.

4.8 Consider a Hamiltonian $H = \frac{T-t}{T}Z + \frac{t}{T}X$, where X and Z are the Pauli matrices, and T is the total evolution time (i.e., $t \in [0, T]$). Write an approximate solution to the Schrödinger equation using Trotterization.

4.9 Find an $\alpha \in \mathbb{R}$ such that $e^{i\alpha Z}Xe^{-i\alpha Z} = Y$, where X, Y, Z are the Pauli matrices. Interpret the meaning of this relation in terms of qubit-state rotations and quantum gates.

II Quantum Gates & Elementary Circuits

5   **Single-Qubit Quantum Gates** 111

6   **Multi-Qubit Systems** 143

7   **Multi-Qubit Quantum Gates** 175

5. Single-Qubit Quantum Gates

Contents

5.1	**Quantum Versus Classical Logic Gates**	112
5.1.1	Quantum Gates as Unitary Operators	113
5.1.2	Reversibility Requirement	113
5.1.3	No-Cloning Requirement	114
5.1.4	Linearity	114
5.1.5	From Quantum Gates to Circuits	115
5.2	**Common Single-Qubit Gates**	115
5.2.1	Approaches of Analysis: X Gate Example	115
5.2.2	Pauli Gates	117
5.2.3	The Hadamard H Gate	118
5.2.4	The Phase Gates	119
5.2.5	The R_x, R_y, and R_z Rotation Gates	120
5.2.6	✳ The Unified Rotation Gate	122
5.3	**From Gate Sequences to Quantum Circuits**	123
5.3.1	The Circuit Model of Quantum Computing	123
5.3.2	Quantum Circuit Diagrams	124
5.3.3	Gate Sequences	125
5.4	**Quantum Random Number Generator**	126
5.5	**The BB84 Quantum Key Distribution (QKD) Protocol**	128
5.5.1	Fundamentals of Cryptography	128
5.5.2	Introduction to the BB84 QKD Protocol	129
5.5.3	Procedure of the BB84 QKD Protocol	130
5.5.4	Detection of Eavesdropping	131
5.5.5	BB84 QKD Implementation with Qubits	132
5.5.6	✳ Further Exploration	134
5.6	**The Quantum Coin Game**	135
5.6.1	The Game Rules	135
5.6.2	Alice Always Wins!	136
5.6.3	Bob's Strategy Shift	137
5.6.4	Alice Overcomes Bob's New Gambit	137

5.7	✳ The No-Cloning Theorem: Proof Outline	138
5.8	Summary and Conclusions	139
	Problem Set 5	140

This chapter marks the beginning of our exploration into the world of quantum gates, the critical building blocks of quantum computing systems.

The principles of quantum computing are deeply intertwined with the fundamental tenets of quantum mechanics. The relationship between the two can be likened to how Newtonian mechanics underpins the launching of satellites. Both fields require an in-depth understanding of the respective theoretical principles and meticulous engineering to ensure precise real-world application.

Just as launching a satellite involves grasping Newtonian principles like gravity and motion to craft the right trajectory and propulsion systems, creating quantum gates and circuits requires understanding the key principles of quantum mechanics, such as superposition and entanglement. This knowledge enables us to control qubits and construct quantum algorithms for tackling complex computational problems. In both contexts, we leverage basic physical theories to devise practical solutions.

In previous discussions, we have laid down the foundational principles of quantum mechanics—including quantum states, measurements, and unitary evolution—and we have developed a solid understanding of qubits. Now, we're ready to extend our exploration into the fascinating world of quantum gates.

We'll commence our exploration with single-qubit gates and their various applications. This groundwork will then pave the way for more advanced topics, including multi-qubit gates, entanglement, quantum teleportation, and the E91 quantum key distribution protocol.

5.1 Quantum Versus Classical Logic Gates

Just as classical logic gates are fundamental to classical computing, quantum gates form the cornerstone of quantum computing. Both types of gates serve as standardized and highly optimized components within their respective computational paradigms. Quantum gates are specifically designed to manipulate quantum states, thereby facilitating the execution of quantum algorithms. These gates can be viewed as specialized solution cases of the Schrödinger equation, tailored for quantum computational tasks. Although implemented differently across various qubit platforms, the operational functionality of quantum gates has remained consistent. This mirrors the evolution of classical gates, which have transitioned from vacuum tubes to transistors and other technologies, yet their core principles of simplified state representations and logical operations have remained unchanged.

5.1 Quantum Versus Classical Logic Gates

5.1.1 Quantum Gates as Unitary Operators

Quantum gates introduce a distinct capability not found in classical computing: the ability to operate on superpositions of states. This unique feature arises from the principles of quantum mechanics that govern qubit behavior. The time evolution of a closed quantum system, as described by the Schrödinger equation, is characterized by unitary operators. Consequently, each quantum gate represents a unitary transformation, capable of acting on superpositions of quantum states and producing superpositions of outcomes.

As explored in § 4.4.1, the action of a quantum gate is analogous to the application of a specially designed Hamiltonian H over a specified time interval Δt. The physical implementation of a quantum gate, thus, depends on the precise engineering of this Hamiltonian.

The evolution of a quantum system's state, in accordance with the Schrödinger equation, is generally expressed as:

$$|\psi(\Delta t)\rangle = U(\Delta t) |\psi(0)\rangle, \tag{5.1}$$

where $U(\Delta t)$ is the unitary operator that describes the gate operation.

In quantum gate and circuit discussions, it's common to omit the Δt term. Thus, Eq. 5.1 is often represented as:

$$|\psi_1\rangle = U |\psi_0\rangle. \tag{5.2}$$

Here, $|\psi_0\rangle \equiv |\psi(0)\rangle$ denotes the system's state prior to the gate operation, $|\psi_1\rangle \equiv |\psi(\Delta t)\rangle$ symbolizes the state post-operation, and $U \equiv U(\Delta t)$ characterizes the operator signifying the gate's action.

5.1.2 Reversibility Requirement

One of the key attributes of quantum gates, stemming from their nature as unitary operators, is reversibility. Specifically, unitary operators possess a notable property: the inverse of a unitary operator is its conjugate transpose, denoted as $U^{-1} = U^\dagger$. Consequently, by applying the inverse operation U^{-1}, which is also unitary, one can effectively reverse the original quantum operation.

This aspect of quantum gates contrasts sharply with classical logic gates. Classical gates are not inherently reversible, as classical mechanics does not enforce the reversibility requirement inherent to quantum mechanics. Because of this not all classical gates have quantum counterparts.

For single-bit classical gates, we can identify four possible ones:

1. Identity (or "do nothing"): 0 maps to 0 and 1 maps to 1.

2. NOT (or bit-flip): 0 maps to 1 and 1 maps to 0.

3. Constant 0: Both 0 and 1 map to 0.

4. Constant 1: Both 0 and 1 map to 1.

It's clear that the third and fourth gates are non-reversible, as they both map two possible inputs to a single output. In quantum mechanics, these gates would violate the principle of unitarity.

Additionally, some well-known multi-bit classical gates are irreversible and hence, cannot be directly implemented as quantum gates:

- AND Gate: This gate has two input bits and one output bit. An output of 0 could arise from inputs 00, 01, or 10. Only the output 1 uniquely corresponds to the input 11. (See Fig. 5.1.)

- OR Gate: Analogous to the AND gate, the OR gate's output of 1 can originate from inputs 01, 10, or 11. An output of 0 is unique to 00.

- Other Gates such as NAND, NOR, and XOR: These gates also exhibit irreversibility for certain outputs.

While these classical gates can't be directly translated to quantum gates, alternative methods exist to emulate their functionality in quantum computing. One prevalent strategy involves using extra output bits (often termed as ancilla bits) to retain all input-bit information. This topic will receive further exploration in subsequent chapters.

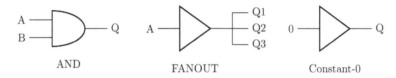

Figure 5.1: Examples of Classical Irreversible Gates

5.1.3 No-Cloning Requirement

Owing to the unitary nature of quantum evolution, there exists a pivotal principle in quantum computing and quantum information: the no-cloning theorem:

> **The No-Cloning Theorem.**
>
> It is fundamentally impossible to create an exact replica of an arbitrary, unknown quantum state solely using any physically realizable quantum operation while retaining the original state.

This theorem implies that one cannot perfectly clone or duplicate a qubit without altering its original state. As a consequence, operations like FANOUT (see Fig. 5.1), which are pervasive in classical digital circuits, are not feasible in quantum computing. On a positive note, the no-cloning principle underscores the robust security in quantum cryptography, exemplified by protocols such as the BB84.

A proof of this theorem is outlined at the end of this chapter (§ 5.7).

5.1.4 Linearity

Quantum gates, represented by complex unitary matrices, act linearly on quantum states. This linearity property, pivotal to quantum mechanics, ensures that any linear combination (superposition) of states is again a quantum state. It allows quantum gates to operate on superpositions, resulting in quantum parallelism and entanglement, which are among the core strengths of quantum computing.

5.1.5 From Quantum Gates to Circuits

Quantum circuits are constructed by composing sequences of quantum gates to implement quantum algorithms. The inherent nature of these circuits is unitary, provided they consist exclusively of quantum gates. This composition guarantees that the resulting operation remains unitary, thereby preserving the quantum state's norm and coherence within a closed system.

However, the scope of quantum circuits extends beyond solely unitary gates. Non-unitary operations, such as measurements and state resets, play a crucial role in quantum computing. While unitary gates facilitate the evolution of quantum states within a closed system, non-unitary operations enable interaction with the external environment. This leads to phenomena such as state collapse or decoherence. Such operations are indispensable for extracting information from quantum systems and are integral to certain steps in quantum algorithms, thereby introducing additional complexity and capabilities to quantum circuits that surpass those of classical logic gates.

The synergy of unitary and non-unitary operations within quantum circuits captures the full breadth of quantum computing capabilities. These range from maintaining coherent superpositions to executing measurements that provide classical information. As we progress in this chapter and in subsequent ones, we will delve into the intricate process of constructing quantum circuits and examine how they form the foundation of complex quantum algorithms.

5.2 Common Single-Qubit Gates

Now we are ready to turn to the common quantum gates involving a single qubit. As shown in Table 5.1, these include the Pauli-X, Y, Z gates, the Hadamard (H) gate, the phase gates, and the rotation gates. By combining these gates in different ways, quantum circuits can solve complex problems and perform tasks that are infeasible for classical computers, such as factoring large numbers or simulating quantum systems.

5.2.1 Approaches of Analysis: X Gate Example

The action of a qubit gate can be represented and analyzed in various ways, which are frequently used in different applications. It's essential to be familiar with all of them. As an illustrative example, we will explore the X gate.

The X gate, also known as the Pauli-X gate, is a fundamental quantum gate, acting as the quantum counterpart to the classical NOT gate. It maps $|0\rangle$ to $|1\rangle$ and vice versa. Therefore, it is also known as the bit-flip gate.

1 Matrix Representation

The X gate is represented by the Pauli-X matrix:

$$X = \begin{bmatrix} 0 & 1 \\ 1 & 0 \end{bmatrix}. \tag{5.3}$$

Given the column vectors

Gate	Symbol	Operator	Description
Identity (I)	$-\boxed{I}-$	$\begin{bmatrix} 1 & 0 \\ 0 & 1 \end{bmatrix}$	Leaves the qubit unchanged.
Pauli-X (X)	$-\boxed{X}-$	$\begin{bmatrix} 0 & 1 \\ 1 & 0 \end{bmatrix}$	NOT or bit-flip gate, maps $\|0\rangle$ to $\|1\rangle$ and $\|1\rangle$ to $\|0\rangle$.
Pauli-Z (Z)	$-\boxed{Z}-$	$\begin{bmatrix} 1 & 0 \\ 0 & -1 \end{bmatrix}$	Phase-flip gate, maps $\|0\rangle$ to $\|0\rangle$ and $\|1\rangle$ to $-\|1\rangle$.
Pauli-Y (Y)	$-\boxed{Y}-$	$\begin{bmatrix} 0 & -i \\ i & 0 \end{bmatrix}$	Combines bit flip and phase flip operations.
Hadamard (H)	$-\boxed{H}-$	$\frac{1}{\sqrt{2}}\begin{bmatrix} 1 & 1 \\ 1 & -1 \end{bmatrix}$	Creates superposition. Maps $\|0\rangle$ to $\|+\rangle$ and $\|1\rangle$ to $\|-\rangle$.
S ($\sqrt{Z}$)	$-\boxed{S}-$	$\begin{bmatrix} 1 & 0 \\ 0 & i \end{bmatrix}$	Applies $\frac{\pi}{2}$ phase shift to the $\|1\rangle$ state.
T ($\sqrt[4]{Z}$)	$-\boxed{T}-$	$\begin{bmatrix} 1 & 0 \\ 0 & e^{i\frac{\pi}{4}} \end{bmatrix}$	Applies $\frac{\pi}{4}$ phase shift to the $\|1\rangle$ state.
Phase ($P(\theta)$)	$-\boxed{P(\theta)}-$	$\begin{bmatrix} 1 & 0 \\ 0 & e^{i\theta} \end{bmatrix}$	Applies θ phase shift to the $\|1\rangle$ state.
$R_x(\theta)$	$-\boxed{R_x(\theta)}-$	$\begin{bmatrix} \cos\frac{\theta}{2} & -i\sin\frac{\theta}{2} \\ -i\sin\frac{\theta}{2} & \cos\frac{\theta}{2} \end{bmatrix}$	Rotates qubit around x-axis of the Bloch sphere by θ.
$R_y(\theta)$	$-\boxed{R_y(\theta)}-$	$\begin{bmatrix} \cos\frac{\theta}{2} & -\sin\frac{\theta}{2} \\ \sin\frac{\theta}{2} & \cos\frac{\theta}{2} \end{bmatrix}$	Rotates qubit around y-axis of the Bloch sphere by θ.
$R_z(\theta)$	$-\boxed{R_z(\theta)}-$	$\begin{bmatrix} e^{-i\theta/2} & 0 \\ 0 & e^{i\theta/2} \end{bmatrix}$	Rotates qubit around z-axis of the Bloch sphere by θ.

Table 5.1: Common Single-Qubit Quantum Gates

$$|0\rangle = \begin{bmatrix} 1 \\ 0 \end{bmatrix} \quad \text{and} \quad |1\rangle = \begin{bmatrix} 0 \\ 1 \end{bmatrix}, \tag{5.4}$$

the bit-flip transformation becomes evident: $X|0\rangle = |1\rangle$ and $X|1\rangle = |0\rangle$.

2 Bra-Ket Representation

A matrix U can be expanded using the outer products of the computational basis states:

$$U = \begin{bmatrix} u_{00} & u_{01} \\ u_{10} & u_{11} \end{bmatrix} = u_{00}|0\rangle\langle 0| + u_{01}|0\rangle\langle 1| + u_{10}|1\rangle\langle 0| + u_{11}|1\rangle\langle 1|. \tag{5.5}$$

Using this expansion, the X gate matrix can be represented in the bra-ket notation:

$$X = |0\rangle\langle 1| + |1\rangle\langle 0|. \tag{5.6}$$

Exercise 5.1 Verify $X|0\rangle = |1\rangle$ and $X|1\rangle = |0\rangle$ using bra-ket notation.

5.2 Common Single-Qubit Gates

3 Applying to Superposition States

Quantum gates, being linear operators, can operate on superposition states. When the X gate acts on

$$|\psi\rangle = \alpha|0\rangle + \beta|1\rangle = \begin{bmatrix} \alpha \\ \beta \end{bmatrix}, \tag{5.7}$$

the resulting state becomes

$$X|\psi\rangle = \beta|0\rangle + \alpha|1\rangle = \begin{bmatrix} \beta \\ \alpha \end{bmatrix}. \tag{5.8}$$

Exercise 5.2 Verify Eq. 5.8 using both the matrix and bra-ket representations of X.

4 Native in $\{|+\rangle, |-\rangle\}$ Basis

The Pauli matrices are both unitary and Hermitian. As such, X possesses eigenvectors that form the $\{|+\rangle, |-\rangle\}$ basis (see also § 3.3.3):

$$|+\rangle = \frac{1}{\sqrt{2}}(|0\rangle + |1\rangle), \tag{5.9a}$$

$$|-\rangle = \frac{1}{\sqrt{2}}(|0\rangle - |1\rangle). \tag{5.9b}$$

Applying the X gate to these vectors yields:

$$X|+\rangle = |+\rangle, \quad X|-\rangle = -|-\rangle. \tag{5.10}$$

5 Rotation on Bloch Sphere

In the Bloch sphere representation, the X gate corresponds to a 180° rotation about the x-axis. As shown in Fig. 2.4, the state vector $|\psi\rangle$ moves 180° along the dotted circle centered around the x-axis.

6 Boolean Representation

In addition to matrix and bra-ket representations, many quantum gates can also be expressed using Boolean representations. We will explore this in detail in § 7.3.

5.2.2 Pauli Gates

We have just delved into a detailed examination of the X gate, introducing various analytical tools in the process. These tools can be extended to the entire family of Pauli gates, which comprise the X, Y, Z, and I gates.

The Z gate, known as the phase-flip gate, imparts a π phase to the $|1\rangle$ state. Additionally, it serves as the flip gate for the $|+\rangle$ and $|-\rangle$ states, transitioning $|+\rangle$ to $|-\rangle$ and vice versa.

The Y gate integrates both the bit flip and phase flip. Interestingly, it can be shown that $ZX = iY$, a derivation we invite the reader to explore in the exercises below.

Gate	Bra-Ket Operators	$\|0\rangle, \|1\rangle \rightarrow$	$\|+\rangle, \|-\rangle \rightarrow$	$\alpha\|0\rangle + \beta\|1\rangle \rightarrow$
I	$\|0\rangle\langle 0\| + \|1\rangle\langle 1\|$ $\|+\rangle\langle +\| + \|-\rangle\langle -\|$	$I\|0\rangle = \|0\rangle$ $I\|1\rangle = \|1\rangle$	$I\|+\rangle = \|+\rangle$ $I\|-\rangle = \|-\rangle$	$\alpha\|0\rangle + \beta\|1\rangle$
X	$\|1\rangle\langle 0\| + \|0\rangle\langle 1\|$ $\|+\rangle\langle +\| - \|-\rangle\langle -\|$	$X\|0\rangle = \|1\rangle$ $X\|1\rangle = \|0\rangle$	$X\|+\rangle = \|+\rangle$ $X\|-\rangle = -\|-\rangle$	$\alpha\|1\rangle + \beta\|0\rangle$
Z	$\|0\rangle\langle 0\| - \|1\rangle\langle 1\|$ $\|-\rangle\langle +\| + \|+\rangle\langle -\|$	$Z\|0\rangle = \|0\rangle$ $Z\|1\rangle = -\|1\rangle$	$X\|+\rangle = \|-\rangle$ $Z\|-\rangle = \|+\rangle$	$\alpha\|0\rangle - \beta\|1\rangle$
Y	$i(\|1\rangle\langle 0\| - \|0\rangle\langle 1\|)$ $i(\|+\rangle\langle -\| - \|-\rangle\langle +\|)$	$Y\|0\rangle = i\|1\rangle$ $Y\|1\rangle = -i\|0\rangle$	$Y\|+\rangle = -i\|-\rangle$ $Y\|-\rangle = i\|+\rangle$	$i\alpha\|1\rangle - i\beta\|0\rangle$
H	$\|+\rangle\langle 0\| + \|-\rangle\langle 1\|$ $\|0\rangle\langle +\| + \|1\rangle\langle -\|$	$H\|0\rangle = \|+\rangle$ $H\|1\rangle = \|-\rangle$	$H\|+\rangle = \|0\rangle$ $H\|-\rangle = \|1\rangle$	$\alpha\|+\rangle + \beta\|-\rangle$
S	$\|0\rangle\langle 0\| + i\|1\rangle\langle 1\|$	$S\|0\rangle = \|0\rangle$ $S\|1\rangle = i\|1\rangle$	-	$\alpha\|0\rangle + i\beta\|1\rangle$
T	$\|0\rangle\langle 0\| + e^{i\frac{\pi}{4}}\|1\rangle\langle 1\|$	$T\|0\rangle = \|0\rangle$ $T\|1\rangle = e^{i\frac{\pi}{4}}\|1\rangle$	-	$\alpha\|0\rangle + e^{i\frac{\pi}{4}}\beta\|1\rangle$
P	$\|0\rangle\langle 0\| + e^{i\theta}\|1\rangle\langle 1\|$	$P(\theta)\|0\rangle = \|0\rangle$ $P(\theta)\|1\rangle = e^{i\theta}\|1\rangle$	-	$\alpha\|0\rangle + e^{i\theta}\beta\|1\rangle$

Table 5.2: Transformations by Common Single-Qubit Quantum Gates

The I gate stands for the identity gate, represented by the identity matrix. While it has no transformative effect on the qubit state and doesn't require physical implementation, its frequent use in circuit analysis justifies its inclusion for completeness.

All Pauli gates are involutory, meaning applying a Pauli gate twice restores the qubit state to its original form. For example, $X^2 = I$.

For quick reference, essential formulas related to these gates and others we will explore subsequently can be found in Tables 5.1 and 5.2. While we won't delve into all the formulas here, we encourage readers to tackle the exercises provided below.

Exercise 5.3 Show that Z acts as the flip gate for $\|+\rangle$ and $\|-\rangle$, meaning $Z\|+\rangle = \|-\rangle$ and $Z\|-\rangle = \|+\rangle$, using both matrix and bra-ket representations.

Exercise 5.4 Show that $Y(\alpha\|0\rangle + \beta\|1\rangle) = i(\alpha\|1\rangle - \beta\|0\rangle)$ employing both matrix and bra-ket representations.

5.2.3 The Hadamard H Gate

The Hadamard (H) gate is fundamental in quantum computing. It is commonly used to create superposition states as it transforms $\|0\rangle$ to $\|+\rangle$, and $\|1\rangle$ to $\|-\rangle$:

5.2 Common Single-Qubit Gates

$$H\,|0\rangle = |+\rangle = \frac{1}{\sqrt{2}}(|0\rangle + |1\rangle), \tag{5.11a}$$

$$H\,|1\rangle = |-\rangle = \frac{1}{\sqrt{2}}(|0\rangle - |1\rangle). \tag{5.11b}$$

For this reason, the $\{|+\rangle, |-\rangle\}$ basis is often referred to as the Hadamard basis.

Being involutory, $H^2 = I$. This means applying H twice restores the qubit state to its original form:

$$H\,|+\rangle = |0\rangle, \quad H\,|-\rangle = |1\rangle. \tag{5.12}$$

Lastly,

$$H(\alpha\,|0\rangle + \beta\,|1\rangle) = \alpha\,|+\rangle + \beta\,|-\rangle. \tag{5.13}$$

> **Exercise 5.5** Show that $H\,|+\rangle = |0\rangle$ and $H\,|-\rangle = |1\rangle$ using both matrix and bra-ket representations.

5.2.4 The Phase Gates

Phase gates are instrumental in introducing a relative phase between the basis states of a qubit. Among these gates, the most commonly utilized are the S, T, and P gates.

The S gate, often termed the "phase gate," adds a phase of $\pi/2$ (or 90°) to the $|1\rangle$ state. It can also be represented as the square root of the Z gate:

$$S\,|0\rangle = |0\rangle, \quad S\,|1\rangle = i\,|1\rangle, \quad S = \sqrt{Z}. \tag{5.14}$$

The T gate introduces a more nuanced phase of $\pi/4$ (or 45°). It is effectively the fourth root of the Z gate:

$$T\,|0\rangle = |0\rangle, \quad T\,|1\rangle = e^{i\pi/4}\,|1\rangle, \quad T = \sqrt[4]{Z}. \tag{5.15}$$

The general phase gate, $P(\phi)$, allows for a custom phase ϕ to be added to the $|1\rangle$ state:

$$P(\phi)\,|0\rangle = |0\rangle, \quad P(\phi)\,|1\rangle = e^{i\phi}\,|1\rangle. \tag{5.16}$$

These gates are foundational for tasks such as phase kickback and quantum Fourier transform. By adjusting the relative phase, they offer an extra degree of freedom and control in quantum algorithms.

> **Exercise 5.6** Show that $S^2 = Z$ and $T^4 = Z$.

> **Exercise 5.7** Demonstrate how the S, T, and $P(\phi)$ gates transform the general qubit state $\alpha\,|0\rangle + \beta\,|1\rangle$ using the bra-ket notation.

5.2.5 The R_x, R_y, and R_z Rotation Gates

Rotation gates play a crucial role in quantum computing, offering a method to rotate qubit states around various axes of the Bloch sphere (see Fig. 5.2). The three primary rotation gates are R_x, R_y, and R_z, corresponding to rotations about the x, y, and z axes, respectively.

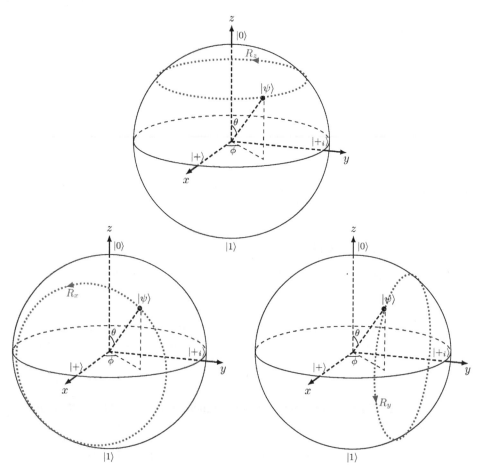

Figure 5.2: R_x, R_y, and R_z Gates as Rotations on the Bloch Sphere

1. The $R_x(\theta)$ gate rotates a qubit state around the x-axis by an angle θ:

$$R_x(\theta) = e^{-i\frac{\theta}{2}X} = \begin{bmatrix} \cos\frac{\theta}{2} & -i\sin\frac{\theta}{2} \\ -i\sin\frac{\theta}{2} & \cos\frac{\theta}{2} \end{bmatrix}. \tag{5.17}$$

Exercise 5.8 Demonstrate that the matrix of $R_x(\theta)$ is unitary. Is it also Hermitian?

2. Similarly, the $R_y(\theta)$ gate rotates a qubit state around the y-axis:

$$R_y(\theta) = e^{-i\frac{\theta}{2}Y} = \begin{bmatrix} \cos\frac{\theta}{2} & -\sin\frac{\theta}{2} \\ \sin\frac{\theta}{2} & \cos\frac{\theta}{2} \end{bmatrix}. \tag{5.18}$$

3. The $R_z(\theta)$ gate performs a rotation around the z-axis:

5.2 Common Single-Qubit Gates

$$R_z(\theta) = e^{-i\frac{\theta}{2}Z} = \begin{bmatrix} e^{-i\frac{\theta}{2}} & 0 \\ 0 & e^{i\frac{\theta}{2}} \end{bmatrix}. \tag{5.19}$$

A key property of these rotation operators is:

$$R_i(\theta)^\alpha = R_i(\alpha\theta), \tag{5.20}$$

where $\alpha \in \mathbb{R}$ and $i \in \{x, y, z\}$.

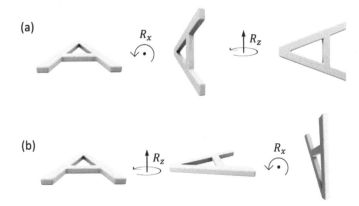

Figure 5.3: Demonstration of Non-Commutative Nature of 3D Rotations

It is imperative to understand that rotations in three-dimensional space are non-commutative when applied in different sequences. As depicted in Fig. 5.3, rotating around the x-axis followed by the z-axis (part a) does not result in the same orientation as when the rotation sequence is reversed (part b). This can be mathematically expressed as $R_z R_x \neq R_x R_z$, or equivalently $[R_z, R_x] \neq 0$.

 The equations for R_x, R_y, and R_z closely resemble the solutions for spin precession (Eq. 4.26) and Rabi oscillations (Eq. 4.36). Hence, both phenomena serve as effective means to implement these gates.

The Pauli gates, Hadamard gate, and Phase gates can all be expressed as special cases of rotation gates, albeit with a potential global phase difference. The relationships between them are given by:

$$X = iR_x(\pi), \tag{5.21a}$$
$$Y = iR_y(\pi), \tag{5.21b}$$
$$Z = -R_z(\pi), \tag{5.21c}$$
$$S = e^{i\frac{\pi}{4}} R_z(\frac{\pi}{2}), \tag{5.21d}$$
$$T = e^{i\frac{\pi}{8}} R_z(\frac{\pi}{4}), \tag{5.21e}$$
$$H = iR_x(\pi)R_y(\frac{\pi}{2}) = iR_y(\frac{\pi}{2})R_z(\pi). \tag{5.21f}$$

Equation 5.21 offers insight into understanding single-qubit gates as rotations on the Bloch sphere. For instance, the Hadamard gate can be interpreted as a rotation of 90° around the y-axis, followed by a 180° rotation around the x-axis. Alternatively, it can also be seen as a 180° rotation around the z-axis, followed by a 90° rotation around the y-axis.

> **Exercise 5.9** Derive the expression for the X gate as a special case of the R_x rotation gate.

> **Exercise 5.10** Among the gates listed in Table 5.1, identify which gates are Hermitian (i.e., $U = U^\dagger$), which are involutory (i.e., $U^2 = I$), and which satisfy both conditions.

5.2.6 ∗The Unified Rotation Gate

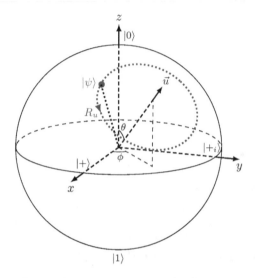

Figure 5.4: The Unified Rotation Gate (R_u) on the Bloch Sphere

We can further unify R_x, R_y, and R_z into a single rotation operator R_u. Below is how it's defined (refer to Fig. 5.4).

Given any real unit vector in three-dimensional space represented by $\hat{u} = (u_x, u_y, u_z)$, where the condition $u_x^2 + u_y^2 + u_z^2 = 1$ holds, we can define the unitary operator

$$\sigma_u = u_x X + u_y Y + u_z Z. \tag{5.22}$$

In the context of spherical coordinates with polar angle θ and azimuthal angle ϕ, the components are given by

$$u_x = \sin\theta \cos\phi, \quad u_y = \sin\theta \sin\phi, \quad u_z = \cos\theta. \tag{5.23}$$

This results in:

$$\sigma_u = \sin\theta \cos\phi\, X + \sin\theta \sin\phi\, Y + \cos\theta\, Z \tag{5.24a}$$

$$= \begin{bmatrix} \cos\theta & \sin\theta e^{-i\phi} \\ \sin\theta e^{i\phi} & -\cos\theta \end{bmatrix}. \tag{5.24b}$$

In this context, we use γ to denote the rotation angle, reserving θ for the polar directional angle.

For a rotation about the axis $\hat{u}$ by an angle γ:

$$R_u(\gamma, \theta, \phi) = e^{-i\frac{\gamma}{2}\sigma_u} \tag{5.25a}$$

$$= \cos\frac{\gamma}{2} I - i \sin\frac{\gamma}{2}\sigma_u \tag{5.25b}$$

$$= \begin{bmatrix} \cos\frac{\gamma}{2} - i\sin\frac{\gamma}{2}\cos\theta & -i\sin\frac{\gamma}{2}\sin\theta e^{-i\phi} \\ -i\sin\frac{\gamma}{2}\sin\theta e^{i\phi} & \cos\frac{\gamma}{2} + i\sin\frac{\gamma}{2}\cos\theta \end{bmatrix}. \tag{5.25c}$$

It might be evident to the astute reader that all single-qubit gates can be represented as a special case of R_u. Then, why not just use R_u and eliminate all others? Primarily because R_u is parameterized by three values (θ, ϕ, and γ) and is challenging to implement physically.

Exercise 5.11 Express R_x, R_y, and R_z in terms of $R_u(\gamma, \theta, \phi)$ and verify that their matrix representations are equivalent.

5.3 From Gate Sequences to Quantum Circuits

Quantum gates, foundational to quantum circuits, dictate the progression of quantum computations on qubits. These circuits visually depict quantum algorithms, offering a clear understanding of their design and sequence. As we've delved into single-qubit gates, our next logical step is to explore sequences of these foundational gates within quantum circuits.

5.3.1 The Circuit Model of Quantum Computing

The circuit model of quantum computing emerges as a leading framework for understanding and implementing quantum algorithms and operations. Drawing parallels with classical circuits, which process information using a series of logical gates, quantum circuits use quantum gates to manipulate qubits. These quantum computations can be visually represented through circuit diagrams, delineating each gate and its sequence, thereby facilitating an intuitive grasp and design of quantum algorithms.

This model is tailored for universal quantum computing, enabling it to theoretically execute any quantum algorithm. However, it's important to recognize that the circuit model isn't the only quantum computational model out there. The quantum Turing machine model serves as another key model, offering a more abstract perspective of quantum computation. While rich in theoretical insights, its high-level nature tends to make it less suited for practical applications compared to the circuit model.

Diverging from these models is the concept of quantum annealing. This paradigm, exemplified by systems like D-Wave, excels in rapidly searching for optimized solutions within vast solution spaces, making it particularly effective for certain optimization problems. However, quantum annealing has its own set of limitations and does not replace the capabilities of universal quantum computers.

A comprehensive study of quantum computing, such as ours, naturally commences with the circuit model for universal computing due to its practical relevance, visual clarity, and wide applicability. This foundational understanding paves the way for diving deeper into other quantum computational paradigms and appreciating their unique strengths and challenges.

5.3.2 Quantum Circuit Diagrams

Quantum circuits are graphically depicted through circuit diagrams. For illustration, Fig. 5.5 will be used to elucidate some salient aspects of quantum circuit diagrams.

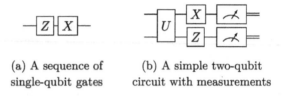

(a) A sequence of single-qubit gates

(b) A simple two-qubit circuit with measurements

Figure 5.5: Examples of Quantum Circuits

1. **Wires**

   Often referred to as "wires," the horizontal lines depict the temporal evolution of individual qubits in the circuit. Each wire denotes a single qubit and charts its state alteration through various gate operations.

2. **Gates**

   Symbols, frequently housed within rectangles or squares, represent the gates and are set upon the wires of the qubits they influence. These symbols signify quantum operations responsible for the metamorphosis of qubit states.

3. **Left-to-right Sequence**

   Gates are arranged from left to right on the wires, encapsulating the chronological sequence of operations. Progressing from left to right mirrors the advancement in time, with the sequence of gates being pivotal, orchestrating the comprehensive state transformation of the qubits.

4. **Order of Operator Multiplication**

   The composition of a sequence of gates is represented as the matrix product of the related gate matrices. Owing to conventional mathematical notation, the sequence of these matrices is in inverse order relative to the sequence of the gates. For instance, in Fig. 5.5 (a), the progression from a Z gate to an X gate results in the matrix product XZ.

5. **Input and Output**

   Positioned on the left of the circuit diagram, the initial states (often initialized to $|0\rangle$) are one form of input to the circuit. However, the input is not solely limited to such initializations; it also encompasses other elements such as query oracles and gate parametrization (to be discussed in Chapter 11). Conversely, the final states on the right side of the circuit diagram represent the output after all quantum gates

5.3 From Gate Sequences to Quantum Circuits

and operations have been applied. Measurement operations, typically represented by rectangles enclosing a meter sign, are introduced at this stage to extract classical data—indicated by double wires—from the quantum states of the qubits.

6 Computation in Place

It is important to understand that quantum circuits differ fundamentally from their classical counterparts, such as printed circuit boards or integrated circuits. In a quantum circuit, the gates and wires are not tangible physical entities in the same way. Instead, only the qubits and their control mechanisms are physically implemented. The wires in the diagram represent the temporal continuity of qubit states, and the quantum gates denote specific transformations of these states, directed by external controls.

> **Exercise 5.12**
>
> $$|0\rangle - \boxed{X} - \boxed{Y} - \boxed{Z} - \boxed{T} - |\psi\rangle$$
>
> (a) What is the output state $|\psi\rangle$ after the above sequence of gates is applied to the initial state $|0\rangle$?
>
> (b) Quantum circuits involving unitary gates are reversible. Draw the inverse circuit for the above sequence. Hint: the inverse of a circuit consists of the inverses of the original gates applied in reverse order.

5.3.3 Gate Sequences

Equipped with our knowledge of quantum gates, we can delve into gate sequences and, crucially, their equivalences. Intriguingly, gate sequences can often be represented by a single, equivalent gate. This capability is vital for optimizing quantum circuits, especially when these circuits must be executed on real quantum hardware with resource constraints. Moreover, mastering the equivalences of gate sequences is foundational for analyzing complex quantum circuits.

In the subsequent discussions, we highlight some pivotal equivalences involving single-qubit gates. This list will be extended as we venture into multi-qubit gates in § 7.4.

1 Pauli Gate Relations

The diagram below demonstrates that the combination of a Z gate followed by an X gate is equivalent to a Y gate, up to a global phase: $ZX = iY$.

$$-\boxed{X}-\boxed{Z}- \quad \Leftrightarrow \quad -\boxed{Y}-$$

Further relations among the Pauli gates include: $XZ = -iY$, $XY = iZ$, $YX = -iZ$, $YZ = iX$, $ZY = -iX$. See Appendix D for more.

2 Involutory Relations

The diagram below shows that applying an H gate twice yields the identity operation (I), leaving the qubit state unchanged. Applying it thrice reduces to just applying it once, as a result of the involutory property of the Hadamard gate: $H^2 = I$.

$$-[H][H]- \Leftrightarrow -[I]-$$

$$-[H][H][H]- \Leftrightarrow -[H]-$$

Other gates exhibit similar involutory relations: $X^2 = I$, $Y^2 = I$, and $Z^2 = I$.

3 Relations Among Hadamard and Pauli Gates

Given the relations $HXH = Z$ and $HX = ZH$, we derive the following gate sequence equivalences:

$$-[X][H]- \Leftrightarrow -[H][Z]-$$

$$-[H][Z][H]- \Leftrightarrow -[X]-$$

$$-[H][X][H]- \Leftrightarrow -[Z]-$$

4 Constructing the Hadamard Gate

In practice, the Hadamard gate is often realized using other basic gates. For instance, $H = iR_x(\pi)R_y(\frac{\pi}{2}) = iR_y(\frac{\pi}{2})R_z(\pi)$. This realization allows us to construct the Hadamard gate as illustrated below:

$$-[R_y(\pi/2)][R_x(\pi)]- \Leftrightarrow -[R_z(\pi)][R_y(\pi/2)]- \Leftrightarrow -[H]-$$

5 Constructing the Y Gate

In practice, the Y gate is often realized using other basic gates. For example, $Y = R_x(-\frac{\pi}{2})ZR_x(\frac{\pi}{2})$.

$$-[R_x(\pi/2)][Z][R_x(-\pi/2)]- \Leftrightarrow -[Y]-$$

Exercise 5.13 Draw circuit diagrams corresponding to the gate relations $HXH = Z$, $HYH = -Y$, and $HY = -YH$, $SXS^\dagger = Y$.

5.4 Quantum Random Number Generator

Quantum applications that use only single-qubit gates are relatively simple, as they do not involve entanglement or complex interactions between multiple qubits. Nonetheless, they can help us build an understanding of basic quantum concepts and operations. Next, we explore are some examples, starting with QRNG.

A Quantum Random Number Generator (QRNG) is a device that generates truly random numbers by exploiting the inherent randomness and unpredictability of quantum mechanical processes. Unlike classical random number generators, which either rely on deterministic algorithms or on external sources of entropy, QRNGs harness the fundamental properties of quantum mechanics to produce genuine randomness that is independent of any underlying patterns or biases.

5.4 Quantum Random Number Generator

QRNG Using Photons

A common method to implement a QRNG is by using a single-photon source and a beam splitter, illustrated in Fig. 5.6(a). In this setup, individual photons are emitted and sent through the beam splitter. The beam splitter has a 50% chance of reflecting the photon and a 50% chance of transmitting it. Detectors are placed at both the reflected and transmitted paths, and the path taken by the photon is measured. At this juncture, we are measuring individual photons, positioning us firmly within the quantum regime, rather than the classical optics regime. Given that the photon's behavior at the beam splitter is inherently random due to quantum mechanics, the detector outcomes generate a truly random binary sequence.

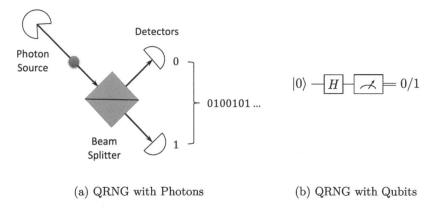

(a) QRNG with Photons (b) QRNG with Qubits

Figure 5.6: Quantum Random Number Generator (QRNG)

QRNG Using Qubits

Another approach to QRNGs is to exploit the quantum superposition and measurement process. A qubit is prepared in a superposition of its basis states, typically by applying a Hadamard gate to the initial state, as shown in Fig. 5.6(b). The qubit is then measured in the computational basis, collapsing the superposition to either 0 or 1 with equal probability. The measurement outcome serves as a random bit. The circuit is run repeated to produce a string of random bits of desired length.

Equivalence

The equivalence between these two methods can be understood by comparing the role of the beam splitter in the photon approach to that of the Hadamard gate in the superposition-measurement approach. The beam splitter places an incoming photon into a superposition of potential paths (reflected or transmitted), introducing quantum uncertainty. Similarly, the Hadamard gate places a qubit into a superposition of its basis states. When the photon or the qubit is observed or measured, this superposition results in inherent quantum randomness due to the unpredictability of the outcome. Both mechanisms, therefore, harness quantum superposition to generate random bits.

> **Exercise 5.14** * The QRNG devices discussed above generate binary sequences that correspond to random numbers with a uniform distribution. Discuss how you

would transform these uniformly distributed random numbers into numbers that follow a Gaussian distribution and a Poisson distribution. Consider the following in your discussion:

(a) Describe any algorithms or mathematical transformations you would use to achieve these distributions from a uniform distribution.

(b) Discuss any practical considerations or limitations that might arise when implementing these transformations, especially in the context of using a quantum device.

(c) Discuss how the properties of quantum randomness impacts the quality or characteristics of the generated numbers compared to classical methods.

5.5 The BB84 Quantum Key Distribution (QKD) Protocol

Another application that primarily uses single-qubit gates is the BB84 Quantum Key Distribution (QKD) Protocol. As we will discuss next, this protocol serves as a beautiful showcase of the real-world application of quantum measurements and quantum gates.

5.5.1 Fundamentals of Cryptography

1 Cryptography

Cryptography comprises techniques for safeguarding communication and data against adversaries. Cryptographic systems fall into two main categories:

- Secret (or symmetric) key cryptography, where both parties utilize a shared key for encryption and decryption.

- Public (or asymmetric) key cryptography, which employs a dual-key system with public and private components. While data encrypted using the public key can only be decrypted using the private key, the significant overhead usually makes this method more suitable for exchanging secret keys.

2 Key Exchange Protocols

Key exchange protocols serve as intermediaries to facilitate the secure sharing of secret keys between parties. Classical key exchange protocols often employ asymmetric cryptographic techniques. Notable examples include:

- Diffie-Hellman key exchange is based on the difficulty of the discrete logarithm problem in a finite field. Both parties use their respective private and shared public parameters to generate an identical shared secret, which is computationally infeasible to deduce from the public parameters.

- RSA key exchange utilizes the RSA algorithm, an asymmetric cryptographic technique. The public key encrypts the shared secret, and only the private key can decrypt it. RSA relies on the computational difficulty of integer factorization.

- Elliptic Curve Diffie-Hellman (ECDH) key exchange operates similarly to the original Diffie-Hellman but employs elliptic curve cryptography, providing equivalent security with shorter key lengths, thus being computationally more efficient.

- Lattice-based key exchange is a more recent approach, considered to be resistant to attacks from quantum computers, hence termed "quantum-safe." These also usually employ asymmetric cryptographic principles.

It is worth mentioning that asymmetric cryptographic techniques, especially in protocols like RSA and Diffie-Hellman, are often used specifically for the key exchange process. Once the key has been securely exchanged, communication might then switch to a symmetric key cryptographic scheme, which is generally less computationally intensive.

3 Quantum Computing and Cryptography

The advent of quantum computing poses significant threats to traditional cryptographic protocols. Shor's algorithm, for example, can efficiently factor large integers and compute discrete logarithms, effectively breaking RSA and Diffie-Hellman protocols. This vulnerability has led to an ongoing effort in the field of post-quantum cryptography, which aims to develop cryptographic protocols that are secure against quantum attacks. Some of the emerging options include lattice-based, code-based, and multivariate polynomial cryptography [14].

4 Quantum Key Distribution (QKD)

QKD leverages quantum mechanics to deliver a categorically secure public-key cryptosystem. QKD protocols can even detect potential eavesdroppers.

Notable QKD protocols include:

- BB84, proposed by Charles Bennett and Gilles Brassard in 1984, serves as one of the foundational QKD protocols.
- E91, which relies on entangled qubit pairs to establish a secure communication channel.
- B92, a less complex yet also less efficient variant of BB84.
- SARG04, an error-resilient protocol that draws inspiration from BB84.
- COW, which is compatible with existing telecommunications infrastructure and also takes inspiration from BB84.

5.5.2 Introduction to the BB84 QKD Protocol

The BB84 QKD protocol is renowned as the inaugural and a quintessential QKD protocol. It facilitates Alice and Bob in establishing a confidential shared key, suitable for secure communication via classical encryption methods. By exploiting quantum mechanics properties, including the no-cloning theorem, the BB84 protocol underscores the potential of quantum physics for bolstering communication security.

Within this protocol, Alice constructs a series of qubits (typically photons) with bases and states determined randomly. She transmits these qubits to Bob through a quantum channel, who then measures the received qubits using randomly opted bases. Subsequently, they utilize a public classical channel to compare their bases, retaining matches and dismissing mismatches. This shared set of measurement results becomes their secret key.

130   Chapter 5. Single-Qubit Quantum Gates

The protocol's security stems from the Heisenberg Uncertainty Principle (HUP, see § 1.6) and the no-cloning theorem (see § 5.1). Owing to the HUP, an eavesdropper cannot simultaneously and precisely measure both bases, given their non-orthogonal nature. The no-cloning theorem affirms the inability to perfectly replicate unknown quantum states, impeding an eavesdropper's attempt to intercept and relay qubits undetected. Consequently, any eavesdropping introduces errors, detectable during the error-checking phase, which alerts Alice and Bob to the need for restarting the transmission.

5.5.3 Procedure of the BB84 QKD Protocol

Typically, employing photon polarization, the BB84 protocol comprises five pivotal steps, elucidated in Fig. 5.7:

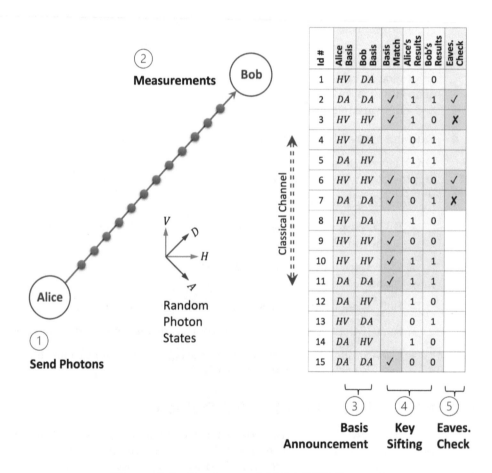

Figure 5.7: Fundamental BB84 QKD Protocol

1 **Photon Generation and Distribution**

Alice uses a QRNG to generate a random bit sequence. For each bit, she also randomly chooses either the rectilinear (HV) or diagonal (DA) basis. She then encodes the bits into photon polarizations based on the chosen bases:

- H polarization for bit 0 in the rectilinear basis

5.5 The BB84 Quantum Key Distribution (QKD) Protocol

- V polarization for bit 1 in the rectilinear basis
- D polarization for bit 0 in the diagonal basis
- A polarization for bit 1 in the diagonal basis

 The rectilinear basis, synonymous with the computational basis, aligns the H and V polarizations with $|0\rangle$ and $|1\rangle$ states, respectively. The diagonal basis mirrors the $\{|+\rangle,|-\rangle\}$ basis, associating the D (45°) polarization with $|+\rangle$ and the A (135°) polarization with $|-\rangle$.

Alice then dispatches the photons to Bob through a quantum medium, which could be a free space light beam or an optical fiber.

2 Measurement

Upon receiving each photon, Bob employs a randomly chosen basis (also either rectilinear or diagonal) to measure each photon's polarization. If Bob's basis aligns with Alice's, he discerns the intended bit. Otherwise, his measurement might diverge. This disparity is illustrated in Fig. 5.7.

3 Basis Announcement

Via a classical public channel, Bob notifies Alice of his measurement bases without disclosing the results.

4 Key Shifting

Comparing bases, Alice and Bob retain results from matching bases and eliminate the rest. They then split the remaining sifted bits for eavesdropping detection and as a shared secret key. Typically, these groups might span 128 and 2048 bits, respectively. However, Fig. 5.7 only illustrates this with 4 bits for each group.

5 Eavesdropping Check

Alice and Bob examine the first sifted key group to identify potential eavesdropping, as demonstrated in the column labeled "Eaves. Check" in Fig. 5.7. Any detected breach compels them to discard the key, initiating a new BB84 QKD session.

5.5.4 Detection of Eavesdropping

Eavesdropping refers to the act of intercepting or copying the quantum states of qubits (typically photons) during a quantum communication process. An eavesdropper, often referred to as Eve, may attempt to gain information about the transmitted key by measuring these qubits. However, the fundamental principles of quantum mechanics naturally limit her ability to do so without detection.

The Heisenberg's Uncertainty Principle (HUP) posits that one cannot measure both the rectilinear and diagonal polarization states of a qubit with arbitrary precision at the same time. This is due to the inherent non-orthogonality of the two bases. Consequently, when Eve measures the qubits using an incorrect basis, she invariably disturbs the states, introducing potential errors in the transmitted key.

Furthermore, the No-Cloning Theorem (see § 5.1.3) emphasizes that it is unfeasible to produce identical copies of an unknown quantum state. This theorem considerably hinders Eve's chances of intercepting and replicating quantum states without alerting the communicating parties.

There are a few notable scenarios wherein Eve might try to eavesdrop:

1. Eve selects the incorrect basis, leading to an altered qubit:
   - Should Bob opt for the correct basis, the introduced error by Eve becomes detectable.
   - On the other hand, if Bob selects an incorrect basis, the result is inherently random, thus keeping Eve concealed.

2. Eve picks the correct basis, leaving the qubit unaffected:
   - If Bob subsequently chooses the correct basis, Eve remains undetected while successfully acquiring one bit of the key.
   - Conversely, if Bob selects the wrong basis, the outcome remains unpredictable, and Eve's presence stays hidden.

Given these conditions, the probability of detecting Eve when using n check bits can be expressed as:

$$P(\text{Detection of Eve with } n \text{ check bits}) = 1 - \left(\frac{3}{4}\right)^n. \tag{5.26}$$

For ensuring a security confidence level of 99.9%, Alice and Bob would need to implement error-checking by comparing a subset of 30 bits from the key to discern any potential eavesdropping activities.

> **Exercise 5.15** Describe the BB84 protocol for Quantum Key Distribution. Your description should include:
>
> (a) The process of key generation and distribution between Alice and Bob.
>
> (b) The role of the quantum and classical channels.
>
> (c) The method used to detect eavesdropping by a third party (Eve).
>
> (d) The procedure for key sifting and reconciliation.

5.5.5 BB84 QKD Implementation with Qubits

The BB84 protocol, which we have previously explained using photon polarization, can also be generalized to employ any qubit system in principle.

1 Key Steps

The key steps of the BB84 protocol using general qubits are outlined below, mirroring the same steps in the photon implementation.

1. Alice prepares qubits randomly in one of four states ($|0\rangle, |1\rangle, |+\rangle, |-\rangle$) and sends them to Bob. The key information is encoded in two bases:
   - Computational basis: $\{|0\rangle, |1\rangle\}$, where $|0\rangle$ encodes bit 0 and $|1\rangle$ encodes bit 1.

5.5 The BB84 Quantum Key Distribution (QKD) Protocol

- Hadamard basis: $\{|+\rangle, |-\rangle\}$, where $|+\rangle$ encodes bit 0 and $|-\rangle$ encodes bit 1.

2. Bob measures the received qubits using a randomly chosen basis.
3. Alice and Bob reveal the bases they used for preparing and measuring the qubits through a classical communication channel.
4. They keep the bits where both Alice and Bob used the same basis and discard the rest. The remaining bits form the shared secret key.
5. They examine a segment of sifted key to identify potential eavesdropping.

2 Implementation with Quantum Gates

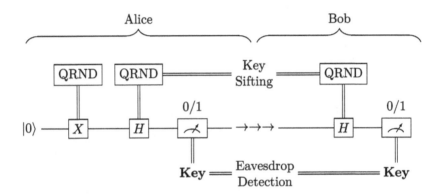

Figure 5.8: BB84 QKD Implementation with Qubits

In this generalized implementation, illustrated in Fig. 5.8, the BB84 protocol uses qubits in four states: $|0\rangle$, $|1\rangle$, $|+\rangle$, and $|-\rangle$. By convention, single lines represent qubits (or quantum channels), while double lines represent classical bits (or classical channels). Alice employs the following methods to prepare her qubit states:

- $|0\rangle$: The qubit is always initialized in the $|0\rangle$ state.
- $|1\rangle$: Apply an X gate after initialization.
- $|+\rangle$: Apply a Hadamard (H) gate after initialization.
- $|-\rangle$: Apply an X gate followed by an H gate after initialization.

Alice's X and H gates are controlled by two independent Quantum Random Number Generator (QRND) devices. A gate is switched into effect if the output bit from its corresponding QRND is 1; otherwise, it does not change the qubit state.

Similarly, when Bob receives a qubit, he selects a random basis using a QRND-controlled H gate. The circuit is meant to run repeatedly to produce a random binary string of desired length.

Exercise 5.16 Assume Eve is trying to eavesdrop on Alice and Bob during their BB84 protocol by performing an intercept-and-resend attack.

(a) Explain Eve's strategy and how it might affect the key generation process.

(b) What is the probability that Eve's interference will be detected by Alice and Bob?

(c) Discuss how Alice and Bob can quantify and mitigate the information gained by Eve.

(d) Express the eavesdrop-detection process using a quantum circuit.

5.5.6 ⁎ Further Exploration

The BB84 QKD protocol leverages the fundamental properties of quantum mechanics to establish a secure communication channel. While classical encryption techniques can be rendered ineffective by the advancement of computing capabilities (especially quantum computers), the security of the BB84 protocol is grounded in the immutable principles of quantum mechanics, making it a future-proof solution for secure communication.

However, practical implementations of the BB84 protocol do face challenges such as photon loss, noise, and equipment imperfections. Continuous research and development efforts are necessary to address these issues and ensure the successful deployment of QKD systems in real-world scenarios. For readers interested in exploring this subject further, the following resources are recommended:

1. General Introduction to Cryptography and Key Distribution Protocols [5]: This textbook provides a comprehensive introduction to modern cryptography, including key distribution protocols.

2. BB84: Original Paper by Charles Bennett and Gilles Brassard [21]: The original paper that introduced the BB84 protocol, laying the foundation for quantum key distribution.

3. B92: A Simpler Yet Less Efficient Variant of BB84 [22]: This paper introduces B92, a simplified and less efficient variant of the BB84 protocol.

4. SARG04: Error-Resilient Quantum Cryptography Inspired by BB84 [79]: A paper that introduces SARG04, an error-resilient quantum key distribution method inspired by BB84.

5. COW: Congruent with Existing Telecom Infrastructure and Inspired by BB84 [86]: A detailed exploration of the COW protocol, compatible with existing telecommunication infrastructure and influenced by BB84.

6. E91: Quantum Cryptography Using Entangled Qubit Pairs [39]: This paper presents the E91 protocol that utilizes entangled qubit pairs for key distribution. We will also discuss the E91 QKD protocol in § 10.6.

7. Quantum Key Distribution in Cryptography: A Survey [13]: This article explores the integration of quantum key distribution (QKD) in cryptographic infrastructures. Highlighting QKD's promise of information-theoretic security, the review underscores its application in renewing symmetric cipher keys and enabling secure key establishment in networks, while addressing inherent challenges and research avenues.

8. Post-Quantum Cryptography [14]: The vulnerabilities of traditional cryptographic protocols, such as RSA and Diffie-Hellman, have led to ongoing efforts in the field of post-quantum cryptography. This field aims to de-

5.6 The Quantum Coin Game

velop cryptographic protocols that are secure against attacks leveraging Shor's algorithm.

5.6 The Quantum Coin Game

Prepare to explore an exciting game: the fair coin game, with Alice and Bob putting a quantum twist on it! This is not merely a game between these two players but an intriguing introduction to the world of quantum algorithms. You, the reader, are invited to follow along as we unravel the fascinating potentials of this quantum challenge.

5.6.1 The Game Rules

To ensure fairness and adaptability in the quantum version of the game, we will use the "Coin in Box" version of the game. Here are the game rules:

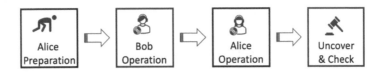

Figure 5.9: Fair Coin Gate - Setup and Rules

1. Alice Preparation: Alice places the coin in a state of her choice (heads or tails on top) in a box that covers the coin. (Note this is a special coin. Even by touching the coin, it is impossible to determine which side is up.)

2. Bob Operation: Bob reaches into the box and has the choice to flip the coin over or not. Alice cannot see what he is doing.

3. Alice Operation: Alice reaches into the box and performs an operation of her choice (flip or not flip). Bob cannot see what she is doing.

4. Uncover and Check: The coin is uncovered, and the result is read.
    - If it is Heads, Alice wins.
    - If it is Tails, Bob wins.

Is it a fair game? In the classical version, you bet.

Now let's consider the quantum version of the game:

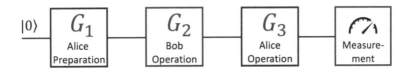

Figure 5.10: Quantum Coin Gate - Setup and Rules

The rules are the same as the classical version mentioned above. However, the following changes are made:

1. The coin is now represented by a qubit, with $|0\rangle$ representing Heads and $|1\rangle$ representing Tails.

2. A flip operation corresponds to the application of a NOT gate, which swaps the states $|0\rangle$ and $|1\rangle$.

3. A non-flip operation corresponds to the application of the Identity (I) gate.

4. In addition to the flip and non-flip operations, Alice and Bob are allowed to use other single-qubit gates (labeled G_1, G_2, and G_3 in Fig. 5.10) in their strategies.

5. In the Preparation step, the qubit is initially in state $|0\rangle$. However, Alice can apply a quantum gate to change it to another state.

6. The final Uncover and Check step is replaced by quantum measurement. If the measurement outcome is $|0\rangle$, Alice wins; if it is $|1\rangle$, Bob wins.

In this quantum version of the game, players can use the principles of quantum mechanics to devise new strategies and possibly gain an advantage over their opponent.

5.6.2 Alice Always Wins!

In the quantum case, Alice can win each time by using the Hadamard (H) gate. Here's the strategy:

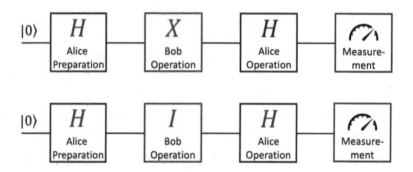

Figure 5.11: Strategy in Which Alice Always Wins

1. Alice applies the Hadamard gate, putting the qubit in an equal superposition state $|+\rangle = \frac{1}{\sqrt{2}}(|0\rangle + |1\rangle)$, namely, $H|0\rangle = |+\rangle$.

2. Regardless of whether Bob applies the I gate or the X gate, the qubit remains in a superposition state. However, in either case, the qubit state is unchanged, because $I|+\rangle = |+\rangle$ and $X|+\rangle = |+\rangle$.

3. Alice applies the Hadamard gate again, converting $|+\rangle$ back to $|0\rangle$ since $H|+\rangle = |0\rangle$.

4. They measure the final state. However, since the final state will always be $|0\rangle$, Alice will win every time.

5.6 The Quantum Coin Game

By introducing the Hadamard gate into the game strategically, Alice ensures that the final state of the qubit is always Heads ($|0\rangle$), allowing her to win the game every time!

5.6.3 Bob's Strategy Shift

Assuming Alice keeps her current strategy of using the Hadamard gate, Bob can adapt his strategy to change the game's outcome in his favor. Here's how Bob can use other gates to beat Alice:

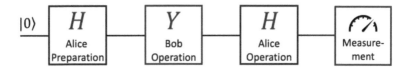

Figure 5.12: Strategy Shift: Bob Beats Alice

1. Alice applies the Hadamard gate, putting the qubit in an equal superposition state $|+\rangle$.

2. Bob applies the Pauli-Y gate to the qubit instead of X or I. After applying the Y gate, the qubit state becomes $-i\frac{1}{\sqrt{2}}(|0\rangle - |1\rangle)$, which is $|-\rangle$, since the global phase $(-i)$ can be ignored.

3. Alice applies the Hadamard gate again. The resulting qubit state is $|1\rangle$ because $H|-\rangle = |1\rangle$.

4. They measure the final state. Since the outcome is $|1\rangle$, Bob wins.

By using the Y gate strategically, Bob can change the outcome of the game and beat Alice!

5.6.4 Alice Overcomes Bob's New Gambit

If Bob continues using his new strategy with the Y gate, Alice can also revise her strategy to beat Bob again. How can that be done? She can compose a new gate G using a gate sequence.

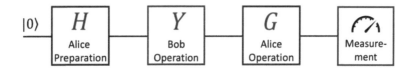

Figure 5.13: Strategy Revised: Alice Wins Again

> **Exercise 5.17** Assuming Bob continues using his new strategy with the Y gate, Alice would revise her strategy to beat Bob again. What gate can she use to achieve this? Here is a hint: $XHYH|0\rangle = i|0\rangle$.

You can imagine that, since they have mastered quantum gates and sequences, Alice and Bob could revise their strategies continuously, playing happily thereafter.

5.7 *The No-Cloning Theorem: Proof Outline

The no-cloning theorem, as introduced in § 5.1.3, asserts that it is impossible to produce an exact replica of an arbitrary unknown quantum state using any physically realizable quantum operation. This section provides a proof of the theorem. Given that the proof involves concepts of multi-qubit gate operation, it is recommended that readers come back to this section after familiarizing themselves with these concepts in Chapter 7.

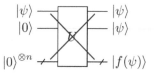

Figure 5.14: Illustration of the No-Cloning Theorem

Let's consider unitary operations first: a hypothetical quantum cloning machine represented by a unitary operator U. This operator acts on a quantum state $|\psi\rangle$, which we wish to clone, and an ancillary state $|0\rangle$, prepared for the cloning process. After applying U, our goal is to obtain two identical copies of $|\psi\rangle$.

Additionally, the cloning system may involve extra qubits. Assuming these qubits are initially in the $|0\rangle$ state without loss of generality, the cloning process can be described by:

$$U(|\psi\rangle \otimes |0\rangle \otimes |0\rangle^{\otimes n}) = |\psi\rangle \otimes |\psi\rangle \otimes |f(\psi)\rangle, \quad (5.27)$$

where $|f(\psi)\rangle$ represents the state of the extra qubits after cloning.

To demonstrate the impossibility of such cloning, we proceed by contradiction. Assume that the same operator U can clone the states $|0\rangle$, $|1\rangle$, and $|+\rangle = \frac{1}{\sqrt{2}}(|0\rangle+|1\rangle)$. This leads to the following equations:

$$U(|0\rangle \otimes |0\rangle \otimes |0\rangle^{\otimes n}) = |0\rangle \otimes |0\rangle \otimes |f(0)\rangle, \quad (5.28a)$$

$$U(|1\rangle \otimes |0\rangle \otimes |0\rangle^{\otimes n}) = |1\rangle \otimes |1\rangle \otimes |f(1)\rangle, \quad (5.28b)$$

$$U(|+\rangle \otimes |0\rangle \otimes |0\rangle^{\otimes n}) = |+\rangle \otimes |+\rangle \otimes |f(+)\rangle. \quad (5.28c)$$

By the linearity of U, adding the first two equations and multiplying by $\frac{1}{2}$ gives:

$$U(|+\rangle \otimes |0\rangle \otimes |0\rangle^{\otimes n}) = \frac{1}{2}(|00\rangle \otimes |f(0)\rangle + |11\rangle \otimes |f(1)\rangle). \quad (5.29)$$

However, the third equation implies:

$$U(|+\rangle \otimes |0\rangle \otimes |0\rangle^{\otimes n}) = \frac{1}{2}(|00\rangle + |01\rangle + |10\rangle + |11\rangle) \otimes |f(+)\rangle. \quad (5.30)$$

The right-hand sides of Eqs. 5.29 and 5.30 represent different quantum states, because measuring the first two qubits in the former yields two possible outcomes

($|00\rangle$ and $|11\rangle$), while the latter yields four different outcomes. This contradiction confirms that it is impossible to create an exact replica of an arbitrary unknown quantum state using only unitary quantum operations.

The proof of the No-Cloning Theorem, demonstrated for unitary transformations above, can be extended to include Completely Positive Trace-Preserving (CPTP) maps, which encompass measurements and other quantum operations. This extension is outlined in § 12.2.9. Thus, the No-Cloning Theorem is not limited to unitary operators alone; it establishes that it is impossible to create an identical copy of an arbitrary unknown quantum state using any physically realizable quantum operation.

5.8 Summary and Conclusions

Understanding Quantum Gates

Throughout this chapter, we navigated the integral concept of quantum gates, which are foundational to quantum computing. Just as Newtonian mechanics underpins satellite launches, core quantum mechanics principles play a pivotal role in shaping and managing quantum gates. We expanded on our prior knowledge of quantum mechanics and qubits, venturing into the world of quantum gates.

Our introduction began with single-qubit gates, giving us a grasp on their varied applications. This groundwork paves the way for deeper insights into multi-qubit gates, entanglement, quantum teleportation, and the E91 quantum key distribution in future discussions.

Distinguishing Quantum from Classical Gates

We emphasized the difference between quantum gates and their classical counterparts. Classical gates modify classical bits, while quantum gates act on qubits, allowing the creation of quantum circuits adept at processing and storing quantum information. These discussions shed light on the divergences and parallels between classical and quantum computing.

Our look into single-qubit gates, like the Pauli and Hadamard gates, showcased the extensive possibilities unlocked when these gates are paired together, enabling quantum circuits to address challenges daunting for classical systems.

Real-World Quantum Applications

We explored applications such as the Quantum Random Number Generator (QRNG) and the BB84 Quantum Key Distribution (QKD) Protocol. These applications highlighted the practical significance of quantum gates, enhancing our understanding of quantum measurements and their uses. The Quantum Coin Game served as a practical demonstration of the potential within quantum algorithms.

The Next Step: Multi-Qubit Systems

After solidifying our understanding of single-qubit gate principles and applications, we are ready to delve into multi-qubit systems. This forthcoming exploration will establish the foundation for multi-qubit gates and the enigmatic concept of quantum entanglement. While classical mechanics employs straightforward addition to describe systems, we will see how quantum mechanics uses the tensor product to represent states of joint quantum systems, offering a more complex interaction framework.

Looking Ahead

The next chapter will spotlight multi-qubit systems, an essential preface to advanced topics such as multi-qubit gates and quantum entanglement. Our focus will be on the composition principle in quantum mechanics, offering clarity on how quantum systems are distinctively represented compared to the classical sum-based approach. We'll aim to understand the nuances of quantum particle interactions and their progression over time, preparing us for an in-depth study of multi-qubit gates in subsequent chapters.

Problem Set 5

5.1 Show that the Pauli Y operator acts as a flip gate for the states $|+\rangle$ and $|-\rangle$, ignoring the global phase (which is inconsequential to the physical state of the qubit). Specifically, demonstrate that $Y|+\rangle = -i|-\rangle$ and $Y|-\rangle = i|+\rangle$. Then, demonstrate this transformation on the Bloch sphere.

5.2 Validate the relation $ZX = iY$ using both matrix and bra-ket representations.

5.3 Validate the following relations. Here $S = \begin{bmatrix} 1 & 0 \\ 0 & i \end{bmatrix}$.

$$XXX = X, \qquad XYX = -Y, \qquad XZX = -Z,$$
$$YXY = -X, \qquad YYY = Y, \qquad YZY = -Z,$$
$$ZXZ = -X, \qquad ZYZ = -Y, \qquad ZZZ = Z,$$
$$SXS^\dagger = Y, \qquad SYS^\dagger = -X, \qquad SZS^\dagger = Z.$$

5.4 Consider $H' = \frac{1}{\sqrt{2}}\begin{bmatrix} 1 & -1 \\ 1 & 1 \end{bmatrix}$.

   (a) If $H' = UH$, where H is the Hadamard operator, find U.

   (b) Find $H'|0\rangle$, $H'|1\rangle$, $H'|+\rangle$, and $H'|-\rangle$.

5.5 Show how the Hadamard gate, H, can be expressed as $iR_x(\pi)R_y(\frac{\pi}{2})$ and $iR_y(\frac{\pi}{2})R_z(\pi)$.

5.6 Show that, up to a global phase factor, $H = X\sqrt{Y}$ and $H = \sqrt{Y}Z$.

5.7 Consider a qubit initially in the state defined by the polar angle α and the azimuthal angle β on the Bloch sphere, expressed as:

$$|\psi\rangle = \cos\frac{\alpha}{2}|0\rangle + e^{i\beta}\sin\frac{\alpha}{2}|1\rangle.$$

The rotation around the x-axis by an angle θ is represented by the matrix $R_x(\theta)$ (as defined in Eq. 5.17). Derive the state vector $|\psi'\rangle$ after the gate $R_x(\theta)$ acts on the state $|\psi\rangle$.

5.8 Verify that the Hadamard gate is equivalent to $R_u(\pi, \frac{\pi}{4}, 0)$, where R_u is defined in Eq. 5.25. Then, show that the Hadamard gate can be visualized as a 180° rotation around the $(x + z)$-axis, which is the angular bisector between the x and z-axes.

5.9 Analyze the quantum circuit below and derive the measured probability P as a function of ϕ. Discuss how this circuit functions similarly to a Mach-Zehnder interferometer (§ 1.5.4.1).

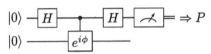

5.10 Investigate the scenario where Alice and Bob are implementing the BB84 protocol, but their quantum channel is noisy, leading to a 1% error rate in the transmission of qubits.

(a) How does noise affect the key generation process?

(b) Describe a method Alice and Bob could use to estimate the error rate in their key bits.

(c) ✶ Explain how Alice and Bob can still establish a secret key using error correction and privacy amplification techniques.

6. Multi-Qubit Systems

Contents

6.1	**Systems of Two Qubits**	144
6.1.1	Review of Single Qubit States	144
6.1.2	Two-Qubit States: Developing Intuition	145
6.1.3	General States vs. Product States	147
6.2	✶**Measurements of Two-Qubit Systems**	148
6.2.1	✶Local Measurements on Subsystems	148
6.2.2	✶Joint Measurements	154
6.3	**Multi-Qubit System States**	156
6.3.1	The Composition Postulate of Quantum Mechanics	157
6.3.2	Basis States	157
6.3.3	Product States, Entangled States, and Correlated States	159
6.3.4	✶Identical Particles	161
6.4	✶**Measurements of Multi-Qubit Systems**	162
6.4.1	✶Measurements on a Single Qubit	162
6.4.2	✶Measurements on a Group of Qubits	163
6.4.3	✶A Partial Measurement Paradigm for Quantum Algorithms	164
6.4.4	✶Measurements in Alternate Bases	166
6.5	✶**Time Evolution of Multi-Qubit States**	167
6.5.1	✶Two-Qubit System	168
6.5.2	✶Multi-Qubit System	168
6.5.3	✶Example: The Ising Model	169
6.5.4	✶Example: CNOT Gate Implementation	170
6.6	**Summary and Conclusions**	171
	Problem Set 6	172

In this chapter, we explore the exciting and complex realm of multi-qubit systems. This crucial understanding is a stepping stone towards more advanced topics in quantum computation, such as multi-qubit gates and quantum entanglement. The defining principle governing these composite quantum systems is the composition principle in quantum mechanics, a concept that markedly separates the quantum world from classical mechanics.

In classical Newtonian mechanics, individual subsystems and their interactions are described with straightforward addition. For instance, the combined effect of individual forces acting on a system can be depicted as the simple sum of these forces, as portrayed by the equation $\sum m\mathbf{a} = \sum \mathbf{F}$.

On the other hand, quantum mechanics approaches this differently. It utilizes state vectors and quantum operators to encapsulate the behavior and interactions of subsystems. According to the composition principle in quantum mechanics, the joint state of multiple quantum subsystems is captured by the tensor product of their individual states. This crucial distinction sets quantum mechanics apart from the additive principle that governs classical systems.

The tensor product principle allows us to understand the intricate dance of quantum particles and their time evolution. Moreover, it is an essential tool to describe entanglement, one of the most fascinating and puzzling phenomena of quantum physics.

In this chapter, we delve into the intriguing world of multiple qubits, setting the stage for a deeper investigation of multi-qubit gates in the subsequent chapter.

6.1 Systems of Two Qubits

Embarking on the journey of multi-qubit systems, we initially focus on the two-qubit system. The understanding gleaned from two-qubit systems will build our intuition and elucidate the more complex principles of multi-qubit systems, while concurrently demonstrating the underlying connection with single-qubit systems.

6.1.1 Review of Single Qubit States

The state of a single qubit can be represented as a two-dimensional complex vector:

$$|\psi\rangle = \alpha |0\rangle + \beta |1\rangle. \tag{6.1}$$

Mathematically, we express this as $|\psi\rangle \in \mathbb{C}^2$, with $\mathbb{C}^2$ symbolizing a complex Hilbert space of two dimensions. In the context of quantum computing, where we focus on finite dimensions, a Hilbert space is essentially a vector space that is equipped with an inner product.

The state vector $|\psi\rangle$ is normalized, which means:

$$\langle\psi|\psi\rangle = |\alpha|^2 + |\beta|^2 = \alpha\alpha^* + \beta\beta^* = 1. \tag{6.2}$$

When we measure the qubit in the basis $\{|0\rangle, |1\rangle\}$, we obtain $|0\rangle$ with probability $|\alpha|^2$ and $|1\rangle$ with probability $|\beta|^2$.

The vectors $|0\rangle$ and $|1\rangle$ form the standard (or computational) basis of the state space $\mathbb{C}^2$. They are orthonormal, which means:

$$\langle i|j\rangle = \delta_{i,j}. \tag{6.3}$$

Here $i, j \in \{0, 1\}$, and $\delta_{i,j}$ is the Kronecker delta, which is defined as $\delta_{i,j} = 1$ if $i = j$ and $\delta_{i,j} = 0$ otherwise.

6.1.2 Two-Qubit States: Developing Intuition

Now consider two independent qubits, A and B. Their state vectors are represented as

$$|\psi_A\rangle = \alpha|0_A\rangle + \beta|1_A\rangle, \quad |\psi_B\rangle = \gamma|0_B\rangle + \delta|1_B\rangle, \tag{6.4}$$

where $\alpha, \beta, \gamma, \delta$ are complex coefficients such that $|\alpha|^2 + |\beta|^2 = 1$ and $|\gamma|^2 + |\delta|^2 = 1$.

We can view the two qubits as a combined quantum system, which should be consistent with our view of them as two separate systems. Let's check this out.

According to the principles of quantum mechanics, the state vector of the combined two-qubit system, given that the two qubits are independent, is represented as the tensor product of the individual states:

$$|\psi_{AB}\rangle = |\psi_A\rangle \otimes |\psi_B\rangle = (\alpha|0_A\rangle + \beta|1_A\rangle) \otimes (\gamma|0_B\rangle + \delta|1_B\rangle). \tag{6.5}$$

By expanding this equation, we find:

$$|\psi_{AB}\rangle = \alpha\gamma|0_A\rangle \otimes |0_B\rangle + \alpha\delta|0_A\rangle \otimes |1_B\rangle + \beta\gamma|1_A\rangle \otimes |0_B\rangle + \beta\delta|1_A\rangle \otimes |1_B\rangle. \tag{6.6}$$

1 Normalization Verification

To validate the quantum state $|\psi_{AB}\rangle$, let's confirm it is normalized:

$$\begin{aligned}\langle\psi_{AB}|\psi_{AB}\rangle &= |\alpha\gamma|^2 + |\alpha\delta|^2 + |\beta\gamma|^2 + |\beta\delta|^2 \\ &= (|\alpha|^2 + |\beta|^2)(|\gamma|^2 + |\delta|^2) \\ &= 1. \end{aligned} \tag{6.7}$$

2 Single-Qubit Measurement Verification

Suppose we measure qubit A and obtain $|0_A\rangle$. This process should select the terms with $|0_A\rangle$ in our state vector, i.e., $\alpha\gamma|0_A\rangle \otimes |0_B\rangle + \alpha\delta|0_A\rangle \otimes |1_B\rangle$. Interpreting the complex coefficients as probability amplitudes for $|0_A\rangle \otimes |0_B\rangle$ and $|0_A\rangle \otimes |1_B\rangle$, the associated probability is

$$\begin{aligned}P_{|0_A\rangle} &= |\alpha\gamma|^2 + |\alpha\delta|^2 \\ &= |\alpha|^2(|\gamma|^2 + |\delta|^2) \\ &= |\alpha|^2. \end{aligned} \tag{6.8}$$

This result matches our expectation when considering qubit A independently. The same method can be applied for measuring $|1_A\rangle$, $|0_B\rangle$, and $|1_B\rangle$. The probability interpretation remains consistent in all cases.

3 Two-Qubit Measurement Verification

Assume we measure qubit A and obtain $|0_A\rangle$, and simultaneously measure qubit B and obtain $|0_B\rangle$. The joint probability of these two measurement outcomes in independent systems would be $|\alpha|^2|\gamma|^2$. In a combined quantum system, these measurements select the term $\alpha\gamma|0_A\rangle \otimes |0_B\rangle$ from the state vector. The probability is given by the square of the coefficient, which also results in $|\alpha\gamma|^2$, thereby providing a consistent representation.

4 Orthonormal Basis Verification

To extend our state vector to represent generic two-qubit states, including interacting qubits, we require the terms $|0_A\rangle \otimes |0_B\rangle$, $|0_A\rangle \otimes |1_B\rangle$, etc., to form an orthonormal basis.

For this, we can express the basis states in matrix form:

$$|0\rangle = \begin{bmatrix} 1 \\ 0 \end{bmatrix}, \quad |1\rangle = \begin{bmatrix} 0 \\ 1 \end{bmatrix}. \tag{6.9}$$

By applying the tensor product rules for matrices, we get:

$$|0_A\rangle \otimes |0_B\rangle = \begin{bmatrix} 1 \\ 0 \\ 0 \\ 0 \end{bmatrix}, \quad |0_A\rangle \otimes |1_B\rangle = \begin{bmatrix} 0 \\ 1 \\ 0 \\ 0 \end{bmatrix},$$

$$|1_A\rangle \otimes |0_B\rangle = \begin{bmatrix} 0 \\ 0 \\ 1 \\ 0 \end{bmatrix}, \quad |1_A\rangle \otimes |1_B\rangle = \begin{bmatrix} 0 \\ 0 \\ 0 \\ 1 \end{bmatrix}. \tag{6.10}$$

These equations clearly indicate that the set $\{|i_A\rangle \otimes |j_B\rangle\}$ ($i,j \in \{0,1\}$) is orthonormal, meaning each pair of vectors is orthogonal, and each vector has unit length.

The two-qubit state space is four-dimensional. Since the above set contains four independent elements, it also spans the entire vector space and hence, it forms a complete orthonormal basis for this state space.

5 Tensor Product of Operators

In our discussion so far, we have successfully established that the state vector of two independent qubits can be represented as the tensor product of their individual state vectors. But how does this formalism extend to the operators that act on these state vectors?

Consider two operators: U_A acting on $|\psi_A\rangle$ and U_B acting on $|\psi_B\rangle$. For our combined two-qubit state, it is natural to anticipate the combined operator as the tensor product $U_A \otimes U_B$, resulting in the state $(U_A \otimes U_B)|\psi_{AB}\rangle$. This expectation is supported by the tensor product property:

$$(U_A \otimes U_B)(|\psi_A\rangle \otimes |\psi_B\rangle) = (U_A |\psi_A\rangle) \otimes (U_B |\psi_B\rangle). \tag{6.11}$$

This relationship confirms the anticipated behavior of tensor product operators on combined state vectors.

6 Shorthand Notation

For succinctness, we will use $|00\rangle$ instead of $|0_A\rangle \otimes |0_B\rangle$ from this point forward, where the first position represents qubit A, and the second position represents qubit B. Other combinations will follow the same convention. Therefore, our earlier state vector in Eq. 6.6 simplifies to:

$$|\psi_{AB}\rangle = \alpha\gamma |00\rangle + \alpha\delta |01\rangle + \beta\gamma |10\rangle + \beta\delta |11\rangle. \tag{6.12}$$

6.1 Systems of Two Qubits

The four basis states are now represented as $|00\rangle$, $|01\rangle$, $|10\rangle$, and $|11\rangle$. Their orthonormal condition can be expressed as:

$$\langle ij|kl\rangle = \delta_{ik}\delta_{jl}, \tag{6.13}$$

where $i,j \in \{0,1\}$, and δ_{ij} is the Kronecker delta function.

Exercise 6.1 Consider the state

$$|\psi\rangle = \frac{1}{\sqrt{2}}(|0\rangle + |1\rangle) \otimes \frac{1}{\sqrt{5}}(|0\rangle - 2i|1\rangle)$$

(a) Express $|\psi\rangle$ in terms of $|00\rangle$, $|01\rangle$, $|10\rangle$, and $|11\rangle$.

(b) Compute the probability of the first qubit being in state $|1\rangle$.

(c) Compute the probability of the first qubit being in state $|0\rangle$ and the second qubit being in state $|1\rangle$.

6.1.3 General States vs. Product States

1 General Two-Qubit States

In our system of two qubits, the state space is four-dimensional with basis states $|00\rangle$, $|01\rangle$, $|10\rangle$, and $|11\rangle$. This allows us to construct a general two-qubit state as a linear combination of these basis states:

$$|\psi\rangle = \sum_{i,j \in \{0,1\}} c_{ij}|ij\rangle = c_{00}|00\rangle + c_{01}|01\rangle + c_{10}|10\rangle + c_{11}|11\rangle. \tag{6.14}$$

Here, c_{ij} are complex coefficients such that $\sum |c_{ij}|^2 = 1$. Therefore, a general state for a two-qubit system is a vector in the four-dimensional complex Hilbert space $\mathbb{C}^4$.

2 Product States

The state $|\psi_{AB}\rangle \equiv |\psi_A\rangle \otimes |\psi_B\rangle$ referred to in Eq. 6.12 is a specific instance known as a product state.

A product state is defined as one where the state vector can be expressed as the tensor product of the state vectors of the individual qubits. It represents a system of independent qubits.

3 Non-Product States

Apart from product states, there exist states that cannot be expressed as a tensor product of individual qubit states. An example of such a state is the Bell state, also known as an EPR pair:

$$|\Phi^+\rangle = \frac{1}{\sqrt{2}}(|00\rangle + |11\rangle). \tag{6.15}$$

This state cannot be written as a tensor product of individual qubit states.

Proof. If we assume that $\frac{1}{\sqrt{2}}(|00\rangle + |11\rangle)$ could be expressed by Eq. 6.12, we find that $\alpha\gamma = \beta\delta = \frac{1}{\sqrt{2}}$ and $\alpha\delta = \beta\gamma = 0$. However, these two conditions are impossible to satisfy simultaneously. □

Non-product states like these exhibit entanglement, a unique quantum phenomenon where the state of one qubit is interconnected with the state of the other, which we will study extensively later.

Exercise 6.2 Demonstrate that $|\Psi^-\rangle = \frac{1}{\sqrt{2}}(|01\rangle - |10\rangle)$ is a non-product state.

Exercise 6.3 Prove that $\frac{1}{2}(|00\rangle - |10\rangle - |01\rangle + |11\rangle)$ is a product state and identify the two component states.

6.2 ∗ Measurements of Two-Qubit Systems

We introduced a broad framework for quantum measurement in § 3.4. This section delves deeper by extending this framework to encompass the measurements of composite quantum systems, specifically those involving multiple qubits.

To provide a clear perspective, we will primarily focus on a composite quantum system comprising two subsystems, A and B. However, it's essential to note that the principles we discuss are readily extendable to more intricate systems consisting of multiple subsystems. These subsystems can be formed by qubits or other advanced quantum particles. When addressing the measurements of such composite systems, the tensor product structure of the state space becomes crucial, as do potential correlations between the subsystems. In the subsequent sections, we will elucidate the two predominant types of quantum measurements suited for these composite systems.

6.2.1 ∗ Local Measurements on Subsystems

In quantum mechanics, a local measurement refers to a measurement performed on one subsystem of a composite system, leaving the other subsystem(s) physically undisturbed. The measurement operation applies only to the subsystem of interest, and the outcome pertains solely to the state of this subsystem.

For a composite system composed of subsystems A and B, the local measurement operators for a measurement on subsystem A or B can be expressed as $M_i \otimes I_B$ and $I_A \otimes M_i$ respectively, where M_i represents the measurement operators of the subsystem in focus and I_A and I_B are identity operators on subsystems A and B.

The sequence in which local measurements are made is not essential, as each measurement operates independently on its respective subsystem. Mathematically, $M_i \otimes I_B$ and $I_A \otimes M_i$ commute. The primary aim of local measurements is to uncover properties of the individual subsystem.

1 Entangled States vs. Product States

Entangled States

For entangled states, the state of the composite system cannot be fully described by the states of its individual subsystems. When a measurement is made on one subsystem of an entangled pair, our description of the overall system, and hence of the other subsystem, must be updated. This can affect the potential outcomes of subsequent measurements on the other subsystem.

For instance, consider a two-qubit entangled state like the Bell state $|\Phi^+\rangle = \frac{1}{\sqrt{2}}(|00\rangle + |11\rangle)$. If a measurement on subsystem A yields the result $|0\rangle$, the entire system collapses into state $|00\rangle$, thus determining that any subsequent measurement on subsystem B would yield $|0\rangle$. Conversely, if the measurement on subsystem A results in $|1\rangle$, the entire system collapses into state $|11\rangle$, and a subsequent measurement on B would yield $|1\rangle$. This reflects the profound correlation that is characteristic of entangled states, underlining the fact that while no direct interaction occurs between the subsystems during a local measurement, the measurement can nevertheless affect our description and future measurement predictions of the other subsystem.

Product States

In contrast to entangled states, for product states such as $|\psi_{AB}\rangle = |\psi_A\rangle \otimes |\psi_B\rangle$, the system can be entirely described by the states of its individual subsystems, and the measurement outcome on one subsystem does not affect our understanding or predictions of the other subsystem. Furthermore, in this case, we can apply measurement operators directly to the individual subsystem states. For instance, $(M_i \otimes I_B)|\psi_{AB}\rangle$ simplifies to $(M_i|\psi_A\rangle) \otimes |\psi_B\rangle$.

If we perform a measurement on subsystem A and find it in some state $|\psi_A\rangle$, this does not impact the state or potential measurement outcomes for subsystem B, which remains in its original state $|\psi_B\rangle$. Consequently, for product states, the measurements on individual subsystems are truly independent, and each subsystem can be analyzed separately without the need to consider their correlations. This decoupling of subsystems is a hallmark of product states. It illustrates that even though the universe is composed of many quantum systems, we can treat each one independently if it is isolated from the rest.

2 Local Measurements in the Computational Basis

To gain a deeper understanding and operational proficiency on measurements on composite quantum systems, it is necessary to study the related mathematics. Let's start with local measurements in the computational (or standard) basis.

When measuring in the computational basis, the single-qubit measurement operators corresponding to the outcomes $|0\rangle$ and $|1\rangle$ are:

$$M_i = |i\rangle\langle i|, \tag{6.16}$$

where $i \in \{0, 1\}$. Note that these are projection operators with the property $M_i^2 = M_i$.

The two-qubit measurement operators for measuring on qubit A and qubit B are:

$$M_{iA} = |i\rangle\langle i| \otimes I, \quad M_{iB} = I \otimes |i\rangle\langle i|. \tag{6.17}$$

When performing local measurements on the general two-qubit state given by

$$|\psi\rangle = \sum_{i,j\in\{0,1\}} c_{ij} |ij\rangle, \qquad (6.14)$$

we first apply the two-qubit measurement operators on $|\psi\rangle$:

$$M_{0A} |\psi\rangle = (|0\rangle\langle 0| \otimes I) \sum_{i,j\in\{0,1\}} c_{ij} |ij\rangle = c_{00} |00\rangle + c_{01} |01\rangle, \qquad (6.18a)$$

$$M_{1A} |\psi\rangle = (|1\rangle\langle 1| \otimes I)... \qquad = c_{10} |10\rangle + c_{11} |11\rangle, \qquad (6.18b)$$

$$M_{0B} |\psi\rangle = (I \otimes |0\rangle\langle 0|)... \qquad = c_{00} |00\rangle + c_{10} |10\rangle, \qquad (6.18c)$$

$$M_{1B} |\psi\rangle = (I \otimes |1\rangle\langle 1|)... \qquad = c_{01} |01\rangle + c_{11} |11\rangle. \qquad (6.18d)$$

Then, we can calculate the associated probabilities according to Eq. 3.38:

$$P_{0A} = \|M_{0A} |\psi\rangle\|^2 = \|c_{00} |00\rangle + c_{01} |01\rangle\|^2 = |c_{00}|^2 + |c_{01}|^2, \qquad (6.19a)$$

$$P_{1A} = ... \qquad = |c_{10}|^2 + |c_{11}|^2, \qquad (6.19b)$$

$$P_{0B} = ... \qquad = |c_{00}|^2 + |c_{10}|^2, \qquad (6.19c)$$

$$P_{1B} = ... \qquad = |c_{01}|^2 + |c_{11}|^2. \qquad (6.19d)$$

Next, we will focus on measuring qubit A and obtaining $|0\rangle$. Other scenarios are similar. The post-measurement state is calculated according to Eq. 3.41:

$$|\psi_{0A}\rangle = \frac{M_{0A} |\psi\rangle}{\sqrt{P_{0A}}} = \frac{c_{00} |00\rangle + c_{01} |01\rangle}{\sqrt{|c_{00}|^2 + |c_{01}|^2}} = \frac{|0\rangle \otimes (c_{00} |0\rangle + c_{01} |1\rangle)}{\sqrt{|c_{00}|^2 + |c_{01}|^2}}. \qquad (6.20)$$

This clearly shows the system collapses to a product state upon measurement, with qubit A in $|0\rangle$ and qubit B in $|\psi_{B|0A}\rangle = \hat{N}(c_{00} |0\rangle + c_{01} |1\rangle)$, where $\hat{N}$ represents normalization, in this case, dividing by $\sqrt{|c_{00}|^2 + |c_{01}|^2}$.

The state of qubit B can also be calculated through the partial product:

$$|\psi_{B|0A}\rangle = \hat{N} \langle 0_A|\psi\rangle, \qquad (6.21)$$

where the index A in $\langle 0_A|$ indicates that it interacts only with the state of qubit A in $|\psi\rangle$ through inner product, while leaving the state of qubit B unchanged. (In matrix operation, $\langle 0_A|\psi\rangle$ is equivalent to $((\langle 0| \otimes I) |\psi\rangle)$.)

At this point, if we measure qubit B and get outcome $|1\rangle$, the system will collapse to $|01\rangle$. The probability for this outcome (A getting $|0\rangle$ and B getting $|1\rangle$) is

$$P_{1B,0A} = P_{0A} \left| c_{01}/\sqrt{|c_{00}|^2 + |c_{01}|^2} \right|^2 = |c_{01}|^2. \qquad (6.22)$$

As you can see, the joint probability for the outcome $|01\rangle$ depends only on c_{01} as expected, and it does not depend on the order of the measurements on qubits A and B.

The results so far is illustrated in Fig. 6.1.

If we assign the value $\lambda_0 = 1$ to $|0\rangle$ and $\lambda_1 = -1$ to $|1\rangle$, as is the case when measuring spin, we can calculate the expected value (statistical average) of λ using Eq. 3.40:

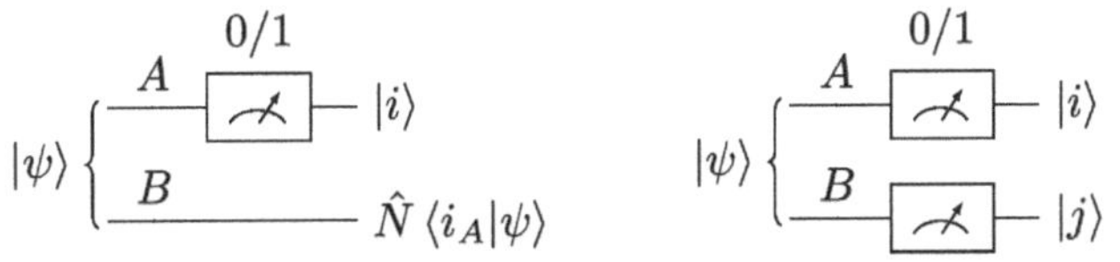

6.2 ∗ Measurements of Two-Qubit Systems

Notations: $i, j \in \{0, 1\}$, and $\hat{N}$ denotes for normalization.

Figure 6.1: Local Measurement Examples

$$\langle \lambda_A \rangle = \sum_i P_{iA} \lambda_i = |c_{00}|^2 + |c_{01}|^2 - |c_{10}|^2 - |c_{11}|^2, \tag{6.23a}$$

$$\langle \lambda_B \rangle = \sum_i P_{iB} \lambda_i = |c_{00}|^2 + |c_{10}|^2 - |c_{01}|^2 - |c_{11}|^2. \tag{6.23b}$$

In this case, the measurement corresponds to the Pauli Z operator observable, allowing us to calculate $\langle \lambda \rangle$ as:

$$\langle \lambda_A \rangle = \langle \psi | Z \otimes I | \psi \rangle, \tag{6.24a}$$
$$\langle \lambda_B \rangle = \langle \psi | I \otimes Z | \psi \rangle. \tag{6.24b}$$

Exercise 6.4 Consider the state

$$|\psi\rangle = c(|00\rangle - 2i\,|01\rangle + 2i\,|10\rangle + 4\,|11\rangle).$$

(a) Determine the coefficient c to normalize the state $|\psi\rangle$.

(b) Compute the probability of the first qubit being in state $|1\rangle$.

(c) Compute the expected value for the observable $\frac{X+Z}{2}$.

3 Local Measurements in Alternate Bases

After familiarizing ourselves with the mechanics of measurements in the computational basis, we will now extend our discussion to local measurements in other bases. We will focus on one of the most common alternative bases: the Hadamard basis, with basis states $|+\rangle$ and $|-\rangle$, defined as:

$$|+\rangle = \frac{1}{\sqrt{2}}(|0\rangle + |1\rangle), \tag{6.25a}$$

$$|-\rangle = \frac{1}{\sqrt{2}}(|0\rangle - |1\rangle). \tag{6.25b}$$

Consider the following observable:

$$M = |+\rangle\langle+| - |-\rangle\langle-|. \tag{6.26}$$

We can verify that M is a Hermitian operator, thereby qualifying it as an observable. When we use M to measure a qubit, the outcome will be either $|+\rangle$ or $|-\rangle$, the eigenstates of M. This observable associates the eigenvalue $+1$ with the outcome $|+\rangle$, and the eigenvalue -1 with the outcome $|-\rangle$.

Let's now use M to measure one of the Bell states, which are entangled states:

$$|\psi\rangle = \frac{1}{\sqrt{2}}(|00\rangle + |11\rangle). \tag{6.27}$$

To facilitate this, we will use the following properties:

$$\langle +|0\rangle = \langle +|1\rangle = \langle -|0\rangle = \frac{1}{\sqrt{2}}, \quad \langle -|1\rangle = -\frac{1}{\sqrt{2}}. \tag{6.28}$$

We apply M to the first qubit. In this process, we first derive the following intermediate quantities (un-normalized states):

$$|\psi_{+A}\rangle \equiv (|+\rangle\langle +| \otimes I)|\psi\rangle \tag{6.29a}$$

$$= \frac{1}{\sqrt{2}}(|+\rangle\langle +||0\rangle I|0\rangle + |+\rangle\langle +||1\rangle I|1\rangle) \tag{6.29b}$$

$$= \frac{1}{2}|+\rangle \otimes (|0\rangle + |1\rangle) \tag{6.29c}$$

$$= \frac{1}{\sqrt{2}}|++\rangle. \tag{6.29d}$$

Similarly,

$$|\psi_{-A}\rangle \equiv (|-\rangle\langle -| \otimes I)|\psi\rangle = \frac{1}{\sqrt{2}}|--\rangle. \tag{6.29e}$$

The probability of obtaining the outcome $|+\rangle$ is $\||\psi_{+A}\rangle\|^2 = \frac{1}{2}$. Upon measuring and obtaining the outcome $|+\rangle$, the state of the first qubit collapses to $|+\rangle$. Simultaneously, the state of the second qubit also collapses to $|+\rangle$, resulting in the two-qubit system being in the product state $|++\rangle$. Likewise for the outcome $|-\rangle$.

The expected value of M when measured on the first qubit is:

$$\langle\psi|M \otimes I|\psi\rangle = \langle\psi|(|\psi_{+A}\rangle - |\psi_{-A}\rangle)\rangle = 0. \tag{6.30}$$

> **Exercise 6.5** Consider $M = a|+\rangle\langle +| + b|-\rangle\langle -|$, and $|\psi\rangle = \cos\alpha|00\rangle + \sin\alpha|11\rangle$, where $a, b, \alpha \in \mathbb{R}$.
>
> (a) Show that M is an observable.
>
> (b) Describe the measurement outcomes, associated probabilities, and post-measurement states of the second qubit when the first qubit is measured with M.
>
> (c) Compute the expected value of M: $\langle\psi|M|\psi\rangle$.

In quantum computing, measurements are typically conducted in the computational basis due to its native support on quantum hardware. To measure $|\psi\rangle$ in the $|+\rangle, |-\rangle$ basis, we first need to convert it to the computational basis. This can be achieved by applying Hadamard gates:

$$H|+\rangle = |0\rangle, \quad H|-\rangle = |1\rangle, \tag{6.31a}$$
$$H|0\rangle = |+\rangle, \quad H|1\rangle = |-\rangle. \tag{6.31b}$$

6.2 * Measurements of Two-Qubit Systems

After this transformation, we perform local measurements on the two qubits. For instance, if the measurement readouts are both 0, we deduce that the original state in the $|+\rangle, |-\rangle$ basis was $|++\rangle$. This logic can be similarly applied to other sets of readouts. The process is illustrated in Fig. 6.2.

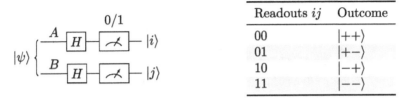

Figure 6.2: Local Measurement in $\{|+\rangle, |-\rangle\}$ Basis

4 * An Advanced Example

(i) This example features advanced mathematical concepts and techniques. While first-time learners may opt to skip it initially, revisiting this content is recommended for a deeper understanding of Bell inequalities discussed in subsequent sections.

Consider qubit A measured along a direction defined by the polar angle α and the azimuthal angle 0 on the Bloch sphere. Effectively, it is measured using the basis $\{|a\rangle, |a_\perp\rangle\}$:

$$|a\rangle = \cos\frac{\alpha}{2}|0\rangle + \sin\frac{\alpha}{2}|1\rangle, \qquad (6.32a)$$

$$|a_\perp\rangle = -\sin\frac{\alpha}{2}|0\rangle + \cos\frac{\alpha}{2}|1\rangle. \qquad (6.32b)$$

The associated observable is:

$$M_a = |a\rangle\langle a| - |a_\perp\rangle\langle a_\perp|. \qquad (6.33)$$

Similarly, qubit B is measured along a different direction defined by the polar angle β on the Bloch sphere. The measurement basis $\{|b\rangle, |b_\perp\rangle\}$ is:

$$|b\rangle = \cos\frac{\beta}{2}|0\rangle + \sin\frac{\beta}{2}|1\rangle, \qquad (6.34a)$$

$$|b_\perp\rangle = -\sin\frac{\beta}{2}|0\rangle + \cos\frac{\beta}{2}|1\rangle. \qquad (6.34b)$$

The associated observable is:

$$M_b = |b\rangle\langle b| - |b_\perp\rangle\langle b_\perp|. \qquad (6.35)$$

Let's consider the combined observable which represents the product (or correlation) of two local measurements:

$$M_{ab} = M_a \otimes M_b. \qquad (6.36)$$

To measure the state $|\psi\rangle = \frac{1}{\sqrt{2}}(|00\rangle + |11\rangle)$, we start with the expected values of M_a across the one-qubit basis states:

$$\langle 0| M_a |0\rangle = \cos\alpha, \tag{6.37a}$$
$$\langle 1| M_a |1\rangle = -\cos\alpha, \tag{6.37b}$$
$$\langle 1| M_a |0\rangle = \sin\alpha, \tag{6.37c}$$
$$\langle 0| M_a |1\rangle = \sin\alpha. \tag{6.37d}$$

Now calculate the expected values of M_{ab} across the two-qubit basis states:

$$\langle 11| (M_a \otimes M_b) |11\rangle = \langle 1| M_a |1\rangle \langle 1| M_b |1\rangle = \cos\alpha \cos\beta, \tag{6.38a}$$
$$\langle 00| (M_a \otimes M_b) |00\rangle = \langle 0| M_a |0\rangle \langle 0| M_b |0\rangle = \cos\alpha \cos\beta, \tag{6.38b}$$
$$\langle 11| (M_a \otimes M_b) |00\rangle = \langle 1| M_a |0\rangle \langle 1| M_b |0\rangle = \sin\alpha \sin\beta, \tag{6.38c}$$
$$\langle 00| (M_a \otimes M_b) |11\rangle = \langle 0| M_a |1\rangle \langle 0| M_b |1\rangle = \sin\alpha \sin\beta. \tag{6.38d}$$

Finally, we obtain:

$$\langle\psi|M_{ab}|\psi\rangle = \cos\alpha \cos\beta + \sin\alpha \sin\beta = \cos(\alpha - \beta). \tag{6.39}$$

6.2.2 ∗ Joint Measurements

In a joint measurement, the measurement operator simultaneously acts on all subsystems under consideration. For instance, in a two-qubit system, we may measure a property that involves both qubits, and the measurement operator is a 4×4 matrix. The probabilities and post-measurement states are calculated by applying the joint measurement operator to the entire composite system state.

1 Parity Measurement

Consider two qubits in a system. They can be in the states $|00\rangle$, $|01\rangle$, $|10\rangle$, or $|11\rangle$. Our task is to measure the system and distinguish $|00\rangle$ and $|11\rangle$ as a group, and $|01\rangle$ and $|10\rangle$ as another, while leaving the state of the system unchanged.

This objective can be achieved using a parity measurement, which determines whether the total count of $|1\rangle$ states is even or odd. To be specific, $|00\rangle$ and $|11\rangle$ have even parity since there are either 0 or 2 $|1\rangle$ states. Conversely, $|01\rangle$ and $|10\rangle$ have odd parity due to the presence of one $|1\rangle$ state.

The parity measurement operator, when applied to the combined state of two qubits, is:

$$P_Z = Z \otimes Z = |00\rangle\langle 00| + |11\rangle\langle 11| - |01\rangle\langle 01| - |10\rangle\langle 10|. \tag{6.40}$$

The states $|00\rangle$ and $|11\rangle$ are eigenvectors of P_Z with an eigenvalue of 1, whereas $|01\rangle$ and $|10\rangle$ are eigenvectors with an eigenvalue of -1. Thus, a measurement reading of 1 indicates even parity and -1 indicates odd parity. Since these states are eigenvectors of P_Z, the measurement leaves them unchanged.

Building on this, the mechanism can also be extended to superposition states. Consider the two superposition states:

$$|\psi_{\text{even}}\rangle = \alpha |00\rangle + \beta |11\rangle, \tag{6.41a}$$
$$|\psi_{\text{odd}}\rangle = \gamma |01\rangle + \delta |10\rangle. \tag{6.41b}$$

6.2 ∗ Measurements of Two-Qubit Systems

These states are also eigenvectors of P_Z:

$$P_Z |\psi_{\text{even}}\rangle = |\psi_{\text{even}}\rangle, \tag{6.42a}$$

$$P_Z |\psi_{\text{odd}}\rangle = -|\psi_{\text{odd}}\rangle. \tag{6.42b}$$

When measuring a general state of the form $|\psi\rangle = \sum_{i,j \in \{0,1\}} c_{ij} |ij\rangle$ using P_Z, the state collapses to either $|\psi_{\text{even}}\rangle$ or $|\psi_{\text{odd}}\rangle$, corresponding to measurement readings of 1 or -1, respectively. However, if the system is already in $|\psi_{\text{even}}\rangle$ or $|\psi_{\text{odd}}\rangle$, the P_Z measurement leaves the state undisturbed, highlighting the essence of parity measurement.

> Parity measurements can discern groups of qubit states because the measurement operator has degenerate eigenvalues, leading to corresponding subspaces in its eigenspace. Specifically, the degenerate eigenvalues of 1 and -1 correspond to two different subspaces: one spanned by the even parity states $|00\rangle$ and $|11\rangle$, and the other by the odd parity states $|01\rangle$ and $|10\rangle$. These subspaces are used to classify the qubits into groups without changing their states.

Analogous to P_Z, there is a phase-parity measurement, which measures in the $\{|+\rangle, |-\rangle\}$ basis and counts the number of $|+\rangle$ states. Its operator is represented as:

$$P_X = X \otimes X = |++\rangle\langle++| + |--\rangle\langle--| - |+-\rangle\langle+-| - |-+\rangle\langle-+|. \tag{6.43}$$

Parity measurements are crucial in numerous quantum algorithms and protocols. For instance, they are foundational in quantum error correction for detecting both bit and phase flip errors. Additionally, they are employed in certain quantum teleportation protocols, wherein the parity of two qubits informs the requisite correction operation for the teleported qubit.

To execute a parity measurement on an actual quantum device, one generally needs the capability to conduct controlled operations (such as a CNOT gate) to generate entanglement between the qubits. Subsequent individual qubit measurements then yield the parity of the initial state. This will be explored further in the context of error correction in § 12.4.

2 Bell Measurement

A Bell measurement, also known as a Bell state measurement or entanglement measurement, is a type of quantum measurement that is particularly important for quantum information processing tasks such as quantum teleportation and superdense coding. In essence, a Bell measurement is a joint measurement on a two-qubit system that distinguishes between the four mutually orthogonal entangled states, called Bell states. More specifically, a Bell measurement results in one of four possible outcomes and transforms the state of the two qubits into one of the four Bell states corresponding to the measurement outcome.

While a Bell measurement can be described theoretically, performing a Bell measurement on a real quantum computer is non-trivial. It requires a sequence of quantum gates that effectively transform the Bell basis into the computational basis. This typically involves a CNOT gate, with the first qubit as the control and the second as the target, followed by a Hadamard gate applied to the first qubit. Finally, individual measurements are performed on each qubit. The measurement results then allow us to determine which of the four Bell states the system was in. We will explore this further in Chapter 8 when we investigate entanglement and Bell states in detail.

> **Exercise 6.6** This is a key exercise that tests your basic understanding of quantum measurements. Ensure you can complete it independently.
>
> Consider a system of three qubits (A, B, and C), in the state
>
> $$|\psi\rangle = \sum_{i,j,k \in \{0,1\}} c_{ijk} |ijk\rangle, \qquad (6.44)$$
>
> where $|ijk\rangle$ represents the joint computational basis states of the qubits A, B, and C, and c_{ijk} are complex coefficients.
>
> (a) Calculate the probability of measuring the third qubit (C) in the state $|0\rangle$.
>
> (b) If qubit C is measured and collapses to $|0\rangle$, determine the resulting joint state of qubits A and B. Also, describe the state of the three-qubit system.
>
> (c) Assuming a Bell measurement is performed on qubits A and B and the outcome is $|\Phi^+\rangle = \frac{1}{\sqrt{2}}(|00\rangle + |11\rangle)$, find the probability of this outcome and describe the post-measurement state of the three-qubit system.

6.3 Multi-Qubit System States

Having introduced the fundamental concepts of composite quantum systems with two-qubit examples, we will now expand our discussion to encompass n-qubit systems, also referred to many-body systems in quantum mechanics. The dimension of the state space for such a system is $N = 2^n$, and correspondingly, an operator in this space would be an $N \times N$ matrix. This is a major factor contributing to the potential exponential computational capacity of quantum computers.

How large can n and N be in practical quantum computing? When $n = 300$, we find that $N = 2^{300}$, which is approximately 10^{90}. This figure surpasses the estimated number of atoms in our observable universe. As of 2024, quantum computers with up to $n = 1000$ qubits already exist. Theoretically, these computers possess immense computational capacity. However, in practice, they are plagued by errors and instability due to qubit decoherence. Facilitating error correction necessitates additional qubits. It is estimated that to have a quantum computer with 1000 error-free logical qubits, we would need approximately a million ($n = 10^6$) physical qubits.

> **The Role of Entanglement in Quantum Computing**
>
> In an n-qubit system, if all qubits are independent, their collective product state can be described using just $2n$ complex numbers. However, when the qubits are entangled, describing their state requires up to 2^n complex numbers. This significant increase in representational complexity is a key to why quantum computers can potentially achieve substantial speedups over classical computers.

> Thus, entanglement is not just a feature of quantum systems; it is a crucial prerequisite for the enhanced computational power of quantum computing.

6.3.1 The Composition Postulate of Quantum Mechanics

Quantum mechanics leverages state vectors and quantum operators to delineate the behavior of quantum systems and their interactions. The Composition Postulate in quantum mechanics stipulates that the collective state of two or more quantum systems is described by the tensor product of their individual states. More succinctly, the joint state space of two or more quantum systems is the tensor product of their individual state spaces.

> **Postulate 5: Composite Systems**
>
> When two or more quantum systems are combined, their joint state space is described by the tensor product of the state spaces of the constituent systems.

The tensor product provides a mechanism to combine two Hilbert spaces, $\mathcal{H}_A$ and $\mathcal{H}_B$, into a larger Hilbert space, denoted as $\mathcal{H}_{AB}$. As per the Composition Postulate, the joint state space of the combined system AB is described by the tensor product of the individual state spaces:

$$\mathcal{H}_{AB} = \mathcal{H}_A \otimes \mathcal{H}_B. \tag{6.45}$$

This implies that the basis states of $\mathcal{H}_{AB}$ are derived from the tensor product of the basis states of the individual spaces. Let's say the basis states for system A are $|a_i\rangle$ and for system B are $|b_j\rangle$. Consequently, the basis states for the combined system AB are:

$$|a_i b_j\rangle = |a_i\rangle \otimes |b_j\rangle. \tag{6.46}$$

If the dimension of $\mathcal{H}_A$ is m, and that of $\mathcal{H}_B$ is n, the resulting dimension of $\mathcal{H}_{AB}$ is mn.

6.3.2 Basis States

For an n-qubit system, the state space is described by the tensor product of the individual qubit state spaces. A single qubit is described by a two-dimensional complex Hilbert space. Therefore, an n-qubit system will be described by an N-dimensional complex Hilbert space, where $N = 2^n$.

The basis for an n-qubit system is a set of orthonormal states that span the Hilbert space of the composite quantum system. A standard choice for the basis of an n-qubit system is the computational basis (also known as the standard basis). The computational basis is formed by taking the tensor product of the individual qubit basis states, as follows:

For a single qubit, the basis states are $|0\rangle$ and $|1\rangle$. For n-qubits, take the tensor product of the individual qubit basis states to create a set of N orthonormal basis states.

1 Binary Representation

In the binary representation, each basis state corresponds to a binary string of length n, with each digit representing a qubit in the state $|0\rangle$ or $|1\rangle$. For example, for a two-qubit system, the basis states are $|00\rangle$, $|01\rangle$, $|10\rangle$, and $|11\rangle$. The computational basis states for an n-qubit system are represented by:

$$|0\cdots00\rangle, |0\cdots01\rangle, |0\cdots10\rangle, \cdots, |1\cdots11\rangle, \tag{6.47}$$

where each binary string corresponds to a unique basis state.

These binary strings can be thought of as names, or tags, of the basis states. The vector of each basis state is a column vector of dimension N, with a '1' at the position indexed by the corresponding binary number. For example,

$$|0\cdots00\rangle = \begin{bmatrix} 1 & 0 & \cdots & 0 & 0 \end{bmatrix}^T,$$
$$|1\cdots11\rangle = \begin{bmatrix} 0 & 0 & \cdots & 0 & 1 \end{bmatrix}^T.$$

These basis states are orthonormal, because each basis state has exactly one '1' in it, each at a different position.

A general n-qubit state can be written as:

$$|\psi\rangle = \sum_{x_1, x_2, \cdots, x_n \in \{0,1\}} c_{x_1 x_2 \cdots x_n} |x_1 x_2 \cdots x_n\rangle, \tag{6.48}$$

or in this compact form:

$$|\psi\rangle = \sum_{x \in \{0,1\}^n} c_x |x\rangle. \tag{6.49}$$

This binary representation is particularly useful when working with the tensor product structure of the n-qubit Hilbert space, as each binary string directly corresponds to the tensor product of individual qubit states.

2 Decimal Representation

There is also the decimal representation which uses decimal numbers ranging from 0 to $N-1$ (where $N = 2^n$) to label the basis states:

$$|0\rangle, |1\rangle, |2\rangle, \cdots, |N-1\rangle, \tag{6.50}$$

and a general n-qubit state can be written as

$$|\psi\rangle = \sum_{k=0}^{N-1} c_k |k\rangle. \tag{6.51}$$

In this representation, each basis state is labeled by a decimal number, which corresponds to the decimal equivalent of the binary string representation. The total number of basis states is still N. For example, for a 2-qubit system, the basis states are $|0\rangle$, $|1\rangle$, $|2\rangle$, and $|3\rangle$, which correspond to the binary notations $|00\rangle$, $|01\rangle$, $|10\rangle$, and $|11\rangle$, respectively.

6.3 Multi-Qubit System States

For those who are reading about this for the first time, it's important to note that the contents inside the $|\cdot\rangle$ are merely labels. In the computational basis, each basis state is a vector of length $N = 2^n$, having exactly one '1' in its structure, with the rest being zeros. See Eq. 6.10 for an example of two qubits.

This decimal representation is convenient when working with the Hilbert space as a whole, especially when dealing with quantum gates or other linear operators that act on the entire Hilbert space rather than individual qubits.

6.3.3 Product States, Entangled States, and Correlated States

We have already introduced the concepts of product and entangled states within the context of two-qubit systems. In the case of composite quantum systems comprising n-qubits, quantum states can be broadly classified into pure states and mixed states. Pure states include product states and entangled states. Mixed states can be correlated or uncorrelated.

1 Product States

A product state for an n-qubit composite quantum system is a state that can be expressed as the tensor product of the individual states of each qubit. Mathematically, an n-qubit product state can be written as:

$$|\psi_1 \psi_2 \cdots \psi_n\rangle = |\psi_1\rangle \otimes |\psi_2\rangle \otimes \cdots \otimes |\psi_n\rangle, \tag{6.52}$$

where $|\psi_i\rangle$ signifies the state of the i-th qubit. In this scenario, the qubits are not entangled, meaning their states can be described independently of each other. Product states are separable, indicating that they can be factored into individual qubit states.

There can also be product states where only some qubits are independent of the rest, such as in the case of $|\psi_1\rangle \otimes |\psi_2 \cdots \psi_n\rangle$.

Verifying Properties of Tensor Products

To confirm the mathematical consistency of product states and operations involving them, consider the following properties of tensor products:

Due to the following property of tensor product,

$$(A_1 \otimes A_2 \cdots \otimes A_n)(B_1 \otimes B_2 \cdots \otimes B_n) = (A_1 B_1) \otimes (A_2 B_2) \cdots \otimes (A_n B_n), \tag{6.53}$$

we know the composite state is normalized if the component states are, because

$$\langle \psi_1 \psi_2 \cdots \psi_n | \psi_1 \psi_2 \cdots \psi_n \rangle = \langle \psi_1 | \psi_1 \rangle \otimes \langle \psi_2 | \psi_2 \rangle \cdots \otimes \langle \psi_n | \psi_n \rangle = 1. \tag{6.54}$$

Furthermore, operators can be 'distributed' into the tensor products:

$$\langle A_1 \otimes A_2 \cdots \otimes A_n | \psi_1 \otimes \psi_2 \cdots \otimes \psi_n \rangle = (A_1 |\psi_1\rangle) \otimes (A_2 |\psi_2\rangle) \cdots \otimes (A_n |\psi_n\rangle). \tag{6.55}$$

2 Entangled States

Entangled states are characterized by the inability to describe the quantum state of one qubit independently of the other qubits in the system. Entangled states exhibit stronger correlations than classical correlations, reflecting their inherently quantum mechanical nature. These states cannot be expressed as a product of individual qubit states and display nonlocal correlations that are fundamental to quantum information processing, including quantum communication, quantum cryptography, and quantum computing.

A prominent example of an entangled state is the GHZ state (Greenberger-Horne-Zeilinger state):

$$|\text{GHZ}\rangle = \frac{1}{\sqrt{2}}(|00\cdots 0\rangle + |11\cdots 1\rangle). \tag{6.56}$$

This state is non-factorizable into a product of individual qubit states, and the measurement outcomes on one qubit demonstrate nonlocal correlations with the measurement outcomes on the other qubits.

As an aside, let's use the following "compact" expressions of the GHZ state to illustrate some interesting notations associated with the tensor product:

$$|\text{GHZ}\rangle = \frac{1}{\sqrt{2}}\left(\bigotimes_{i=1}^{n}|0\rangle + \bigotimes_{i=1}^{n}|1\rangle\right) \tag{6.57a}$$

$$= \frac{1}{\sqrt{2}}\left(|0\rangle^{\otimes n} + |1\rangle^{\otimes n}\right). \tag{6.57b}$$

These expressions illustrate the use of the tensor product symbol, $\otimes$, in the context of a state space composed of multiple qubits. The use of $\bigotimes_{i=1}^{n}$ and $\otimes^n$ effectively conveys the tensor product of n qubits, each in the state $|0\rangle$ or $|1\rangle$.

Exercise 6.7 Consider an n-qubit GHZ state given by Eq. 6.57, where $n > 2$. Determine the state of the rest of the system after the following measurements:

(a) The first qubit is measured with an outcome $|0\rangle$.

(b) The first qubit is measured with an outcome $|+\rangle$, where $|+\rangle = \frac{1}{\sqrt{2}}(|0\rangle + |1\rangle)$.

(c) The first two qubits are measured with an outcome corresponding to the projection onto the state $\frac{1}{\sqrt{2}}(|00\rangle + |11\rangle)$.

(d) The first two qubits are measured with an outcome corresponding to the projection onto the state $\frac{1}{\sqrt{2}}(|01\rangle + |10\rangle)$.

3 ✳ Correlated States

Correlated states describe composite system states wherein the properties of one qubit are statistically linked to the properties of other qubits in the system. These correlations can arise due to interactions between the qubits, shared environments, or common preparation procedures. Correlated states can be either entangled or classically correlated.

6.3 Multi-Qubit System States

Classically correlated states, while not entangled, still exhibit correlations between the qubits. These correlations are akin to classical correlations in statistical systems, wherein knowledge of one qubit provides information about other qubits, but each qubit can still be described by individual quantum states.

An example of a classically correlated n-qubit state is (expressed as a density matrix):

$$\rho_{1\cdots n} = p_1 \, |00\cdots 0\rangle\langle 0\cdots 0| + p_2 \, |11\cdots 1\rangle\langle 1\cdots 1|, \tag{6.58}$$

where p_1 and p_2 are probabilities that satisfy $p_1 + p_2 = 1$.

Correlated states are a special case of mixed states, which we will explore in § 12.2.

6.3.4 ∗ Identical Particles

While the discussion on fermions and bosons is not broadly applicable to general quantum computing, it does hold relevance in specialized scenarios such as quantum chemistry simulations. Furthermore, anyons directly link to emerging platforms in quantum computing. This section is thus intended to provide both context and an overview of particle statistics in quantum systems.

The Composition Postulate in quantum mechanics, as previously described, applies to distinguishable quantum systems. In these systems, like qubits, each unit can be uniquely identified, and the joint state space is the tensor product of the individual state spaces.

For unbound, indistinguishable particles, the joint state space is still constructed using the tensor product of the individual state spaces. However, the overall state (wavefunction) of the combined system must satisfy certain symmetry requirements, which depend on whether the particles are fermions or bosons.

Fermions and bosons are elementary particles that follow different statistical behaviors due to their intrinsic properties. Fermions, particles with half-integer spin such as electrons, protons, and neutrons, follow Fermi-Dirac statistics. Bosons, which have integer spin, like photons and certain atomic nuclei, follow Bose-Einstein statistics.

In contrast, qubits, even if they are based on half or integer spin particles, are localized in space, making them distinguishable from one another. Hence, their combined quantum states are formed according to the regular Composition Postulate, without needing to account for the symmetrization (for bosons) or antisymmetrization (for fermions) rules that apply to free, indistinguishable particles.

When combining indistinguishable quantum systems, the Composition Postulate is modified as follows:

- Fermions: Fermions adhere to the Pauli Exclusion Principle, which prevents two fermions from simultaneously occupying the same quantum state. Therefore, the combined state of multiple identical fermions must be antisymmetric under the exchange of any pair of fermions. This can be mathematically expressed using a Slater determinant or through second quantization formalism, which ensures that the overall wavefunction changes sign upon the exchange of any two fermions.

- Bosons: Bosons, unlike fermions, can simultaneously occupy the same quantum state. The combined state of multiple identical bosons must be symmetric under the exchange of any pair of bosons. This symmetry can be expressed using symmetrization operators or second quantization formalism, which guarantees that the overall wavefunction remains unchanged upon the exchange of any two bosons.

- Anyons: Anyons are a class of quasiparticles in two-dimensional systems that do not strictly follow Fermi-Dirac or Bose-Einstein statistics. Instead, their wave functions acquire a complex phase upon exchanging positions, providing a unique behavior that lies between fermionic and bosonic statistics. This property makes them particularly intriguing for quantum computing. Anyonic systems, such as those realized in certain quantum Hall systems, are actively researched as platforms for topological quantum computing. In this computational paradigm, the quantum information is stored in the topological features of the system, making it robust against local errors. The anyonic braiding operations, where anyons are moved around each other, can serve as fault-tolerant quantum gates, significantly contributing to the robustness and scalability of a quantum computer.

6.4 ∗Measurements of Multi-Qubit Systems

The principles developed through two-qubit systems extend to multi-qubit systems for measurements, as outlined in § 6.2. We now enrich this understanding with additional insights and generalized formulae.

6.4.1 ∗Measurements on a Single Qubit

Assume we have an n-qubit system in a general superposition state:

$$|\psi\rangle = \sum_{x_1,x_2,\cdots,x_n \in \{0,1\}} c_{x_1 x_2 \cdots x_n} |x_1 x_2 \cdots x_n\rangle. \qquad \text{(Copy of 6.48)}$$

If we measure the first qubit in the computational basis, the post-measurement state of the entire system will exist as a superposition of states where the first qubit is either in state $|0\rangle$ or in state $|1\rangle$, depending on the measurement outcome.

Assuming $|0\rangle$ is the measurement result, the associated probability is

$$P_0^{(1)} = \sum_{x_2,\cdots,x_n \in \{0,1\}} |c_{0 x_2 \cdots x_n}|^2. \qquad (6.59)$$

The post-measurement state of the entire system would comprise a superposition of all states with the first qubit in state $|0\rangle$, normalized by the probability of obtaining $|0\rangle$:

$$|\psi'\rangle = \frac{1}{\sqrt{P_0^{(1)}}} \sum_{x_2,\cdots,x_n \in \{0,1\}} c_{0 x_2 \cdots x_n} |0 x_2 \cdots x_n\rangle \qquad (6.60a)$$

$$= |0\rangle \otimes |\psi'_{n-1}\rangle. \qquad (6.60b)$$

6.4 ∗ Measurements of Multi-Qubit Systems

After the measurement, the first qubit becomes disentangled from the rest of the system. Here $|\psi'_{n-1}\rangle$ represents the post-measurement state of the remaining $n-1$ qubits:

$$|\psi'_{n-1}\rangle = \frac{1}{\sqrt{P_0^{(1)}}} \sum_{x_2,\cdots,x_n \in \{0,1\}} c_{0x_2\cdots x_n} |x_2 \cdots x_n\rangle. \tag{6.61}$$

Let's introduce the *partial product notation*:

$$\langle y^{(j)}|x_1 x_2 \cdots x_j \cdots x_n\rangle \equiv \langle y|x_j\rangle |x_1 x_2 \cdots x_{j-1} x_{j+1} \cdots x_n\rangle. \tag{6.62}$$

Note that in this notation, the bra $\langle y^{(j)}|$ does not match in length with the ket $|x_1 x_2 \cdots x_n\rangle$. Instead, the index j in $\langle y^{(j)}|$ indicates that it interacts only with the corresponding $|x_j\rangle$ through the inner product, while leaving the states of all other qubits unchanged. This concept can be generalized to situations where there are multiple y's.

Then the probability of measuring the first qubit with outcome $|0\rangle$ can be expressed in this shorthand form:

$$P_0^{(1)} = \left\| \langle 0^{(1)}|\psi\rangle \right\|^2. \tag{6.63}$$

And the post-measurement state $|\psi'_{n-1}\rangle$ is given by:

$$|\psi'_{n-1}\rangle = \frac{1}{\sqrt{P_0^{(1)}}} \langle 0^{(1)}|\psi\rangle. \tag{6.64}$$

An analogous treatment can be done for the case where we obtained $|1\rangle$ as the measurement result, but it won't be elaborated here.

> **Exercise 6.8** Consider the methodology outlined in Eqs. 6.60 to 6.64 for the case of measuring the first qubit and obtaining the outcome $|0\rangle$.
>
> Your task is to replicate this analysis, but now assume that it's the last qubit being measured, and the measurement outcome is $|1\rangle$.

6.4.2 ∗ Measurements on a Group of Qubits

To extend our previous discussion, where only the first qubit was measured, let's consider dividing the system into two sections, each referred to as a 'register': one with m qubits and the other with $k \equiv n - m$ qubits. If we perform a measurement on the m-qubit register, the post-measurement state of the entire system, depending on the measurement outcome, will be a superposition of states corresponding to that outcome, normalized appropriately. This process is often referred to as *partial measurement*.

For instance, the probability of obtaining a certain outcome $|a_1 a_2 \cdots a_m\rangle$ for measuring the first m qubits is given by, in the partial product notation:

$$P_m = \left\| \langle a_1^{(1)} a_2^{(2)} \cdots a_m^{(m)} | \psi \rangle \right\|^2 \tag{6.65a}$$

$$= \sum_{x_{m+1},\cdots,x_n \in \{0,1\}} |c_{a_1 a_2 \cdots a_m x_{m+1} \cdots x_n}|^2. \tag{6.65b}$$

And the post-measurement state of the remaining $n - m$ qubits is:

$$|\psi'_{n-m}\rangle = \frac{1}{\sqrt{P_m}} \langle a_1^{(1)} a_2^{(2)} \cdots a_m^{(m)} | \psi \rangle \quad (6.66a)$$

$$= \frac{1}{\sqrt{P_m}} \sum_{x_{m+1}, \cdots, x_n \in \{0,1\}} c_{a_1 a_2 \cdots a_m x_{m+1} \cdots x_n} |x_{m+1} \cdots x_n\rangle. \quad (6.66b)$$

The state $|\psi'_{n-m}\rangle$ resides within a subspace of the complete Hilbert space, spanned by the basis states $\{|x_{m+1} \cdots x_n\rangle\}$. The overall system state is given by:

$$|\psi'\rangle = |a_1 a_2 \cdots a_m\rangle \otimes |\psi'_{n-m}\rangle. \quad (6.67)$$

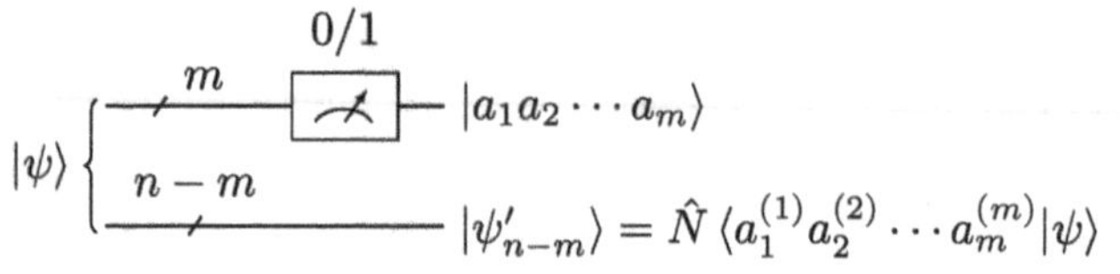

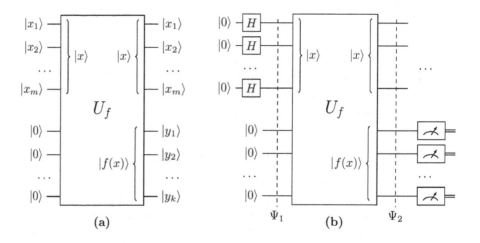

Notations: $a_1 a_2 \cdots a_m \in \{0,1\}$ represents a series of binary values, $\hat{N}$ denotes normalization, and $\langle a_1^{(1)} a_2^{(2)} \cdots a_m^{(m)} | \psi \rangle$ makes use of the partial product notation as defined in Eq. 6.62.

Figure 6.3: Measurement on a Group of Qubits

6.4.3 ✶ A Partial Measurement Paradigm for Quantum Algorithms

Figure 6.4: A Partial Measurement Paradigm for Quantum Algorithms

As an application of the partial measurement formulation discussed above, we examine a foundational model of measurement frequently employed in quantum algorithms.

1   Implementation of a Function

In Fig. 6.4(a), we show how a function $f(x)$ is commonly implemented as a unitary transformation U_f. The first m-qubit register, labeled $|x_1\rangle, |x_2\rangle, \ldots, |x_m\rangle$, encodes

6.4 ✳ Measurements of Multi-Qubit Systems

the input x. You can think of x as an m-bit string or a number made of the m binary digits.

The output of the function is the second register of k qubits, labeled $|y_1\rangle$, $|y_2\rangle$, ..., $|y_k\rangle$. Similarly, you can regard these as representing the binary digits of the numerical output of $f(x)$. The function can also be understood in vector form:

$$[y_1, y_2, \ldots, y_k]^T = f([x_1, x_2, \ldots, x_m]^T). \tag{6.68}$$

The function f can be a one-to-one or many-to-one function. Below are some examples:

1. IsPrime Function: The output is one bit. $y = 1$ if x is a prime number, and $y = 0$ otherwise.

2. Mod-N Function: The output needs at least $\log_2 N$ qubits, representing $y = x$ mod N. Here, N is an integer smaller than 2^k.

Note that the output is put into a separate second register, while the input is preserved in the first register. It is designed this way so that the transformation is U_f invertible (in fact, unitary), as required by quantum mechanics.

2 Uniform Superposition Input

Employing a superposition state as the input is a powerful tool of quantum computation, as demonstrated in Fig. 6.4(b). The Hadamard gate, denoted by H, transforms $|0\rangle$ into $|+\rangle = \frac{1}{\sqrt{2}}(|0\rangle + |1\rangle)$. This operation equips the input register at Ψ_1 with a uniform superposition of *all m-qubit basis states*, which is mathematically formulated as:

$$|\text{in}\rangle = |+\rangle^{\otimes m} = \sum_{x \in \{0,1\}^m} |x\rangle, \tag{6.69}$$

and the combined system state at the input is given by:

$$|\Psi_1\rangle = \sum_{x \in \{0,1\}^m} |x\rangle \otimes |0\rangle^{\otimes k}. \tag{6.70}$$

Due to the linearity of the transform U_f, the output register becomes:

$$|\text{out}\rangle = \sum_{x \in \{0,1\}^m} |f(x)\rangle, \tag{6.71}$$

and the combined system state at the output is given by:

$$|\Psi_2\rangle = \sum_{x \in \{0,1\}^m} |x\rangle \otimes |f(x)\rangle. \tag{6.72}$$

3 Measuring the Output

When we measure the output register, we obtain one of the possible values of $f(x)$, which we denote as $\tilde{y}$. Note that each measurement yields a $\tilde{y}$ randomly. According to Eq. 6.66, the system collapses into a state where the second register is $|\tilde{y}\rangle$. Simultaneously, the first register is projected onto a superposition of only those states $|x\rangle$ for which $f(x) = \tilde{y}$. This effectively singles out, within the first register, the superposition of all states corresponding to the pre-image of $\tilde{y}$ under f. This superposition, which mathematically represents the solution set to the equation $f(x) = \tilde{y}$, is denoted by $|\tilde{x}\rangle$:

$$|\tilde{x}\rangle = \sum_{\substack{x \in \{0,1\}^m \\ f(x)=\tilde{y}}} |x\rangle. \qquad (6.73)$$

Below are some examples:

1. IsPrime: If $f(x)$ is the IsPrime function given above, and the measurement outcome is 1, then $|\tilde{x}\rangle$ contains all the prime numbers less than 2^m. Conversely, if the measurement outcome is 0, $|\tilde{x}\rangle$ contains all the non-prime numbers less than 2^m.

2. Modular Exponentiation: $f(x) = p^x \mod N$, where N and p are coprime integers with $p < N$. This function is periodic with some period r, such that $f(x) = f(x + nr)$ for any integer n.

   Upon measuring the output and obtaining a value $\tilde{y}$, if $f(x_0) = \tilde{y}$ for some x_0, then $|\tilde{x}\rangle$ is a superposition of the form $|x_0\rangle + |x_0 + r\rangle + |x_0 + 2r\rangle + |x_0 + 3r\rangle + \cdots$, due to the periodicity of $f(x)$. Applying the quantum Fourier transform to $|\tilde{x}\rangle$ can reveal the period r. Discovering this period r enables an efficient method to factorize N, which is the foundational principle of Shor's algorithm.

This measurement paradigm is a key component in many quantum algorithms, including those of Simon and Shor, as well as in schemes for public-key quantum money (§ 11.3.2).

Exercise 6.9 Consider the Mod-N function with $N = 5$ and $m = 5$, i.e., $y = x \mod 5$ with $0 \leq x < 32$. Assuming you have already prepared the system in a state of uniform superposition over all possible x values, apply the partial measurement paradigm as detailed in § 6.4.3.

(a) What are the possible measurement outcomes?

(b) If the measurement outcome is $\tilde{y} = 1$, what is the set of values for $\tilde{x}$?

6.4.4 * Measurements in Alternate Bases

While most quantum computing devices natively support measurements in the computational basis, there are scenarios demanding measurements in other bases. In such cases, one would typically apply a basis rotation just before the measurement.

Consider an n-qubit state, $|\psi\rangle$. If we wish to measure it in an orthonormal basis, denoted by $\{|\phi_i\rangle\}$, the necessary unitary transformation U can be constructed as:

$$U = \sum_{i=0}^{2^n - 1} |i\rangle\langle\phi_i|, \qquad (6.74)$$

where $|i\rangle$ refers to an n-qubit computational basis state in decimal notation.

6.5 ∗ Time Evolution of Multi-Qubit States

As demonstrated in § 3.4.5, a measurement of $U|\psi\rangle$ in the computational basis is equivalent to measuring $|\psi\rangle$ in the basis $\{|\phi_i\rangle\}$. This concept is illustrated in Fig. 6.5.

The probabilities from the measurement form a distribution over the possible outcomes, which represents the likelihood of the state $|\psi\rangle$ being projected onto each of the basis states, $\{|\phi_i\rangle\}$:

$$P_i = |\langle\phi_i|\psi\rangle|^2. \tag{6.75}$$

An equivalent perspective is to consider the probabilities associated with $U|\psi\rangle$ with respect to the computational basis:

$$P_i = |\langle i|U|\psi\rangle|^2. \tag{6.76}$$

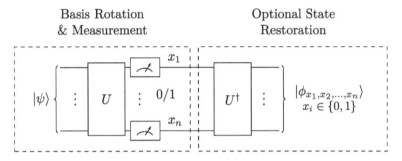

Figure 6.5: Multi-Qubit Measurement in Alternate Basis

Exercise 6.10 In many quantum machine learning algorithms, a specific circuit structure is observed. The depiction below showcases this typical circuit. Notably, the measurement is conducted against the $|00...0\rangle$ state.

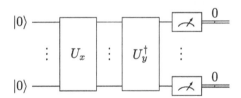

Consider the scenario where this measurement is performed repeatedly on the circuit. After accumulating a significant number of repetitions, one can derive a statistical average of the readouts.

Given this context, your tasks are as follows:

(a) Determine the expression for the measurement result, articulating it in terms of an inner product.

(b) Discuss the interpretation of the derived measurement result.

6.5 ∗Time Evolution of Multi-Qubit States

We have learned that Schrödinger equation governs the time evolution of the state of a quantum system, such as a qubit or an atom. In the context of multi-qubit systems, the Schrödinger equation governs the time evolution of the joint state vector of the system, given by Eq. 6.48.

6.5.1 *Two-Qubit System

The Hamiltonian of a system of two qubits, A and B, is composed of individual Hamiltonians for each qubit and a possible interaction term:

$$H_{AB} = H_A \otimes I_B + I_A \otimes H_B + H_{\text{int}}. \tag{6.77}$$

Here, H_A and H_B are the Hamiltonians of qubits A and B, and I_A and I_B are the identity operators in their respective Hilbert spaces. The term H_{int} takes into account of the interaction between the two qubits, which depends on the specific physical implementation of the qubits and the type of interaction between them.

The Schrödinger equation for the two-qubit system is given by:

$$i\hbar \frac{\partial}{\partial t} |\psi_{AB}(t)\rangle = H_{AB} |\psi_{AB}(t)\rangle. \tag{6.78}$$

This equation governs the time evolution of the joint state vector $|\psi_{AB}(t)\rangle$ in the presence of interactions between the two qubits.

The solution to this time-dependent Schrödinger equation can be expressed formally in terms of a time-ordered exponential:

$$|\psi_{AB}(t)\rangle = e^{-iH_{AB}t/\hbar} |\psi_{AB}(0)\rangle. \tag{6.79}$$

This is a direct application of the time-evolution operator. Note that this solution assumes that the Hamiltonian does not explicitly depend on time.

6.5.2 *Multi-Qubit System

For an n-qubit system, the Schrödinger equation and Hamiltonian generalize naturally. The Hamiltonian of an n-qubit system is composed of individual Hamiltonians for each qubit and all possible interaction terms:

$$H = \sum_{i=1}^{n} H_i \otimes I_{\neg i} + \sum_{i<j} H_{ij} \otimes I_{\neg ij} + \dots. \tag{6.80}$$

Here, H_i are the individual Hamiltonians associated with the i-th qubit. These Hamiltonians govern the local dynamics of each qubit, which could be due to an external field or some local potential. $I_{\neg i}$ denotes the identity operators on all other qubits except the i-th qubit. By using the identity operator on these other qubits, we ensure that the Hamiltonian H_i only acts on the i-th qubit and leaves the other qubits unaffected.

H_{ij} are the interaction Hamiltonians between the i-th and j-th qubits. These represent the influence of one qubit on another and could describe a variety of interactions, such as spin-spin interactions or couplings mediated by some external field. Again, $I_{\neg ij}$ are the identity operators on all other qubits except the i,j-th qubits, ensuring that the interaction Hamiltonian H_{ij} only affects the qubits it is meant to.

Note that this is a general form of a multi-qubit Hamiltonian and the exact form of H_i and H_{ij} will depend on the specific quantum system being described. For example, in the Ising model discussed below, the H_{ij} terms are given by $J_{ij}Z_iZ_j$.

The ellipsis at the end of the equation indicates that this structure can be extended to incorporate interactions among larger groups of qubits, should your specific system require it. However, in many practical cases, one often restricts to single-qubit and two-qubit interactions due to their physical realisability and computational manageability.

The Schrödinger equation for the n-qubit system becomes:

$$i\hbar \frac{\partial}{\partial t}|\psi(t)\rangle = H|\psi(t)\rangle. \tag{6.81}$$

If the Hamiltonian does not explicitly depend on time, the above equation has a solution given by:

$$|\psi(t)\rangle = U(t)|\psi(0)\rangle, \tag{6.82}$$

where $U(t)$ is the time-evolution operator:

$$U(t) = e^{-iHt/\hbar}. \tag{6.83}$$

6.5.3 ∗ Example: The Ising Model

The transverse field Ising model (TFIM), Ising model for short, is a well-known model in condensed matter physics that describes a system of spins under the influence of a transverse magnetic field. In the context of quantum computing, it serves as a canonical example of a multi-qubit system whose ground state can be efficiently prepared on a quantum computer, and therefore, it can be used for various quantum optimization and simulation algorithms.

The Hamiltonian of the TFIM is given by:

$$H = \sum_{i<j} J_{ij}Z_iZ_j - \Gamma \sum_i X_i. \tag{6.84}$$

There are two terms that make up the Hamiltonian of the transverse field Ising model:

(1) $\sum_{i<j} J_{ij}Z_iZ_j$: This is the Ising interaction term. Here the sum is taken over all pairs of neighboring spins i and j on a lattice. The $i < j$ in the sum ensures the pairs are not double counted. J_{ij} denotes the strength of interaction between the ith and j^{th} spins. Z_i and Z_j are the Pauli Z operators that operate on the spins at the ith and j^{th} locations, respectively. Thus, Z_iZ_j represents the operation performed on the spins at the ith and j^{th} locations, i.e., it operates on the state $|i\rangle|j\rangle$. See Fig. 6.6.

In the above notation, Z_iZ_j is a shorthand for $Z_i \otimes Z_j$, acting on spins i and j.

Similarly, X_i is a shorthand for $X_i \otimes I_{\neg i}$, where $I_{\neg i}$ denotes the identity operators on all other qubits except the i-th qubit.

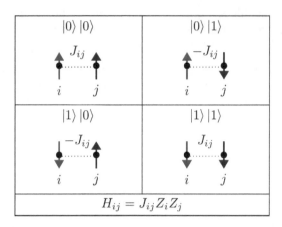

Figure 6.6: Illustration of Ising Interaction

(2) $-\Gamma \sum_i X_i$: This is the transverse field term, where the sum runs over all spins i on the lattice. The parameter Γ denotes the strength of the transverse magnetic field. The Pauli X operator, represented by X_i, operates on the spin at the ith location and flips it. The negative sign indicates that the transverse field tries to align the spins along the x-axis, opposing the Ising interaction.

The two terms do not commute, making the exact time evolution of this system non-trivial. However, using methods such as Trotterization, one can efficiently simulate the time evolution of the TFIM on a quantum computer.

The Ising model provides a bridge between physical systems and computational problems, allowing the use of quantum devices to solve problems of practical interest. We will explore its applications in § 11.2.

6.5.4 ✳ Example: CNOT Gate Implementation

The CNOT gate is a two-qubit operation. In many high-level descriptions, the intricacies of this operation are simplified using quantum gates depicted by their unitary transformations. However, for professionals engaged in quantum hardware implementation, comprehending the foundational concepts dictated by the Schrödinger equation remains essential.

The formal definition of the CNOT operator is provided in § 7.1.1:

$$\text{CNOT} = \begin{bmatrix} I & 0 \\ 0 & X \end{bmatrix} \tag{6.85a}$$

$$= \frac{I+Z}{2} \otimes I + \frac{I-Z}{2} \otimes X \tag{6.85b}$$

$$= \frac{1}{2}(I \otimes I + Z \otimes I + I \otimes X - Z \otimes X). \tag{6.85c}$$

In the context of Hamiltonian evolution, the key element in the CNOT operator is the term $\frac{1}{2}(Z \otimes X)$, symbolizing the interaction between the two qubits.

The typical Hamiltonian, instrumental in realizing the CNOT gate, is represented as:

$$H_{\text{CNOT}} = -\frac{\pi\hbar}{4\Delta t}(Z \otimes X). \tag{6.86}$$

Utilizing this Hamiltonian, we can substantiate that during a time evolution spanning Δt and accompanied by singular qubit rotations, the two-qubit system undergoes a transformation analogous to the CNOT gate action.

To validate this, let us revisit a generic matrix property: for any normal matrix A with $A^2 = I$, it holds that $e^{i\theta A} = \cos\theta I + i\sin\theta A$. In the context of our scenario, $(Z \otimes X)^2 = I \otimes I$, where $I \otimes I$ is the 4×4 identity matrix. Thus, we can infer:

$$U(\Delta t) = e^{-iH_{\text{CNOT}}\Delta t/\hbar} \tag{6.87a}$$

$$= e^{i\frac{\pi}{4}(Z \otimes X)} \tag{6.87b}$$

$$= \cos\frac{\pi}{4}(I \otimes I) + i\sin\frac{\pi}{4}(Z \otimes X) \tag{6.87c}$$

$$= \frac{1}{\sqrt{2}}\begin{bmatrix} 1 & i & 0 & 0 \\ i & 1 & 0 & 0 \\ 0 & 0 & 1 & -i \\ 0 & 0 & -i & 1 \end{bmatrix}. \tag{6.87d}$$

While the above matrix is not a precise replica of the CNOT operator, aligning it with the CNOT gate becomes feasible upon introducing a Z-rotation for the first qubit and an X-rotation for the second qubit:

$$\text{CNOT} = e^{i\frac{\pi}{4}}U(\Delta t)\left(R_z(\frac{\pi}{2}) \otimes R_x(\frac{\pi}{2})\right), \tag{6.88}$$

where the rotation gates R_x and R_z are elaborated in § 5.2.5.

> **Exercise 6.11** Validate Eq. 6.88.
>
> Hint: $U(\Delta t) = \begin{bmatrix} R_x(-\frac{\pi}{2}) & 0 \\ 0 & R_x(\frac{\pi}{2}) \end{bmatrix}$, $\quad \text{CNOT} = U(\Delta t)\begin{bmatrix} R_x(\frac{\pi}{2}) & 0 \\ 0 & iR_x(\frac{\pi}{2}) \end{bmatrix}$.

> **Exercise 6.12** Illustrate that, if H_{CNOT} is directly represented as $\frac{\pi\hbar}{2\Delta t}\text{CNOT}$ (often deemed impractical), then the resulting $U(\Delta t) = -i\,\text{CNOT}$.
>
> Hint: $\text{CNOT}^2 = I$.

6.6 Summary and Conclusions

Conceptual Progression

The exploration of multi-qubit systems in this chapter serves as a cornerstone for more advanced topics in quantum computation. As we transitioned from the familiar realm of classical Newtonian mechanics with its additive principles, we dove into the quantum domain governed by the tensor product principle. This principle emerges as the defining aspect of quantum mechanics that distinguishes it from classical mechanics.

Our journey began with an examination of the two-qubit system, gradually expanding our scope to systems of n qubits. By doing so, we highlighted the exponential complexity in terms of state space and computational capabilities. While practical applications remain limited due to challenges such as qubit decoherence, the theoretical underpinnings suggest immense potential, especially when error correction methods mature in the future.

Measurements in Multi-Qubit Systems

An integral part of this chapter was dedicated to understanding measurements in multi-qubit systems. We revisited the quantum measurement framework, extending its application to composite quantum systems, especially those with multiple qubits. The distinction between local measurements on subsystems and joint measurements was elaborated upon. Emphasis was placed on the tensor product structure of state space and the associated correlations between subsystems.

Time Evolution and Quantum Interactions

By invoking the Schrödinger equation, we shed light on the time evolution of multi-qubit systems. Exploring both two-qubit and n-qubit systems, we grasped the intricacies of their evolution. Practical applications and examples, such as the Ising model and the CNOT gate implementation, further illustrated these abstract principles.

Practical Implications and Challenges

While the theoretical implications of multi-qubit systems are profound, practical challenges such as qubit decoherence pose hurdles. This interplay between theory and practicality will be significant as quantum computing progresses, with error correction emerging as a fundamental area of research.

Upcoming Topics

As we venture into the subsequent chapters, our focus will shift to multi-qubit quantum gates. These gates, vital for quantum computing, operate on multiple qubits simultaneously and embody properties distinct from single-qubit gates. We will explore commonly used multi-qubit gates, universal gate sets, gate sequences, and introduce new tools like the Boolean representation of quantum gates and circuits. This transition will lay a firm foundation for advanced topics in quantum computing and information processing.

Problem Set 6

6.1 Consider $|\psi\rangle = \frac{1}{\sqrt{2}}(|00\rangle + |11\rangle)$, $|\alpha\rangle = \cos\alpha |0\rangle + \sin\alpha |1\rangle$, and $|\beta\rangle = \cos\beta |0\rangle + \sin\beta |1\rangle$, where $\alpha, \beta \in \mathbb{R}$.

Compute the probability $P(\alpha, \beta) = |\langle\alpha| \langle\beta| |\psi\rangle |^2$.

6.2 Determine if the state $c(|00\rangle - 2i|01\rangle + 2i|10\rangle + 4|11\rangle)$ is a product state. If it is, identify the two component states.

6.3 (a) Define $\Pi_a = \frac{1}{2}(I - X)$ and $\Pi_b = \frac{1}{2}(I - Z)$. Show that Π_a and Π_b are projection operators, i.e., $\Pi^2 = \Pi$. Here, X and Z are Pauli operators.

(b) Given $|\psi\rangle = \frac{1}{\sqrt{2}}(|01\rangle - |10\rangle)$, calculate $\langle\psi|\Pi_a \otimes \Pi_b|\psi\rangle$.

6.4 Find a 4×4 Hamiltonian H such that e^{-iH} equals the CNOT matrix.

6.5 Consider a general n-qubit system in a superposition state as defined in Eq. 6.48. We will conduct a measurement on the first two qubits and obtain the outcome corresponding to one of the Bell states, specifically $|\Psi^-\rangle = \frac{1}{\sqrt{2}}(|01\rangle - |10\rangle)$.

Your task is to construct equations that determine the probability of this measurement outcome and the resulting post-measurement state of the system. You should follow the process outlined in Eqs. 6.66 and 6.67.

6.6 Consider a function $y = f(x)$ with a pairing property that $f(x) = f(x \oplus s)$ for any x, where $\oplus$ denotes the bitwise XOR operation, and x, y, and s are all m-bit binary strings. This function is a component of Simon's algorithm.

Assuming you have already prepared the system in a state of uniform superposition over all possible x values, apply the partial measurement paradigm as detailed in § 6.4.3. Describe the resulting state $|\tilde{x}\rangle$ upon measurement of the system. (The Simon's algorithm hinges on getting information of s from $|\tilde{x}\rangle$ using a subsequent Hadamard transform.)

6.7 The following quantum circuit is run repeatedly for N times. For each run, each of the n qubits is measured in the computational basis; the k-th run yields n measurement values $z_1^{(k)}, z_2^{(k)}, \ldots, z_n^{(k)}$, where $z_i^{(k)} = 1$ for measurement outcome $|0\rangle$, and $z_i^{(k)} = -1$ for $|1\rangle$. Essentially, we are measuring each qubit with the Z observable.

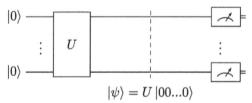

$|\psi\rangle = U|00\ldots 0\rangle$

From the measurement results, do the following:

(a) Calculate the empirical probability of measuring the i-th qubit and obtaining $|0\rangle$. Estimate the sampling error.

(b) Calculate the empirical probability of measuring all n qubits and obtaining $|00\ldots 0\rangle$. Estimate the sampling error.

(c) Calculate the empirical expectation value $\langle Z_i \rangle$ for the i-th qubit. Estimate the sampling error.

(d) Calculate the empirical expectation value $\langle Z_i \otimes Z_j \rangle$ for the i-th and j-th qubits. Estimate the sampling error.

(e) Can you obtain the empirical expectation values $\langle X_i \rangle$ and $\langle Z_i \otimes X_j \rangle$? Explain. If not, how would you revise the circuit to obtain these values?

6.8 The single-qubit Hadamard transform is the unitary transformation defined by:

$$H|0\rangle = \frac{1}{\sqrt{2}}(|0\rangle + |1\rangle), \quad H|1\rangle = \frac{1}{\sqrt{2}}(|0\rangle - |1\rangle).$$

The n-qubit Hadamard transform is defined as the tensor product of n single-qubit Hadamard transforms:

$$H_n = H^{\otimes n} = H \otimes H \otimes \cdots \otimes H \quad (n \text{ times}).$$

(a) Write out the matrix form of H_3.

(b) Prove that H_n is a unitary transformation.

(c) Calculate the state vector $H_n |00\cdots 0\rangle$.

(d) Calculate the state vector $H_n |11\cdots 1\rangle$.

(e) *Derive a general formula for $H_n |x_1 x_2 \cdots x_n\rangle$.

(f) *Derive a general formula for $H_n \sum_{x_1, x_2, \cdots, x_n \in \{0,1\}} c_{x_1 x_2 \cdots x_n} |x_1 x_2 \cdots x_n\rangle$.

6.9 The n-qubit XOR operation is a transformation often used in quantum algorithms such as Simon's algorithm. Given a binary string $s = a_1 a_2 \cdots a_n$ (for example $0101\cdots$), the XOR transform is defined as follows:

$$\text{XOR}_n |x_1 x_2 \cdots x_n\rangle = |y_1 y_2 \cdots y_n\rangle,$$

where $y_j = x_j \oplus a_j$, for all j from 1 to n.

(a) Prove that XOR_n is a unitary transformation.

(b) Calculate the state vector $\text{XOR}_n |00\cdots 0\rangle$.

(c) Calculate the state vector $\text{XOR}_n |11\cdots 1\rangle$.

(d) *Derive a general formula for $\text{XOR}_n |x_1 x_2 \cdots x_n\rangle$.

(e) *Derive a general formula for
$\text{XOR}_n \sum_{x_1, x_2, \cdots, x_n \in \{0,1\}} c_{x_1 x_2 \cdots x_n} |x_1 x_2 \cdots x_n\rangle$.

(f) *Write out the matrix form of XOR_n.

6.10 The N-dimensional quantum Fourier tranform (QFT) is given by

$$\text{QFT} |x\rangle = |\tilde{x}\rangle = \frac{1}{\sqrt{N}} \sum_{k=0}^{N-1} \omega_N^{kx} |k\rangle, \quad \text{where } x = 0, 1, 2, \ldots, N-1. \quad (6.89)$$

It can be viewed as a change from the computational basis $\{|x\rangle\}$ to the Fourier basis $\{|\tilde{x}\rangle\}$. Here $\omega_N = e^{\frac{2\pi i}{N}}$, $i = \sqrt{-1}$, and $\{\omega_N^k\}$ are the N-th complex roots of 1.

(a) Find the basis transformation matrix QFT for $N = 4$.

(b) Find the matrix form of QFT and QFT† for any N.

(c) *Calculate the representation of $|\psi\rangle = \sum_k c_k |k\rangle$ in the Fourier basis.

(d) *Compute $\langle A \rangle = \langle \psi | A | \psi \rangle$ in both the computational and Fourier bases and demonstrate they are equal, for $N = 4$ and $A = \sum_{k=0}^{3} (k-2) |k\rangle\langle k|$.

7. Multi-Qubit Quantum Gates

Contents

7.1	**Common Multi-Qubit Gates**	**176**
7.1.1	CNOT Gate	178
7.1.2	Flipped CNOT Gate	179
7.1.3	✶ General Controlled-U Gate	180
7.1.4	Classically Controlled-U Gate	182
7.1.5	✶ Two-Qubit Rotation Gates	183
7.1.6	✶ Parallel Gates	184
7.2	**Universal Sets of Qubit Gates**	**185**
7.2.1	CNOT + Arbitrary Single-Qubit Rotations	185
7.2.2	Clifford Group + T	186
7.2.3	Toffoli + H	186
7.3	**Boolean Representation of Quantum Gates**	**187**
7.3.1	Properties of XOR Operations	187
7.3.2	Boolean Representation of Common Quantum Gates	188
7.3.3	Understanding Boolean Representation	189
7.4	**Equivalent Gate Sequences**	**191**
7.4.1	Order Independence in Separate Qubit Operations	192
7.4.2	Gate Sequence: A First Example	192
7.4.3	Sequences Involving X, Z, and CNOT	194
7.4.4	Sequences Involving H	195
7.4.5	✶ Deferred Measurement Principle	197
7.4.6	✶ Operator Sum Relations	199
7.5	**✶ Exploratory Topics**	**200**
7.5.1	✶ Expressing Common Gates with the Toffoli Gate	200
7.5.2	✶ Advanced Quantum Gate Examples	201
7.6	**Summary and Conclusions**	**202**
	Problem Set 7	**203**

Multi-qubit gates serve a critical function in quantum computing, primarily due to their ability to create and manipulate entangled states. These states, which are essential for executing universal quantum algorithms, cannot be generated using single-qubit gates alone. Unlike single-qubit gates that apply unitary transformations to individual qubits, multi-qubit gates extend their action to multiple qubits simultaneously. They act on an n-qubit state space, making their corresponding matrix representation a $2^n \times 2^n$ matrix, where n denotes the number of qubits involved.

Despite the unique and critical role they play, multi-qubit gates maintain the key property of being unitary operators. They preserve the inner product between state vectors, ensuring that the laws of quantum mechanics, particularly the conservation of probability, remain unbroken after their application. This unitary nature makes them consistent with the overall framework of quantum mechanics and quantum computation.

This chapter starts by examining commonly used multi-qubit gates in quantum computing and universal gate sets. Building on these foundational concepts, we delve into the study of gate sequences, including a rigorous exploration of sequences that produce equivalent transformations on qubits. To aid this exploration, we introduce a novel tool: the Boolean representation of quantum gates and circuits. These topics are fundamental to the field of quantum computing and information processing and set the stage for further inquiry.

7.1 Common Multi-Qubit Gates

We will begin with an in-depth exploration of multi-qubit gates commonly used in quantum computing. Subsequently, we will delve into related topics such as parallel quantum gates and classically controlled gates.

Multi-qubit gates often encountered in quantum computing include the Controlled NOT (CNOT), Controlled Z (CZ), SWAP, Toffoli (also known as CCNOT), Fredkin (also known as CSWAP) gates, the general Controlled-U (CU) gate, and the two-qubit rotation gates. In the CU gate, U represents any single-qubit unitary transformation. These gates play essential roles in quantum circuits and quantum algorithms, enabling the creation of entangled states and execution of conditional operations based on the state of one or more control qubits.

 The highlighted terms in the operators in Table 7.1 denote the changes in the basis mappings induced by the gates.

The matrix representation of the gates in Table 7.1 is provided in Eq. 7.1.

$$\text{CNOT} = \begin{bmatrix} 1 & 0 & 0 & 0 \\ 0 & 1 & 0 & 0 \\ 0 & 0 & 0 & 1 \\ 0 & 0 & 1 & 0 \end{bmatrix} = \begin{bmatrix} I & 0 \\ 0 & X \end{bmatrix}, \quad (7.1a)$$

$$\text{CZ} = \begin{bmatrix} 1 & 0 & 0 & 0 \\ 0 & 1 & 0 & 0 \\ 0 & 0 & 1 & 0 \\ 0 & 0 & 0 & -1 \end{bmatrix} = \begin{bmatrix} I & 0 \\ 0 & Z \end{bmatrix}, \quad (7.1b)$$

7.1 Common Multi-Qubit Gates

Gate	Symbol	Operator	Description
CNOT (CX)		$\|00\rangle\langle00\| + \|01\rangle\langle01\|$ $+ \|10\rangle\langle11\| + \|11\rangle\langle10\|$	Applies X to the target qubit if the control qubit is $\|1\rangle$.
CZ (ZZ)		$\|00\rangle\langle00\| + \|01\rangle\langle01\|$ $+ \|10\rangle\langle10\| - \|11\rangle\langle11\|$	Applies Z to the target qubit if the control qubit is $\|1\rangle$.
SWAP		$\|00\rangle\langle00\| + \|01\rangle\langle10\|$ $+ \|10\rangle\langle01\| + \|11\rangle\langle11\|$	Exchanges the states of the two qubits it acts upon.
CCNOT (Toffoli)		$\|000\rangle\langle000\| + \|001\rangle\langle001\|$ $+ \|010\rangle\langle010\| + \|011\rangle\langle011\|$ $+ \|100\rangle\langle100\| + \|101\rangle\langle101\|$ $+ \|110\rangle\langle111\| + \|111\rangle\langle110\|$	Applies X to the target qubit if both control qubits are $\|1\rangle$.
CSWAP (Fredkin)		$\|000\rangle\langle000\| + \|001\rangle\langle001\|$ $+ \|010\rangle\langle010\| + \|011\rangle\langle011\|$ $+ \|100\rangle\langle100\| + \|101\rangle\langle101\|$ $+ \|110\rangle\langle110\| + \|111\rangle\langle111\|$	Swaps the states of the second and third qubits if the control qubit is $\|1\rangle$.
Ctrl-U (CU)		$\|00\rangle\langle00\| + \|01\rangle\langle01\|$ $+ u_{00}\|10\rangle\langle10\| + u_{01}\|10\rangle\langle11\|$ $+ u_{10}\|11\rangle\langle10\| + u_{11}\|11\rangle\langle11\|$	Applies U to the target qubit if the control qubit is $\|1\rangle$.
$R_{zz}(\theta)$	$R_{zz}(\theta)$		Two-qubit rotation gates. Also, R_{zx}, R_{xy}, etc.

Note: The CZ gate exhibits a unique property: it does not matter whether qubit 1 or 2 serves as the control qubit (see Exercise 7.4). This symmetry in its behavior leads to its alternative name, the ZZ gate, which is represented by the symbol shown to the right.

Table 7.1: Common Multi-Qubit Quantum Gates

$$CU = \begin{bmatrix} 1 & 0 & 0 & 0 \\ 0 & 1 & 0 & 0 \\ 0 & 0 & u_{00} & u_{01} \\ 0 & 0 & u_{10} & u_{11} \end{bmatrix} = \begin{bmatrix} I & 0 \\ 0 & U \end{bmatrix}, \qquad (7.1c)$$

$$SWAP = \begin{bmatrix} 1 & 0 & 0 & 0 \\ 0 & 0 & 1 & 0 \\ 0 & 1 & 0 & 0 \\ 0 & 0 & 0 & 1 \end{bmatrix} = \frac{1}{2}\begin{bmatrix} I+Z & X-iY \\ X+iY & I-Z \end{bmatrix}, \qquad (7.1d)$$

$$CCNOT = \begin{bmatrix} 8 \times 8 \\ \text{matrix} \end{bmatrix} = \begin{bmatrix} I_6 & 0 \\ 0 & X \end{bmatrix}, \qquad (7.1e)$$

$$CSWAP = \begin{bmatrix} 8 \times 8 \\ \text{matrix} \end{bmatrix} = \begin{bmatrix} I_4 & 0 \\ 0 & SWAP \end{bmatrix}. \qquad (7.1f)$$

Exercise 7.1 Verify that the operator matrices in Eq. 7.1 are unitary.

Exercise 7.2 Investigate why Controlled-H is not a commonly used quantum gate.

7.1.1 CNOT Gate

Let's use the CNOT gate, which is the most useful multi-qubit gate, to examine how its operator and matrix work. For the CNOT gate shown in Table 7.1, qubit 1 is the control qubit, and qubit 2 is the target qubit. When the control qubit is $|1\rangle$, an X (i.e., NOT) gate is applied to the target qubit, flipping its 0 and 1.

For example, the basis state $|10\rangle$ is transformed to $|11\rangle$. This corresponds to the term $|11\rangle\langle 10|$ in the operator of CNOT. For the CNOT matrix (see Eq. 7.2), we label the rows with $|00\rangle$, $|01\rangle$, $|10\rangle$, and $|11\rangle$, and the columns with $\langle 00|$, $\langle 01|$, $\langle 10|$, and $\langle 11|$. Then $|11\rangle\langle 10|$ corresponds to the 1 at the row $|11\rangle$ and column $\langle 10|$. This is summarized as the third row in the following Table:

Basis Mapping	Projection Operator	Matrix Element
$\|00\rangle \mapsto \|00\rangle$	$\|00\rangle\langle 00\|$	1 at row $\|00\rangle$, col $\langle 00\|$
$\|01\rangle \mapsto \|01\rangle$	$\|01\rangle\langle 01\|$	1 at row $\|01\rangle$, col $\langle 01\|$
$\|10\rangle \mapsto \|11\rangle$	$\|11\rangle\langle 10\|$	1 at row $\|11\rangle$, col $\langle 10\|$
$\|11\rangle \mapsto \|10\rangle$	$\|10\rangle\langle 11\|$	1 at row $\|10\rangle$, col $\langle 11\|$

$$\text{CNOT} = \begin{array}{c} \\ |00\rangle \\ |01\rangle \\ |10\rangle \\ |11\rangle \end{array} \begin{array}{c} \begin{array}{cccc} \langle 00| & \langle 01| & \langle 10| & \langle 11| \end{array} \\ \left[\begin{array}{cccc} 1 & 0 & 0 & 0 \\ 0 & 1 & 0 & 0 \\ 0 & 0 & 0 & 1 \\ 0 & 0 & 1 & 0 \end{array} \right] \end{array}. \quad (7.2)$$

The CNOT gate is idempotent, i.e., $\text{CNOT}^2 = I_4$, where $I_4 = I \otimes I$ is the 4×4 identity matrix. This means applying the CNOT gate twice in succession results in no change in the system.

A distinctive feature of quantum gates compared to classical logic gates is that quantum gates can work with superposition states. Let's examine a general two-qubit state:

$$|\psi\rangle = \alpha |00\rangle + \beta |01\rangle + \gamma |10\rangle + \delta |11\rangle. \quad (7.3)$$

The term $|11\rangle\langle 10|$ in the operator of CNOT picks out $\gamma |10\rangle$ from $|\psi\rangle$ and transforms it to $\gamma |11\rangle$. Similarly, for other terms in the operator. Thus

$$\text{CNOT} |\psi\rangle = \alpha |00\rangle + \beta |01\rangle + \gamma |11\rangle + \delta |10\rangle. \quad (7.4)$$

The corresponding matrix operation is:

$$\begin{bmatrix} 1 & 0 & 0 & 0 \\ 0 & 1 & 0 & 0 \\ 0 & 0 & 0 & 1 \\ 0 & 0 & 1 & 0 \end{bmatrix} \begin{bmatrix} \alpha \\ \beta \\ \gamma \\ \delta \end{bmatrix} = \begin{bmatrix} \alpha \\ \beta \\ \delta \\ \gamma \end{bmatrix}. \quad (7.5)$$

7.1 Common Multi-Qubit Gates

 The term "control qubit" in the context of the CNOT gate might suggest that the state of the control qubit remains invariant. However, this is not always true, especially when the control qubit is in a superposition. Consider the $\{|+\rangle, |-\rangle\}$ basis. When applying the CNOT gate to the basis states, we observe:

$$\text{CNOT} |++\rangle = |++\rangle, \tag{7.6a}$$
$$\text{CNOT} |+-\rangle = |--\rangle, \tag{7.6b}$$
$$\text{CNOT} |-+\rangle = |-+\rangle, \tag{7.6c}$$
$$\text{CNOT} |--\rangle = |+-\rangle. \tag{7.6d}$$

It is evident from these results that, in the $\{|+\rangle, |-\rangle\}$ basis, the state of the "target qubit" remains unchanged, while the "control qubit" can change based on the operation. Therefore, *multi-qubit gates should always be regarded as operating on the all the qubits as an integrated system, rather than affecting them independently.*

Exercise 7.3 Verify Eq. 7.6.

7.1.2 Flipped CNOT Gate

Let's consider the CNOT gate with qubit 2 as control qubit and qubit 1 target qubit. We name it CNOT' or CX'. Now when the second qubit is $|1\rangle$, an X (i.e., NOT) gate is applied to the first qubit, flipping its 0 and 1.

Symbol, operator, and description:

Gate	Symbol	Operator	Description
CNOT' (CX')		$\|00\rangle\langle 00\| + \|11\rangle\langle 01\|$ $+ \|10\rangle\langle 10\| + \|01\rangle\langle 11\|$	Applies X to the first qubit if the second qubit is $\|1\rangle$.

Basis mapping:

Basis Mapping	Projection Operator	Matrix Element
$\|00\rangle \mapsto \|00\rangle$	$\|00\rangle\langle 00\|$	1 at row $\|00\rangle$, col $\langle 00\|$
$\|01\rangle \mapsto \|11\rangle$	$\|11\rangle\langle 01\|$	1 at row $\|11\rangle$, col $\langle 01\|$
$\|10\rangle \mapsto \|10\rangle$	$\|10\rangle\langle 10\|$	1 at row $\|10\rangle$, col $\langle 10\|$
$\|11\rangle \mapsto \|01\rangle$	$\|01\rangle\langle 11\|$	1 at row $\|01\rangle$, col $\langle 11\|$

Matrix:

$$\text{CNOT}' = \begin{array}{c} \\ |00\rangle \\ |01\rangle \\ |10\rangle \\ |11\rangle \end{array} \begin{array}{c} \begin{array}{cccc} \langle 00| & \langle 01| & \langle 10| & \langle 11| \end{array} \\ \left[\begin{array}{cccc} 1 & 0 & 0 & 0 \\ 0 & 0 & 0 & 1 \\ 0 & 0 & 1 & 0 \\ 0 & 1 & 0 & 0 \end{array} \right] \end{array}. \quad (7.7)$$

The CNOT′ matrix can be written in 2×2 block form as:

$$\text{CNOT}' = \frac{1}{2} \begin{bmatrix} I+Z & I-Z \\ I-Z & I+Z \end{bmatrix}. \quad (7.8)$$

Exercise 7.4 Demonstrate that, in contrast to the fact that CNOT is not equal to CNOT′, the gate CZ is equal to CZ′. Here, CZ′ represents the CZ gate with qubit 2 serving as the control qubit and qubit 1 as the target qubit. In fact, because of this property, the CZ gate is also referred as the ZZ gate.

7.1.3 ∗ General Controlled-U Gate

A Controlled-U (or CU) gate applies a generic unitary gate U to the target qubit if the control qubit is $|1\rangle$. The CNOT, CZ, CCNOT, and CSWAP gates listed in Table 7.1 are special cases of Controlled-U gates. The target and control can also be more than one qubit. For example, the CCNOT gate has two qubits for the control, and the CSWAP gate has two qubits for the target.

1 Mathematical Representations

Mathematically, the Controlled-U gate can be represented as:

$$\text{CU} = \begin{bmatrix} I & 0 \\ 0 & U \end{bmatrix}, \quad (7.9)$$

where U is a generic unitary matrix, and I is the identity matrix.

In the following, for simplicity, we only consider the case where the target and control are single qubits, so U and I are 2×2 matrices.

The above matrix for CU has a corresponding operator form:

$$\text{CU} = |0\rangle\langle 0| \otimes I + |1\rangle\langle 1| \otimes U, \quad (7.10)$$

which, when expanded, yields the operator in Table 7.1.

The fact that the CU gate applies U to the target qubit when the control qubit is $|1\rangle$ can be expressed as:

$$\text{CU} |0\rangle |\psi\rangle = |0\rangle |\psi\rangle, \quad (7.11a)$$
$$\text{CU} |1\rangle |\psi\rangle = |1\rangle U |\psi\rangle. \quad (7.11b)$$

7.1 Common Multi-Qubit Gates

Exercise 7.5 Demonstrate that the representations of the CU gate are equivalent: Eq. 7.9, Eq. 7.10, Eq. 7.11, and the operator form in Table 7.1.

The skills involved in this exercise is foundational for working with quantum algorithms. Make sure you can do this exercise independently.

2 Properties

The Controlled-U gate has a number of important properties:

Unitarity

The Controlled-U gate is unitary if U is unitary.

Since Pauli operators are unitary, so are CNOT and CZ.

Proof. The matrix representation of CU is given by Eq. 7.9. The Hermitian conjugate of CU, denoted by $CU^\dagger$, is:

$$CU^\dagger = \begin{bmatrix} I & 0 \\ 0 & U^\dagger \end{bmatrix}. \tag{7.12}$$

Since U is unitary, $UU^\dagger = I$. Therefore,

$$CU \cdot CU^\dagger = \begin{bmatrix} I & 0 \\ 0 & U \end{bmatrix} \begin{bmatrix} I & 0 \\ 0 & U^\dagger \end{bmatrix} = \begin{bmatrix} I & 0 \\ 0 & I \end{bmatrix} = I, \tag{7.13}$$

which proves that CU is unitary if U is unitary. □

Not a Parallel Gate

The CU gate cannot be decomposed as a tensor product $W \otimes U$ where U and W are non-identity unitary matrices.

This property implies that the application of a Controlled-U gate involves interaction between qubits.

A gate with an operator of the form $W \otimes U$ is a parallel gate, which will be discussed in § 7.1.6.

Proof.

The matrix representation of CU is given by Eq. 7.9. Now, assume for the sake of contradiction that we could write CU as a product operator, $W \otimes U$. Then, this would imply that

$$W \otimes U = \begin{bmatrix} w_{00}U & w_{01}U \\ w_{10}U & w_{11}U \end{bmatrix}, \tag{7.14}$$

for some matrix W. Comparing this with the matrix of CU in Eq. 7.9, we have the following system of equations:

$$w_{00}U = I, \tag{7.15a}$$
$$w_{01}U = 0, \tag{7.15b}$$
$$w_{10}U = 0, \tag{7.15c}$$
$$w_{11}U = U. \tag{7.15d}$$

For $w_{01}U$ to be the zero matrix, w_{01} must be zero since U is a unitary matrix. Similarly, w_{10} must be zero. This leads us to

$$W \otimes U = \begin{bmatrix} w_{00}U & 0 \\ 0 & w_{11}U \end{bmatrix}. \tag{7.16}$$

Comparing this with the matrix of CU, we can see that this form can represent the CU gate only if $w_{00} = 1$, $w_{11} = 1$, and $U = I$.

Therefore, we conclude that the CU gate is not a simple product operator like $W \otimes U$. □

Involutory Property

If $U^2 = I$, then $\text{CU}^2 = I_{2\times 2}$.

Proof.

Given that $U^2 = I$, we can write

$$\text{CU}^2 = \begin{bmatrix} I & 0 \\ 0 & U \end{bmatrix} \begin{bmatrix} I & 0 \\ 0 & U \end{bmatrix} = \begin{bmatrix} I & 0 \\ 0 & U^2 \end{bmatrix} = \begin{bmatrix} I & 0 \\ 0 & I \end{bmatrix} = I_{2\times 2}. \tag{7.17}$$

□

3 Controlled-U with Phase

For a standalone U gate, its global phase does not have any physical effect; that is, U and $e^{i\phi}U$ are equivalent as quantum gates. However, for a controlled-U gate, the global phase of U does matter, because

$$\begin{bmatrix} I & 0 \\ 0 & e^{i\phi}U \end{bmatrix} \neq e^{i\phi} \begin{bmatrix} I & 0 \\ 0 & U \end{bmatrix}. \tag{7.18}$$

In particular, CNOT and CZ have their "negative" counterparts, which are distinct from their original versions:

$$\overline{\text{CNOT}} = \begin{bmatrix} I & 0 \\ 0 & -X \end{bmatrix}, \tag{7.19a}$$

$$\overline{\text{CZ}} = \begin{bmatrix} I & 0 \\ 0 & -Z \end{bmatrix}. \tag{7.19b}$$

> **Exercise 7.6** Evaluate $\overline{\text{CNOT}}\,|++\rangle$, $\overline{\text{CNOT}}\,|+-\rangle$, $\overline{\text{CNOT}}\,|-+\rangle$, and $\overline{\text{CNOT}}\,|--\rangle$.

7.1.4 Classically Controlled-U Gate

Controlled gates like CNOT, CZ, and CU are quantum controlled gates that operate based on the states of control qubits. However, there also exist classically controlled gates, which perform actions dictated by classical logic. These gates involve the inclusion or exclusion of certain gates in the instruction sequence for the quantum computer based on classical control signals.

Classically controlled gates function similarly to their quantum controlled counterparts. However, there is a key difference. In a quantum controlled gate, the control qubit can be in a superposition state, whereas a classically controlled gate is

7.1 Common Multi-Qubit Gates

driven by a classical bit (0 or 1), often determined by the measurement outcome of a qubit. The operator for a quantum controlled U gate is given by Eq. 7.10, whereas for a classically controlled U gate,

$$\text{cCU} = U^i = \begin{cases} I & \text{for } i = 0, \\ U & \text{for } i = 1, \end{cases} \tag{7.20}$$

where i denotes the control bit.

The symbol for a classically controlled gate employs a double line for the control wire, as shown in Fig. 7.1.

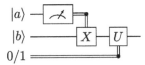

Figure 7.1: Examples of Classically Controlled Gates

7.1.5 ∗Two-Qubit Rotation Gates

Two-qubit rotation gates are essential components in quantum algorithms, facilitating complex interactions between qubits. These gates generalize the notion of single-qubit rotations and find applications in quantum algorithms like VQE and VAOA, which will be discussed in § 11.2. They can be represented as follows:

$$R_{pq}(\theta) = e^{-i\frac{\theta}{2}P \otimes Q}, \quad P, Q \in \{X, Y, Z\}. \tag{7.21}$$

Recall that for any normal matrix A with $A^2 = I$, it holds that $e^{i\theta A} = \cos\theta I + i\sin\theta A$. In the present case, $(P \otimes Q)^2 = I \otimes I$, where $I \otimes I$ is the 4×4 identity matrix. Thus, we can infer:

$$R_{pq}(\theta) = \cos\frac{\theta}{2}(I \otimes I) - i\sin\frac{\theta}{2}(P \otimes Q). \tag{7.22}$$

For example, the $R_{zz}(\theta)$ gate, also commonly termed the parameterized ZZ gate, is defined as:

$$R_{zz}(\theta) \equiv ZZ(\theta) = \cos\frac{\theta}{2}(I \otimes I) - i\sin\frac{\theta}{2}(Z \otimes Z) \tag{7.23a}$$

$$= \begin{bmatrix} R_z(\theta) & 0 \\ 0 & R_z(-\theta) \end{bmatrix} \tag{7.23b}$$

$$= \begin{bmatrix} e^{-i\frac{\theta}{2}} & 0 & 0 & 0 \\ 0 & e^{i\frac{\theta}{2}} & 0 & 0 \\ 0 & 0 & e^{i\frac{\theta}{2}} & 0 \\ 0 & 0 & 0 & e^{-i\frac{\theta}{2}} \end{bmatrix}. \tag{7.23c}$$

Exercise 7.7 Explain why $R_{zz}(\theta) \neq R_z(\theta) \otimes R_z(\theta)$.

$$R_{zx}(\theta) = \cos\frac{\theta}{2}(I \otimes I) - i\sin\frac{\theta}{2}(Z \otimes X) \tag{7.24a}$$

$$= \begin{bmatrix} \cos\frac{\theta}{2} & -i\sin\frac{\theta}{2} & 0 & 0 \\ -i\sin\frac{\theta}{2} & \cos\frac{\theta}{2} & 0 & 0 \\ 0 & 0 & \cos\frac{\theta}{2} & i\sin\frac{\theta}{2} \\ 0 & 0 & i\sin\frac{\theta}{2} & \cos\frac{\theta}{2} \end{bmatrix}. \tag{7.24b}$$

7.1.6 * Parallel Gates

A gate with an operator of the form $W \otimes U$ is referred to as a parallel gate. When acting on unentangled qubits, W and U act in parallel on their respective qubits, behaving as two independent gates, as illustrated in Eq. 7.25 and Fig. 7.2. In this case, if W or U equals the identity operator I, the parallel gate reduces to a single-qubit gate.

$$(W \otimes U)(|\psi\rangle \otimes |\phi\rangle) = (W|\psi\rangle) \otimes (U|\phi\rangle). \tag{7.25}$$

Equation 7.25 is applicable for product states, where $|\psi\rangle$ and $|\phi\rangle$ are unentangled.

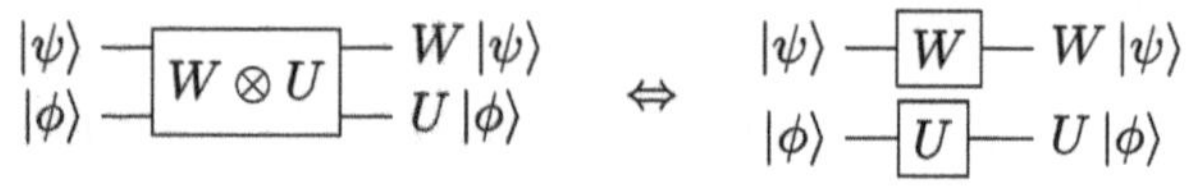

Figure 7.2: Parallel Gate on Unentangled State

Parallel gates are considered "trivial" multi-qubit gates, which is why the common multi-qubit gate discussed in § 7.1 has not included any of those. However, there is an important non-trivial application of parallel gates: when we want to apply a gate to part of an entangled state $|\psi\rangle$, for example, applying U to the second qubit of a Bell state, we are effectively applying the parallel gate $I \otimes U$ to the combined system, even though we may still draw a single qubit gate on the second qubit, as illustrated in Fig. 7.3.

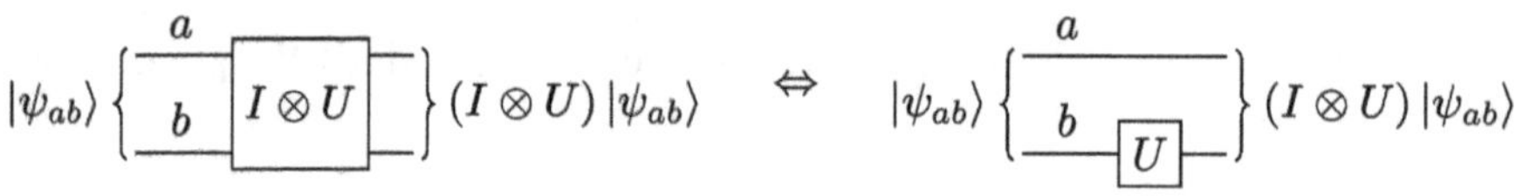

Figure 7.3: Gate on Part of an Entangled State

We often use a shorthand notation for $(I \otimes U)|\psi_{ab}\rangle$: $U_b|\psi_{ab}\rangle$, where U_b stands for U applied to qubit-b. This notation is convenient when we work with basis states, for example,

$$(I \otimes X)\frac{1}{\sqrt{2}}(|00\rangle + |11\rangle) = X_2\frac{1}{\sqrt{2}}(|00\rangle + |11\rangle) \tag{7.26a}$$

$$= \frac{1}{\sqrt{2}}(|0\rangle X|0\rangle + |1\rangle X|1\rangle) \tag{7.26b}$$

$$= \frac{1}{\sqrt{2}}(|01\rangle + |10\rangle). \tag{7.26c}$$

7.2 Universal Sets of Qubit Gates

In classical computing, complex operations are implemented as sequences of simpler operations. A combination of the NAND gate and the FANOUT circuit element can construct digital circuits capable of performing all computational tasks, forming what is known as a universal set of gates for classical computing. However, the NAND and FANOUT gates are not reversible. If we restrict ourselves to reversible gates, achieving universality with only one- and two-bit gates becomes impossible. Notably, a set consisting solely of the Toffoli gate can serve as a universal gate set for classical reversible computation.

Quantum computing employs gates as unitary transformations, which leads to an infinite number of possible gates for single, two, and multi-qubit operations on a quantum computer. This is in stark contrast to classical computing, which only has four functions mapping one bit to another, for example. Despite this apparent infinity, there exist universal sets of quantum gates, composed of a few specific gate types, that are capable of implementing any quantum computation. As a result, we can construct complex quantum circuits using only a limited number of gates from these universal sets to execute non-trivial quantum computations. This explains why the sets of common quantum gates typically include only a select few types.

In real-world quantum computing, we often focus on approximating desired computations to a controllable degree of accuracy. Hence, we define a universal gate set as follows:

> **Definition** A *universal gate set* is a set of quantum gates that can approximate any unitary transformation on any number of qubits to an arbitrary degree of accuracy.

In simpler terms, a universal gate set allows us to approximate any desired quantum operation with a sequence of gates from the set, up to a predetermined error bound.

Universality is crucial in quantum computing because it allows us to solve any computational problem by using a fixed set of quantum gates, just as we can solve any computational problem in classical computing using a fixed set of classical gates.

So, what are the requirements for a universal set of quantum gates? At a minimum, they must allow us to reach any point on the Bloch sphere (including those with complex amplitudes), and they must be able to create both superposition states and entanglement states.

Below are a few examples of universal gate sets. Importantly, the choice of gate set depends on specific factors such as the quantum computing architecture, physical implementation, and system error rates. Although a universal gate set can approximate any quantum computation, it might not always provide the most efficient solution for certain tasks. Each gate set presents its own set of advantages and challenges. As such, research is ongoing to identify the optimal gate sets for various quantum computing platforms and applications.

7.2.1 CNOT + Arbitrary Single-Qubit Rotations

The combination of the CNOT gate and single-qubit unitary rotation gates accomplishes the requirements for a universal set. This gate set is universal because it can generate any single-qubit unitary transformation and create entanglement between qubits. (For a rigorous proof, see Refs. [18, 7].)

Single-qubit unitary rotations allow us to reach any point on the Bloch sphere, corresponding to any state the qubit can be in. Typically, two non-commuting rotation gates (representing rotations around different axes of the Bloch sphere) are sufficient to generate arbitrary rotations. A common choice involves the rotation gates around the x and z axes, denoted as R_x and R_z respectively, as defined in § 5.2.5.

In addition to the CNOT gate, frequently used gates in this universal set include X, Z, and H. While these are essential, other single-qubit gates may also be required for specific applications or implementations. The versatility and power of this universal set lie in its ability to perform both fundamental quantum operations and more complex algorithms across various quantum computing platforms.

7.2.2 Clifford Group + T

The Clifford Group, T gate set is a commonly utilized universal gate set, especially important in the context of fault-tolerant quantum computing, where fault-tolerance refers to the ability of a system to correct errors that occur during quantum computations. The Clifford group here encompasses the Pauli gates (X, Y, Z), the Hadamard gate (H), the phase gate (S), and the CNOT gate. These gates can be implemented efficiently and with fault-tolerance in many quantum error-correcting codes.

The T gate, representing a $\pi/4$ rotation around the z-axis of the Bloch sphere, is not a part of the Clifford group. It is specifically included in this set to efficiently approximate all single-qubit rotations, thus ensuring universality. In comparison to the Clifford gates, the T gate is more challenging to implement fault-tolerantly, thereby making it a major contributor to the computational cost of a quantum algorithm in fault-tolerant settings.

The T gate is harder to implement fault-tolerantly because it requires precise control over the qubit state, which can be challenging due to issues such as quantum decoherence and gate errors. This distinction has led to the concept of the T-count, which measures the number of T gates used in a quantum algorithm, serving as an indicator of the algorithm's computational expense in a fault-tolerant quantum computing paradigm. The universality of this gate set, coupled with its fault-tolerant characteristics, underpins its widespread usage.

7.2.3 Toffoli + H

The gate set comprised of the Toffoli gate (also known as the CCNOT gate) and Hadamard (H) gate forms a universal gate set for quantum computation. The Toffoli gate is a three-qubit gate, and is the most elementary quantum gate that is capable of executing deterministic universal classical computation. See § 7.5.1 for more details.

This gate set can produce any unitary transformation on any number of qubits, making it universal for quantum computation. It allows for the effective utilization of all inherent quantum mechanical properties such as superposition, entanglement, and quantum interference, providing the computational advantage sought in quantum computing.

The main practical challenge with this gate set lies in the implementation of the Toffoli gate. Because the Toffoli gate is a three-qubit gate, it often requires more complex quantum operations and higher error rates compared to single- and two-qubit gates. To implement a Toffoli gate, one might need to use a combination of simpler gates, which increases the gate count and thereby the possibility of errors.

7.3 Boolean Representation of Quantum Gates

In this section, we introduce a powerful tool for analyzing quantum gates and circuits: the Boolean representation of quantum gates and circuits. At its core, this representation offers an efficient method for us to examine a quantum circuit for each of the basis states separately. The results for each basis state are then combined to form a complete picture of the circuit's behavior for a general quantum state, leveraging the linearity property of quantum transformations. We also refer to this approach the logic operation approach.

7.3.1 Properties of XOR Operations

Consider x as a Boolean variable where $x \in \{0, 1\}$, and let $\bar{x} \equiv 1 - x$. When $x = 0$, $|x\rangle$ represents $|0\rangle$, and $|\bar{x}\rangle$ represents $|1\rangle$. When $x = 1$, $|x\rangle$ represents $|1\rangle$, and $|\bar{x}\rangle$ represents $|0\rangle$.

The XOR operator, denoted as $\oplus$, corresponds to a bitwise modulo 2 addition:

$$x \oplus y \equiv (x + y) \bmod 2. \tag{7.27}$$

Because $2x = 2y = 0 \bmod 2$, we also have

$$x \oplus y = (x - y) \bmod 2 = (y - x) \bmod 2. \tag{7.28}$$

The XOR operation results in '1' when the number of '1's in the operands is odd; otherwise, it returns '0'. (For example, $1 \oplus 1 \oplus 0 \oplus 1 = 1$, and $1 \oplus 1 \oplus 1 \oplus 1 = 0$.)

The following lists some key properties of XOR operations:

$$x \oplus 0 = x, \tag{7.29a}$$
$$x \oplus 1 = \bar{x}, \tag{7.29b}$$
$$\bar{x} \oplus 1 = x, \tag{7.29c}$$
$$x \oplus x = 0, \tag{7.29d}$$
$$x \oplus \bar{x} = 1, \tag{7.29e}$$
$$\overline{x \oplus y} = \bar{x} \oplus y = x \oplus \bar{y} = 1 \oplus x \oplus y, \tag{7.29f}$$
$$(-1)^{x \oplus y} = (-1)^{\bar{x} \oplus \bar{y}} = (-1)^{x+y}, \tag{7.29g}$$
$$(-1)^{x \oplus \bar{y}} = (-1)^{\bar{x} \oplus y} = (-1)^{1+x+y}, \tag{7.29h}$$
$$x \oplus y = x + y - 2xy. \tag{7.29i}$$

Exercise 7.8 Verify the last three relations in Eq. 7.29 for all combinations of $x = 0, 1$ and $y = 0, 1$.

Exercise 7.9 Simplify $\overline{x \oplus \overline{y} \oplus 1}$.

Exercise 7.10 Simplify $(-1)^{\overline{1 \oplus x \oplus \overline{y}}}$.

7.3.2 Boolean Representation of Common Quantum Gates

The Boolean representation of a quantum gate can be thought of as a formula that succinctly characterizes the transformation applied by the gate to all possible basis states. This is similar to the use of Boolean algebra and truth tables in classical computing, with significant differences which we will explain in § 7.3.3.

The Boolean representation of some common quantum gates is given in Table 7.2.

Gate	Symbol	Boolean Representation
X (NOT)	$\lvert x\rangle - \boxed{X} - \lvert \bar{x}\rangle$	$\lvert x\rangle \mapsto \lvert \bar{x}\rangle$
CNOT	$\lvert x\rangle \longrightarrow \lvert x\rangle$ $\lvert y\rangle \longrightarrow \oplus \longrightarrow \lvert x \oplus y\rangle$	$\lvert xy\rangle \mapsto \lvert x(x \oplus y)\rangle$
Z	$\lvert x\rangle - \boxed{Z} - (-1)^x \lvert x\rangle$	$\lvert x\rangle \mapsto (-1)^x \lvert x\rangle$
CZ	$\lvert x\rangle \longrightarrow \lvert x\rangle$ $\lvert y\rangle - \boxed{Z} - (-1)^{xy} \lvert y\rangle$	$\lvert xy\rangle \mapsto (-1)^{xy} \lvert xy\rangle$
H	$\lvert x\rangle - \boxed{H} - \frac{1}{\sqrt{2}}(\lvert 0\rangle + (-1)^x \lvert 1\rangle)$	$\lvert x\rangle \mapsto \frac{1}{\sqrt{2}}(\lvert 0\rangle + (-1)^x \lvert 1\rangle)$
Y	$\lvert x\rangle - \boxed{Y} - i(-1)^x \lvert \bar{x}\rangle$	$\lvert x\rangle \mapsto i(-1)^x \lvert \bar{x}\rangle$
$P(\phi)$	$\lvert x\rangle - \boxed{e^{i\phi}} - e^{ix\phi} \lvert x\rangle$	$\lvert x\rangle \mapsto e^{ix\phi} \lvert x\rangle$
U	$\lvert x\rangle - \boxed{U} - u_{xx} \lvert x\rangle + u_{\bar{x}x} \lvert \bar{x}\rangle$	$\lvert x\rangle \mapsto u_{xx} \lvert x\rangle + u_{\bar{x}x} \lvert \bar{x}\rangle$
CU	$\lvert x\rangle \longrightarrow \lvert x\rangle$ $\lvert y\rangle - \boxed{U} - \begin{array}{l}U^x \lvert y\rangle, \text{ or}\\ u_{yy}^x \lvert y\rangle + xu_{\bar{y}y} \lvert \bar{y}\rangle\end{array}$	$\lvert xy\rangle \mapsto \lvert x\rangle U^x \lvert y\rangle$ $\lvert xy\rangle \mapsto u_{yy}^x \lvert xy\rangle + xu_{\bar{y}y} \lvert x\bar{y}\rangle$
Toffoli (CCNOT)	$\lvert x\rangle \longrightarrow \lvert x\rangle$ $\lvert y\rangle \longrightarrow \lvert y\rangle$ $\lvert z\rangle \longrightarrow \oplus \longrightarrow \lvert xy \oplus z\rangle$	$\lvert xyz\rangle \mapsto \lvert xy(xy \oplus z)\rangle$
SWAP	$\lvert x\rangle \longrightarrow \lvert y\rangle$ $\lvert y\rangle \longrightarrow \lvert x\rangle$	$\lvert xy\rangle \mapsto \lvert yx\rangle$
Fredkin (CSWAP)	$\lvert x\rangle \longrightarrow \lvert x\rangle$ $\lvert y\rangle \longrightarrow \lvert \bar{x}y \oplus xz\rangle$ $\lvert z\rangle \longrightarrow \lvert xy \oplus \bar{x}z\rangle$	$\lvert xyz\rangle \mapsto \lvert x(\bar{x}y \oplus xz)(xy \oplus \bar{x}z)\rangle$

Table 7.2: Boolean Representation of Common Qubit Gates

7.3 Boolean Representation of Quantum Gates

 A note on the notation in Table 7.2. In the case of the CNOT gate, the expression $|x(x \oplus y)\rangle$ should be interpreted as $|x\rangle \otimes |(x \oplus y)\rangle$. This might appear unusual due to the omission of the tensor product operator for brevity. The juxtaposition of x and $(x \oplus y)$ within the ket notation does not imply multiplication. In this context, $|x(x \oplus y)\rangle$ represents states such as $|00\rangle$, $|11\rangle$, etc., with the first and second qubit values being x and $x \oplus y$, respectively.

In the Toffoli gate representation, the output of the third qubit is $|xy \oplus z\rangle$ in $|xyz\rangle \mapsto |xy(xy \oplus z)\rangle$. Here, the operation xy within the "()" denotes standard multiplication, not a tensor product. This is because $|xy \oplus z\rangle$ in this case represents a single qubit state.

Consider the Z gate as an example. Its Boolean representation is $|x\rangle \mapsto (-1)^x |x\rangle$. This formula means that for $x = 0$, we have $|0\rangle \mapsto |0\rangle$; and for $x = 1$, $|1\rangle \mapsto -|1\rangle$. Given the superposition state $\alpha |0\rangle + \beta |1\rangle$, we can apply the Z gate by leveraging the linearity of quantum transformations, which allows us to handle each term separately. Therefore, we obtain $\alpha |0\rangle + \beta |1\rangle \mapsto \alpha |0\rangle - \beta |1\rangle$, accurately reflecting the transformation function of the Z gate.

Exercise 7.11 Demonstrate that the Boolean representation of gate CZ is equal to that of CZ$'$. Here, CZ$'$ represents the CZ gate with qubit 2 serving as the control.

Exercise 7.12 Formulate the Boolean representation of the parallel gate $I \otimes X$ and compare with the CNOT gate.

Exercise 7.13 Formulate the Boolean representation of the parallel gate $I \otimes Z$ and compare with the CZ gate.

7.3.3 Understanding Boolean Representation

While Boolean representation is a powerful technique for analyzing quantum circuits, it can sometimes lead to confusion during interpretation. To elucidate some key aspects and avoid potential pitfalls, we will examine specific examples next.

1 Forming Superposition

Boolean representation informs us the transformation of each basis state by a quantum gate or circuit. The results for each basis state are then combined to form a complete picture of the circuit's behavior for a general quantum state. Note that when multiple qubits are involved, the basis states are tensor product states.

■ **Example 7.1** Let's use CNOT as an example to illustrate how the above process works. The symbol and formula for CNOT in Table 7.2 capture the transformation of CNOT on all combinations of the basis states as follows:

Combining these cases in a superposition, we obtain the general transformation of the CNOT gate:

$$\alpha |00\rangle + \beta |01\rangle + \gamma |10\rangle + \delta |11\rangle \mapsto \alpha |00\rangle + \beta |01\rangle + \gamma |11\rangle + \delta |10\rangle.$$

■

$$|00\rangle \mapsto |00\rangle \qquad |01\rangle \mapsto |01\rangle \qquad |10\rangle \mapsto |11\rangle \qquad |11\rangle \mapsto |10\rangle$$

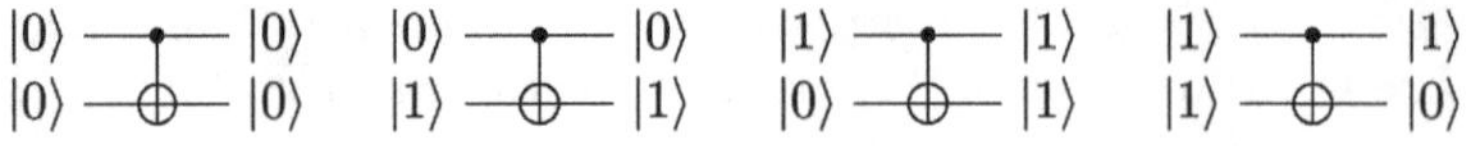

Figure 7.4: Expansion of the Boolean Representation of CNOT

2 $|x\rangle$ vs. $|\psi\rangle$

If a qubit has an output $|x\rangle$, we can treat $|x\rangle$ as the general quantum state $|\psi\rangle$, provided that qubit is disentangled from the rest of the system.

Care should be taken when dealing with multi-qubit systems. If $|x\rangle$ is entangled with another qubit, then we cannot directly substitute $|x\rangle$ with the general state $|\psi\rangle$ as $|x\rangle$ must be treated as a part of the multi-qubit basis states. Let's illustrate this caveat with the following example.

■ **Example 7.2** A common question for the Boolean representation of CNOT gate is: If $|x\rangle \equiv |\psi\rangle$, doesn't the following diagram imply that we can duplicate an arbitrary state $|\psi\rangle$, thereby violating the no-cloning theorem?

The answer is NO. The replacement of $|x\rangle$ with $|\psi\rangle$ is invalid here since the two qubits are entangled. Thus, we need to instantiate the left part of the diagram twice, once with $x = 0$ and again with $x = 1$, and then perform a linear combination. Therefore, the diagram should be corrected as follows:

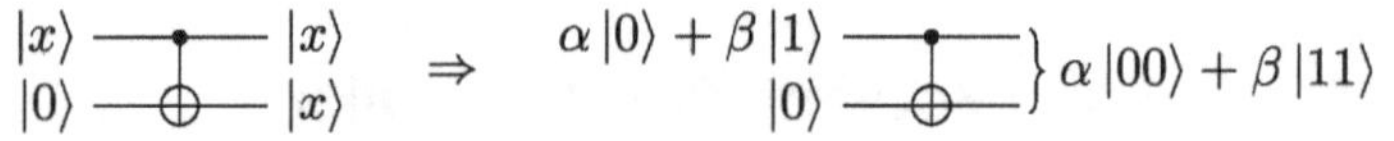

■

3 ✻ Identifying Product States

In the context of Boolean representation, where qubits are denoted as $|x\rangle$, $|y\rangle$, etc., we need to be cautious in discerning product states. A quantum state, even if it seems to present as a product state, may not be a true product state, as shown in the following example. The crux lies in understanding that only if a quantum state consistently preserves its form as a product state across all combinations of x and y values, inclusive of their signs, can it be treated as a genuine product state. In the absence of this uniformity across all basis states, the seeming product state is not a true one.

■ **Example 7.3** This example comes from the discussion of quantum teleportation in § 10.3.3. At some stage in the process, the three-qubit system is in the state:

$$|\Psi_1\rangle = \frac{1}{\sqrt{2}}(|xx0\rangle + |x\bar{x}1\rangle). \tag{7.30}$$

7.4 Equivalent Gate Sequences

At a first glance, you might think that the first qubit is separated from the system, because it seems that $|\Psi_1\rangle$ can be written as a product state $|x\rangle \otimes \frac{1}{\sqrt{2}}(|x0\rangle + |\bar{x}1\rangle)$.

But this conclusion is wrong. Because of the presence of $\bar{x}$, the seemingly product state does not keep the same form for $x = 0$ and 1.

To interpret Eq. 7.30 correctly, we need to run it twice, once for with $|x\rangle = |0\rangle$ and the other with $|x\rangle = |1\rangle$, and then do a linear combination of the two resulting equations with weights α and β. Thus, Eq. 7.30 represents

$$|\Psi_1\rangle = \frac{1}{\sqrt{2}} \left(\alpha(|000\rangle + |011\rangle) + \beta(|110\rangle + |101\rangle) \right), \qquad (7.31)$$

which cannot be decomposed as the tensor product of $\alpha|0\rangle + \beta|1\rangle$ with another state. ∎

Exercise 7.14 What is the output state from the following circuit when the input state is $|x\rangle = \alpha|0\rangle + \beta|1\rangle$ and $|y\rangle = \gamma|0\rangle + \delta|1\rangle$?

$$
\begin{array}{l}
|x\rangle \longrightarrow \bullet \longrightarrow \boxed{Z} \longmapsto (-1)^x |x\rangle \\
|y\rangle \longrightarrow \oplus \longrightarrow \phantom{\boxed{Z}} \longmapsto |x \oplus y\rangle
\end{array}
$$

4 ✻ Working with Coefficients

The following example illustrates how coefficients (of $|x\rangle$, etc.) work in Boolean representation.

■ **Example 7.4** Let's examine the CU gate in the last row of Table 7.2.

$$|xy\rangle \mapsto u_{yy}^x |xy\rangle + x u_{\bar{y}y} |x\bar{y}\rangle$$

$$= \begin{cases} |0y\rangle & \text{for } x = 0 \\ u_{yy}|1y\rangle + u_{\bar{y}y}|1\bar{y}\rangle & \text{for } x = 1 \end{cases}$$

$$= \begin{cases} |00\rangle & \text{for } x = 0,\ y = 0 \\ |01\rangle & \text{for } x = 0,\ y = 1 \\ u_{00}|10\rangle + u_{10}|11\rangle & \text{for } x = 1,\ y = 0 \\ u_{11}|11\rangle + u_{01}|10\rangle & \text{for } x = 1,\ y = 1. \end{cases}$$

∎

Exercise 7.15 Verify the Boolean representation of other gates in Table 7.2 as we have done for CU.

7.4 Equivalent Gate Sequences

In this section, we delve into a fundamental aspect of quantum circuits: gate sequences, and more importantly, their equivalence. The sequence of gates, much akin to the building blocks of a structure, forms the simplest embodiment of a quantum circuit. A fascinating attribute of these sequences is their ability to combine into a single gate equivalent, paving the path for simplification and optimization of quantum circuits. This becomes crucial in practical scenarios when the circuits need to be executed on actual quantum hardware where resources are limited and hence, optimization is key. Moreover, the understanding of equivalent gate sequences forms the bedrock for the analysis of more intricate quantum circuits, for instance, quantum gate teleportation. Mastering the techniques and skills required to dissect and interpret these sequences is a pivotal part of one's journey in quantum computing. As we proceed, we'll learn more about these sequences, their inherent properties, and the tools needed to comprehend their nuances.

7.4.1 Order Independence in Separate Qubit Operations

The figure below shows a sequence often implied without proof. Two linear operations, W and V, pertaining to two qubits a and b, respectively, are order independent. Here, W and V can represent either quantum gates or measurements.

$$
\begin{array}{ccc}
a-\boxed{W} & \boxed{W} & \boxed{W} \\
b\boxed{V} \quad\Leftrightarrow\quad & \boxed{V} \quad\Leftrightarrow\quad & \boxed{V} \\
(W\otimes I)(I\otimes V) & (I\otimes V)(W\otimes I) & W\otimes V
\end{array}
$$

Mathematically, applying W to qubit a without affecting qubit b is equivalent to applying $W \otimes I$ to the joint state of the two qubits. Similarly, applying V to qubit b only is equivalent to $I \otimes V$. In this case, W and V commute regardless of their commutation properties within their respective systems. This is because the operations act on different Hilbert spaces and do not interfere with each other. In other words,

$$(W \otimes I)(I \otimes V) = (I \otimes V)(W \otimes I), \text{ or } [W \otimes I, I \otimes V] = 0. \tag{7.32}$$

Because of this property, we often express the whole operation as $W \otimes V$ or $W_a V_b$.

Proof. Since

$$(W \otimes I)(I \otimes V) = WI \otimes IV = W \otimes V, \tag{7.33a}$$
$$(I \otimes V)(W \otimes I) = IW \otimes VI = W \otimes V, \tag{7.33b}$$

we have

$$[W \otimes I, I \otimes V] = (W \otimes I)(I \otimes V) - (I \otimes V)(W \otimes I) \tag{7.34a}$$
$$= W \otimes V - W \otimes V \tag{7.34b}$$
$$= 0. \tag{7.34c}$$

□

7.4.2 Gate Sequence: A First Example

Gate equivalence can be analyzed using three approaches: matrix product, operator product, and logic operation. Mastering these approaches can greatly aid in optimizing quantum circuits and algorithms. The choice of approach, whether more intuitive or effective, hinges on the nature of the gate sequence.

7.4 Equivalent Gate Sequences

We will illustrate these approaches with a prominent example of gate equivalence, depicted in the diagram below. When sequentially applied, the trio of CNOT gates (with the central one flipped) effectively emulates the operation of a SWAP gate, which interchanges the states of two qubits.

$$
\begin{array}{c}
\text{CNOT} \quad \text{CNOT}' \quad \text{CNOT} \\
|x\rangle \!-\!\!\bullet\!-\!\!\oplus\!-\!\!\bullet\!- \\
|y\rangle \!-\!\!\oplus\!-\!\!\bullet\!-\!\!\oplus\!-
\end{array}
\quad \Leftrightarrow \quad
\begin{array}{c}
\text{SWAP} \\
|x\rangle \!-\!\!\times\!- \\
|y\rangle \!-\!\!\times\!-
\end{array}
$$

1 Matrix Product Approach

The matrix product representing the gate sequence:

$$\text{CNOT} \cdot \text{CNOT}' \cdot \text{CNOT} \tag{7.35a}$$

$$= \begin{bmatrix} 1 & 0 & 0 & 0 \\ 0 & 1 & 0 & 0 \\ 0 & 0 & 0 & 1 \\ 0 & 0 & 1 & 0 \end{bmatrix} \begin{bmatrix} 1 & 0 & 0 & 0 \\ 0 & 0 & 0 & 1 \\ 0 & 0 & 1 & 0 \\ 0 & 1 & 0 & 0 \end{bmatrix} \begin{bmatrix} 1 & 0 & 0 & 0 \\ 0 & 1 & 0 & 0 \\ 0 & 0 & 0 & 1 \\ 0 & 0 & 1 & 0 \end{bmatrix} \tag{7.35b}$$

$$= \begin{bmatrix} 1 & 0 & 0 & 0 \\ 0 & 0 & 1 & 0 \\ 0 & 1 & 0 & 0 \\ 0 & 0 & 0 & 1 \end{bmatrix}. \tag{7.35c}$$

We recognize that the final product matrix corresponds to the SWAP gate.

2 Operator Product Approach

Recall from §§ 7.1.1 and 7.1.2 that:

$$\text{CNOT} = |00\rangle\langle 00| + |01\rangle\langle 01| + |11\rangle\langle 10| + |10\rangle\langle 11|, \tag{7.36a}$$

$$\text{CNOT}' = |00\rangle\langle 00| + |11\rangle\langle 01| + |10\rangle\langle 10| + |01\rangle\langle 11|. \tag{7.36b}$$

The operator product representing the gate sequence:

$$\text{CNOT} \cdot \text{CNOT}' \cdot \text{CNOT} \tag{7.37a}$$
$$= (|00\rangle\langle 00| + |01\rangle\langle 01| + |11\rangle\langle 10| + |10\rangle\langle 11|) \tag{7.37b}$$
$$ (|00\rangle\langle 00| + |11\rangle\langle 01| + |10\rangle\langle 10| + |01\rangle\langle 11|) \tag{7.37c}$$
$$ (|00\rangle\langle 00| + |01\rangle\langle 01| + |11\rangle\langle 10| + |10\rangle\langle 11|) \tag{7.37d}$$
$$= (|00\rangle\langle 00| + |01\rangle\langle 01| + |11\rangle\langle 10| + |10\rangle\langle 11|) \tag{7.37e}$$
$$ (|00\rangle\langle 00| + |11\rangle\langle 01| + |10\rangle\langle 11| + |01\rangle\langle 10|) \tag{7.37f}$$
$$= |00\rangle\langle 00| + |01\rangle\langle 10| + |10\rangle\langle 01| + |11\rangle\langle 11| \tag{7.37g}$$
$$= \text{SWAP}. \tag{7.37h}$$

3 Logic Operation Approach

Using the Boolean representation tool developed in § 7.3 we can derive the same equivalence as follows.

Recall from § 7.3 that:

$$\text{CNOT} : |x, y\rangle \mapsto |x, x \oplus y\rangle, \qquad (7.38a)$$
$$\text{CNOT}' : |x, y\rangle \mapsto |x \oplus y, y\rangle, \qquad (7.38b)$$

where $x, y \in \{0, 1\}$.

Now we go through the three quantum gates in the sequence:

$$|x, y\rangle \xrightarrow{\text{first CNOT}} |x, x \oplus y\rangle \qquad (7.39a)$$
$$\xrightarrow{\text{mid CNOT'}} |x \oplus (x \oplus y), x \oplus y\rangle = |y, x \oplus y\rangle \qquad (7.39b)$$
$$\xrightarrow{\text{second CNOT}} |y, (x \oplus y) \oplus y\rangle = |y, x\rangle. \qquad (7.39c)$$

We can see $|x, y\rangle$ maps to $|y, x\rangle$, which represents the SWAP operation.

> **Exercise 7.16** Replace CNOT with CZ in the above analysis, and investigate the equivalence of the following gate sequence. Identify if this sequence emulates any well-known gate operation, similar to how the previous sequence performed a SWAP operation:
>

7.4.3 Sequences Involving X, Z, and CNOT

The ordered gate sequences in this subsection, involving X, Z, and CNOT, are essential for understand CNOT gate transportation discussed in § 10.5.2.

Seq.	Gate Sequence	Equivalent Sequence	Proof
1	$\|x\rangle$ ─●─, $\|y\rangle$ ─X─⊕─ $\}\|\Psi_1\rangle$	$\|x\rangle$ ─●─, $\|y\rangle$ ─⊕─X─ $\}\|\Psi_2\rangle$	Eq. 7.40
2	$\|x\rangle$ ─X─●─, $\|y\rangle$ ───⊕─ $\}\|\Psi_1\rangle$	$\|x\rangle$ ─●─X─, $\|y\rangle$ ─⊕─X─ $\}\|\Psi_2\rangle$	Eq. 7.41
3	$\|x\rangle$ ───●─, $\|y\rangle$ ─Z─⊕─ $\}\|\Psi_1\rangle$	$\|x\rangle$ ─●─Z─, $\|y\rangle$ ─⊕─Z─ $\}\|\Psi_2\rangle$	Eq. 7.42
4	$\|x\rangle$ ─Z─●─, $\|y\rangle$ ───⊕─ $\}\|\Psi_1\rangle$	$\|x\rangle$ ─●─Z─, $\|y\rangle$ ─⊕───$\}\|\Psi_2\rangle$	Eq. 7.43

Table 7.3: Gate Sequence Equivalence Involving X, Z and CNOT

The following equations provide proofs for the gate sequence equivalences referenced in Table 7.3. Here, C denotes the operator of the CNOT gate. We have used the logic operation approach for these proofs. Of course, we can also prove these equivalences using matrix multiplication. Please reference § 7.3 for the properties of the XOR ($\oplus$) operation and Boolean representations of quantum gates.

7.4 Equivalent Gate Sequences

Gate Sequence 1

$$|\Psi_1\rangle = CX_2 |x\rangle |y\rangle = C |x\rangle |\bar{y}\rangle = |x\rangle |x \oplus \bar{y}\rangle, \tag{7.40a}$$

$$|\Psi_2\rangle = X_2 C |x\rangle |y\rangle = X_2 |x\rangle |x \oplus y\rangle = |x\rangle |\overline{x \oplus y}\rangle. \tag{7.40b}$$

Here we have used $\overline{x \oplus y} = x \oplus \bar{y} = \bar{x} \oplus y = 1 \oplus x \oplus y$.

Given that $|\Psi_1\rangle$ and $|\Psi_2\rangle$ equal to $|x\rangle |\bar{x} \oplus y\rangle$, these two gate sequences also equivalent to $\overline{\text{CNOT}}$, the inverted CNOT where the NOT gate is activated when the control qubit is $|0\rangle$, as opposed to $|1\rangle$. See also Problem 7.1. This equivalence can also be derived through matrix multiplication.

Gate Sequence 2

$$|\Psi_1\rangle = CX_1 |x\rangle |y\rangle = C |\bar{x}\rangle |y\rangle = |\bar{x}\rangle |\bar{x} \oplus y\rangle, \tag{7.41a}$$

$$|\Psi_2\rangle = X_2 X_1 C |x\rangle |y\rangle = X_2 X_1 |x\rangle |x \oplus y\rangle = |\bar{x}\rangle |\overline{x \oplus y}\rangle. \tag{7.41b}$$

Here we have used $\overline{x \oplus y} = \bar{x} \oplus y = 1 \oplus x \oplus y$.

Gate Sequence 3

$$|\Psi_1\rangle = CZ_2 |x\rangle |y\rangle = (-1)^y C |x\rangle |y\rangle = (-1)^y |x\rangle |x \oplus y\rangle, \tag{7.42a}$$

$$|\Psi_2\rangle = Z_2 Z_1 C |x\rangle |y\rangle = Z_2 Z_1 |x\rangle |x \oplus y\rangle = (-1)^{x+x\oplus y} |x\rangle |x \oplus y\rangle. \tag{7.42b}$$

Here we have used $(-1)^{x+x\oplus y} = (-1)^{x+x+y} = (-1)^{2x+y} = (-1)^y$.

Gate Sequence 4

$$|\Psi_1\rangle = CZ_1 |x\rangle |y\rangle = (-1)^x C |x\rangle |y\rangle = (-1)^x |x\rangle |x \oplus y\rangle, \tag{7.43a}$$

$$|\Psi_2\rangle = Z_1 C |x\rangle |y\rangle = Z_1 |x\rangle |x \oplus y\rangle = (-1)^x |x\rangle |x \oplus y\rangle. \tag{7.43b}$$

Exercise 7.17 ✳ The diagram below depicts a gate sequence similar to the sequence 1 in Table 7.3, except CNOT is replaced with $\overline{\text{CNOT}}$. The $\overline{\text{CNOT}}$ gate is an inverted CNOT gate: it applies an X gate to the target qubit when the control qubit is in state $|0\rangle$, instead of $|1\rangle$. Find an equivalent sequence to this sequence, similar to the ones in Table 7.3.

7.4.4 Sequences Involving H

The gate sequences detailed in Table 7.4 incorporate the Hadamard gate H. For these sequences, analysis via matrix multiplication tends to be the most straightforward approach.

The subsequent equations furnish proofs for the gate sequence equivalences cited in Table 7.4. For each case, we use U to denote the unitary matrix representing the gate sequence.

Seq.	Gate Sequence	Equivalent Sequence	Proof
5	$\|x\rangle$ ──●── , $\|y\rangle$ ─H─⊕─H─ $\}\|\Psi_1\rangle$	$\|x\rangle$ ──●── , $\|y\rangle$ ──Z── $\}\|\Psi_2\rangle$	Eq. 7.44
6	$\|x\rangle$ ─H─●─H─ , $\|y\rangle$ ──Z── $\}\|\Psi_1\rangle$	$\|x\rangle$ ──⊕── , $\|y\rangle$ ──●── $\}\|\Psi_2\rangle$	Eq. 7.45
7	$\|x\rangle$ ─H─●─H─ , $\|y\rangle$ ─H─⊕─H─ $\}\|\Psi_1\rangle$	$\|x\rangle$ ──⊕── , $\|y\rangle$ ──●── $\}\|\Psi_2\rangle$	Eq. 7.46
8	$\|x\rangle$ ─H─●─H─ , $\|y\rangle$ ──⊕── $\}\|\Psi_1\rangle$	$\|x\rangle$ ──⊕── , $\|y\rangle$ ─H─●─H─ $\}\|\Psi_2\rangle$	Eq. 7.47

Table 7.4: Gate Sequence Equivalence Involving H

Gate Sequence 5

$$U = (I \otimes H)\,\text{CNOT}\,(I \otimes H) \tag{7.44a}$$

$$= \begin{bmatrix} H & 0 \\ 0 & H \end{bmatrix} \begin{bmatrix} I & 0 \\ 0 & X \end{bmatrix} \begin{bmatrix} H & 0 \\ 0 & H \end{bmatrix} \tag{7.44b}$$

$$= \begin{bmatrix} H^2 & 0 \\ 0 & HXH \end{bmatrix} \tag{7.44c}$$

$$= \begin{bmatrix} I & 0 \\ 0 & Z \end{bmatrix} \tag{7.44d}$$

$$= \text{CZ}. \tag{7.44e}$$

Here we have used $H^2 = I$ and $HXH = Z$.

Gate Sequence 6

$$U = (H \otimes I)\,\text{CZ}\,(H \otimes I) \tag{7.45a}$$

$$= \frac{1}{\sqrt{2}} \begin{bmatrix} I & I \\ I & -I \end{bmatrix} \begin{bmatrix} I & 0 \\ 0 & Z \end{bmatrix} \frac{1}{\sqrt{2}} \begin{bmatrix} I & I \\ I & -I \end{bmatrix} \tag{7.45b}$$

$$= \frac{1}{2} \begin{bmatrix} I+Z & I-Z \\ I-Z & I+Z \end{bmatrix} \tag{7.45c}$$

$$= \text{CNOT}'. \tag{7.45d}$$

Here we have used Eq. 7.8 for the matrix of CNOT$'$.

Gate Sequence 7

$$U = (H \otimes H)\,\text{CNOT}\,(H \otimes H) \tag{7.46a}$$

$$= \frac{1}{\sqrt{2}} \begin{bmatrix} H & H \\ H & -H \end{bmatrix} \begin{bmatrix} I & 0 \\ 0 & X \end{bmatrix} \frac{1}{\sqrt{2}} \begin{bmatrix} H & H \\ H & -H \end{bmatrix} \tag{7.46b}$$

$$= \frac{1}{2} \begin{bmatrix} I+Z & I-Z \\ I-Z & I+Z \end{bmatrix} \tag{7.46c}$$

$$= \text{CNOT}'. \tag{7.46d}$$

7.4 Equivalent Gate Sequences

Here we have used Eq. 7.8 for the matrix of CNOT′.

You may have observed that gate sequences 6 and 7 are equivalent, both corresponding to CNOT′. Interestingly, another method to deduce sequence 7 is by combining sequences 5 and 6, and applying the identity $H^2 = I$. We leave this derivation to the reader.

Gate Sequence 8

As it turns out, gate sequence 8, even though appearing very similar to sequence 6, does not correspond to a single equivalent gate within our current repertoire. Its equivalent form shown in Table 7.4 can be derived from sequence 7. Its operator is given by:

$$U = (H \otimes I)\,\text{CNOT}\,(H \otimes I) \tag{7.47a}$$

$$= \frac{1}{\sqrt{2}} \begin{bmatrix} I & I \\ I & -I \end{bmatrix} \begin{bmatrix} I & 0 \\ 0 & X \end{bmatrix} \frac{1}{\sqrt{2}} \begin{bmatrix} I & I \\ I & -I \end{bmatrix} \tag{7.47b}$$

$$= \frac{1}{2} \begin{bmatrix} I+X & I-X \\ I-X & I+X \end{bmatrix}. \tag{7.47c}$$

Exercise 7.18 You're given the following gate sequence, which is similar to sequence 7 in Table 7.4 except the CNOT is replaced by CZ.

Your tasks are two-fold:

(a) Simplify this sequence using the existing equivalent sequences found in Table 7.4. Provide at least two different ways to simplify this sequence.

(b) Verify your simplifications by performing matrix multiplication. The aim is to demonstrate that your simplified sequence is indeed equivalent to the original one.

7.4.5 ∗ Deferred Measurement Principle

In this section, we will explore a type of gate sequence equivalence that involves measurements. This is enabled by the deferred measurement principle, which allows us to measure control qubits early in a quantum computation. As a result, quantum controlled gates can often be replaced with their classically controlled analogs. Fig. 7.5 provides an example of equivalent circuits that make use of this principle.

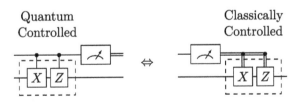

Figure 7.5: Example of Deferred Measurements

> **The Deferred Measurement Principle.**
>
> Measurements can be moved from an intermediate stage of a quantum circuit to the end. If the results of these measurements are used at any point in the circuit, classically controlled operations can be replaced by quantum controlled operations.

Mathematically, the deferred measurement principle signifies that the two circuits in Fig. 7.6 are equivalent, indicating that control (for a quantum Controlled-U gate) and measurement (whose outcome is utilized in the corresponding classically controlled U gate) commute.

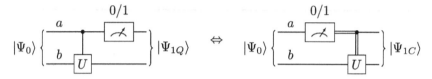

Figure 7.6: Deferred Measurement Principle

Proof. $*$ To prove the equivalence of the two circuits in Fig. 7.6, we need to show that their output states, i.e., $|\Psi_{1C}\rangle$ and $|\Psi_{1Q}\rangle$, are identical.

The measurement operator for both circuits is:

$$M_i = |i\rangle\langle i| \otimes I, \tag{7.48}$$

where i is the measurement outcome, 0 or 1.

The operator for a quantum controlled U gate, CU, is provided in Eq. 7.10; and cCU for a classically controlled U gate is in Eq. 7.20.

For the circuit on the left with a quantum controlled U gate:

$$|\Psi_{1Q}\rangle = \text{CU} \, M_i \, |\Psi_0\rangle \tag{7.49a}$$
$$= (|0\rangle\langle 0| \otimes I + |1\rangle\langle 1| \otimes U)(|i\rangle\langle i| \otimes I) |\Psi_0\rangle \tag{7.49b}$$
$$= (|0\rangle\langle 0| \, |i\rangle\langle i|) \otimes (II) |\Psi_0\rangle + (|1\rangle\langle 1| \, |i\rangle\langle i|) \otimes (UI) |\Psi_0\rangle \tag{7.49c}$$
$$= \begin{cases} (|0\rangle\langle 0| \otimes I) |\Psi_0\rangle & \text{for } i = 0, \\ (|1\rangle\langle 1| \otimes U) |\Psi_0\rangle & \text{for } i = 1. \end{cases} \tag{7.49d}$$

For the circuit on the right with a classically controlled U gate:

$$|\Psi_{1C}\rangle = M_i \, \text{cCU} \, |\Psi_0\rangle \tag{7.50a}$$
$$= (|i\rangle\langle i| \otimes I)(I \otimes U^i) |\Psi_0\rangle \tag{7.50b}$$
$$= (|i\rangle\langle i| \, I) \otimes (IU^i) |\Psi_0\rangle \tag{7.50c}$$
$$= (|i\rangle\langle i| \otimes U^i) |\Psi_0\rangle \tag{7.50d}$$
$$= \begin{cases} (|0\rangle\langle 0| \otimes I) |\Psi_0\rangle & \text{for } i = 0, \\ (|1\rangle\langle 1| \otimes U) |\Psi_0\rangle & \text{for } i = 1. \end{cases} \tag{7.50e}$$

Therefore, we conclude that $|\Psi_{1C}\rangle = |\Psi_{1Q}\rangle$. ∎

7.4 Equivalent Gate Sequences

Exercise 7.19 * The following circuit is a basic building block in error correction codes.

$$\begin{array}{c} \alpha|0\rangle + \beta|1\rangle - \boxed{H} - \bullet - \boxed{H} - \\ |\psi\rangle \hspace{1.2cm} - \boxed{U} \hspace{0.6cm} - \end{array} \Bigg\} |\Psi_1\rangle$$

(a) Demonstrate that the output state can be written as:

$$|\Psi_1\rangle = \frac{\alpha+\beta}{\sqrt{2}}\left(|0\rangle|\psi\rangle + |1\rangle|\psi\rangle\right) + \frac{\alpha-\beta}{\sqrt{2}}\left(|0\rangle U|\psi\rangle - |1\rangle U|\psi\rangle\right).$$

(b) If the first qubit is measured and the outcome is $|0\rangle$ (or $|1\rangle$), what is the state of the second qubit upon measurement?

7.4.6 * Operator Sum Relations

In quantum circuits, gates are applied sequentially, corresponding to the multiplication of the respective matrices, not their addition. This distinction means that certain equivalences between gates, while mathematically accurate, cannot be translated directly into equivalent sequences of quantum gates.

A notable example of this is that some quantum gates can be expressed in terms of Pauli operators:

$$H = \frac{1}{\sqrt{2}}(X+Z), \tag{7.51a}$$

$$\text{CNOT} = \frac{1}{2}(I \otimes I + I \otimes X + Z \otimes I - Z \otimes X), \tag{7.51b}$$

$$\text{CZ} = \frac{1}{2}(I \otimes I + I \otimes Z + Z \otimes I - Z \otimes Z), \tag{7.51c}$$

$$\text{SWAP} = \frac{1}{2}(I \otimes I + X \otimes X + Y \otimes Y + Z \otimes Z). \tag{7.51d}$$

These relations are not directly translatable into quantum circuits in general as they are sums of operators, not compositions. However, They can still provide useful insights into the underlying operation of the gates. As an example, let's find an equivalent for the following gate sequence using the above equations:

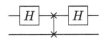

$$(H \otimes I) \cdot \text{SWAP} \cdot (H \otimes I) \tag{7.52a}$$

$$= (H \otimes I)\frac{1}{2}(I \otimes I + X \otimes X + Y \otimes Y + Z \otimes Z)(H \otimes I) \tag{7.52b}$$

$$= \frac{1}{2}(HIH \otimes I + HXH \otimes X + HYH \otimes Y + HZH \otimes Z) \tag{7.52c}$$

$$= \frac{1}{2}(I \otimes I + Z \otimes X - Y \otimes Y + X \otimes Z) \tag{7.52d}$$

$$= \frac{1}{2}\begin{bmatrix} I+X & Z+iY \\ Z-iY & I-X \end{bmatrix} \tag{7.52e}$$

$$= \frac{1}{2}\begin{bmatrix} 1 & 1 & 1 & 1 \\ 1 & 1 & -1 & -1 \\ 1 & -1 & 1 & -1 \\ 1 & -1 & -1 & 1 \end{bmatrix}. \tag{7.52f}$$

Here we have used the following identities:

$$HXH = Z, \tag{7.53a}$$
$$HZH = X, \tag{7.53b}$$
$$HYH = -Y, \tag{7.53c}$$

and

$$(A \otimes B)(C \otimes D) = AC \otimes BD. \tag{7.54}$$

7.5 *Exploratory Topics

In this section, we delve into advanced topics that might not be essential for those new to quantum computing but can provide additional insights and depth for more advanced readers.

7.5.1 *Expressing Common Gates with the Toffoli Gate

In this subsection, we investigate the properties of the Toffoli (CCNOT) gate, as shown in Table 7.2. The Boolean representation of the Toffoli gate is given as

$$\text{Toffoli}: |xyz\rangle \mapsto |xy(xy \oplus z)\rangle. \tag{7.55}$$

The classical logic functions such as NOT, AND, and FANOUT can be expressed in terms of the Toffoli gate alone as demonstrated in Table 7.5. An OR gate can be represented using several stages of Toffoli gates following the relation $\text{OR}(x,y) = \text{NOT}(\text{AND}(\text{NOT}(x), \text{NOT}(y)))$. This demonstrates that we have a quantum equivalent for basic classical logic functions. An additional ancillary qubit, introduced in the Toffoli gate, makes this translation possible.

By establishing a connection with the universal set of classical logic gates—NOT, AND, and FANOUT—we can translate any classical logic circuit into a quantum circuit using only the Toffoli gate. This implies that a quantum computer can emulate the performance of a classical computer, at minimum, if hardware is available.

7.5 ⁎ Exploratory Topics

However, to utilize the full potential of quantum computing, we need gates that facilitate superposition and entanglement. The Toffoli gate can act as a CNOT gate, making it suitable for entanglement operations. When coupled with the Hadamard (H) gate, which allows for superposition, we obtain a universal set of quantum gates. This powerful combination, discussed in § 7.2.3, signifies that the Toffoli gate and Hadamard gate together can be utilized for any quantum computation.

Gate	Symbol	Boolean Representation
NOT (X)	$\|1\rangle \longrightarrow \|1\rangle$ $\|1\rangle \longrightarrow \|1\rangle$ $\|z\rangle \longrightarrow \|\bar{z}\rangle$	$\|z\rangle \mapsto \|\bar{z}\rangle$
AND	$\|x\rangle \longrightarrow \|x\rangle$ $\|y\rangle \longrightarrow \|y\rangle$ $\|0\rangle \longrightarrow \|(xy)\rangle$	$\|xy\rangle \mapsto z = \|(xy)\rangle$
FANOUT	$\|1\rangle \longrightarrow \|1\rangle$ $\|y\rangle \longrightarrow \|y\rangle$ $\|0\rangle \longrightarrow \|y\rangle$	Only for classical bits represented by basis states; not for superposition states.
CNOT	$\|1\rangle \longrightarrow \|1\rangle$ $\|y\rangle \longrightarrow \|y\rangle$ $\|z\rangle \longrightarrow \|(y \oplus z)\rangle$	$\|yz\rangle \mapsto \|y(y \oplus z)\rangle$

Table 7.5: Expressing Common Gates Using Toffoli Gate

Exercise 7.20 Express NOT, AND, and classical FANOUT using the Fredkin (CSWAP) gate exclusively, similar to the Toffoli gate. Evaluate if the Fredkin gate alone can construct a CNOT or H gate.

7.5.2 ⁎ Advanced Quantum Gate Examples

Suppose we need a quantum gate to perform for the following function:

$$\begin{array}{r} |x\rangle \longrightarrow |x\rangle \\ |y\rangle \longrightarrow G \longrightarrow |y\rangle \\ |z\rangle \longrightarrow |z \oplus x\bar{y}\rangle \end{array}$$

Is this gate realizable as a quantum gate? In order to examine this, let's work out the mappings of the basis states:

$$|000\rangle \mapsto |000\rangle, \quad |001\rangle \mapsto |001\rangle, \quad |010\rangle \mapsto |010\rangle, \quad |011\rangle \mapsto |011\rangle,$$
$$|100\rangle \mapsto |101\rangle, \quad |101\rangle \mapsto |100\rangle, \quad |110\rangle \mapsto |110\rangle, \quad |111\rangle \mapsto |111\rangle.$$

We notice the basis states $|100\rangle$ and $|101\rangle$ are changed, while the others remain unchanged. This mapping corresponds to the following matrix for the gate:

$$G = \begin{bmatrix} I_4 & 0 & 0 \\ 0 & U & 0 \\ 0 & 0 & I_2 \end{bmatrix},$$

where I_4 is the 4×4 identity matrix, I_2 is the 2×2 identity matrix, and U is a 2×2 matrix defined by:

$$U = \begin{bmatrix} 0 & 1 \\ 1 & 0 \end{bmatrix}. \qquad (7.56)$$

We recognize $U = X$. Apparently G is a unitary matrix. So it is realizable as a quantum gate.

What kind of gate is G? The changed basis states, $|100\rangle \mapsto |101\rangle$ and $|101\rangle \mapsto |100\rangle$, imply that the third qubit z is flipped only when the first qubit $x = 1$ and the second qubit $y = 1$. From this we conclude that G can be named the $\overline{\text{CCN}}\text{OT}$ gate, with a symbol:

$$\begin{array}{rl} |x\rangle & \longrightarrow |x\rangle \\ |y\rangle & \longrightarrow |y\rangle \\ |z\rangle & \longrightarrow |z \oplus x\bar{y}\rangle \end{array}$$

This gate can be regarded as a variation of the original CCNOT (Toffoli) gate in Table 7.2.

Exercise 7.21 Given the matrix representation of a gate

$$U = \begin{bmatrix} 0 & 1 & 0 & 0 \\ 0 & 0 & 0 & 1 \\ 0 & 0 & 1 & 0 \\ 1 & 0 & 0 & 0 \end{bmatrix}.$$

Find its corresponding Boolean representation.

7.6 Summary and Conclusions

Multi-Qubit Operations

The heart of this chapter revolved around multi-qubit gates, the operational bedrock upon which complex quantum circuits are constructed. These gates, acting on the tensor product of state spaces of the involved qubits, stand distinct from their single-qubit counterparts. One prominent aspect is their ability to generate and manipulate entangled states. While these gates possess the power to intricately intertwine qubits, they, like single-qubit gates, still represent unitary evolutions, ensuring conservation of total probability post-application.

Delving into the specifics, we embarked on a comprehensive study of familiar multi-qubit gates such as CNOT, CZ, and CU, highlighting their pivotal roles in quantum circuit design and algorithms. The structure and application of these gates were emphasized, setting the groundwork for intricate quantum computing tasks.

Universal Quantum Computing Framework

Drawing a parallel to classical computing, where complex operations evolve from simpler ones, the chapter transitioned to universal sets of qubit gates. The universality concept in quantum computing stems from the necessity to approximate any quantum operation using a restricted gate set. By introducing the reader to diverse universal gate sets and elaborating on their prerequisites, the chapter emphasized the importance of adaptability in the quantum realm, especially when harnessing limited gate types for non-trivial quantum computations.

Analytical Tools and Techniques

A central tenet of the chapter was the introduction of the Boolean representation of quantum gates and circuits. This methodology allows for efficient analysis of quantum circuits, considering each basis state independently, then combining them for a holistic understanding, thereby harnessing the linearity of quantum transformations. This technique not only aids in conceptual clarity but also proves instrumental in dissecting complex quantum circuits.

Venturing into gate sequences, we underlined the notion of equivalence in gate sequences. Recognizing sequences that combine to form a singular gate equivalent is essential for understanding complex quantum circuits, as well as for optimization, particularly when addressing real-world quantum hardware constraints.

Advanced Explorations

In this chapter, we also introduced some exploratory topics to provide a deeper understanding of quantum gates and their nuances. While these discussions might be seen as advanced, they aim to provide a foundational understanding for readers interested in further exploring the subject of quantum mechanics.

Upcoming Topics

In the forthcoming chapter, we journey into the realm of Bell states. These states, defined amidst the backdrop of the intriguing EPR paradox and Bell's theorem, stand as beacons elucidating quantum mechanics' complexities. As quintessential examples of quantum entanglement, Bell states not only accentuate quantum superposition and nonlocal correlations but also form the backbone of numerous quantum information processing tasks. This sets the stage for an enriching discourse on the EPR paradox and the myriad applications of quantum entanglement.

Problem Set 7

7.1 Work out the operator and matrix for a variant of the CNOT gate, denoted as $\overline{\text{CNOT}}$. In this gate, qubit 1 serves as the control qubit and qubit 2 as the target qubit. However, in contrast to the regular CNOT gate, an X (i.e., NOT) gate is applied to the target qubit, flipping its states 0 and 1, when the control qubit is in state $|0\rangle$ rather than $|1\rangle$.

7.2 Intuitively, you should be able to achieve the function of $\overline{\text{CNOT}}$ in Problem 7.1 by adding an X gate on the control qubit of a regular CNOT to flip the $|0\rangle$ and $|1\rangle$, and appending another X gate to restore the flip:

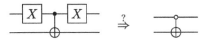

In terms of operators, this means

$$\overline{\text{CNOT}} = (X \otimes I) \cdot \text{CNOT} \cdot (X \otimes I).$$

Prove or disprove the above equation using the $\overline{\text{CNOT}}$ operator and/or matrix you have obtained in Problem 7.1.

7.3 A CU gate applies a generic unitary gate U to the target qubit if the control qubit is $|1\rangle$. A $\overline{\text{CU}}$ gate applies U to the target qubit if the control qubit is $|0\rangle$. Work out the operator and matrix for $\overline{\text{CU}}$.

7.4 Derive the operators corresponding to the quantum gates defined by the following block matrices, where U is a 2×2 unitary matrix, and I the 2×2 identify matrix. For each gate, discuss its properties similarly to the approach used for CU in § 7.1.3. Analyze the patterns within each operator, formulate a description analogous to that of CU, and devise a corresponding circuit diagram.

$$G_1 = \begin{bmatrix} 0 & U \\ I & 0 \end{bmatrix}, \quad G_2 = \begin{bmatrix} 0 & I \\ U & 0 \end{bmatrix}.$$

7.5 Derive the matrix of $R_{xy}(\theta)$, similar to Eq. 7.24.

7.6 (CNOT Mirror) The circuit equivalence shown below demonstrates that a pair of CNOT gates on three qubits—where the target qubit of the first CNOT is the control qubit of the second CNOT—can be rearranged. By reversing the sequence of these two CNOT gates and adding a CNOT at the beginning, the overall action remains unchanged. Prove this equivalence.

7.7 (Distributed CNOT) In the distributed CNOT protocol, the CNOT operation is performed between two qubits without direct interaction by using an intermediate qubit and four CNOT gates, as illustrated by the equivalent circuits below. Prove this equivalence.

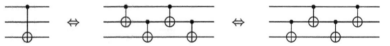

7.8 Examine the following circuit, which represents a basic encoder in error correction codes designed to rectify phase-inversion errors.

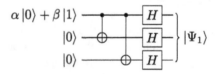

Demonstrate that the output state is:

$$|\Psi_1\rangle = \alpha |+++\rangle + \beta |---\rangle.$$

7.9 The following gate sequence is often used to decompose three-qubit gates into two-qubit gates. Prove its equivalence.

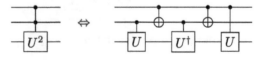

III Quantum Entanglement

8	Bell States	207
9	Entanglement and Bell Inequalities	223
10	Key Applications of Entanglement	251

8. Bell States

Contents

8.1	**Maximally Entangled States**	208
8.1.1	Definitions	208
8.1.2	Maximum Entanglement	209
8.2	✻ **Bell Basis**	209
8.2.1	✻ Orthonormality and Completeness	210
8.2.2	✻ Two-Qubit States in Bell Basis	210
8.3	**Bell State Creation**	211
8.3.1	Basic Bell State Generator	211
8.3.2	Generating All Bell States from $\|00\rangle$	212
8.3.3	Photon Pair Creation with SPDC	213
8.3.4	✻ Boolean Representation	214
8.4	**Bell Measurement**	214
8.4.1	Joint Measurement	214
8.4.2	Basic Implementation: Bell Analyzer	215
8.4.3	Post-Measurement State	215
8.5	**Bell State Conversion**	215
8.5.1	Converting $\|\beta_{00}\rangle$ to Other Bell States	216
8.5.2	Converting Other Bell States to $\|\beta_{00}\rangle$	216
8.5.3	Additional Conversions Related to $\|\beta_{11}\rangle$	217
8.6	✻ **Deferred Proofs**	217
8.7	✻ **Generalization: GHZ States**	218
8.8	**Summary and Conclusions**	219
	Problem Set 8	220

Bell states were first defined in the context of the famous "EPR paradox," a thought experiment devised by Einstein, Podolsky, and Rosen in 1935 to argue that quantum mechanics was incomplete. It wasn't until 1964 that John Bell formulated what is now known as Bell's theorem, which showed that the kind of "hidden variables" suggested by EPR could not explain the predictions of quantum mechanics, and therefore quantum mechanics is a complete theory.

The Bell states were then used to test Bell's theorem experimentally, and the results of these tests have been overwhelmingly in favor of quantum mechanics. These tests, and the Bell states they rely on, are a major reason why physicists today accept quantum mechanics as a correct and complete description of the microscopic world.

Bell states are a set of four distinct entangled quantum states of two qubits. They are the simplest and most widely recognized examples of quantum entanglement — a remarkable quantum phenomenon wherein the properties of one qubit become intricately connected to the properties of another, irrespective of their spatial separation.

Bell states, being maximally entangled, serve to elucidate fundamental aspects of quantum mechanics such as quantum superposition and nonlocal correlations. Furthermore, Bell states play an indispensable role in a myriad of quantum computing and quantum information processing applications, including quantum teleportation, quantum cryptography, and quantum error correction.

In this chapter, we delve into a detailed exploration of Bell states, setting the stage for our discussion on the EPR paradox and applications of quantum entanglement in subsequent chapters. This also allows us to put into practice the concepts and techniques we have thus far studied concerning quantum gates.

8.1 Maximally Entangled States

8.1.1 Definitions

The four Bell states are:

$$|\Phi^+\rangle \equiv |\beta_{00}\rangle = \frac{1}{\sqrt{2}}(|00\rangle + |11\rangle), \tag{8.1a}$$

$$|\Psi^+\rangle \equiv |\beta_{01}\rangle = \frac{1}{\sqrt{2}}(|01\rangle + |10\rangle), \tag{8.1b}$$

$$|\Phi^-\rangle \equiv |\beta_{10}\rangle = \frac{1}{\sqrt{2}}(|00\rangle - |11\rangle), \tag{8.1c}$$

$$|\Psi^-\rangle \equiv |\beta_{11}\rangle = \frac{1}{\sqrt{2}}(|01\rangle - |10\rangle). \tag{8.1d}$$

These states are all non-product states, representing different types of correlations between the qubits.

In the context of quantum physics, the $|\Psi^-\rangle$ state is often referred to as a "spin singlet" while the other three states, "spin triplets". The reason for this terminology is that, when we interpret the qubits as spin-1/2 particles, the $|\Psi^-\rangle$ state corresponds to a pair of particles with a total combined spin of 0. Conversely, the other three states represent a combined particle pair with a total spin of 1.

Exercise 8.1 Show that $\frac{1}{2}(|00\rangle + |01\rangle + |10\rangle + |11\rangle)$ is a product state.

Exercise 8.2 Show that $|\Phi^+\rangle = \frac{1}{\sqrt{2}}(|00\rangle + |11\rangle)$ is a non-product state.

8.1.2 Maximum Entanglement

Bell states are maximally entangled states of two qubits. Consequently, when we perform local measurements on the qubits in a Bell state, the measurement outcomes are perfectly correlated (or anti-correlated) due to their entanglement.

The measurement operators acting on the first qubit in the standard basis:

$$M_0 = |0\rangle\langle 0| \otimes I \quad \text{for measurement outcome } |0\rangle, \tag{8.2a}$$

$$M_1 = |1\rangle\langle 1| \otimes I \quad \text{for measurement outcome } |1\rangle. \tag{8.2b}$$

Let's take the $|\Phi^+\rangle$ Bell state as an example. If we measure the first qubit and find it in state $|0\rangle$, we have effectively applied the M_0 measurement operator:

$$M_0 |\Phi^+\rangle = (|0\rangle\langle 0| \otimes I) \frac{1}{\sqrt{2}}(|00\rangle + |11\rangle). \tag{8.3}$$

This results in the post-measurement state $|00\rangle$:

$$\frac{1}{\sqrt{2}}(|00\rangle + |11\rangle) \xrightarrow{\text{measure 1st qubit to be } |0\rangle} |00\rangle. \tag{8.4}$$

As a result, we can be certain that the second qubit will also be in the state $|0\rangle$.

Similarly, if we measure the first qubit and find it in the state $|1\rangle$, the second qubit will also be in the state $|1\rangle$, because:

$$\frac{1}{\sqrt{2}}(|00\rangle + |11\rangle) \xrightarrow{\text{measure 1st qubit to be } |1\rangle} |11\rangle. \tag{8.5}$$

This demonstrates perfect correlation between the measurement outcomes of the individual qubits in state $|\Phi^+\rangle$, measured along the z direction. Similarly, $|\Phi^-\rangle$ also exhibits perfect correlation, while $|\Psi^+\rangle$ and $|\Psi^-\rangle$ exhibit perfect anti-correlation. Later on (in § 9.8), you will discover that this strong correlation also exists in other directions.

8.2 ∗ Bell Basis

In this section, we delve deeper into the mathematical properties of Bell states, specifically examining their orthonormality and completeness. The orthonormality of Bell states stems from the fact that the inner product of distinct Bell states is zero, whereas the inner product of a Bell state with itself is one. Their completeness, on the other hand, ensures that any two-qubit state can be represented as a linear combination of Bell states. As such, the set of Bell states forms an orthonormal basis for the two-qubit Hilbert space, a fundamental concept in quantum information theory and quantum computing. In the subsequent subsections, we'll explore these properties and their implications in more detail.

8.2.1 ⁎Orthonormality and Completeness

The four Bell states are orthonormal because their inner products are either zero (when the states are different) or one (when the states are the same), which is proved in § 8.6. In the $|\beta_{ij}\rangle$ notation, this is expressed as:

$$\langle \beta_{ij}|\beta_{kl}\rangle = \delta_{ik}\delta_{jl}, \tag{8.6}$$

where δ_{ik} and δ_{jl} are Kronecker deltas and $i, j, k, l \in \{0, 1\}$.

Furthermore, $\{|\beta_{ij}\rangle\}$ is a complete basis set, as demonstrated by

$$\sum_{i,j \in \{0,1\}} |\beta_{ij}\rangle\langle \beta_{ij}| = I, \tag{8.7}$$

where I is the 4×4 identity matrix. This means that the Bell states span the entire two-qubit Hilbert space. Thus, the Bell states form an orthonormal basis for the two-qubit Hilbert space.

> The set of Bell states serves as an orthonormal basis for two-qubit states, much like the set $|+\rangle, |-\rangle$ forms an orthonormal basis for single-qubit states.

8.2.2 ⁎Two-Qubit States in Bell Basis

Any arbitrary two-qubit quantum state can be expressed as

$$|\psi\rangle = \alpha |00\rangle + \beta |01\rangle + \gamma |10\rangle + \delta |11\rangle. \tag{8.8}$$

Given that the Bell states form a complete basis for two-qubit states, we can also express $|\psi\rangle$ as a linear combination of these states:

$$|\psi\rangle = \sum_{i,j \in \{0,1\}} c_{ij} |\beta_{ij}\rangle, \tag{8.9}$$

where c_{ij} are complex coefficients. These coefficients must satisfy the normalization condition:

$$\sum_{i,j \in \{0,1\}} |c_{ij}|^2 = 1. \tag{8.10}$$

We can calculate the coefficients c_{ij} using the inner product between the Bell states and $|\psi\rangle$:

$$c_{ij} = \langle \beta_{ij}|\psi\rangle. \tag{8.11}$$

We can combine Eqs. 8.9 and 8.11 as a single equation using projectors $|\beta_{ij}\rangle\langle\beta_{ij}|$:

$$|\psi\rangle = \sum_{i,j \in \{0,1\}} |\beta_{ij}\rangle \langle \beta_{ij}|\psi\rangle. \tag{8.12}$$

In terms of the coefficients α, β, γ, and δ, the coefficients c_{ij} are:

$$c_{00} = \frac{1}{\sqrt{2}}(\alpha + \delta), \tag{8.13a}$$

$$c_{01} = \frac{1}{\sqrt{2}}(\beta + \gamma), \tag{8.13b}$$

$$c_{10} = \frac{1}{\sqrt{2}}(\beta - \gamma), \tag{8.13c}$$

$$c_{11} = \frac{1}{\sqrt{2}}(\alpha - \delta). \tag{8.13d}$$

This is equivalent to a basis-rotation using the "Bell matrix":

$$\begin{bmatrix} c_{00} \\ c_{00} \\ c_{10} \\ c_{11} \end{bmatrix} = \frac{1}{\sqrt{2}} \begin{bmatrix} 1 & 0 & 0 & 1 \\ 0 & 1 & 1 & 0 \\ 0 & 1 & -1 & 0 \\ 1 & 0 & 0 & -1 \end{bmatrix} \begin{bmatrix} \alpha \\ \beta \\ \gamma \\ \delta \end{bmatrix}. \tag{8.14}$$

> ⓘ The Bell state generator we'll discuss in subsequent sections can be understood as a circuit that transforms the computational basis into the Bell basis. Conversely, the Bell state analyzer can be seen as performing the reverse transformation, from the Bell basis back to the computational basis.

Exercise 8.3 Derive the equations for the coefficients α, β, γ, and δ in terms of the amplitudes c_{ij}. This will provide the inverse relation of Eq. 8.13.

8.3 Bell State Creation

This section focuses on the practical creation of Bell states, which is a fundamental operation in quantum computing and quantum information theory. In an ideal quantum computer, Bell states can be readily generated using universal quantum gates like the Hadamard (H) and the controlled NOT (CNOT) gates. We will first investigate the basic methods of generating Bell states using these quantum gates. However, in practice, especially when working with systems like photons, we often resort to physical phenomena such as spontaneous parametric down-conversion (SPDC) to create entangled pairs of particles, which form the Bell states. Throughout this section, we will explore these methods and the underlying physics, providing a comprehensive understanding of Bell state creation in both theoretical and experimental scenarios.

8.3.1 Basic Bell State Generator

Bell states can be generated using qubit gates. For example, the Hadamard gate and the CNOT gate can be used together to create the Bell state Φ^+ from an initial state $|00\rangle$ as follows.

Apply a Hadamard gate (H) to the first qubit. This creates a superposition state:

$$(H \otimes I)|00\rangle = \frac{1}{\sqrt{2}}(|00\rangle + |10\rangle). \tag{8.15}$$

Apply a CNOT gate to the two-qubit state, with the first qubit as the control and the second qubit as the target. This entangles the qubits and generates the Bell state:

$$\text{CNOT} \frac{1}{\sqrt{2}}(|00\rangle + |10\rangle) = \frac{1}{\sqrt{2}}(|00\rangle + |11\rangle). \tag{8.16}$$

All four Bell states can be created using qubit gates as in Fig. 8.1:

In	Out	
$\|00\rangle$	$\|\beta_{00}\rangle \equiv \|\Phi^+\rangle = \frac{1}{\sqrt{2}}(\|00\rangle + \|11\rangle)$	
$\|01\rangle$	$\|\beta_{01}\rangle \equiv \|\Psi^+\rangle = \frac{1}{\sqrt{2}}(\|01\rangle + \|10\rangle)$	
$\|10\rangle$	$\|\beta_{10}\rangle \equiv \|\Phi^-\rangle = \frac{1}{\sqrt{2}}(\|00\rangle - \|11\rangle)$	
$\|11\rangle$	$\|\beta_{11}\rangle \equiv \|\Psi^-\rangle = \frac{1}{\sqrt{2}}(\|01\rangle - \|10\rangle)$	

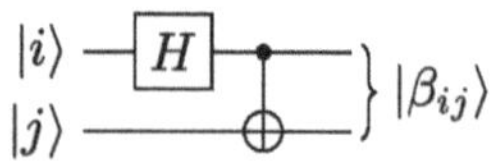

$|i\rangle$ —[H]—•—
$|j\rangle$ ———⊕— $\}|\beta_{ij}\rangle$

Bell-State Generator

Figure 8.1: Bell State Creation using H and CNOT Gates

Clarification of Quantum Circuits and Math Formulas

In quantum circuit diagrams, time generally flows from left to right, and operations (gates) are applied to qubits as you move along this direction. However, when you represent the operations mathematically as matrices, the sequence of operations follows the convention of matrix multiplication, which is right-to-left.

This discrepancy can sometimes be confusing for those new to quantum computing. But it's an important aspect to understand for accurate interpretation of quantum circuits and corresponding mathematical expressions.

Exercise 8.4 Verify the generation of $|\Psi^+\rangle$, $|\Phi^-\rangle$, and $|\Psi^-\rangle$ in Fig. 8.1.

8.3.2 Generating All Bell States from $|00\rangle$

Adding X gates for the input state preparation, we can create all four Bell states directly from $|00\rangle$, as shown in Fig. 8.2. Remember that an X gate transforms a qubit from $|0\rangle$ to $|1\rangle$.

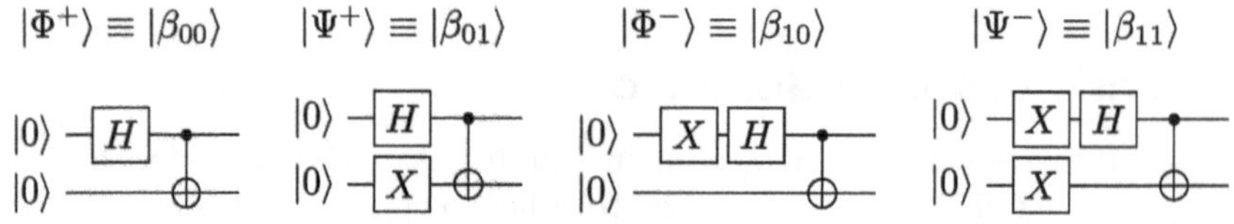

$|\Phi^+\rangle \equiv |\beta_{00}\rangle$ $|\Psi^+\rangle \equiv |\beta_{01}\rangle$ $|\Phi^-\rangle \equiv |\beta_{10}\rangle$ $|\Psi^-\rangle \equiv |\beta_{11}\rangle$

$|0\rangle$ —[H]—•— $|0\rangle$ —[H]—•— $|0\rangle$ —[X]—[H]—•— $|0\rangle$ —[X]—[H]—•—
$|0\rangle$ ———⊕— $|0\rangle$ —[X]—⊕— $|0\rangle$ ———————⊕— $|0\rangle$ —[X]———⊕—

Figure 8.2: Bell State Creation from Initial State $|00\rangle$

Exercise 8.5 Verify the generation of $|\Phi^-\rangle$ and $|\Psi^-\rangle$ in Fig. 8.2 using operator algebra.

8.3 Bell State Creation

Alternatively, we can use the Z gate to generate a π phase shift (i.e., negative sign) in the output to create $|\Phi^-\rangle$ and $|\Psi^-\rangle$ as shown in Fig. 8.3. Remember that applying a Z gate to a qubit changes its $|1\rangle$ to $-|1\rangle$. (Applying the Z gate to either the first or the second qubit yields the same Bell states, as the difference is only in global phase which is not measurable in a quantum system.)

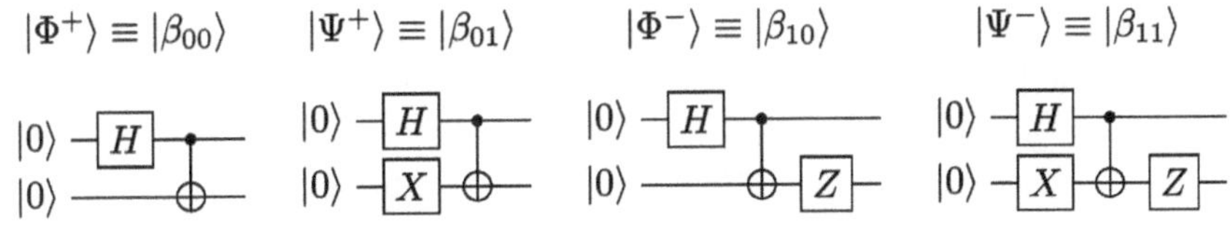

$$|\Phi^+\rangle \equiv |\beta_{00}\rangle \quad |\Psi^+\rangle \equiv |\beta_{01}\rangle \quad |\Phi^-\rangle \equiv |\beta_{10}\rangle \quad |\Psi^-\rangle \equiv |\beta_{11}\rangle$$

Figure 8.3: Bell State Creation Using Output Phase Shift

Exercise 8.6 Verify the generation of $|\Phi^-\rangle$ and $|\Psi^-\rangle$ in Fig. 8.3 using operator algebra.

8.3.3 Photon Pair Creation with SPDC

Direct application of Hadamard and CNOT gates to photons is not readily feasible. Instead, entangled photons are generated through spontaneous parametric down-conversion (SPDC) or other nonlinear optical processes. SPDC is a widely used method for producing pairs of entangled photons for a variety of quantum communication and computation experiments.

In SPDC, as illustrated in Fig. 8.4 a single high-energy photon passes through a nonlinear crystal and occasionally splits into two lower-energy photons that are entangled. The entangled photons can be created in various Bell states, depending on the experimental setup and the properties of the nonlinear crystal.

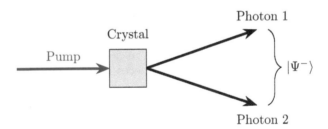

Figure 8.4: Photon Pair Creation with SPDC

For polarization-entangled photon pairs created by SPDC, the Bell state is

$$|\Psi^-\rangle = \frac{1}{\sqrt{2}} \left(|\leftrightarrow\rangle_1 \otimes |\updownarrow\rangle_2 - |\updownarrow\rangle_1 \otimes |\leftrightarrow\rangle_2 \right), \quad (8.17)$$

where $|\leftrightarrow\rangle_1$, $|\updownarrow\rangle_1$, $|\leftrightarrow\rangle_2$ and $|\updownarrow\rangle_2$ represent the horizontal and vertical polarizations of the two photons in the pair.

If we interpret $|\leftrightarrow\rangle$ as $|0\rangle$ and $|\updownarrow\rangle$ as $|1\rangle$, then Eq. 8.17 yields the same $|\Psi^-\rangle$ as in the standard notation presented in Eq. 8.1.

8.3.4 ∗ Boolean Representation

Using the Boolean representation introduced in § 7.3, the four Bell states can be succinctly expressed in a single equation:

$$|\beta_{xy}\rangle = \frac{1}{\sqrt{2}} \left(|0\rangle |y\rangle + (-1)^x |1\rangle |\bar{y}\rangle\right). \tag{8.18}$$

The Bell state creation circuit can be presented as in Fig. 8.5.

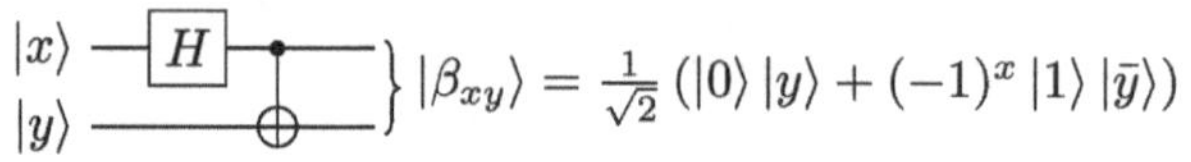

$$\begin{array}{c}|x\rangle - \boxed{H} - \bullet - \\ |y\rangle ----\oplus-\end{array}\Big\} |\beta_{xy}\rangle = \frac{1}{\sqrt{2}}\left(|0\rangle|y\rangle + (-1)^x|1\rangle|\bar{y}\rangle\right)$$

Figure 8.5: Boolean Representation of Bell States

8.4 Bell Measurement

This section delves into Bell measurements, which hold a pivotal position in quantum information processing due to their ability to distinguish between different Bell states. These measurements operate by projecting a two-qubit state onto the Bell basis, effectively resulting in one of the four Bell states. Such a process forms the core of numerous quantum protocols like quantum teleportation, superdense coding, and quantum key distribution. Moreover, we will take a look at how Bell measurements can be practically implemented using a quantum circuit known as a "Bell analyzer."

8.4.1 Joint Measurement

The orthonormality of the Bell states (discussed in § 8.2) allows us to distinguish between different Bell states through measurements. If a quantum state is prepared in one of the Bell states, it can be unambiguously determined through a joint measurement on both qubits, known as a Bell measurement or a Bell state analysis.

A Bell measurement differs from the local measurements discussed in § 8.1.2 in that it is a type of joint quantum measurement (see § 6.2.2) performed on a two-qubit system. This implies the necessity of having both qubits at the same location or some form of quantum communication between the two qubits, which characterizes the non-local nature of Bell measurements.

If we perform a Bell measurement on a general two-qubit state $|\psi\rangle$ given by

$$|\psi\rangle = \alpha|00\rangle + \beta|01\rangle + \gamma|10\rangle + \delta|11\rangle \tag{8.19a}$$

$$= \sum_{i,j \in \{0,1\}} c_{ij} |\beta_{ij}\rangle, \tag{8.19b}$$

we will obtain one of the Bell states $|\beta_{ij}\rangle$ with probability $|c_{ij}|^2$, where $i,j \in \{0,1\}$. The quantum state will then collapse into that Bell state. The coefficients c_{ij} are given by Eq. 8.13 in terms of α, β, γ, and δ.

Effectively, a Bell measurement can be thought of as projecting the two-qubit state $|\psi\rangle$ onto one of the four Bell states and performing the measurement in the Bell basis. The measurement operator for outcome $|\beta_{ij}\rangle$ is the projection operator:

8.5 Bell State Conversion

$$M_{ij} = |\beta_{ij}\rangle\langle\beta_{ij}|. \tag{8.20}$$

The set of all measurement operators $\{M_{ij} \,|\, i,j \in \{0,1\}\}$ forms a complete measurement set. That is, the sum of all measurement operators is the identity operator on the two-qubit Hilbert space (see Eq. 8.7). Compare these operators to the single-qubit local operators (e.g., $|0\rangle\langle 0| \otimes I$) used in § 8.1.2.

After a Bell measurement, all subsequent measurements on the unaltered system will yield the same outcome, as the state has been collapsed into the corresponding Bell state.

8.4.2 Basic Implementation: Bell Analyzer

A Bell measurement can be implemented by the circuit in Fig. 8.6. The Bell measurement circuit (or Bell analyzer) is the inverse of the Bell state generation circuit shown in Fig. 8.1. After the CNOT and Hadamard gates, we measure the two qubits in the computational basis. The first qubit gives us i and second j. Together we obtain one of four labels 00, 01, 10, or 11 indicating the measurement outcome $|\beta_{ij}\rangle$.

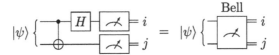

Figure 8.6: Bell Measurement

It's worth noting that the Bell analyzer can be utilized with any two-qubit state, not just Bell states. In the case of an arbitrary two-qubit state, the measurement results could be any combination of i and j, with each outcome occurring with a probability of $|c_{ij}|^2$. Here, i and j correspond to the indices of the Bell state to which the system collapses.

8.4.3 Post-Measurement State

In Fig. 8.6, we presume that our primary concern is the classical measurement results, i.e., the two-bit outcomes 00, 01, 10, or 11, and the resultant quantum state is disregarded. Nevertheless, there might be circumstances in which we want to use the Bell state $|\beta_{ij}\rangle$ that the system collapses into after the measurement in subsequent portions of the quantum circuit. In such cases, we would need to recreate that state, as illustrated in Fig. 8.7.

Note with this full implementation, the Bell measurement not only provides insight into the state of the system but also engenders a state change. If the qubits were initially not entangled, the measurement procedure will bring about their entanglement.

8.5 Bell State Conversion

Bell states can be converted from one to another using Pauli gates. In fact, this conversion can be achieved by applying a Pauli gate to one of the qubits locally. The ability to convert between Bell states forms an important aspect of quantum

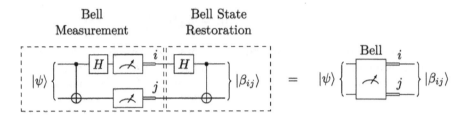

Figure 8.7: Bell Measurement with Post-Measurement State Restored

information processing and serves as a fundamental skill for anyone seeking to work in this field. This section sheds light on such conversions. It details converting $|\beta_{00}\rangle$ to other Bell states and vice versa.

8.5.1 Converting $|\beta_{00}\rangle$ to Other Bell States

Let's take $|\beta_{00}\rangle \equiv |\Phi^+\rangle$ as our reference state. Fig. 8.8 shows several ways to convert $|\beta_{00}\rangle$ to other Bell states.

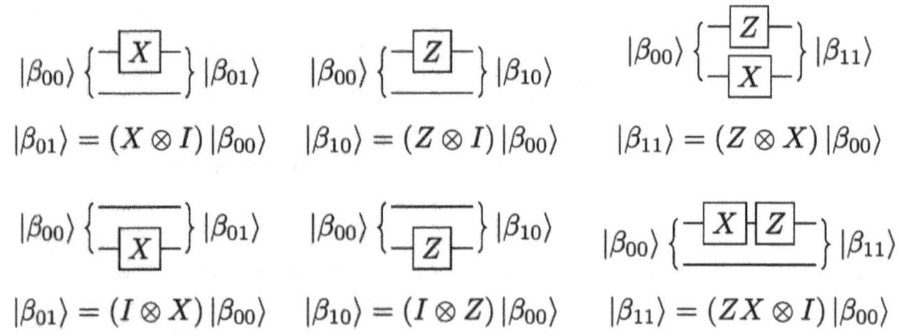

$|\beta_{00}\rangle \{\boxed{X}\} |\beta_{01}\rangle \qquad |\beta_{00}\rangle \{\boxed{Z}\} |\beta_{10}\rangle \qquad |\beta_{00}\rangle \{\boxed{\begin{array}{c} Z \\ X \end{array}}\} |\beta_{11}\rangle$

$|\beta_{01}\rangle = (X \otimes I) |\beta_{00}\rangle \qquad |\beta_{10}\rangle = (Z \otimes I) |\beta_{00}\rangle \qquad |\beta_{11}\rangle = (Z \otimes X) |\beta_{00}\rangle$

$|\beta_{00}\rangle \{\boxed{X}\} |\beta_{01}\rangle \qquad |\beta_{00}\rangle \{\boxed{Z}\} |\beta_{10}\rangle \qquad |\beta_{00}\rangle \{\boxed{X\ Z}\} |\beta_{11}\rangle$

$|\beta_{01}\rangle = (I \otimes X) |\beta_{00}\rangle \qquad |\beta_{10}\rangle = (I \otimes Z) |\beta_{00}\rangle \qquad |\beta_{11}\rangle = (ZX \otimes I) |\beta_{00}\rangle$

Figure 8.8: Converting $|\beta_{00}\rangle \equiv |\Phi^+\rangle$ to Other Bell States

> (i) The expression $|\beta_{11}\rangle = (Z \otimes X) |\beta_{00}\rangle$ is sometimes written in shorthand form as $|\beta_{11}\rangle = Z_1 X_2 |\beta_{00}\rangle$, where Z_1 refers to Z applied to the first qubit, and X_2 meaning X applied to the second qubit. This convention also applies to other similar expressions.

> **Exercise 8.7** Demonstrate $|\beta_{11}\rangle = (ZX \otimes I) |\beta_{00}\rangle$ which underlines the last conversion circuit in Fig. 8.8.

8.5.2 Converting Other Bell States to $|\beta_{00}\rangle$

Similarly, Fig. 8.9 shows several ways to convert other Bell states to $|\beta_{00}\rangle$. These are the inverse operations of the transformations in Fig. 8.8. You notice the forward conversion and reverse conversion are the same for the first five cases. This is because $X^2 = I$ and $Z^2 = I$. In the last case, the order of X and Z is different between forward and reverse conversion, because $XZZX = I$.

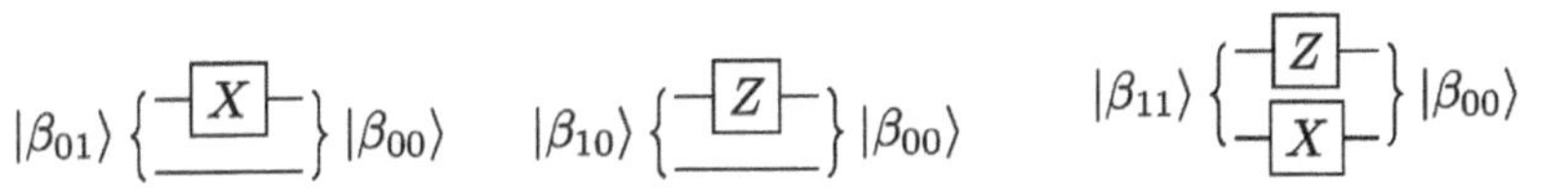

$|\beta_{01}\rangle \left\{\boxed{X}\right\} |\beta_{00}\rangle \qquad |\beta_{10}\rangle \left\{\boxed{Z}\right\} |\beta_{00}\rangle \qquad |\beta_{11}\rangle \left\{\boxed{\begin{array}{c}Z\\X\end{array}}\right\} |\beta_{00}\rangle$

$|\beta_{00}\rangle = (X \otimes I)|\beta_{01}\rangle \qquad |\beta_{00}\rangle = (Z \otimes I)|\beta_{10}\rangle \qquad |\beta_{00}\rangle = (Z \otimes X)|\beta_{11}\rangle$

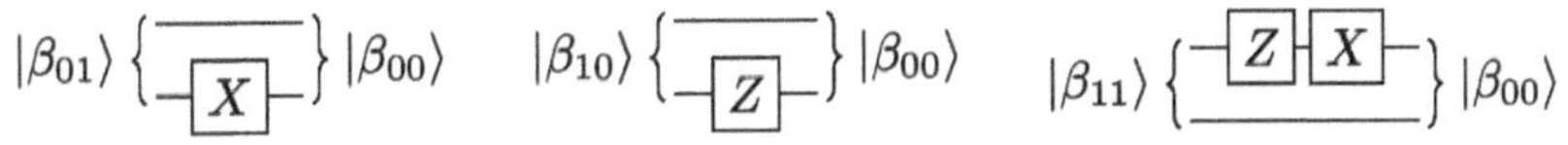

$|\beta_{01}\rangle \left\{\boxed{X}\right\} |\beta_{00}\rangle \qquad |\beta_{10}\rangle \left\{\boxed{Z}\right\} |\beta_{00}\rangle \qquad |\beta_{11}\rangle \left\{\boxed{Z|X}\right\} |\beta_{00}\rangle$

$|\beta_{00}\rangle = (I \otimes X)|\beta_{01}\rangle \qquad |\beta_{00}\rangle = (I \otimes Z)|\beta_{10}\rangle \qquad |\beta_{00}\rangle = (XZ \otimes I)|\beta_{11}\rangle$

Figure 8.9: Converting Other Bell States to $|\beta_{00}\rangle \equiv |\Phi^+\rangle$

Exercise 8.8 In Figs. 8.8 and 8.9, we have adopted $|\beta_{00}\rangle \equiv |\Phi^+\rangle$ as the standard Bell state, following the usual convention in quantum computing. However, in quantum communication, photon pairs generated via Spontaneous Parametric Down-Conversion (SPDC) often exhibit the state $|\beta_{11}\rangle \equiv |\Psi^-\rangle$. Recreate comparable figures using $|\Psi^-\rangle$ as the reference state.

8.5.3 Additional Conversions Related to $|\beta_{11}\rangle$

Some acute readers may have discovered that some combinations for $|\beta_{11}\rangle$ are missing in Figs. 8.8 and 8.9. Well, some combinations introduce a global phase, which in practical sense, is not a problem. These are shown in Fig. 8.10.

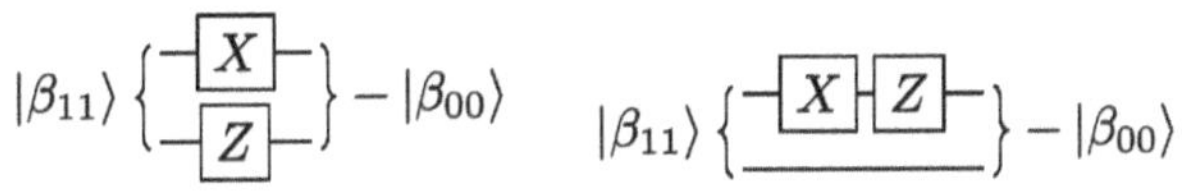

$|\beta_{11}\rangle \left\{\boxed{\begin{array}{c}X\\Z\end{array}}\right\} -|\beta_{00}\rangle \qquad |\beta_{11}\rangle \left\{\boxed{X|Z}\right\} -|\beta_{00}\rangle$

$|\beta_{00}\rangle = -(X \otimes Z)|\beta_{11}\rangle \qquad |\beta_{00}\rangle = -(ZX \otimes I)|\beta_{11}\rangle$

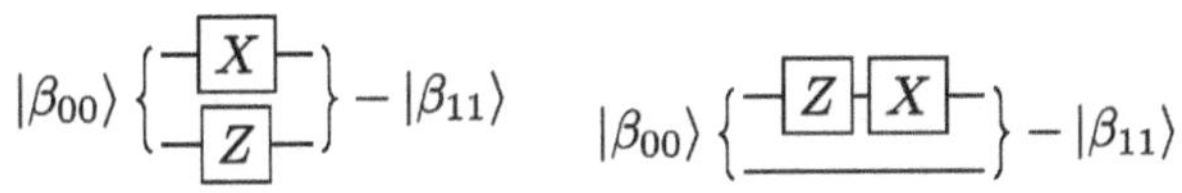

$|\beta_{00}\rangle \left\{\boxed{\begin{array}{c}X\\Z\end{array}}\right\} -|\beta_{11}\rangle \qquad |\beta_{00}\rangle \left\{\boxed{Z|X}\right\} -|\beta_{11}\rangle$

$|\beta_{11}\rangle = -(X \otimes Z)|\beta_{00}\rangle \qquad |\beta_{11}\rangle = -(XZ \otimes I)|\beta_{00}\rangle$

Figure 8.10: Additional Conversions Related to $|\beta_{11}\rangle \equiv |\Psi^-\rangle$

8.6 ∗ Deferred Proofs

Equation 8.11 can be proved as follows:

$$\langle \beta_{ij}|\psi\rangle = \langle \beta_{ij}| \sum_{k,l\in\{0,1\}} c_{kl}|\beta_{kl}\rangle \rangle \quad (8.21\text{a})$$

$$= \sum_{k,l\in\{0,1\}} c_{kl} \langle \beta_{ij}|\beta_{kl}\rangle \quad (8.21\text{b})$$

$$= \sum_{k,l\in\{0,1\}} c_{kl} \delta_{ik}\delta_{jl} \quad (8.21\text{c})$$

$$= c_{ij}. \quad (8.21\text{d})$$

Now, let's prove Eq. 8.6, the orthonormality of the Bell states.

For normalization, let's calculate the inner product of each Bell state with itself:

$$\langle \beta_{00}|\beta_{00}\rangle = \frac{1}{2}(\langle 00|00\rangle + \langle 00|11\rangle + \langle 11|00\rangle + \langle 11|11\rangle) = \frac{1}{2}(1+0+0+1) = 1,$$

$$\langle \beta_{01}|\beta_{01}\rangle = \frac{1}{2}(\langle 00|00\rangle - \langle 00|11\rangle - \langle 11|00\rangle + \langle 11|11\rangle) = \frac{1}{2}(1-0-0+1) = 1,$$

$$\langle \beta_{10}|\beta_{10}\rangle = \frac{1}{2}(\langle 01|01\rangle + \langle 01|10\rangle + \langle 10|01\rangle + \langle 10|10\rangle) = \frac{1}{2}(1+0+0+1) = 1,$$

$$\langle \beta_{11}|\beta_{11}\rangle = \frac{1}{2}(\langle 01|01\rangle - \langle 01|10\rangle - \langle 10|01\rangle + \langle 10|10\rangle) = \frac{1}{2}(1-0-0+1) = 1.$$

For orthogonality, let's calculate the inner product of distinct Bell states:

$$\langle \beta_{01}|\beta_{00}\rangle = \frac{1}{2}(\langle 01|00\rangle + \langle 01|11\rangle + \langle 10|00\rangle + \langle 10|11\rangle) = \frac{1}{2}(0+0+0+0) = 0,$$

$$\langle \beta_{01}|\beta_{10}\rangle = \frac{1}{2}(\langle 01|00\rangle - \langle 01|11\rangle + \langle 10|00\rangle - \langle 10|11\rangle) = \frac{1}{2}(0-0+0-0) = 0,$$

$$\langle \beta_{01}|\beta_{11}\rangle = \frac{1}{2}(\langle 01|01\rangle - \langle 01|10\rangle + \langle 10|01\rangle - \langle 10|10\rangle) = \frac{1}{2}(1-0+0-1) = 0,$$

$$\langle \beta_{10}|\beta_{00}\rangle = \frac{1}{2}(\langle 00|00\rangle + \langle 00|11\rangle - \langle 11|00\rangle - \langle 11|11\rangle) = \frac{1}{2}(1+0-0-1) = 0,$$

$$\langle \beta_{10}|\beta_{11}\rangle = \frac{1}{2}(\langle 00|01\rangle - \langle 00|10\rangle + \langle 11|01\rangle - \langle 11|10\rangle) = \frac{1}{2}(0-0+0-0) = 0,$$

$$\langle \beta_{00}|\beta_{11}\rangle = \frac{1}{2}(\langle 00|01\rangle - \langle 00|10\rangle + \langle 11|01\rangle - \langle 11|10\rangle) = \frac{1}{2}(0-0+0-0) = 0.$$

All the inner products of distinct Bell states are equal to 0, which indicates that the Bell states are orthogonal to each other.

Exercise 8.9 Prove the completeness of Bell basis, i.e., Eq. 8.7.

8.7 ∗ Generalization: GHZ States

Bell states, or EPR pairs, involve two entangled qubits. Generalizing Bell states to more than two qubits can lead to more complex types of entanglement. There are numerous ways to entangle more than two qubits, and these lead to different classes of multi-qubit entangled states.

GHZ (Greenberger-Horne-Zeilinger) states are a class of multi-qubit states that demonstrate stronger forms of quantum entanglement than can be seen with two qubits. The simplest GHZ state involves three qubits and is given by the superposition

$\frac{1}{\sqrt{2}}(|000\rangle + |111\rangle)$, where each term in the superposition involves all qubits being in the same state.

When we increase the number of qubits, we encounter more types of multipartite entangled states, such as the W states, e.g., $\frac{1}{\sqrt{3}}(|001\rangle + |010\rangle + |100\rangle)$. It's worth noting that these different classes of entangled states cannot always be converted into each other without some probability of failure, which suggests they demonstrate fundamentally different types of quantum correlation.

※ **Further Exploration**

The concept of GHZ states, as well as their applications in quantum information theory and quantum computation, extends beyond the basic explanation given in this text. For those seeking to deepen their understanding, the following topics are suggested for further exploration:

1. Going Beyond Bell's Theorem [48]: In this seminal paper, Greenberger, Horne, and Zeilinger introduce the concept of GHZ states, which exemplify a more profound form of quantum entanglement than Bell states.

2. Quantum Computation and Quantum Information [7]: This comprehensive textbook by Nielsen and Chuang covers the entire field of quantum information, including the detailed theory of multi-qubit entangled states such as GHZ and W states.

3. Observation of Three-Photon Greenberger-Horne-Zeilinger Entanglement [27]: In this experimental paper, Bouwmeester et al. demonstrate the creation of a three-qubit GHZ state, putting the theoretical predictions into practice.

8.8 Summary and Conclusions

Bell States and Quantum Entanglement

Throughout this chapter, we have ventured into the realm of Bell states and their profound implications. By studying Bell states, we were able to demystify complex quantum phenomena like quantum entanglement and to understand them in a precise mathematical framework. We emphasized the indispensable nature of Bell states in the realm of quantum computation and information processing.

Theoretical Foundations and Practical Implementations

We subsequently dissected the properties of Bell states, establishing their role as the quintessential representation of two-qubit entanglement. The concepts of orthonormality and completeness of Bell states were meticulously explored, highlighting their significance in quantum computing and information.

Our expedition into the practical creation of Bell states encompassed both theoretical constructs, employing quantum gates, and real-world scenarios utilizing physical phenomena like spontaneous parametric down-conversion (SPDC).

Measurement and Conversion

The chapter further unveiled the intricacies of Bell measurements, which are essential for many quantum protocols. Their ability to discern between different Bell states and the projection mechanism onto the Bell basis was elaborated upon. Moreover, the concept of Bell state conversion was introduced, showcasing the flexibility and adaptability of these entangled states in quantum operations.

Generalizations and Future Directions

In closing the chapter, we introduced the reader to the more general entangled states such as the GHZ states. These states serve as a natural extension of the Bell states, embracing more than two qubits. Such generalizations enrich our understanding of multi-qubit entanglement and open doors for further research and applications.

Upcoming Topics

In the succeeding chapter, we will venture deeper into the enigmatic domain of quantum entanglement, specifically focusing on Bell inequalities. These inequalities will be our tool to probe the disparity between quantum mechanics and the principles of local realism. By delving into experimental tests that validate quantum entanglement and the consequent negation of local hidden variables, we shall further unravel the mysteries of the quantum realm. Additionally, we will explore the pragmatic applications of quantum entanglement, particularly in the domains of secure communication and cryptographic protocols.

Problem Set 8

8.1 A general form of two-qubit Bell states is

$$|\Phi\rangle = \frac{1}{\sqrt{2}}(|ab\rangle + |cd\rangle),$$

with $\langle a|c\rangle = \langle b|d\rangle = 0$.

(a) Prove that $|\Phi\rangle$ is not a product state.

(b) Confirm this general form encompasses the definitions in Eq. 8.1.

8.2 (a) Using the general form of Bell states in Problem 8.1, show that if $|\Psi\rangle$ and $|\Phi\rangle$ are two Bell states shared by Alice and Bob, then there is a unitary operator local to Alice (i.e., of the form $U \otimes I$) which takes $|\Psi\rangle$ to $|\Phi\rangle$:

$$(U \otimes I)|\Psi\rangle = |\Phi\rangle.$$

Similarly, there is such a unitary operator local to Bob. Thus, all Bell states are equivalent in this sense.

(b) Find U for the specific converstions in § 8.5.

8.3 Derive the unitary matrix U_β for the basis change from the computational basis to the Bell basis. This transformation is important in quantum computing for encoding and decoding information in Bell states. Specifically, find U_β such that

$$[|\beta_{00}\rangle, |\beta_{01}\rangle, |\beta_{10}\rangle, |\beta_{11}\rangle] = U_\beta[|00\rangle, |01\rangle, |10\rangle, |11\rangle],$$

Problem Set 8

where each ket $(|...\rangle)$ represents a column vector.

Relate U_β to the specific cases in § 8.3.

8.4 For some qubit platforms, it is more natural to implement the Bell state

$$|\Phi_i^+\rangle = \frac{1}{\sqrt{2}}(|00\rangle + i|11\rangle)$$

than the standard Bell state $|\Phi^+\rangle = \frac{1}{\sqrt{2}}(|00\rangle + |11\rangle)$.

Find the other three Bell states ($|\Phi_i^-\rangle, |\Psi_i^+\rangle, |\Psi_i^-\rangle$) such that, together with $|\Phi_i^+\rangle$, the four states form a complete orthonormal basis.

8.5 Consider the three-qubit state given by:

$$|\Psi_0\rangle = \frac{1}{\sqrt{2}}(\alpha|0\rangle + \beta|1\rangle) \otimes (|00\rangle + |11\rangle).$$

Assume we measure the first two qubits in the Bell basis $|\beta_{ij}\rangle$ (where $i,j \in \{0,1\}$).

(a) What is the probability for each of the Bell states $|\beta_{ij}\rangle$ to occur?

(b) If the measurement yields $|\beta_{ij}\rangle$, what is the state of the third qubit after the measurement?

8.6 Consider the four-qubit state given by:

$$|\Psi_0\rangle = \frac{1}{2}(|0000\rangle + |0101\rangle + |1010\rangle + |1111\rangle).$$

Assume we measure the first two qubits in the Bell basis $|\beta_{ij}\rangle$ (where $i,j \in \{0,1\}$).

(a) What is the probability for each of the Bell states $|\beta_{ij}\rangle$ to occur?

(b) If the measurement yields $|\beta_{ij}\rangle$, what is the state of the last two qubits after the measurement?

8.7 ✶ Consider the Bell state

$$|\Psi^-\rangle = \frac{1}{\sqrt{2}}(|01\rangle - |10\rangle).$$

Let **n** and **m** be unit vectors in $\mathbb{R}^3$. Calculate the expectation value

$$E(\mathbf{n}, \mathbf{m}) = \langle \Psi^-|(\mathbf{n}\cdot\boldsymbol{\sigma}) \otimes (\mathbf{m}\cdot\boldsymbol{\sigma})|\Psi^-\rangle$$

where $\boldsymbol{\sigma}$ represents the vector of Pauli matrices. Show that

$$E(\mathbf{n}, \mathbf{m}) = -n_1 m_1 - n_2 m_2 - n_3 m_3 = -\mathbf{n}\cdot\mathbf{m} = -\cos\theta_{\mathbf{n},\mathbf{m}}.$$

8.8 ✶ **Multipartite Bell States**

Consider the extension of the Bell state concept to a system of four qubits, or equivalently, a pair of qudits with $d = 4$. We define the uniform Bell state for such a system as follows:

$$|\beta_{0000}\rangle = \frac{1}{2} \sum_{i,j \in \{0,1\}} |ij\rangle \otimes |ij\rangle = \frac{1}{2}(|0000\rangle + |0101\rangle + |1010\rangle + |1111\rangle).$$

This state can also be represented in decimal notation as:

$$|\beta_0\rangle = \frac{1}{2} \sum_{i=0}^{3} |i\rangle \otimes |i\rangle = \frac{1}{2}(|0\rangle + |5\rangle + |10\rangle + |15\rangle),$$

with $|i\rangle$ now ranging over the set $\{0,1,2,3\}$, corresponding to the decimal equivalent of the binary states.

The other Bell states can be generated using the transformation:

$$|\beta_{ijkl}\rangle = U_{ijkl}|\beta_{0000}\rangle \quad (i,j,k,l \in \{0,1\}),$$

with U_{ijkl} defined as the application of specific Pauli and identity operators on the base Bell state. The operators are:

$U_{0000} = II\,II$ $U_{0001} = IX\,II$ $U_{0010} = IZ\,II$ $U_{0011} = IY\,II$
$U_{0100} = XI\,II$ $U_{0001} = XX\,II$ $U_{0010} = XZ\,II$ $U_{0011} = XY\,II$
$U_{0100} = ZI\,II$ $U_{0101} = ZX\,II$ $U_{0110} = ZZ\,II$ $U_{0111} = ZY\,II$
$U_{1100} = YI\,II$ $U_{1101} = YX\,II$ $U_{1110} = YZ\,II$ $U_{1111} = YY\,II$

Your tasks are as follows:

(a) Show that these Bell states are orthonormal, i.e.,

$$\langle \beta_{ijkl}|\beta_{i'j'k'l'}\rangle = \delta_{ii'}\delta_{jj'}\delta_{kk'}\delta_{ll'}.$$

(b) Demonstrate that these Bell states form a complete basis for the space of four qubits, i.e.,

$$\sum_{i,j,k,l \in \{0,1\}} |\beta_{ijkl}\rangle\langle\beta_{ijkl}| = I.$$

(c) Extend this construction to describe Bell states for systems comprising $2n$ qubits. (In the above example $n = 2$.)

9. Entanglement and Bell Inequalities

Contents

9.1	**Classical Correlation vs. Quantum Entanglement**	224
9.1.1	Classical Correlations	224
9.1.2	Quantum Entanglement	225
9.2	**The EPR Paradox**	227
9.2.1	Hidden Variable Theory, Realism, and Locality	228
9.3	**Bell Inequalities**	229
9.3.1	Bell-CHSH Inequality	229
9.4	**Bell-CHSH Inequality with Classical Correlation**	231
9.4.1	Balls with Contrasting Colors	231
9.4.2	Gray Balls with Shared Ink	232
9.4.3	Adaptable Intelligent Balls	232
9.5	**Bell-CHSH Inequality with Quantum Entanglement**	233
9.5.1	Violation of the Bell CHSH Inequality: A Special Case	233
9.5.2	Violation of the Bell CHSH Inequality: General Case	235
9.6	**Experimental Verification**	236
9.6.1	✶ Basic Experimental Setup	236
9.6.2	Early Experiments	237
9.6.3	Loophole-free Measurements	238
9.6.4	Key Experimental Findings and Implications	238
9.6.5	2022 Nobel Prize in Physics	239
9.7	**The No-Communication Theorem**	240
9.7.1	Myths and Misconceptions	240
9.8	**✶ Derivation of Bell-CHSH Quantity for Bell States**	241
9.8.1	✶ Derivation via Measurement Bases	241
9.8.2	✶ Derivation Using Pauli Operators	244
9.9	**✶ Further Exploration**	246
9.10	**Summary and Conclusions**	247
	Problem Set 9	248

In this chapter, we are going to embark on an exploration of the captivating world of quantum entanglement and Bell inequalities. Quantum entanglement is a phenomenon that challenges our conventional wisdom, especially the notion of local realism, which postulates that physical systems have predetermined properties and cannot interact faster than light.

One of the most famous examples that highlight the puzzling nature of quantum entanglement is the century-old Einstein-Podolsky-Rosen (EPR) paradox. This thought experiment raised questions about how complete and accurate quantum mechanics is, and sparked many discussions among scientists.

In recent years, scientists have tested quantum entanglement using Bell inequalities. These inequalities are tools used to distinguish between predictions made by quantum mechanics and the principles of local realism. Experiments that have confirmed quantum entanglement have greatly expanded our knowledge of the quantum world.

Beyond being a mere intellectual puzzle, quantum entanglement boasts tangible applications, becoming integral to emerging technologies. The E91 protocol, for instance, harnesses entanglement for secure communication through cryptographic keys and employs Bell inequalities to detect eavesdropping.

This chapter will thoroughly examine the historical background, theories, and experimental achievements of quantum entanglement and Bell inequalities. In the next chapter, we will study their real-world applications.

9.1 Classical Correlation vs. Quantum Entanglement

Quantum entanglement is a fundamental phenomenon in quantum mechanics where particles become correlated in such a way that the state of one particle cannot be described independently of the state of the other, even when they are separated by large distances. The physics underlying quantum entanglement is distinct from classical correlation.

To illustrate this distinction, consider a machine that emits pairs of balls in opposite directions, as depicted in Fig. 9.1. Envisage two individuals, Alice and Bob, each catching one of these balls. The balls are arranged such that when Bob catches a black ball, he knows that Alice must have caught a white one, and vice versa. This correlation can manifest in both classical and quantum realms but arises for different reasons.

9.1.1 Classical Correlations

In the classical domain, correlations can occur in various scenarios:

- Balls with contrasting colors: The machine dispenses pairs of balls with contrasting colors that are predetermined. In this scenario, Alice and Bob always catch balls of opposite colors because they were already arranged that way. This is similar to the predetermined coin-pair scenario: If a pair of coins is tossed such that when one shows heads, the other must show tails.

9.1 Classical Correlation vs. Quantum Entanglement

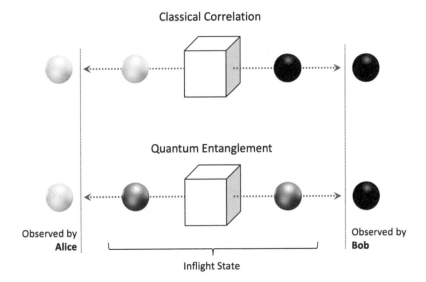

Figure 9.1: Classical Correlation versus Quantum Entanglement

- Gray balls with shared ink: Initially, the balls contain an equal mixture of black and white pigments, hence appearing gray. Upon observation, the pigments separate, flowing to one of the balls, with the direction of the flow being random. As a result, one ball randomly turns black, and the other turns white. In this scenario, the colors are not predetermined, and a transfer of pigment between the balls is necessary for them to assume their respective colors.

- Communicating intelligent balls: Consider two balls capable of sensing their surroundings and communicating wirelessly. They can coordinate to display different colors based on the environment (e.g., who catches them and how they are caught).

 The examples above illustrate anti-correlation. Correlation can also be direct. For instance, with fair coins, if Alice's coin shows heads, so does Bob's, and the same applies for tails.

9.1.2 Quantum Entanglement

Transitioning to the quantum realm, let's consider the two balls as particles, or qubits. The scenario in Fig. 9.1 can occur if the qubits are in a Bell state, represented as $|\Psi^-\rangle = \frac{1}{\sqrt{2}}(|01\rangle - |10\rangle)$. Here, we associate the white color with state $|0\rangle$ and black with state $|1\rangle$.

When Alice measures her qubit, she employs the projection operator

$$M_{0a} = |0\rangle\langle 0| \otimes I, \tag{9.1}$$

which collapses the Bell state to $M_{0a}|\Psi^-\rangle = |01\rangle$. (The normalization factor is not relevant here and is therefore omitted.) Subsequently, Bob's qubit is in state $|1\rangle$.

Likewise, if Bob measures his qubit first and observes $|0\rangle$, he uses the projection operator

$$M_{0b} = I \otimes |0\rangle\langle 0|, \tag{9.2}$$

which collapses the Bell state to $M_{0b}|\Psi^-\rangle = |10\rangle$. At this point, Alice's qubit is in state $|1\rangle$.

This demonstrates that the measurement outcomes of Alice and Bob are anti-correlated. However, a key difference from classical correlation lies in the quantum measurement principle. The states of the individual qubits in the entangled pair are not defined until they are measured. This is shown in Fig. 9.1, where the balls are gray before being caught, symbolizing their undefined states. When Alice measures her qubit, she obtains either $|0\rangle$ or $|1\rangle$ with equal probability, but whichever she gets, Bob will always observe the opposite. This resembles the "grey balls with shared ink" classical scenario, except that no communication or transfer of matter is needed between the qubits.

Similar logic applies to other Bell states. For $|\Phi^+\rangle$ and $|\Phi^-\rangle$, the measurement outcomes are correlated; for $|\Psi^+\rangle$ and $|\Psi^-\rangle$, they are anti-correlated.

The above discussion on entanglement is quite abstract. You might be wondering, how are these "gray balls" created in the phyisical world? In quantum computing, entanglement between qubits is achieved through a variety of physical implementations, each exhibiting the behavior described earlier. The following examples should help provide a more tangible understanding.

- Photon Pairs Generated With SPDC: Spontaneous parametric down-conversion (SPDC) is a process where a single photon decays into two entangled photons with lower energy. These entangled photon pairs show quantum correlations and can be sent to distant locations.

- Entangled Atomic States: Quantum entanglement can also include other particles, such as atoms. Entangled atomic states can be produced by changing the internal energy levels of atoms using techniques like stimulated Raman adiabatic passage (STIRAP).

- Superconducting Qubits: In quantum computing, superconducting qubits can be entangled through interactions controlled by microwave resonators or other coupling methods. This enables the creation of quantum logic gates and algorithms.

- Trapped Ions: Trapped ion systems create entangled states by controlling the motion and internal energy levels of ions held in electromagnetic traps. These systems are used for quantum computing and simulation.

- NV Centers: Nitrogen-vacancy (NV) centers in diamond are defect structures that can hold electron spins for a long time. These spins can be entangled using methods such as resonant microwave driving and optical control.

- Quantum Dots: Quantum dots are tiny semiconductor structures that can hold and control individual electrons or excitons. Entangled states can be created in quantum dots through methods like parametric fluorescence or electron spin interactions.

9.2 The EPR Paradox

Quantum mechanics posits that particles were not predetermined to be in a certain state but were actually in a superposition of multiple states until observed. For instance, in the context of the example given earlier, before observation, each particle could be thought of as being both black and white at the same time (analogous to being gray). The moment they are observed, they randomly collapse into one of the states, instantaneously determining the state of the other particle, no matter how far apart they are.

This seems to contradict our classical intuitions about cause and effect and the nature of reality. How can one particle be influenced by an observation made on another particle at a distance without any signal passing between them? According to relativity, a signal cannot travel faster than light, but quantum mechanics doesn't seem to require a signal for this entangled behavior. Moreover, how can it be that the outcomes of our observations are random and probabilistic?

Albert Einstein, along with his colleagues Boris Podolsky and Nathan Rosen, found this counterintuitive. In 1935, they argued that quantum mechanics does not provide a complete description of reality. This argument came to be known as the EPR paradox, named after the initials of the three scientists.

> " I cannot seriously believe in quantum theory because the theory cannot be reconciled with the idea that physics should represent a reality in time and space, free from *spooky actions at a distance*.
>
> Do you really believe that *the moon isn't there when nobody looks?*
>
> Quantum mechanics is very worthy of regard. But an inner voice tells me that this is not yet the right track. The theory yields much, but it hardly brings us closer to the Old One's secrets. I, in any case, am convinced that *He does not play dice.*"

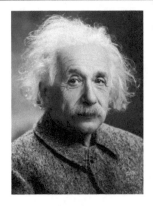

Albert Einstein
Credit: Library of Congress

Interpretations and Philosophical Implications of Quantum Mechanics

Quantum mechanics is both an experimentally verified and mathematically rigorous theory, yet it opens avenues for philosophical debates. It questions our understanding of reality, the role of observers, and concepts like determinism and causality. Prominent physicists like Niels Bohr and Richard Feynman have noted the inherent challenges in fully comprehending the theory. There are currently several interpretations of quantum mechanics leading to different philosophical implications.

Copenhagen Interpretation: The Copenhagen interpretation, which we have implicitly followed so far, postulates that after a quantum measurement, the state vector collapses to the measured eigenstate. In this view, quantum mechanics is fundamentally a theory of probability amplitudes, allowing for phenomena like superposition and entanglement.

Realist Interpretation: Contrastingly, the realist interpretation suggests that quantum mechanics is deterministic, and any uncertainties arise due to hidden variables. Albert Einstein was a strong advocate of this view, encapsulated in his phrase "God does not play dice." However, modern experiments, including those recognized by the 2022 Nobel Prize, have largely ruled out this interpretation.

Other Interpretations: Various other interpretations exist, including many-worlds, information-based interpretations, and decoherence-based views. Each offers a unique perspective on the underlying principles of quantum mechanics.

9.2.1 Hidden Variable Theory, Realism, and Locality

EPR proposed an alternative to standard quantum mechanics that employs hidden variables. They assumed that particles always contained information, which might be hidden, about their states and that this information determined their states even before being observed. Essentially, this information caused the correlation between particles.

In this context, "hidden variables" refer to unobservable factors hypothesized to exist to account for the behavior of quantum systems in a deterministic way. The concept of hidden variables is associated with local realism, which encompasses two philosophical ideas: realism and locality.

Realism: This is the belief that objects possess definite properties and values even when they are not being observed or measured. In the context of hidden variables, realism implies that the properties of a quantum system (such as position, spin, or polarization) have well-defined values at all times, even when not being measured. These values are thought to be determined by hidden variables.

Locality: This refers to the idea that physical processes occurring at one location do not depend on the properties of objects at other locations that are spacelike separated (i.e., separated by a distance greater than the speed of light multiplied by the time it takes for information to travel between them). In the context of hidden variables, locality means that the outcomes of measurements on spatially separated particles are independent of each other and are determined solely by the local hidden variables associated with each particle.

In § 9.1.1, two examples - balls with contrasting colors and predetermined coin pairs - adhere to the principle of local realism. However, the other two examples - gray balls with shared ink and communicating intelligent balls - violate this principle.

However, the notion of local hidden variables faced challenges from Bell's inequalities. Proposed by physicist John Bell in 1964, Bell's inequalities are a set of mathematical inequalities that, if violated by experimental data, indicate that the underlying theory cannot be explained by local hidden variables. The violation of Bell's inequalities by experimental data suggests that entangled quantum systems exhibit correlations that cannot be explained by classical physics.

Subsequent experiments consistently demonstrated violations of Bell's inequalities, suggesting that the behavior of entangled particles does not align with the concept of local hidden variables. This, in turn, provides strong support for the predictions of quantum mechanics and has profound implications for our understanding of the fundamental nature of reality.

> **Key Takeaways**
>
> Classical correlations and quantum entanglement both involve connections between separate systems, but the nature of these connections is fundamentally different. Classical correlations under the assumption of local realism can be explained by predetermined states, whereas quantum entanglement involves an inherent, non-classical correlation that exists even without prior conditions or communication.

In the next section, we will explore in more detail how Bell's inequalities have shaped the modern understanding of quantum mechanics and its intriguing properties.

9.3 Bell Inequalities

Bell inequalities provide a mathematical framework that allows experiments to differentiate between the predictions of local hidden variable theories and those of quantum mechanics in the context of the EPR paradox. There are several forms of Bell inequalities, each applicable to various scenarios and types of observables. The original Bell inequality was defined for dichotomic observables (i.e., observables with two possible outcomes). The Bell-CHSH inequality is a specific case derived by John Clauser, Michael Horne, Abner Shimony, and Richard Holt in 1969, which is an extension of the original Bell inequality and is designed for scenarios with two entangled particles and dichotomic observables.

John Bell
Credit: CERN Science Photo

In this text, we will focus on the Bell-CHSH inequality, as it is one of the most widely studied and experimentally tested forms of Bell inequalities. It serves as a cornerstone in our understanding of the fundamental differences between classical and quantum correlations, and it provides a clear demonstration of the nonlocal and counter-intuitive nature of quantum entanglement.

9.3.1 Bell-CHSH Inequality

 First time readers and undergraduate students may skip the following derivation until Eq. 9.7.

1 Simplified Form

The Bell-CHSH inequality is derived based on the assumptions of local realism, using classical probability theory. The derivation starts by defining measurement settings for a pair of objects (colored balls, or particles in the hidden variable theory), A sent to Alice and B sent to Bob.

Each object can be measured along two different measurement settings or bases, such as polarization directions, represented by a and a' for Alice, and b and b' for Bob. Assume that Alice randomly chooses between a and a', and Bob randomly chooses between b and b', to measure each pair of objects.

Let's define $A(a)$ and $B(b)$ as the outcomes of the measurements for Alice and Bob, respectively, with respect to the settings specified by a and b. Similarly, $A(a')$ and $B(b')$ for a' and b'.

We assume the outcomes of these measurements are either $+1$ or -1, that is:

$$A(a), A(a'), B(b), B(b') \in \{-1, 1\}. \tag{9.3}$$

Now, let's consider the following quantities:

$$A(a)\left[B(b) - B(b')\right],$$
$$A(a')\left[B(b) + B(b')\right].$$

Since the outcomes of the measurements are either $+1$ or -1, depending on if $B(b)$ and $B(b')$ have the same sign or opposite signs, one of the above quantities is 0 while the other is bounded to $[-2, 2]$.

Thus, the sum of the above quantities is also bounded to $[-2, 2]$:

$$-2 \leq A(a)\left[B(b) - B(b')\right] + A(a')\left[B(b) + B(b')\right] \leq 2. \tag{9.4}$$

2 Statistically Verifiable Version

However, Inequality 9.4 is not directly verifiable. This is because for each pair of objects, Alice randomly chooses between a and a' to measure, and Bob between b and b'. There are four combinations, and they need to measure at least four pairs to test the inequality thoroughly. To address this, we can derive a statistically verifiable version as follows.

Let's introduce λ, the "hidden variables," which encompass all factors, such as initial conditions and environmental influences, that determine the states and properties of the objects. These factors may be inherently random. In classical probability theory, this is represented as a probability distribution function $P(\lambda)$. It is also assumed that the choice of measurement setting (such as a or a' for Alice) is independent of the hidden variables and does not influence the properties of the objects being measured, thus there is a single $P(\lambda)$ function.

Now $A(a)$, $A(a')$, $B(b)$, and $B(b')$ all depend on λ, and we denote them as $A(a, \lambda)$, $A(a', \lambda)$, $B(b, \lambda)$, and $B(b', \lambda)$.

The expected value of the product of two measurement outcomes, which describes the correlation of measurements under two settings, is given by:

$$E(a, b) = \sum_{\lambda} P(\lambda) A(a, \lambda) B(b, \lambda). \tag{9.5}$$

These expected values can be obtained through repeated measurements on many identically prepared pairs of particles and averaging the outcomes.

9.4 Bell-CHSH Inequality with Classical Correlation

Now, we sum inequality (9.4) over the hidden variable λ with its probability distribution $P(\lambda)$:

$$-2 \leq \sum_\lambda P(\lambda)\left[A(a,\lambda)\left[B(b,\lambda) - B(b',\lambda)\right] + A(a',\lambda)\left[B(b,\lambda) + B(b',\lambda)\right]\right] \leq 2. \tag{9.6}$$

Using the definition of the expected value, we can rewrite the inequality as:

$$-2 \leq S \leq 2, \text{ or, } |S| \leq 2, \tag{9.7}$$

which is the Bell-CHSH inequality, where

$$S = E(a,b) - E(a,b') + E(a',b) + E(a',b') \tag{9.8}$$

is referred to as the Bell-CHSH quantity.

> **Points to Note**
>
> Quantum mechanics is not involved in the derivation of the Bell-CHSH inequality. It is purely based on classical probability theory and the hidden variable (i.e., local realism) assumption. As such, the inequality sets constraints on the correlations between measurements of two particles under any local hidden variable theory. If the inequality is violated, it suggests that the system in question does not adhere to local realism but instead exhibits the nonlocal correlations that are characteristic of quantum entanglement.

9.4 Bell-CHSH Inequality with Classical Correlation

9.4.1 Balls with Contrasting Colors

To build intuition, let's start with a classical analogy using "balls with contrasting colors." We will consider pairs of balls where each ball is either black or white. We define $+1$ for white, and -1 for black.

Imagine that Alice and Bob each receive one ball from a pair, and they have some method to measure or analyze the colors of the balls. We denote Alice's measurement settings as a and a', and Bob's settings as b and b'.

Assume that Alice and Bob randomly choose between their respective settings to measure each pair of balls.

Here, the hidden variable λ represents the set of inherent properties of the balls which determine their color. For instance, it could be some internal parameters set before Alice and Bob receive them.

We assume perfect anti-correlation, meaning if one ball in the pair is white, the other is guaranteed to be black. Mathematically:

$$P(\lambda) = \begin{cases} 1, & \text{if } \lambda \Rightarrow \text{one ball white and the other black,} \\ 0, & \text{otherwise.} \end{cases} \tag{9.9}$$

The correlation functions in the Bell-CHSH quantity all equal -1, because only the terms involving $AB = -1$ are non-zero in the following calculation:

$$E(a,b) = \sum_\lambda P(\lambda) A(a,\lambda) B(b,\lambda) = -1, \qquad (9.10a)$$

$$E(a',b') = \sum_\lambda P(\lambda) A(a',\lambda) B(b',\lambda) = -1, \qquad (9.10b)$$

$$E(a,b') = \sum_\lambda P(\lambda) A(a,\lambda) B(b',\lambda) = -1, \qquad (9.10c)$$

$$E(a',b) = \sum_\lambda P(\lambda) A(a',\lambda) B(b,\lambda) = -1. \qquad (9.10d)$$

The Bell-CHSH quantity becomes

$$S = -1 + 1 - 1 - 1 = -2. \qquad (9.11)$$

This value of $|S|$ satisfies the Bell-CHSH inequality, $|S| \leq 2$.

Note that in this classical example, the outcome can be explained using local realism, where the color correlation is predetermined by some hidden variable (λ). This is in contrast to the quantum case, which we will discuss next, where the Bell-CHSH inequality can be violated, indicating the presence of nonlocal correlations or entanglement.

Exercise 9.1 Derive S for the perfectly correlated classical case, where the color of Bob's ball always matches that of Alice's for the same pair.

9.4.2 Gray Balls with Shared Ink

The Bell-CHSH inequality is not violated as long as the colors of the balls are independent of the measurement settings. However, this scenario becomes problematic when considering simultaneous measurements. If Alice and Bob make measurements at the exact same time while being separated by a distance, the pigments would not have enough time to transfer between the balls to determine their colors. This means that the outcomes are not available at the time of measurement, rendering this scenario unsuitable for testing the Bell-CHSH inequality. It fails to account for the type of correlations observed in entangled quantum systems, as it relies on a physical transfer of pigment that cannot occur instantaneously over a distance.

9.4.3 Adaptable Intelligent Balls

Consider a scenario where the balls possess robotic-like features, allowing them to sense their environment and independently choose which color to display based on the measurement settings they observe.

We still have the measurement settings (a, a', b, b') as different ways Alice and Bob decide to measure or analyze the colors of the balls (e.g., using a net, an umbrella, gloves, or bare hands.) However, in this scenario, Alice and Bob can delay their choice of measurement settings until after the balls are dispensed and just before they make the measurements.

As the balls are intelligent and can adapt independently, they observe how Alice and Bob prepare to measure them and individually set their colors to create the following outcomes:

- $E(a, b) = 1$ (Alice: bare hands; Bob: net $\Rightarrow$ Balls: both white.)
- $E(a, b') = -1$ (Alice: bare hands; Bob: umbrella $\Rightarrow$ Balls: black and white.)
- $E(a', b) = 1$ (Alice: gloves; Bob: net $\Rightarrow$ Balls: both white.)
- $E(a', b') = 1$ (Alice: gloves; Bob: umbrella $\Rightarrow$ Balls: both black.)

Now, the Bell-CHSH quantity becomes:

$$|S| = |1 - (-1) + 1 + 1| = 4. \tag{9.12}$$

As you can see, in this example, the value of $|S|$ is 4, which violates the Bell-CHSH inequality ($|S| \leq 2$). This violation occurs because the intelligent balls do not adhere to the principle of realism; their colors are not predetermined but change based on the measurement settings at the time of the measurement. Furthermore, the balls need to communicate in order to decide whether they will display identical or opposite colors, which violates the locality assumption and makes it impossible to test a pair of balls simultaneously.

This analogy helps us understand the puzzling nature of quantum entanglement, where particles seemingly violate local realism in a similar way. However, as you will see in the subsequent sections, in the quantum case, the Bell inequality is violated under simultaneous measurements, even when the particles are separated by large distances. This violation is considered evidence of the non-classical nature of quantum entanglement, which cannot be explained by any classical theory based on local hidden variables.

In the subsequent discussion on experimental verification, we will explore Bell test experiments that are 'loophole-free', and are specifically designed to eliminate possibilities similar to those of the adaptable intelligent balls.

9.5 Bell-CHSH Inequality with Quantum Entanglement

9.5.1 Violation of the Bell CHSH Inequality: A Special Case

In this section, we will illustrate how the Bell CHSH inequality can be violated with a specific arrangement of measurement settings for Alice and Bob using entangled qubits. A comprehensive mathematical treatment of Bell CHSH inequality for Bell states will follow in the next section.

One key feature that separates quantum mechanics from classical probability theory is the way expected values of measurements are determined in quantum mechanics – they are governed by the inner products of measurement observables.

Let's consider two entangled qubits, A and B, in the Bell state $|\Psi^-\rangle = \frac{1}{\sqrt{2}}(|01\rangle - |10\rangle)$. This state is often used in experiments testing Bell inequalities with photons. If Alice employs M_a to measure her qubit and Bob uses M_b, the expected value of the joint measurement, analogous to the classical counterpart in Eq. 9.5, is given by:

$$E(a, b) = \langle \Psi^- | (M_a \otimes M_b) | \Psi^- \rangle. \tag{9.13}$$

Now, let's consider the following measurement observables for Alice and Bob:

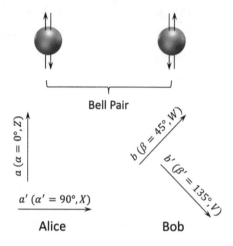

Figure 9.2: Settings for Violating the Bell-CHSH Inequality

$$M_a = Z, \qquad (9.14a)$$
$$M_{a'} = X, \qquad (9.14b)$$
$$M_b = W \equiv \frac{1}{\sqrt{2}}(X + Z), \qquad (9.14c)$$
$$M_{b'} = V \equiv \frac{1}{\sqrt{2}}(X - Z). \qquad (9.14d)$$

These measurement observables are associated with measurement angles on the Bloch sphere $\alpha = 0$, $\alpha' = \frac{\pi}{2}$, $\beta = \frac{\pi}{4}$, and $\beta' = \frac{3\pi}{4}$, as depicted in Fig. 9.2.

Observe that

$$\langle 0|Z|0\rangle = 1, \quad \langle 1|Z|1\rangle = -1, \quad \langle 0|Z|1\rangle = 0, \quad \langle 1|Z|0\rangle = 0, \qquad (9.15a)$$
$$\langle 0|X|0\rangle = 0, \quad \langle 1|X|1\rangle = 0, \quad \langle 0|X|1\rangle = 1, \quad \langle 1|X|0\rangle = 1. \qquad (9.15b)$$

We can then compute

$$E(a,b) = \frac{1}{\sqrt{2}}(\langle 01| - \langle 10|)Z \otimes \frac{1}{\sqrt{2}}(X+Z)\frac{1}{\sqrt{2}}(|01\rangle - |10\rangle) \qquad (9.16a)$$
$$= \frac{1}{2\sqrt{2}}(\langle 0|Z|0\rangle\langle 1|Z|1\rangle + \langle 1|Z|1\rangle\langle 0|Z|0\rangle) \qquad (9.16b)$$
$$= -\frac{1}{\sqrt{2}}, \qquad (9.16c)$$

$$E(a,b') = \frac{1}{\sqrt{2}}(\langle 01| - \langle 10|)Z \otimes \frac{1}{\sqrt{2}}(X-Z)\frac{1}{\sqrt{2}}(|01\rangle - |10\rangle) \qquad (9.17a)$$
$$= \frac{1}{2\sqrt{2}}(\langle 0|Z|0\rangle\langle 1|(-Z)|1\rangle + \langle 1|Z|1\rangle\langle 0|(-Z)|0\rangle) \qquad (9.17b)$$
$$= \frac{1}{\sqrt{2}}, \qquad (9.17c)$$

9.5 Bell-CHSH Inequality with Quantum Entanglement

$$E(a',b) = \frac{1}{\sqrt{2}}(\langle 01| - \langle 10|)X \otimes \frac{1}{\sqrt{2}}(X+Z)\frac{1}{\sqrt{2}}(|01\rangle - |10\rangle) \tag{9.18a}$$

$$= \frac{1}{2\sqrt{2}}\left(-\langle 0|X|1\rangle\langle 1|X|0\rangle - \langle 1|X|0\rangle\langle 0|X|1\rangle\right) \tag{9.18b}$$

$$= -\frac{1}{\sqrt{2}}, \tag{9.18c}$$

$$E(a',b') = \frac{1}{\sqrt{2}}(\langle 01| - \langle 10|)X \otimes \frac{1}{\sqrt{2}}(X-Z)\frac{1}{\sqrt{2}}(|01\rangle - |10\rangle) \tag{9.19a}$$

$$= \frac{1}{2\sqrt{2}}\left(-\langle 0|X|1\rangle\langle 1|X|0\rangle - \langle 1|X|0\rangle\langle 0|X|1\rangle\right) \tag{9.19b}$$

$$= -\frac{1}{\sqrt{2}}. \tag{9.19c}$$

Hence, the Bell CHSH quantity is:

$$S = E(a,b) - E(a,b') + E(a',b) + E(a',b') = -2\sqrt{2}. \tag{9.20}$$

And

$$|S| > 2, \tag{9.21}$$

demonstrating that the Bell-CHSH inequality is violated. This violation is a signature of quantum entanglement and shows that local hidden-variable theories are insufficient to account for the correlations observed in quantum systems.

An Analog for the Bell Inequality Violation

Consider a triangle with sides a, b, and c. Naturally, the sum of any two sides, say $a + b$, is greater than the third side, c. However, the relationship $a^2 + b^2 > c^2$ is not universally true.

In the classical domain, if we associate probabilities with the lengths a, b, and c, the inequality $a + b > c$ represents the Bell inequality holding true.

Contrastingly, in quantum mechanics, the linearly additive quantities are probability amplitudes instead of probabilities. Thus, when probabilities are associated with a^2, b^2, and c^2, the inequality $a^2 + b^2 \gtrless c^2$ showcases the potential violation of the Bell inequality.

Exercise 9.2 Derive the Bell CHSH quantity S for the state $|01\rangle$ with the measurement settings specified in this subsection. Note this state also gives rise to perfect anti-correlation when measured in the computational basis. Check if the Bell CHSH inequality is violated.

9.5.2 Violation of the Bell CHSH Inequality: General Case

We will defer the more general aspects of the Bell CHSH inequality and its violation in Bell states to § 9.8, given its mathematical complexities.

> **Key Takeaways**
>
> Bell inequalities provide a mathematical framework that allows for experiments to differentiate between these two cases: hidden variable theories and quantum theory of entanglement. These inequalities establish limits on the correlations that can be observed if the entangled particles followed local hidden variable theories. Experiments testing Bell inequalities have consistently shown that the predictions of quantum mechanics are correct, and the correlations between entangled particles are better explained by the non-local, instantaneous nature of quantum mechanics rather than local hidden variable theories.

9.6 Experimental Verification

Experimental verification of Bell's inequality, also known as Bell tests, refers to tests of local realism against the predictions of quantum mechanics using entangled particles. These experiments are designed to measure the correlations between the outcomes of spatially separated particles to check whether they obey the constraints set by Bell inequalities, such as the Bell-CHSH inequality. If the inequalities are violated, it implies that local realism is incompatible with the observed behavior, and the particles exhibit nonlocal correlations characteristic of quantum entanglement.

9.6.1 * Basic Experimental Setup

A basic experimental setup to test the EPR paradox is illustrated in Fig. 9.3. The entangled photon pairs are generated through spontaneous parametric down-conversion (SPDC), which is a widely-used method to produce pairs of entangled photons for various quantum communication and computation experiments. In SPDC, a single high-energy photon (labeled as pump laser) passes through a nonlinear BBO crystal, and occasionally, it splits into two lower-energy photons that are entangled. The entangled photons can be created in various Bell states, depending on the experiment setup and the properties of the nonlinear crystal. The state is commonly $\Psi^- = \frac{1}{\sqrt{2}}(|01\rangle - |10\rangle)$, which means in terms of photon polarization:

$$\Psi^- = \frac{1}{\sqrt{2}}\left(|\leftrightarrow\rangle_1 \otimes |\updownarrow\rangle_2 - |\updownarrow\rangle_1 \otimes |\leftrightarrow\rangle_2\right), \tag{9.22}$$

where $|\leftrightarrow\rangle_1$, $|\updownarrow\rangle_1$, $|\leftrightarrow\rangle_2$, and $|\updownarrow\rangle_2$ represent the horizontal and vertical polarizations of the two photons in the pair.

The two photons in each pair are sent to Alice and Bob, respectively, who are located at spatially separated locations. The transmission media can be free space or optical fiber, spanning distances up to hundreds of kilometers.

Both Alice and Bob pass their respective photons through a beam splitter (labeled BS in Fig. 9.3), which is essentially a half mirror. The beam splitter sends the photon in two different paths, corresponding to measurement bases a and a' for Alice, and b and b' for Bob, randomly. The splitter can also be replaced by an optical switch, which can be controlled by Alice or Bob to select the measurement bases.

The angles of the measurement bases (α, α' for Alice and β, β' for Bob) are adjusted using half-wave plates (HWP). The photon then arrives at a polarization beam splitter (PBS). The PBS sends the photon to different detectors according to

9.6 Experimental Verification

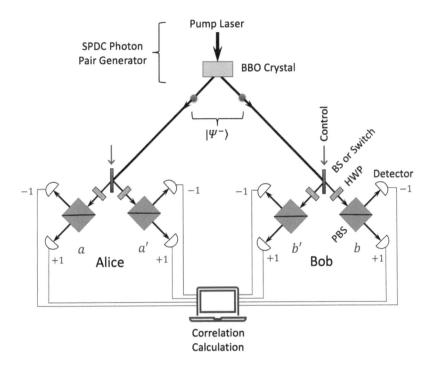

Figure 9.3: Basic Experimental Setup for Bell Test

its polarization. The measurement results (always +1 or −1) are fed into a computer to analyze the correlations and calculate the Bell-CHSH quantity.

The experimental setup described above allows testing the Bell-CHSH inequality by measuring correlations between the polarization states of entangled photon pairs. By varying the settings of the measurement bases, one can obtain different values of the Bell-CHSH quantity S and check for violation of the Bell-CHSH inequality.

9.6.2 Early Experiments

Early experiments aimed at verifying Bell's inequality played a crucial role in establishing the empirical basis for quantum entanglement. Notable among these are:

Aspect experiments (1981-1982) [16], [17]: Alain Aspect and his team conducted a series of groundbreaking experiments using entangled photons to test the Bell-CHSH inequality. They employed the technique of time-coincidence measurements, which involves ensuring that detections are only counted when they could be paired as originating from the same event, and observed strong correlations between the polarization states of the photons, which violated the Bell-CHSH inequality. These results provided strong evidence in favor of quantum mechanics over local hidden variable theories.

Weihs et al. (1998) [89]: In this experiment, Gregor Weihs and his team tested the Bell-CHSH inequality with entangled photon pairs over a distance of about 400 meters. They used a high-efficiency source of entangled photon pairs and fast-switching polarization analyzers. Their results showed a significant violation of the Bell-CHSH inequality, ruling out local hidden variable theories even for particles separated by large distances.

9.6.3 Loophole-free Measurements

There are several loopholes, or measurement flaws, that have been identified in earlier Bell test experiments:

Detection (or Fair Sampling) loophole: This loophole arises when the detectors used in the experiment have low efficiency or when there is a bias in the detection process. It suggests that the observed correlations could be a result of selective sampling rather than genuine quantum entanglement.

Communication (or Locality) loophole: This loophole arises when there is insufficient space or time separation between the measurement events, allowing for the possibility of classical communication between the detectors or the particles during the experiment.

Freedom-of-choice loophole: This loophole arises when the choice of measurement settings is not truly random or when there is a possibility that the measurement settings could be influenced by some unknown variables correlated with the particles.

"Loophole-free" Bell test experiments are designed to test the predictions of quantum mechanics against local hidden variable theories without being affected by any of the known loopholes.

Hensen et al. (2015) [54]: This experiment, also known as the "loophole-free Bell test," addressed some of the potential loopholes in previous experiments, such as the detection loophole and the locality loophole. Unlike the previous experiments that used entangled photons, Hensen and his team used entangled electron spins in diamond nitrogen-vacancy centers separated by 1.3 kilometers, which offered different experimental advantages. They observed a statistically significant violation of the Bell-CHSH inequality, providing strong evidence against local realism.

Giustina et al. (2015) [45] and Shalm et al. (2015) [80]: These two experiments, conducted independently by different research groups, addressed the detection loophole and the locality loophole simultaneously. Both experiments used entangled photons and highly efficient detectors, ensuring that the observed correlations could not be explained by local hidden variable theories. The results of both experiments showed a clear violation of the Bell-CHSH inequality.

By separating the entangled particles by a significant distance and switching the measurement settings quickly, these experiments aimed to rule out the possibility of any local interaction or communication between the particles that could influence their correlated outcomes. This addresses the locality loophole, which suggests that entangled particles might exchange information or be influenced by a common cause in their local environment, thereby explaining the observed correlations.

9.6.4 Key Experimental Findings and Implications

The loophole-free experiments offer compelling evidence for the nonlocal character of quantum entanglement, a fundamental aspect in the comprehension of quantum

9.6 Experimental Verification

phenomena. The consistent experimental confirmations of Bell's inequalities are an important cornerstone in the development of quantum mechanics.

These experiments have profound implications for our understanding of the fundamental nature of the universe. They challenge the premise of local realism, which posits that physical properties are predetermined by the local environment and unaffected by measurement, and that no physical influence can travel faster than light. In contrast, the experimental results align with quantum mechanics, which predicts nonlocal correlations and intrinsic randomness.

Rejecting local realism leads to two possibilities: either accepting the concept of superluminal communication, akin to speculative notions found in science fiction but lacking scientific basis, or embracing the idea of quantum entanglement being inherently nonlocal. The latter is widely accepted in the scientific community, reinforcing the paradigm shift in our understanding of quantum interactions and their implications for the nature of reality.

Furthermore, the intrinsic randomness observed in quantum measurements, as quantified by Born's rule (§ 1.5.1), underscores a departure from classical determinism. Born's rule, which assigns probabilities to the outcomes of quantum measurements, is not a reflection of our ignorance about the system's state but a fundamental aspect of quantum mechanics. This probabilistic nature reveals that, unlike classical systems where uncertainty often stems from incomplete knowledge, quantum randomness is an inherent feature of the universe.

9.6.5 2022 Nobel Prize in Physics

The Nobel Prize in Physics 2022 was awarded jointly to Alain Aspect, John F. Clauser, and Anton Zeilinger for experiments with entangled photons, establishing the violation of Bell inequalities and pioneering quantum information science.

Alain Aspect, John Clauser, and Anton Zeilinger are three prominent physicists who have made significant contributions to the experimental verification of quantum entanglement and the testing of local realism through Bell inequalities. Their work earned them the Nobel Prize in Physics in 2022. Aspect's groundbreaking experiments in the 1980s provided strong evidence in favor of quantum mechanics over local hidden variable theories. Clauser, along with Stuart Freedman, performed one of the first experimental tests of Bell's inequality in the early 1970s, which showed results consistent with the predictions of quantum mechanics. Zeilinger has been a pioneer in the field of quantum information and has conducted numerous experiments on entanglement, pushing the boundaries of our understanding of quantum mechanics. The groundbreaking work of these physicists has been instrumental in shaping our current understanding of the nonlocal nature of quantum entanglement and the fundamental nature of the universe.

As highlighted by the 2022 Nobel Prize, the experimental verification of Bell's inequalities has been central to the development and confirmation of quantum theory. It has challenged our classical intuitions and has opened new avenues for understanding the fundamental aspects of the physical world. As our understanding and technological capabilities evolve, future experiments may offer new insights into the nature of quantum entanglement and its implications for our understanding of the physical world. Additionally, the practical applications of entanglement in the fields of quantum computing, quantum communication, and quantum cryptography make this area of research not only fundamentally interesting but also highly relevant to emerging technologies.

9.7 The No-Communication Theorem

Quantum entanglement and the violation of Bell's inequalities raise questions regarding the possibility of transmitting information faster than the speed of light, a concept often dubbed "superluminal communication." While entangled particles exhibit nonlocal correlations, it's essential to recognize that these correlations do not allow for instantaneous information transfer between distant observers. Nonlocal correlations cannot be exploited for superluminal communication, preserving the fundamental principle that information cannot travel faster than light, and thereby maintaining the core tenets of special relativity and causality. This is formally stated as the no-communication theorem which is a key result in quantum information theory:

> **The No-Communication Theorem.**
>
> During the measurement of an entangled quantum state, it is fundamentally impossible for one observer to communicate information to another solely through the act of measurement.

Let's consider a scenario where two observers, Alice and Bob, each hold one qubit of a two-qubit entangled state. If Alice measures her qubit, the state of her qubit collapses to a specific value. However, this does not immediately change the state of Bob's qubit in any way that Bob can observe.

The no-communication theorem, despite the nonlocality of quantum mechanics, holds true primarily because the measurement outcome in quantum mechanics is intrinsically random. While Alice's measurement collapses the overall state from her perspective, Bob, without additional information, can only see a random outcome when he measures his qubit. He cannot distinguish whether the randomness is due to quantum uncertainty or because of Alice's prior measurement.

To illustrate this further, consider the Ekert 91 (E91) quantum key distribution (QKD) protocol (see § 10.6 for details). In this protocol, Alice and Bob use entangled pairs of qubits to generate a shared secret key. At first glance, it might seem as if Alice can instantaneously communicate secret key bits to Bob, especially if they have previously agreed on a sequence of measurement bases. However, the key they generate is a string of random bits, which carries no information about Alice's state. It's only after Alice communicates with Bob through a classical channel (which cannot exceed the speed of light) to discuss the measurement results that the key becomes useful.

A proof of the No-Communication Theorem is given in § 12.2.9.

9.7.1 Myths and Misconceptions

It is important to note that quantum entanglement is a well-defined concept within the realm of quantum mechanics, and it should not be conflated with unrelated phenomena. There are several myths and misconceptions surrounding quantum entanglement, often stemming from a misunderstanding of the underlying principles of quantum mechanics or from misapplying quantum phenomena to everyday life. Here are a few examples:

- Instantaneous communication through entanglement: Although entanglement can create strong correlations between particles regardless of the distance between them, it cannot be used to transmit information instantaneously, as this would violate the principle of causality and the speed of light limit imposed by special relativity. The process of measurement and the collapse of entangled states still follow the constraints of relativistic causality.

- Macroscopic objects can be easily entangled: Entanglement is typically observed at the level of subatomic particles, and it becomes increasingly difficult to maintain and observe entanglement as the size of the system increases. While there have been some experimental demonstrations of entanglement in larger systems, entangling macroscopic objects like everyday items or living organisms is extremely challenging due to decoherence and environmental factors.

- Entanglement as a source of paranormal phenomena: Some people may claim that quantum entanglement is responsible for various paranormal phenomena, such as telepathy or psychic powers. There are also claims of quantum entanglement between people and their spirit, between dreams and reality, and so on. However, there is no scientific evidence to support such claims. Entanglement is a well-established concept in quantum mechanics, but it has not been shown to have any connection to paranormal phenomena.

9.8 ∗ Derivation of Bell-CHSH Quantity for Bell States

In this section, we present a more general, step by step analysis of the Bell CHSH inequality. We use the Bell state $|\Psi^-\rangle$ for this purpose; other Bell states follow similarly.

9.8.1 ∗ Derivation via Measurement Bases

1 Measurement Bases and Observables

Suppose a qubit is measured along a direction defined by the polar angle α and the azimuthal angle 0 on the Bloch sphere. Effectively, it is measured using the basis $\{|a\rangle, |a_\perp\rangle\}$:

$$|a\rangle = \cos\frac{\alpha}{2}|0\rangle + \sin\frac{\alpha}{2}|1\rangle, \tag{9.23a}$$

$$|a_\perp\rangle = -\sin\frac{\alpha}{2}|0\rangle + \cos\frac{\alpha}{2}|1\rangle. \tag{9.23b}$$

The associated measurement observables can be represented as:

$$M_a = |a\rangle\langle a| - |a_\perp\rangle\langle a_\perp|. \tag{9.24}$$

In the Bell CHSH inequality, there are four bases and four measurement observables involved, similar to the above. For each pair of qubits, Alice selects one of the two bases, $\{|a\rangle, |a_\perp\rangle\}$ or $\{|a'\rangle, |a'_\perp\rangle\}$, to measure qubit A:

$$|a\rangle = \cos\frac{\alpha}{2}|0\rangle + \sin\frac{\alpha}{2}|1\rangle, \qquad |a_\perp\rangle = -\sin\frac{\alpha}{2}|0\rangle + \cos\frac{\alpha}{2}|1\rangle, \qquad (9.25a)$$

$$|a'\rangle = \cos\frac{\alpha'}{2}|0\rangle + \sin\frac{\alpha'}{2}|1\rangle, \qquad |a'_\perp\rangle = -\sin\frac{\alpha'}{2}|0\rangle + \cos\frac{\alpha'}{2}|1\rangle. \qquad (9.25b)$$

Similarly, Bob chooses between the bases $\{|b\rangle, |b_\perp\rangle\}$ and $\{|b'\rangle, |b'_\perp\rangle\}$ for measuring qubit B:

$$|b\rangle = \cos\frac{\beta}{2}|0\rangle + \sin\frac{\beta}{2}|1\rangle, \qquad |b_\perp\rangle = -\sin\frac{\beta}{2}|0\rangle + \cos\frac{\beta}{2}|1\rangle, \qquad (9.26a)$$

$$|b'\rangle = \cos\frac{\beta'}{2}|0\rangle + \sin\frac{\beta'}{2}|1\rangle, \qquad |b'_\perp\rangle = -\sin\frac{\beta'}{2}|0\rangle + \cos\frac{\beta'}{2}|1\rangle. \qquad (9.26b)$$

These four measurement bases correspond to the angles α, α', β, and β' on the Bloch sphere.

The corresponding measurement observables are:

$$M_a = |a\rangle\langle a| - |a_\perp\rangle\langle a_\perp|, \qquad (9.27a)$$
$$M_{a'} = |a'\rangle\langle a'| - |a'_\perp\rangle\langle a'_\perp|, \qquad (9.27b)$$
$$M_b = |b\rangle\langle b| - |b_\perp\rangle\langle b_\perp|, \qquad (9.27c)$$
$$M_{b'} = |b'\rangle\langle b'| - |b'_\perp\rangle\langle b'_\perp|. \qquad (9.27d)$$

When Alice measures qubit A and Bob measures qubit B, they employ joint measurement observables like the following:

$$M_{ab} = M_a \otimes M_b. \qquad (9.28)$$

2 Measurement Expected Values

To check the Bell-CHSH inequality, we will calculate the expected values of the products of the outcomes, or correlation functions, $E(a,b)$, $E(a,b')$, $E(a',b)$, and $E(a',b')$. Note in quantum mechanics, expected values are computed using inner products, which is a fundamental distinction from classical mechanics:

$$E(a,b) = \langle\Psi^-|(M_a \otimes M_b)|\Psi^-\rangle, \qquad (9.29a)$$
$$E(a,b') = \langle\Psi^-|(M_a \otimes M_{b'})|\Psi^-\rangle, \qquad (9.29b)$$
$$E(a',b) = \langle\Psi^-|(M_{a'} \otimes M_b)|\Psi^-\rangle, \qquad (9.29c)$$
$$E(a',b') = \langle\Psi^-|(M_{a'} \otimes M_{b'})|\Psi^-\rangle. \qquad (9.29d)$$

After some algebraic manipulations (see § 9.8.1.5), we obtain:

$$E(a,b) = -\cos(\alpha - \beta), \qquad (9.30a)$$
$$E(a,b') = -\cos(\alpha - \beta'), \qquad (9.30b)$$
$$E(a',b) = -\cos(\alpha' - \beta), \qquad (9.30c)$$
$$E(a',b') = -\cos(\alpha' - \beta'). \qquad (9.30d)$$

The result in Eq. 9.30 is remarkable in its own right. It implies that the measurement correlations solely depend on the difference between the measurement angles, rather than their individual values. For example, if $\alpha = \beta$, then $E(a,b) = -1$, signifying that the state $|\Psi^-\rangle$ exhibits isotropic anti-correlations. This is further supported by the equation:

$$|\Psi^-\rangle = \frac{1}{\sqrt{2}}(|01\rangle - |10\rangle)$$
$$= \frac{1}{\sqrt{2}}(|aa_\perp\rangle - |a_\perp a\rangle).$$

This is quite different from correlated classical spins. In fact, this is an underlying reason for the violation of the Bell inequality.

3 The Bell CHSH Quantity

Inserting the expressions from Eq. 9.30 into the Bell-CHSH quantity in Eq. 9.8 we obtain:

$$S = E(a,b) - E(a,b') + E(a',b) + E(a',b') \tag{9.31a}$$
$$= -\cos(\alpha - \beta) + \cos(\alpha - \beta') - \cos(\alpha' - \beta) - \cos(\alpha' - \beta'). \tag{9.31b}$$

Exercise 9.3 Derive expressions of the Bell-CHSH quantity S (similar to Eq. 9.31) for the other three Bell states.

4 Violation of the Bell CHSH Inequality

Upon examining Eq. 9.31, we observe that it is feasible to select angles α, α', β, and β' such that $|S|$ exceeds 2, thereby violating the Bell-CHSH inequality. Specifically, by setting $\alpha = 0$, $\beta = \frac{\pi}{4}$, $\alpha' = \frac{\pi}{2}$, and $\beta' = \frac{3\pi}{4}$ (as depicted in Fig. 9.2), we find that:

$$|S| = \frac{1}{2}\left|\sqrt{2} - (-\sqrt{2}) + \sqrt{2} + \sqrt{2}\right| = 2\sqrt{2} > 2. \tag{9.32}$$

This result illustrates that the Bell-CHSH inequality can be violated by entangled qubits in a quantum scenario, which implies that local hidden-variable theories are insufficient in fully accounting for the correlations in quantum systems.

The measurement observables M_a, M_a', M_b, and M_b' can be expressed as linear combinations of Pauli operators. For the angles specified above, they correspond to Pauli operators or their simple linear combinations given by Eq. 9.14.

Exercise 9.4 Consider the product state

$$|++\rangle \equiv |+\rangle \otimes |+\rangle = \frac{1}{2}(|00\rangle + |11\rangle + |01\rangle + |10\rangle).$$

Derive S for this state as a function of α and β. Investigate if the Bell CHSH inequality can be violated with this state.

5 Derivation of Eq. 9.30

Start with the expected values of M_a across the one-qubit basis states:

$$\langle 0| M_a |0\rangle = \cos\alpha, \tag{9.33a}$$
$$\langle 1| M_a |1\rangle = -\cos\alpha, \tag{9.33b}$$
$$\langle 1| M_a |0\rangle = \sin\alpha, \tag{9.33c}$$
$$\langle 0| M_a |1\rangle = \sin\alpha. \tag{9.33d}$$

Now calculate the expected values of $M_a \otimes M_b$ across the two-qubit basis states:

$$\langle 11| (M_a \otimes M_b) |11\rangle = \langle 1| M_a |1\rangle \langle 1| M_b |1\rangle = \cos\alpha\cos\beta, \tag{9.34a}$$
$$\langle 00| (M_a \otimes M_b) |00\rangle = \langle 0| M_a |0\rangle \langle 0| M_b |0\rangle = \cos\alpha\cos\beta, \tag{9.34b}$$
$$\langle 11| (M_a \otimes M_b) |00\rangle = \langle 1| M_a |0\rangle \langle 1| M_b |0\rangle = \sin\alpha\sin\beta, \tag{9.34c}$$
$$\langle 00| (M_a \otimes M_b) |11\rangle = \langle 0| M_a |1\rangle \langle 0| M_b |1\rangle = \sin\alpha\sin\beta, \tag{9.34d}$$
$$\langle 01| (M_a \otimes M_b) |01\rangle = \langle 0| M_a |0\rangle \langle 1| M_b |1\rangle = -\cos\alpha\cos\beta, \tag{9.34e}$$
$$\langle 10| (M_a \otimes M_b) |10\rangle = \langle 1| M_a |1\rangle \langle 0| M_b |0\rangle = -\cos\alpha\cos\beta, \tag{9.34f}$$
$$\langle 01| (M_a \otimes M_b) |10\rangle = \langle 0| M_a |1\rangle \langle 1| M_b |0\rangle = \sin\alpha\sin\beta, \tag{9.34g}$$
$$\langle 10| (M_a \otimes M_b) |01\rangle = \langle 1| M_a |0\rangle \langle 0| M_b |1\rangle = \sin\alpha\sin\beta. \tag{9.34h}$$

Finally, calculate the expected values of $M_a \otimes M_b$ for the Bell states:

$$\langle \Phi^+| (M_a \otimes M_b) |\Phi^+\rangle = \cos\alpha\cos\beta + \sin\alpha\sin\beta = \cos(\alpha-\beta), \tag{9.35a}$$
$$\langle \Phi^-| (M_a \otimes M_b) |\Phi^-\rangle = \cos\alpha\cos\beta - \sin\alpha\sin\beta = \cos(\alpha+\beta), \tag{9.35b}$$
$$\langle \Psi^+| (M_a \otimes M_b) |\Psi^+\rangle = -\cos\alpha\cos\beta + \sin\alpha\sin\beta = -\cos(\alpha+\beta), \tag{9.35c}$$
$$\langle \Psi^-| (M_a \otimes M_b) |\Psi^-\rangle = -\cos\alpha\cos\beta - \sin\alpha\sin\beta = -\cos(\alpha-\beta). \tag{9.35d}$$

9.8.2 * Derivation Using Pauli Operators

This subsection outlines a general derivation of the Bell CHSH inequality using Pauli matrices (X, Y, Z). This approach is based on expressing measurement operators as linear combinations of the Pauli matrices. Even though more abstract, this approach is a powerful technique that can be used to analyze more general scenarios beyond the Bell inequalities.

1 Outline of the Derivation

The measurement operator in the direction specified by a 3D unit vector $\mathbf{a} = (a_1, a_2, a_3)$ is given by:

$$M_{\mathbf{a}} = \mathbf{a} \cdot \boldsymbol{\sigma} \tag{9.36a}$$
$$\equiv a_1\sigma_1 + a_2\sigma_2 + a_3\sigma_3, \tag{9.36b}$$

where $\sigma_1 \equiv X$, $\sigma_2 \equiv Y$, and $\sigma_3 \equiv Z$ are the Pauli operators.

Suppose Alice measures her qubit along directions $\mathbf{a}$ and $\mathbf{a}'$, and Bob $\mathbf{b}$ and $\mathbf{b}'$. Then the measurement correlation functions are given by:

$$E(\mathbf{a}, \mathbf{b}) = \langle \Psi^-| (M_{\mathbf{a}} \otimes M_{\mathbf{b}}) |\Psi^-\rangle, \tag{9.37a}$$
$$E(\mathbf{a}, \mathbf{b}') = \langle \Psi^-| (M_{\mathbf{a}} \otimes M_{\mathbf{b}'}) |\Psi^-\rangle, \tag{9.37b}$$
$$E(\mathbf{a}', \mathbf{b}) = \langle \Psi^-| (M_{\mathbf{a}'} \otimes M_{\mathbf{b}}) |\Psi^-\rangle, \tag{9.37c}$$
$$E(\mathbf{a}', \mathbf{b}') = \langle \Psi^-| (M_{\mathbf{a}'} \otimes M_{\mathbf{b}'}) |\Psi^-\rangle. \tag{9.37d}$$

9.8 * Derivation of Bell-CHSH Quantity for Bell States

With much algebra (see subsection below), it can be shown:

$$E(\mathbf{a}, \mathbf{b}) = -\mathbf{a} \cdot \mathbf{b} \tag{9.38a}$$
$$= -(a_1 b_1 + a_2 b_2 + a_3 b_3) \tag{9.38b}$$
$$= -\cos\phi_{\mathbf{ab}}, \tag{9.38c}$$

where $\phi_{\mathbf{ab}}$ represents the angle between $\mathbf{a}$ and $\mathbf{b}$.

And similarly for the other correlation functions. These results mirror Eq. 9.30.

The Bell CHSH quantity is:

$$S = E(\mathbf{a}, \mathbf{b}) - E(\mathbf{a}, \mathbf{b}') + E(\mathbf{a}', \mathbf{b}) + E(\mathbf{a}', \mathbf{b}') \tag{9.39a}$$
$$= -\cos\phi_{\mathbf{ab}} + \cos\phi_{\mathbf{ab}'} - \cos\phi_{\mathbf{a'b}} - \cos\phi_{\mathbf{a'b'}}. \tag{9.39b}$$

Now we can pick non-coplanar measurement directions to demonstrate the violation of Bell CHSH inequality. A particular choice is:

$$\mathbf{a} = (1, 0, 0), \tag{9.40a}$$
$$\mathbf{a}' = (0, 1, 0), \tag{9.40b}$$
$$\mathbf{b} = \left(\frac{1}{\sqrt{3}}, \frac{1}{\sqrt{3}}, \frac{1}{\sqrt{3}}\right), \tag{9.40c}$$
$$\mathbf{b}' = \left(-\frac{1}{\sqrt{3}}, \frac{1}{\sqrt{3}}, \frac{1}{\sqrt{3}}\right). \tag{9.40d}$$

This yields $S = \frac{4}{\sqrt{3}}$ which is smaller than $2\sqrt{2}$ but still greater than 2.

2 Derivation of Eq. 9.38

Let us consider the correlation function $E(\mathbf{a}, \mathbf{b})$:

$$E(\mathbf{a}, \mathbf{b}) = \langle \Psi^- | (M_\mathbf{a} \otimes M_\mathbf{b}) | \Psi^- \rangle \tag{9.41a}$$
$$= \langle \Psi^- | ((a_1\sigma_1 + a_2\sigma_2 + a_3\sigma_3) \otimes (b_1\sigma_1 + b_2\sigma_2 + b_3\sigma_3)) | \Psi^- \rangle \tag{9.41b}$$
$$= \sum_{i=1}^{3}\sum_{j=1}^{3} a_i b_j \langle \Psi^- | (\sigma_i \otimes \sigma_j) | \Psi^- \rangle. \tag{9.41c}$$

Now, it can be shown that

$$\langle \Psi^- | (\sigma_i \otimes \sigma_j) | \Psi^- \rangle = -\delta_{ij}. \tag{9.42}$$

While we will not delve into the detailed algebra required to prove this equation here, it is important to note the following insight. This equation indicates that there is a perfect anti-correlation in the Bell state $|\Psi^-\rangle$ when Alice and Bob measure their qubits along the same direction, meaning they will always obtain opposite outcomes. However, if the measurement directions are orthogonal, the correlation function is zero.

Substituting this result back into Eq. 9.41, we obtain:

$$E(\mathbf{a}, \mathbf{b}) = \sum_{i=1}^{3} a_i b_i \langle \Psi^- | (\sigma_i \otimes \sigma_i) | \Psi^- \rangle \quad (9.43a)$$

$$= -a_1 b_1 - a_2 b_2 - a_3 b_3, \quad (9.43b)$$

which matches Eq. 9.38.

We can derive $E(\mathbf{a}, \mathbf{b})$ for other Bell states similarly. Here is a summary:

$$\langle \Phi^+ | (M_\mathbf{a} \otimes M_\mathbf{b}) | \Phi^+ \rangle = a_1 b_1 - a_2 b_2 + a_3 b_3, \quad (9.44a)$$
$$\langle \Psi^+ | (M_\mathbf{a} \otimes M_\mathbf{b}) | \Psi^+ \rangle = a_1 b_1 + a_2 b_2 - a_3 b_3, \quad (9.44b)$$
$$\langle \Phi^- | (M_\mathbf{a} \otimes M_\mathbf{b}) | \Phi^- \rangle = -a_1 b_1 + a_2 b_2 + a_3 b_3, \quad (9.44c)$$
$$\langle \Psi^- | (M_\mathbf{a} \otimes M_\mathbf{b}) | \Psi^- \rangle = -a_1 b_1 - a_2 b_2 - a_3 b_3. \quad (9.44d)$$

9.9 *Further Exploration

For readers interested in deepening their understanding of entanglement and Bell inequalities, the following research papers and books are suggested:

1. Can Quantum-Mechanical Description of Physical Reality Be Considered Complete? [38]: In this famous paper, Einstein, Podolsky, and Rosen question the completeness of quantum mechanics and introduce the concept of "elements of reality".

2. On the Einstein Podolsky Rosen Paradox [20]: This is John Bell's seminal paper where he introduces Bell's inequalities and provides a way to test the validity of local hidden variable theories against quantum mechanics.

3. Quantum Entanglement and Bell Inequalities [72]: This book chapter by Asher Peres provides an in-depth treatment of quantum entanglement and Bell inequalities, along with various proofs and experimental considerations.

4. Foundations and Applications of Quantum Entanglement [55]: This review by Horodecki et al. provides a comprehensive overview of the theoretical foundations and various applications of quantum entanglement. The authors cover entanglement measures, entanglement transformations, and the use of entanglement in quantum information processing.

5. Experimental Realization of Einstein-Podolsky-Rosen-Bohm Gedankenexperiment [15]: This landmark paper by Alain Aspect and colleagues reports the experimental realization of the Einstein-Podolsky-Rosen-Bohm gedankenexperiment and the observation of Bell's inequality violation with high statistical significance.

6. Loophole-free Bell inequality violation using electron spins [54]: In this paper, Hensen et al. present a loophole-free Bell test that addresses the detection and locality loopholes by using entangled electron spins in diamond nitrogen-vacancy centers.

7. Loopholes in Bell Inequality Tests of Local Realism [65]: This review by Larsson discusses the various loopholes that can occur in experimental tests of Bell inequalities and how they can be addressed. The article provides insights into the challenges of performing loophole-free Bell tests.

8. Challenging Local Realism with Human Choices [53]: In this paper, Handsteiner et al. present an experiment where human choices are used to set the measurement settings in a Bell test. This experiment addresses the freedom-of-choice loophole, making it one of the most stringent tests of local realism.

9.10 Summary and Conclusions

The Enigma of Quantum Entanglement

The chapter set forth a profound exploration into the realm of quantum entanglement, an inherently quantum phenomenon that sets quantum systems apart from their classical counterparts. The inseparable correlation exhibited by entangled particles challenges classical notions, particularly local realism, leading to far-reaching implications in both the theoretical and experimental domains of quantum mechanics. Through an examination of the Einstein-Podolsky-Rosen (EPR) paradox, the chapter underscored the tension between quantum mechanics and classically intuitive perspectives on reality.

Bell Inequalities: Bridging Theory and Experiment

Central to the discourse on quantum entanglement are the Bell inequalities, which serve as invaluable mathematical tools to distinguish quantum correlations from classical ones. Specifically, the Bell-CHSH inequality, a particular form of Bell inequalities, emerged as a focal point in our discussions. Experiments testing these inequalities have consistently highlighted the nonlocal correlations between entangled particles, emphasizing that these correlations are intrinsically quantum and cannot be explained by local hidden variable theories. The reaffirmation of the nonlocal nature of quantum entanglement by loophole-free measurements has further solidified the understanding that quantum mechanics provides a more accurate description of reality than classical intuitions might suggest.

The No-Communication Theorem

While quantum entanglement might seem like a conduit for faster-than-light communication given its nonlocal characteristics, the no-communication theorem elucidates that this is not the case. This theorem effectively ensures that quantum mechanics remains consistent with relativity, barring the instantaneous transmission of information through quantum measurements.

A Methodical Delve into Bell-CHSH Inequality

For those academically inclined, the chapter probed deeper into the mathematical intricacies of the Bell-CHSH inequality using Bell states, exemplifying the conceptual richness and rigor of the subject.

Upcoming Topics

As we progress into the next chapter, our emphasis will shift from the theoretical underpinnings of quantum entanglement to its versatile applications. We will journey through key concepts like superdense coding and quantum teleportation, unravelling how quantum entanglement can be harnessed to facilitate groundbreaking advancements in quantum communication and computation. The indispensable role of quantum teleportation, in particular, will be accentuated, given its foundational importance in the realm of quantum technologies.

This transition from the foundational understanding of entanglement to its practical applications in the next chapter will provide a holistic view, bridging the gap between abstract quantum principles and their real-world implications.

Problem Set 9

9.1 Investigate the following questions and elaborate your answers:

(a) Do entangled particles affect each other instantaneously over vast distances?

(b) How do "hidden variables" relate to Bell inequalities?

(c) What is the significance of "loophole-free" Bell tests?

(d) How do the findings of Bell-test experiments relate to the no-superluminal communication constraint?

(e) Have there been experiments confirming the violation of Bell inequalities using qubits other than photons?

(f) Are there practical applications for the violation of Bell inequalities, or is the phenomenon only of theoretical interest?

(g) The violation of Bell inequalities suggests that either realism or locality is not applicable in the quantum realm. Which one is it, or are both concepts challenged?

(h) Are there any interpretations of quantum mechanics that are consistent with both Bell inequality violations and a form of realism?

(i) Is it possible to have a local but non-realistic theory that accurately describes quantum mechanics?

(j) Does assigning a probability to a measurement outcome, as per Born's rule, indicate a lack of information about the system, or are quantum measurements intrinsically random?

(k) If Bob holds a qubit from a Bell pair, and Alice holds the other, and they are separated by vast distances, say Mars and Jupiter, will Alice be able to detect if Bob measures his qubit?

(l) How does quantum entanglement contribute to quantum computing and communication?

9.2 Show that, for the Bell state $|\Phi^-\rangle$, the maximum of $|S|$ is $2\sqrt{2}$. Find all combinations of α, α', β, and β' such that $|S| = 2\sqrt{2}$.

9.3 Consider the product state

$$|++\rangle \equiv |+\rangle \otimes |+\rangle = \frac{1}{2}(|00\rangle + |11\rangle + |01\rangle + |10\rangle),$$

where $|+\rangle = \frac{1}{\sqrt{2}}(|0\rangle + |1\rangle)$. Derive the Bell-CHSH quantity S (Eq. 9.8) for this state using the measurement settings specified in Eq. 9.14. Check if the Bell CHSH inequality is violated.

9.4 The CHSH Game is a thought experiment involving two players, Alice and Bob, who are separated by a great distance and cannot communicate with each other. The game is played with a referee who provides inputs and determines the outcome.

Rules of the Game:

- Inputs: The referee randomly gives Alice and Bob each a single bit (0 or 1) as input.
- Outputs: Alice and Bob each provide a single bit (0 or 1) as output, without knowing the other player's input.
- Winning Condition: Alice and Bob win the round if the following rule is satisfied: The XOR of their outputs must equal the AND of their inputs. In other words, their outputs should be different (01 or 10) only when both received a 1 for inputs.

Strategies and Winning Probabilities:

- Classical Strategy (Local Hidden Variables): In the classical world, Alice and Bob could pre-agree on a strategy or have some locally shared information ("hidden variables") to coordinate their answers.
- Quantum Strategy (Entanglement): Alice and Bob share an entangled pair of qubits and perform specific measurements based on their inputs.

Your tasks are to demostrate:

(a) The best classical strategy leads to a maximum winning probability of 75%.

(b) With the quantum strategy, they can achieve a winning probability of approximately 85%!

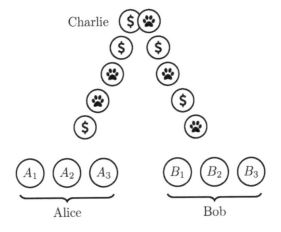

Figure 9.4: Illustration for Problem 9.5

9.5 Three-Qubit Bell Inequality. Charlie prepares many pairs of coins and sends one coin of each pair to Alice and the other to Bob. These coins are set to land randomly as Heads (H) or Tails (T), each maintaining perfect correlation or anticorrelation with its paired coin.

Alice and Bob each have three landing pads for the coins, labeled A_1, A_2, A_3 for Alice and B_1, B_2, B_3 for Bob, respectively. Each coin lands on a pad matched with the pad of its pair of the other participant (i.e., a coin on A_2 is matched with the coin on B_2), revealing either Heads (H) or Tails (T).

Alice and Bob verify the perfect correlation of their coin pairs by confirming that:

$$P(A_1 = B_1) = P(A_2 = B_2) = P(A_3 = B_3) = 1,$$

and perfect anticorrelation by confirming that:

$$P(A_1 \neq B_1) = P(A_2 \neq B_2) = P(A_3 \neq B_3) = 1.$$

They also compute a correlation C from many observations:

$$C = P(A_1 = B_2) + P(A_2 = B_3) + P(A_3 = B_1).$$

Here, $P(A_i = B_j)$ denotes the probability that Alice's coin on pad A_i and Bob's coin on pad B_j show the same face, i.e., both are Heads (H, H) or both are Tails (T, T). Show that:

(a) If different pairs of coins are also perfectly correlated, then $C = 3$.

(b) If the coins alternate between being perfectly correlated and perfectly anticorrelated across the pairs, then $C = 1$.

(c) In the general case, C satisfies $C \geq 1$.

Here, $C \geq 1$ represents a version of Bell's inequality for three qubits. In classical systems, the outcomes for pairs of coins are bound by the statistical constraint $C \geq 1$. Yet, quantum entanglement allows for the violation of this classical limit. Show that:

(d) If Alice and Bob use entangled qubit pairs instead of classical coins, and if Alice's measurements are set at angles $0°$, $120°$, and $-120°$ (corresponding to A_1, A_2, and A_3), while Bob's are at $180°$, $-60°$, and $60°$ (corresponding to B_1, B_2, and B_3), they can achieve $C = 0.75$, violating the classical bound of $C \geq 1$.

10. Key Applications of Entanglement

Contents

10.1	**Review of Preliminaries**	**252**
10.1.1	Bell State Generation and Measurement	252
10.1.2	Conversion of Single-Qubit States	253
10.1.3	Conversion of Bell States	253
10.2	**Superdense Coding**	**254**
10.2.1	Introduction	254
10.2.2	Step-by-Step Analysis	254
10.2.3	✻ Further Exploration	256
10.3	**Quantum Teleportation**	**257**
10.3.1	Introduction	257
10.3.2	Basis State Approach	258
10.3.3	✻ Logic Operation Approach	260
10.3.4	✻ Bell Measurement Approach	262
10.3.5	✻ Further Exploration	265
10.4	**Entanglement Swapping**	**266**
10.4.1	Introduction	266
10.4.2	✻ Step-by-Step Analysis	267
10.4.3	✻ Further Exploration	270
10.5	**Quantum Gate Teleportation**	**270**
10.5.1	✻ Single-Qubit Gate Teleportation	271
10.5.2	✻ Teleportation of Two-Qubit Gates	277
10.5.3	✻ Further Exploration	282
10.6	**E91 Quantum Key Distribution Protocol**	**283**
10.6.1	Procedure of the E91 QKD Protocol	284
10.6.2	Comparison to the BB84 Protocol	286
10.6.3	✻ Further Exploration	287
10.7	**Summary and Conclusions**	**288**
	Problem Set 10	**289**

Quantum entanglement underpins a myriad of applications in quantum computing and quantum communication. These applications range from efficient quantum teleportation and superdense coding, which facilitate communication, to resilient quantum information processing using error-correcting block codes. As research in these areas advances, we can expect a surge of even more innovative applications that leverage the remarkable capabilities of quantum entanglement.

Application	Purpose	Practical Uses
Superdense coding	Allows the transmission of two classical bits using a single entangled quantum bit.	Efficient quantum communication.
Quantum Teleportation	Allows the transfer of quantum information from one location to another, with the help of two classical bits and entanglement.	Quantum communication. Foundation for other entanglement-based applications.
Entanglement swapping	Allows entanglement to be shared between systems that have never interacted.	Creation of quantum networks and long-distance quantum communication.
Quantum gate teleportation	Facilitates the transfer of quantum gates or operations rather than states.	Extends the range of quantum operations. Applications in fault-tolerant computing.
E91 QKD	A quantum key distribution protocol that exploits the principles of quantum entanglement to guarantee secure communication.	Cryptographically secure communication in quantum networks.

Table 10.1: Summary of Key Applications of Quantum Entanglement

In this long chapter, we will explore a number of fundamental applications of quantum entanglement as listed in Table 10.1. Among these, quantum teleportation is the most fundamental, and we will delve into it in extensive detail. However, we will begin our journey with superdense coding, as it is the most straightforward to understand given its operation on the basis states of only two qubits.

10.1 Review of Preliminaries

This section provides a review of some preliminaries necessary for the exploration of quantum entanglement-based applications.

10.1.1 Bell State Generation and Measurement

Bell states represent maximally entangled two-qubit systems and provide a complete, orthogonal basis for the space of two-qubit states. For a detailed discussion of Bell

10.1 Review of Preliminaries

states, refer to Chapter 8.

For any $i, j \in \{0, 1\}$, these states are defined as follows:

$$|\beta_{00}\rangle \equiv |\Phi^+\rangle = \frac{1}{\sqrt{2}}(|00\rangle + |11\rangle), \tag{10.1a}$$

$$|\beta_{01}\rangle \equiv |\Psi^+\rangle = \frac{1}{\sqrt{2}}(|01\rangle + |10\rangle), \tag{10.1b}$$

$$|\beta_{10}\rangle \equiv |\Phi^-\rangle = \frac{1}{\sqrt{2}}(|00\rangle - |11\rangle), \tag{10.1c}$$

$$|\beta_{11}\rangle \equiv |\Psi^-\rangle = \frac{1}{\sqrt{2}}(|01\rangle - |10\rangle). \tag{10.1d}$$

The generation and analysis of Bell states form the basis of many quantum algorithms and protocols. The following figure illustrates the quantum circuits for a Bell state generator and a Bell state analyzer (also referred to as Bell measurement), respectively.

(a) Bell State Generator **(b)** Bell State Analyzer

Figure 10.1: Bell State Generator and Analyzer

10.1.2 Conversion of Single-Qubit States

Table 10.2 shows how "correction gates" convert a qubit state $|\psi_{ij}\rangle$ into the state $|\psi\rangle = \alpha|0\rangle + \beta|1\rangle$. These conversions are reversible with the same gates, enabling state interchangeability, ignoring a global phase factor.

Index ij	Qubit State $|\psi_{ij}\rangle$	Correction Gates
00	$|\psi_{00}\rangle = \alpha|0\rangle + \beta|1\rangle$	I (identity, or none)
01	$|\psi_{01}\rangle = \alpha|1\rangle + \beta|0\rangle$	X (bit-flip)
10	$|\psi_{10}\rangle = \alpha|0\rangle - \beta|1\rangle$	Z (phase-flip)
11	$|\psi_{11}\rangle = \alpha|1\rangle - \beta|0\rangle$	X & Z or iY (bit and phase flip)

Table 10.2: Correction Gates for General Qubit States

These relations can be represented by the following equations:

$$|\psi_{00}\rangle \equiv |\psi\rangle, \tag{10.2a}$$
$$|\psi_{01}\rangle = X|\psi\rangle, \quad |\psi\rangle = X|\psi_{01}\rangle, \tag{10.2b}$$
$$|\psi_{10}\rangle = Z|\psi\rangle, \quad |\psi\rangle = Z|\psi_{10}\rangle, \tag{10.2c}$$
$$|\psi_{11}\rangle = XZ|\psi\rangle, \quad |\psi\rangle = ZX|\psi_{11}\rangle. \tag{10.2d}$$

10.1.3 Conversion of Bell States

Similarly, Pauli gates applied to *either one* of the qubits in a Bell pair can transform the state $|\beta_{ij}\rangle$ into $|\beta_{00}\rangle$ or vice versa, up to a global phase factor. The transformation depends on the index ij, as summarized in Table 10.3. For a full treatment of this topic, see § 8.5.

Index ij	Bell State	Conversion Gates
00	$\|\beta_{00}\rangle \equiv \|\Phi^+\rangle = \frac{1}{\sqrt{2}}(\|00\rangle + \|11\rangle)$	I (identity, or none)
01	$\|\beta_{01}\rangle \equiv \|\Psi^+\rangle = \frac{1}{\sqrt{2}}(\|01\rangle + \|10\rangle)$	X (bit-flip)
10	$\|\beta_{10}\rangle \equiv \|\Phi^-\rangle = \frac{1}{\sqrt{2}}(\|00\rangle - \|11\rangle)$	Z (phase-flip)
11	$\|\beta_{11}\rangle \equiv \|\Psi^-\rangle = \frac{1}{\sqrt{2}}(\|01\rangle - \|10\rangle)$	X & Z or iY (bit and phase flip)

Table 10.3: Conversion Gates for Bell States

These relations can be represented by the following equations:

$$|\beta_{01}\rangle = (I \otimes X)|\beta_{00}\rangle, \qquad |\beta_{00}\rangle = (I \otimes X)|\beta_{01}\rangle, \tag{10.3a}$$

$$|\beta_{01}\rangle = (X \otimes I)|\beta_{00}\rangle, \qquad |\beta_{00}\rangle = (X \otimes I)|\beta_{01}\rangle, \tag{10.3b}$$

$$|\beta_{10}\rangle = (I \otimes Z)|\beta_{00}\rangle, \qquad |\beta_{00}\rangle = (I \otimes Z)|\beta_{10}\rangle, \tag{10.3c}$$

$$|\beta_{10}\rangle = (Z \otimes I)|\beta_{00}\rangle, \qquad |\beta_{00}\rangle = (Z \otimes I)|\beta_{10}\rangle, \tag{10.3d}$$

$$|\beta_{11}\rangle = (I \otimes XZ)|\beta_{00}\rangle, \qquad |\beta_{00}\rangle = (I \otimes ZX)|\beta_{11}\rangle, \tag{10.3e}$$

$$|\beta_{11}\rangle = (ZX \otimes I)|\beta_{00}\rangle, \qquad |\beta_{00}\rangle = (XZ \otimes I)|\beta_{11}\rangle. \tag{10.3f}$$

10.2 Superdense Coding

10.2.1 Introduction

Basic information theory dictates that by sending n bits, one cannot communicate more than n bits of information. Superdense coding, however, is a quantum communication protocol that defies this limit by enabling the transmission of two classical bits of information using only a single qubit. This remarkable achievement is made possible through the use of quantum entanglement, local operations, and classical communication channels. In the absence of prior entanglement, the sender is limited to conveying no more than one bit of information per qubit, a constraint fundamentally outlined in Holevo's Theorem (§ 13.3.2). Superdense coding opens the door to potential advancements in communication efficiency, secure data transmission, and distributed quantum computing.

> **Superdense Coding: Core Intuition**
>
> Superdense coding utilizes: (1) the transformation of any Bell pair into another by applying a Pauli gate to one of the qubits locally (see § 10.1.3); and (2) the orthogonality of Bell states, which enables their perfect discrimination through measurement, as discussed in § 8.2.

10.2.2 Step-by-Step Analysis

1 Bell State Preparation

The procedure of supserdense coding encompasses four primary steps, as depicted in

10.2 Superdense Coding

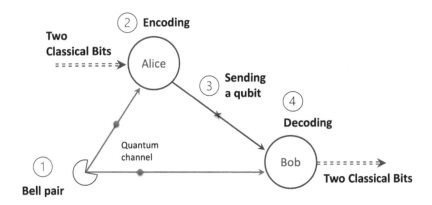

Figure 10.2: Basic Process of Superdense Coding

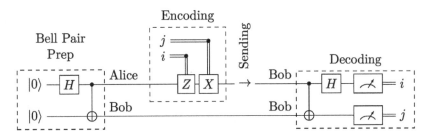

Figure 10.3: Basic Quantum Circuit for Superdense Coding

Fig. 10.2. The corresponding quantum circuit is shown in Fig. 10.3. Alice and Bob share an entangled pair of qubits, pre-established by Alice, Bob, or a third party. The initial state vector of the system is the Bell state:

$$|\Psi_0\rangle = |\Phi^+\rangle = \frac{1}{\sqrt{2}}(|00\rangle + |11\rangle). \tag{10.4}$$

2 Encoding

Alice wants to send two classical bits of information, ij, to Bob. Depending on the two bits she wants to send, Alice applies some Pauli gates to her qubit, according to Table 10.3. For '00', no gates are applied. For '01', the Z gate is applied. For '10', the X gate is applied. For '11', both the X and Z gates are applied. These gates transform the original Bell state $|\Phi^+\rangle \equiv |\beta_{00}\rangle$ to another Bell state, which is the system state after encoding:

Bits to Send (ij)	Encoding Gates	Encoded State (New Bell State)
00	I (identity)	$|\Phi^+\rangle \equiv |\beta_{00}\rangle = \frac{1}{\sqrt{2}}(|00\rangle + |11\rangle)$
01	X (bit-flip)	$|\Psi^+\rangle \equiv |\beta_{01}\rangle = \frac{1}{\sqrt{2}}(|10\rangle + |01\rangle)$
10	Z (phase-flip)	$|\Phi^-\rangle \equiv |\beta_{10}\rangle = \frac{1}{\sqrt{2}}(|00\rangle - |11\rangle)$
11	X & Z or iY	$|\Psi^-\rangle \equiv |\beta_{11}\rangle = \frac{1}{\sqrt{2}}(|01\rangle - |10\rangle)$

3 Decoding

Alice sends her qubit to Bob. Bob now has both qubits. He applies a CNOT gate with the received qubit as control and his own as the target, followed by a Hadamard gate on the control qubit. Bob then measures both qubits in the computational basis. Bob thus effectively performs a Bell measurement, and recovers the two classical bits Alice intended to send.

Alice's Bits	Encoded Bell State	Bob's Bits
00	$\|\Phi^+\rangle \equiv \|\beta_{00}\rangle = \frac{1}{\sqrt{2}}(\|00\rangle + \|11\rangle)$	00
01	$\|\Psi^+\rangle \equiv \|\beta_{01}\rangle = \frac{1}{\sqrt{2}}(\|10\rangle + \|01\rangle)$	01
10	$\|\Phi^-\rangle \equiv \|\beta_{10}\rangle = \frac{1}{\sqrt{2}}(\|00\rangle - \|11\rangle)$	10
11	$\|\Psi^-\rangle \equiv \|\beta_{11}\rangle = \frac{1}{\sqrt{2}}(\|01\rangle - \|10\rangle)$	11

Exercise 10.1 Explain how to perform superdense coding if Alice and Bob share a different Bell state at the start (e.g., $|\Phi^-\rangle$, $|\Psi^+\rangle$, or $|\Psi^-\rangle$). What operations should Alice use to encode her two bits of information?

Why is superdense coding potentially more efficient than classical transmission if both methods involve two bits? In superdense coding, one of the qubits can be shared with the receiver before the message is decided. So, when it's time to communicate the message, only one qubit needs to be sent. This method converts high-latency bandwidth into low-latency bandwidth. The receiver measures two qubits, but the first one could have been received long before the message was decided. This technique can effectively double the communication rate, given enough pre-shared entanglement.

10.2.3 ∗ Further Exploration

There are variations and extensions to the concept of superdense coding. These variations encompass the type and dimension of the quantum states utilized, the number of participants, and the operations applied to the states. Aspired learners are encouraged to delve into these topics to gain a comprehensive understanding.

1. Higher-dimensional Systems and Quantum Networks [50]: The protocol of superdense coding has been extended beyond mere qubits. Its evolution is discussed from discrete and continuous variables to quantum networks. This progress report elucidates the foundational theoretical principles behind quantum dense coding and spans its variant protocols to its applications in quantum secure communication.

2. Probabilistic Superdense Coding [71]: The feasibility of performing superdense coding with nonmaximally entangled states is examined in this study. The authors highlight that it's possible to transmit two classical bits probabilistically by sending just one qubit.

3. Distributed Quantum Dense Coding [32]: In this paper, the notion of distributed quantum dense coding is introduced. This is a generalization of quantum dense coding that incorporates more than one sender and receiver. The research proposes a classification scheme of quantum states based on their

utility in dense coding. The study particularly emphasizes that in the bipartite scenario, bound entanglement is not conducive for this endeavor.

10.3 Quantum Teleportation

Quantum teleportation facilitates the transfer of quantum information (the state of a qubit) from one point to another without physically transmitting the qubit itself. This remarkable accomplishment is realized via entanglement, classical communication channels, and local operations. Quantum teleportation has potential applications in secure communication and distributed quantum computing.

Quantum teleportation is the only known method capable of transferring quantum information between systems without data loss. Due to quantum superposition, it's impossible to measure and replicate all properties of a quantum system for reconstruction elsewhere. In quantum mechanics, a system exists in a combination of multiple potential states. Upon measurement, the system collapses to a single state, erasing information about other possibilities. Quantum teleportation, however, enables the replication of a qubit state, including its superposition, in another qubit. This process results in the destruction of the original state in the source qubit, thus adhering to the no-cloning theorem.

Quantum teleportation not only forms a key foundation in quantum computing, but also serves as an exemplary case study for contrasting and comprehending different analytical methods. If you aim to pioneer in the field of quantum computation or potentially conceive innovative algorithms similar to quantum teleportation, it is essential to have a comprehensive understanding of these varied methods. Through this exploration, our goal is to arm you with the analytical tools necessary for further pioneering and advancement in this domain.

10.3.1 Introduction

Quantum teleportation is a complex process that facilitates the transfer of quantum information from one point to another, without physically moving the quantum system itself. At its heart, this procedure encompasses four primary steps, as depicted in Fig. 10.4.

1. Establishing Entanglement: The sender (Alice) and the receiver (Bob) share an entangled quantum state, typically a pair of qubits in a Bell state.

2. Bell Measurement: Alice conducts a Bell measurement on her part of the entangled pair and the qubit she wishes to teleport, collapsing these qubits into one of four Bell states. This action annihilates the original state of the qubit Alice intends to teleport.

3. Classical Communication: Alice conveys the outcomes of her measurement to Bob via a classical communication channel.

4. State Reconstruction: On receiving this information, Bob applies a specific quantum gate to his part of the entangled pair, contingent on Alice's measurement result. This step reproduces the original quantum state of the qubit that Alice intended to teleport on Bob's qubit.

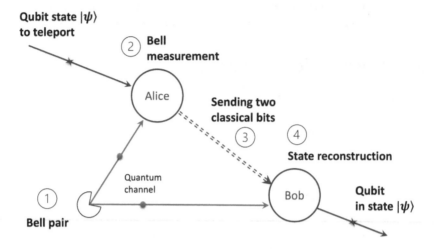

Figure 10.4: Basic Process of Quantum Teleportation

The fundamental procedure for quantum teleportation can be enacted by the quantum circuit shown in Fig. 10.5, which we will dissect in the forthcoming discussion.

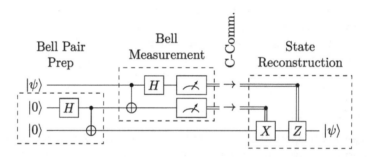

Figure 10.5: Fundamental Quantum Circuit for Teleportation

Several strategies exist for analyzing and comprehending quantum teleportation, and our objective is to explore three separate but interrelated approaches: the basis state approach, the logic operation approach, and the Bell measurement approach. Each of these methods showcases essential toolsets and provides unique insights into this foundational concept, thereby proving invaluable in the analysis of more advanced quantum algorithms.

10.3.2 Basis State Approach

This approach involves examining the teleportation circuit separately for each of the basis states $|0\rangle$ and $|1\rangle$. Utilizing the principle of linearity inherent to quantum transformations, we can subsequently combine these separate analyses to understand the teleportation of a general quantum state $|\psi\rangle = \alpha |0\rangle + \beta |1\rangle$, which Alice wishes to transmit.

This method allows us to derive the state of all three qubits at each step without invoking advanced concepts such as Bell measurements. However, a downside of this

10.3 Quantum Teleportation

approach is that the fundamental physical principles become somewhat obscured within routine algebraic manipulations.

1 Deferred Measurement Principle

To simplify the analysis, we will employ the principle of deferred measurement (introduced in § 7.4.5). This principle states that measurements can be shifted from an intermediate stage of a quantum circuit to the end. Furthermore, if the measurement results are used at any point in the circuit, the classically controlled operations can be replaced by quantum controlled operations.

Leveraging this principle, we can substitute classically controlled X and Z gates with quantum CX (i.e., CNOT) and CZ gates, as illustrated in Fig. 10.6. The subsequent analysis uses this version of the quantum teleportation circuit.

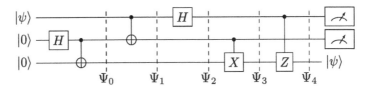

Figure 10.6: Quantum Circuit for Teleportation with Deferred Measurement

2 Establishing Entanglement

The initial state vector of the system after the Bell state preparation is:

$$|\Psi_0\rangle = \begin{cases} \frac{1}{\sqrt{2}}(|000\rangle + |011\rangle) & \text{for basis state } |0\rangle, \\ \frac{1}{\sqrt{2}}(|100\rangle + |111\rangle) & \text{for basis state } |1\rangle. \end{cases} \tag{10.5}$$

3 Alice Applies CNOT Gate

Now Alice applies a CNOT gate with her qubit (first qubit) as the control and her entangled qubit (second qubit) as the target:

$$|\Psi_1\rangle = \begin{cases} \frac{1}{\sqrt{2}}(|000\rangle + |011\rangle) & \text{for basis state } |0\rangle, \\ \frac{1}{\sqrt{2}}(|110\rangle + |101\rangle) & \text{for basis state } |1\rangle. \end{cases} \tag{10.6}$$

4 Alice Applies H Gate

Then Alice applies a Hadamard gate to her qubit:

$$|\Psi_2\rangle = \begin{cases} \frac{1}{2}(|000\rangle + |011\rangle + |100\rangle + |111\rangle) & \text{for basis state } |0\rangle, \\ \frac{1}{2}(|010\rangle + |001\rangle - |110\rangle - |101\rangle) & \text{for basis state } |1\rangle. \end{cases} \tag{10.7}$$

5 Bob Applies CX Gate

Then Alice measures the second qubit and Bob applies an X gate to the third qubit if the measurement result is 1. This is equivalent to applying a controlled-X (same as CNOT) gate, which flips the third qubit if the second qubit is 1.

$$|\Psi_3\rangle = \begin{cases} \frac{1}{2}(|000\rangle + |010\rangle + |100\rangle + |110\rangle) & \text{for basis state } |0\rangle, \\ \frac{1}{2}(|011\rangle + |001\rangle - |111\rangle - |101\rangle) & \text{for basis state } |1\rangle. \end{cases} \tag{10.8}$$

6 Bob Applies CZ Gate

Finally, Alice measures the first qubit and Bob applies a Z gate to the third qubit if the measurement result is 1. This is equivalent to applying a controlled-Z gate, which changes the sign of the state vector if both the second and third qubits are 1.

$$|\Psi_4\rangle = \begin{cases} \frac{1}{2}(|000\rangle + |010\rangle + |100\rangle + |110\rangle) & \text{for basis state } |0\rangle, \\ \frac{1}{2}(|011\rangle + |001\rangle + |111\rangle + |101\rangle) & \text{for basis state } |1\rangle, \end{cases} \quad (10.9)$$

$$= \begin{cases} \frac{1}{2}(|0\rangle + |1\rangle) \otimes (|0\rangle + |1\rangle) \otimes |0\rangle & \text{for basis state } |0\rangle, \\ \frac{1}{2}(|0\rangle + |1\rangle) \otimes (|0\rangle + |1\rangle) \otimes |1\rangle & \text{for basis state } |1\rangle. \end{cases} \quad (10.10)$$

If we use $|x\rangle$ to represent the basis states, where $x \in \{0,1\}$, we can combine the above equation as:

$$|\Psi_4\rangle = \frac{1}{2}(|0\rangle + |1\rangle) \otimes (|0\rangle + |1\rangle) \otimes |x\rangle. \quad (10.11)$$

At this stage, it becomes evident that the third qubit has become disentangled from the first two qubits, since its state is now a component of a product state. Specifically, it is in the state $|x\rangle$. Since $|x\rangle$ can represent any basis state and quantum transformations are linear by nature, Eq. 10.11 will still hold if we replace $|x\rangle$ with a general qubit state $|\psi\rangle$. In fact, $|\Psi_4\rangle = |+\rangle |+\rangle |\psi\rangle$. This indicates that Alice has successfully teleported her qubit state $|\psi\rangle$ to Bob.

10.3.3 ∗ Logic Operation Approach

This method builds on the basis state approach but deviates by directly deriving equations for the combined state $|x\rangle$, rather than considering individual basis states. This is achieved by employing mathematical operations using the Boolean representation technique, similar to Boolean algebra used for classical logic gates.

The Boolean representation of quantum gates and circuits is detailed in § 7.3. Below is a brief summary.

With $|x\rangle$ and $|y\rangle$ representing $|0\rangle$ and $|1\rangle$, i.e., $x, y \in \{0, 1\}$, and $\bar{x} \equiv 1 - x$, $\bar{y} \equiv 1 - y$, the Boolean representation of some popular gates is summarized in Table 10.4.

Gate	Boolean Representation
X or NOT	$|x\rangle \longrightarrow |\bar{x}\rangle$
CNOT or CX	$|x,y\rangle \longrightarrow |x, x \oplus y\rangle$
Z	$|x\rangle \longrightarrow (-1)^x |x\rangle$
CZ	$|x,y\rangle \longrightarrow (-1)^{x \cdot y} |x,y\rangle$
H	$|x\rangle \longrightarrow \frac{1}{\sqrt{2}}(|0\rangle + (-1)^x |1\rangle)$
Y	$|x\rangle \longrightarrow i(-1)^x |\bar{x}\rangle$

Table 10.4: Boolean Representation of Selected Qubit Gates

Here, $\oplus$ denotes the exclusive OR (or XOR) Boolean operator, which corresponds to bitwise modulo 2 addition. Below are some useful identities related to XOR:

10.3 Quantum Teleportation

$$x \oplus y = x + y \mod 2 \tag{10.12a}$$
$$x \oplus 0 = 0 \oplus x = x \tag{10.12b}$$
$$x \oplus 1 = 1 \oplus x = \bar{x} \tag{10.12c}$$
$$\bar{x} \oplus 1 = 1 \oplus \bar{x} = x \tag{10.12d}$$
$$x \oplus x = 0 \tag{10.12e}$$
$$x \oplus \bar{x} = \bar{x} \oplus x = 1 \tag{10.12f}$$
$$\overline{x \oplus y} = \bar{x} \oplus y = x \oplus \bar{y} = 1 \oplus x \oplus y \tag{10.12g}$$
$$(-1)^{x \oplus y} = (-1)^{\bar{x} \oplus \bar{y}} = (-1)^{x+y} \tag{10.12h}$$
$$(-1)^{x \oplus \bar{y}} = (-1)^{\bar{x} \oplus y} = (-1)^{1+x+y}. \tag{10.12i}$$

With the above mathematical tool, let's revisit the steps from the basis states approach, referring to Fig. 10.6.

1 Establishing Entanglement

The initial state vector of the system is:

$$|\Psi_0\rangle = \frac{1}{\sqrt{2}}(|x00\rangle + |x11\rangle). \tag{10.13}$$

2 Alice Applies CNOT Gate

Now Alice applies a CNOT gate with her qubit (first qubit) as the control and her entangled qubit (second qubit) as the target:

$$|\Psi_1\rangle = \frac{1}{\sqrt{2}}(|x(0 \oplus x)0\rangle + |x(1 \oplus x)1\rangle) \tag{10.14a}$$
$$= \frac{1}{\sqrt{2}}(|xx0\rangle + |x\bar{x}1\rangle). \tag{10.14b}$$

> ⓘ Did the CNOT operation entangle the first and second qubit? Yes. Even though $|\Psi_1\rangle = |x\rangle \otimes \frac{1}{\sqrt{2}}(|x0\rangle + |\bar{x}1\rangle)$ appears to be a product state, it is not. It represents $\frac{1}{\sqrt{2}}(\alpha |0\rangle (|00\rangle + |11\rangle) + \beta |1\rangle (|10\rangle + |01\rangle))$ which cannot be decomposed as the tensor product of $\alpha |0\rangle + \beta |1\rangle$ with another state.

3 Alice Applies H Gate

Then Alice applies a Hadamard gate to her qubit (first qubit). To make the equations shorter, let's rewrite $|\Psi_1\rangle$ as:

$$|\Psi_1\rangle = |x\rangle \otimes \frac{1}{\sqrt{2}}(|x0\rangle + |\bar{x}1\rangle). \tag{10.15}$$

After the application of the H gate:

$$|\Psi_2\rangle = \frac{1}{\sqrt{2}}(|0\rangle + (-1)^x |1\rangle) \otimes \frac{1}{\sqrt{2}}(|x0\rangle + |\bar{x}1\rangle) \tag{10.16a}$$
$$= \frac{1}{2}(|0x0\rangle + |0\bar{x}1\rangle + (-1)^x |1x0\rangle + (-1)^x |1\bar{x}1\rangle). \tag{10.16b}$$

How many qubits are currently entangled? Three. Although Eq. 10.16a seems to represent a product state, the inclusion of the $(-1)^x$ factor renders it not a genuine product state.

4 Bob Applies CX Gate

Then Alice measures the second qubit and Bob applies an X gate to the third qubit if the measurement result is 1. Due to the deferred measurement theorem, this is equivalent to applying a CX (i.e., CNOT) gate.

$$|\Psi_3\rangle = \frac{1}{2}(|0x(x \oplus 0)\rangle + |0\bar{x}(\bar{x} \oplus 1)\rangle$$
$$+ (-1)^x |1x(x \oplus 0)\rangle + (-1)^x |1\bar{x}(\bar{x} \oplus 1)\rangle) \quad (10.17\text{a})$$
$$= \frac{1}{2}(|0xx\rangle + |0\bar{x}x\rangle + (-1)^x |1xx\rangle + (-1)^x |1\bar{x}x\rangle). \quad (10.17\text{b})$$

5 Bob Applies CZ Gate

Finally, Alice measures the first qubit and Bob applies a Z gate to the third qubit if the measurement result is 1:

$$|\Psi_4\rangle = \frac{1}{2}(|0xx\rangle + |0\bar{x}x\rangle + (-1)^x(-1)^x |1xx\rangle$$
$$+ (-1)^x(-1)^x |1\bar{x}x\rangle) \quad (10.18\text{a})$$
$$= \frac{1}{2}(|0xx\rangle + |0\bar{x}x\rangle + |1xx\rangle + |1\bar{x}x\rangle) \quad (10.18\text{b})$$
$$= \frac{1}{2}(|0\rangle + |1\rangle) \otimes (|x\rangle + |\bar{x}\rangle) \otimes |x\rangle \quad (10.18\text{c})$$
$$= \frac{1}{2}(|0\rangle + |1\rangle) \otimes (|0\rangle + |1\rangle) \otimes |x\rangle. \quad (10.18\text{d})$$

At this stage, it becomes evident that the three qubits become disentangled, and the third qubit is $|x\rangle$. Alice has successfully teleported her qubit state $|\psi\rangle$ to Bob.

Exercise 10.2 Try to replicate the above analysis steps. This time, however, assume that Alice and Bob share a different Bell state, $|\Psi^-\rangle$, at the start.

10.3.4 ∗ Bell Measurement Approach

We will now examine the quantum teleportation process by utilizing Bell measurement, discussed § 8.4, as an analytical tool. Despite incorporating the advanced concept of Bell measurement, this approach provides the clearest physical intuition about the construction of the teleportation circuit in Fig. 10.5.

1 Establishing Entanglement

Alice wants to send a quantum state $|\psi\rangle = \alpha|0\rangle + \beta|1\rangle$ to Bob. They share an entangled pair of qubits in the Bell state $|\Phi^+\rangle = \frac{1}{\sqrt{2}}(|00\rangle + |11\rangle)$. The initial state vector of the system can be written as:

10.3 Quantum Teleportation

$$|\Psi_0\rangle = \frac{1}{\sqrt{2}}(\alpha|0\rangle + \beta|1\rangle) \otimes (|00\rangle + |11\rangle) \tag{10.19a}$$

$$= \frac{1}{\sqrt{2}}(\alpha|000\rangle + \alpha|011\rangle + \beta|100\rangle + \beta|111\rangle) \tag{10.19b}$$

$$= \frac{1}{\sqrt{2}}(\alpha|00\rangle + \beta|10\rangle)|0\rangle + \frac{1}{\sqrt{2}}(\alpha|01\rangle + \beta|11\rangle)|1\rangle. \tag{10.19c}$$

In the last step, we have singled out the 3rd qubit since we are going to measure the first two qubit in the next step.

2 Bell Measurement

Alice applies a CNOT gate with her qubit as the control and the entangled qubit as the target. Then she applies a Hadamard gate to her qubit. Finally, she measures the two qubits, effectively performing a Bell measurement on them.

The four Bell states form an orthonormal basis for two-qubit states. By performing Bell measurements, Alice projects the two-qubit states onto the Bell states.

The measurement operators, for the overall system of three qubits, are given by:

$$M_{ij} = |\beta_{ij}\rangle\langle\beta_{ij}| \otimes I, \tag{10.20}$$

where $i,j \in \{0,1\}$, β_{ij} are the Bell states given by Eq. 10.1, and I is the identity operator applied to the 3rd qubit.

After the measurement, depending on the measurement result (i,j), the system will be in the state:

$$|\Psi_1\rangle = M_{ij}|\Psi_0\rangle \tag{10.21a}$$

$$= \frac{1}{\sqrt{2}}(|\beta_{ij}\rangle\langle\beta_{ij}| \otimes I)((\alpha|00\rangle + \beta|10\rangle)|0\rangle + (\alpha|01\rangle + \beta|11\rangle)|1\rangle) \tag{10.21b}$$

$$= \frac{1}{\sqrt{2}}|\beta_{ij}\rangle\langle\beta_{ij}|(\alpha|00\rangle + \beta|10\rangle)|0\rangle$$
$$+ \frac{1}{\sqrt{2}}|\beta_{ij}\rangle\langle\beta_{ij}|(\alpha|01\rangle + \beta|11\rangle)|1\rangle \tag{10.21c}$$

$$= |\beta_{ij}\rangle \otimes |\psi_{ij}\rangle, \tag{10.21d}$$

where $|\psi_{ij}\rangle$ is defined in Table 10.2, copied below for convenience:

$$|\psi_{00}\rangle = \alpha|0\rangle + \beta|1\rangle, \tag{10.22a}$$
$$|\psi_{01}\rangle = \alpha|1\rangle + \beta|0\rangle, \tag{10.22b}$$
$$|\psi_{10}\rangle = \alpha|0\rangle - \beta|1\rangle, \tag{10.22c}$$
$$|\psi_{11}\rangle = \alpha|1\rangle - \beta|0\rangle. \tag{10.22d}$$

 How is $|\Psi_1\rangle$ in this equation related to the $|\Psi_1\rangle$ in Eq. 10.6? The latter is a linear combination of the four cases of the former - see Eq. 10.23.

3 Classical Communication

Alice transmits the measurement outcomes (i, j, representing two bits of information) to Bob through a classical communication channel. It is important to note that the original qubit state encapsulates more information than merely two bits; it's characterized by two complex numbers, α and β. Furthermore, exact values cannot be obtained from a single measurement, as performing a measurement invariably disrupts the original quantum state.

4 State Reconstruction

Based on the two bits (i, j) resulting from Alice's measurements, Bob will apply the necessary correction gates to his qubit: a Z gate is applied if $i = 1$, and an X gate is applied if $j = 1$.

Outcome ij	Intermediate Qubit State	Correction Gates
00	$\alpha\,\|0\rangle + \beta\,\|1\rangle$	I (None)
01	$\alpha\,\|1\rangle + \beta\,\|0\rangle$	X
10	$\alpha\,\|0\rangle - \beta\,\|1\rangle$	Z
11	$\alpha\,\|1\rangle - \beta\,\|0\rangle$	X & Z, or iY

The final state vector of Bob's qubit becomes $|\psi\rangle = \alpha\,|0\rangle + \beta\,|1\rangle$. Now Bob's qubit is in the same state as Alice's initial qubit, completing the quantum teleportation process.

> For measurement outcome '11', both an X and a Z gate are applied. Their sequence does not matter in this case, as swapping their order only incurs a global phase shift ($XZ = -ZX$), which is unobservable in quantum mechanics. However, if subsequent stages involve interference with other qubits, the order may become crucial due to the potential impact on relative phases. In such scenarios, the correct sequence is X before Z, i.e., in operator terms, ZX.

> **Exercise 10.3** Try to replicate the above analysis steps. This time, however, assume that Alice and Bob share a different Bell state, specifically $|\Psi^-\rangle$, at the outset. What differences do you observe from the initial scenario?

Quantum Teleportation: Core Summary

This analysis captures the essence of quantum teleportation. The initial state vector can be decomposed into the orthonormal Bell basis as follows:

$$|\Psi_0\rangle = \frac{1}{\sqrt{2}}(\alpha\,|0\rangle + \beta\,|1\rangle) \otimes (|00\rangle + |11\rangle),$$

$$= \frac{1}{2} \sum_{i,j \in \{0,1\}} |\beta_{ij}\rangle \otimes |\psi_{ij}\rangle. \quad (10.23)$$

where $|\beta_{ij}\rangle$ represents the Bell states, as defined in Eq. 10.1, and $|\psi_{ij}\rangle$ is defined in Table 10.2 and Eq. 10.22a.

When Alice measures the first two qubits in the Bell basis $\{\beta_{ij}\}$, each of the Bell states occurs with equal probability. If she obtains $|\beta_{ij}\rangle$ (where $i,j \in \{0,1\}$), the state of the system collapses to $|\beta_{ij}\rangle \otimes |\psi_{ij}\rangle$, and the third qubit takes on the state $|\psi_{ij}\rangle$.

The third qubit can then be adjusted based on Alice's measurement results (i,j) to reconstruct the original state $\alpha|0\rangle + \beta|1\rangle$.

10.3.5 ∗ Further Exploration

The concept of quantum teleportation extends beyond the basic protocol discussed in this text. Many variations and enhancements center around the nature and dimensionality of the quantum states used, the number of parties involved, and the types of operations executed on the states. For those seeking to deepen their understanding, the following topics are suggested for further exploration:

1. Quantum Teleportation of Composite Systems and Entanglement [7]: : This topic deals with the teleportation of composite systems, which can involve multiple entangled qubits. Understanding this concept is fundamental in advancing complex quantum information processing protocols.

2. Superdense Teleportation [47]: : This variant combines principles of quantum teleportation and superdense coding to enable the transfer of more information than either standalone protocol. This involves the teleportation of a two-qubit state using two classical bits and a shared entangled pair of qubits.

3. Higher-dimensional Quantum Teleportation [26]: : While the fundamental quantum teleportation protocol involves qubits, researchers have broadened this framework to include qudits. Higher-dimensional quantum systems can potentially provide richer quantum information and more robust entanglement.

4. Multiparty Quantum Teleportation [0]: : Beyond the traditional two-party setup, quantum teleportation can be extended to multiple parties. This development could involve numerous participants each transmitting information through a shared multipartite entangled state.

5. Teleportation of an Arbitrary Two-Qubit State [75]: : This work expands the scope of quantum teleportation to two-qubit states, linking this to multipartite entanglement. The insights provided enhance the understanding of complex quantum information transfer and potential improvements in teleportation protocols.

6. Continuous-Variable Quantum Teleportation [44]: : This variation uses continuous-variable systems, such as the position and momentum of a particle. As these systems can assume any value within a certain range, this approach often involves quantum states of light and can potentially provide higher fidelity in teleportation.

10.4 Entanglement Swapping

Entanglement swapping, also known as quantum relay, is a fundamental procedure in quantum information science that allows two quantum systems to become entangled without direct interaction, purely through the use of entanglement and local operations. It plays an instrumental role in quantum repeaters and quantum networks, enhancing secure long-distance communication and creating a web of quantum devices.

Entanglement swapping is a powerful technique in overcoming the physical limitations inherent in creating direct entanglement between distant particles. Entanglement is typically established by allowing two quantum systems to interact directly. However, as the distance between the two systems grows, the feasibility of this interaction rapidly diminishes due to environmental interference. Entanglement swapping breaks down the barrier of distance, allowing quantum systems to become entangled without the necessity for direct interaction.

Understanding entanglement swapping is essential for those delving into the field of quantum networking and quantum cryptography. Its concept, combined with other quantum protocols such as quantum teleportation, enables the creation of secure and efficient quantum communication systems. In this section, we aim to shed light on the intricacies of entanglement swapping and its pivotal role in quantum information science.

10.4.1 Introduction

Entanglement swapping is a procedure that leads to the entanglement of two quantum systems that have never interacted before. It is executed through a sequence of steps as outlined in Fig. 10.7.

1. Initial Entanglement: Two pairs of entangled quantum states are created, typically qubits in a Bell state. The first pair is shared between Alice and a mediator (Charlie), and the second pair is shared between Charlie and Bob.

2. Bell Measurement: Charlie conducts a Bell measurement on his two qubits from the entangled pairs, collapsing these qubits into one of four Bell states.

3. Classical Communication: Charlie communicates the outcomes of his measurement to Bob via a classical communication channel.

4. State Reconstruction: Upon receiving Charlie's measurement result, Bob performs certain quantum operations on his qubit depending on the result. This step results in Alice's and Bob's qubits becoming entangled, thereby completing the entanglement swapping process.

 In our example, the state reconstruction is done on Bob's qubit. Depending on the specific requirements of the quantum operation or communication, the state reconstruction can be performed on Alice's qubit, Bob's, or both. The choice of where to perform the state reconstruction can have implications for the efficiency, security, and other characteristics of these quantum protocols.

10.4 Entanglement Swapping

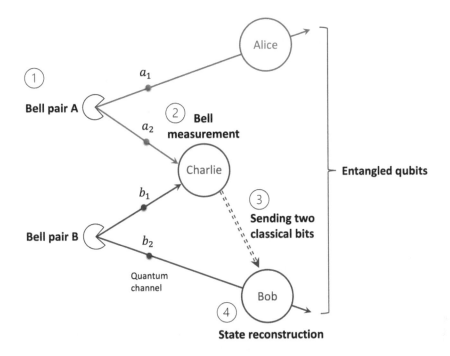

Figure 10.7: Basic Process of Entanglement Swapping

The fundamental procedure for entanglement swapping can be represented by the quantum circuit shown in Fig. 10.8, which we will examine in more detail in the following discussion.

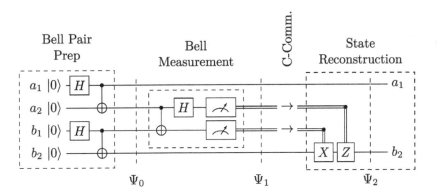

Figure 10.8: Fundamental Quantum Circuit for Entanglement Swapping

10.4.2 ✻ Step-by-Step Analysis

Let's delve into the entanglement swapping protocol by going through each step. This process involves creating entanglement between two qubits without direct interaction, utilizing local operations, and classical communication.

1 Initial Entanglement

Consider two entangled pairs of qubits: Alice's pair A consisting of qubits a_1 and a_2, and Bob's pair B consisting of qubits b_1 and b_2. These pairs can be in any entangled state, but for simplicity, we'll consider them to be in the Bell state $|\Phi^+\rangle = \frac{1}{\sqrt{2}}(|00\rangle + |11\rangle)$.

The initial state vector of the system can be written as:

$$|\Psi_0\rangle_{a_1 a_2 b_1 b_2} = |\Phi^+\rangle_{a_1 a_2} \otimes |\Phi^+\rangle_{b_1 b_2} \tag{10.24a}$$

$$= \frac{1}{\sqrt{2}}(|00\rangle + |11\rangle) \otimes \frac{1}{\sqrt{2}}(|00\rangle + |11\rangle) \tag{10.24b}$$

$$= \frac{1}{2}(|0000\rangle + |0011\rangle + |1100\rangle + |1111\rangle). \tag{10.24c}$$

For the purpose of the entanglement swapping protocol, we will treat a_2 and b_1 as one group, and a_1 and b_2 as another. This grouping aligns with the steps of the protocol, where a_2 and b_1 are sent to Charlie for a Bell measurement, while a_1 and b_2 become the final entangled pair. To reflect this in our notation, we will rearrange the order of the qubits in our state vector representation. Note that this is merely a mathematical adjustment and does not involve any physical changes to the qubits. In the new representation, the state vector is:

$$|\Psi_0\rangle_{a_2 b_1 a_1 b_2} = \frac{1}{2}(|0000\rangle + |0101\rangle + |1010\rangle + |1111\rangle). \tag{10.25}$$

2 Bell Measurement

Alice and Bob each send one qubit, a_2 and b_1 respectively, to Charlie. Upon receiving the qubits a_2 and b_1, Charlie performs a Bell measurement. This measurement process is achieved by first applying a CNOT gate with a_2 as the control qubit and b_1 as the target, followed by a Hadamard gate on a_2. This transforms the two-qubit state into a superposition of the Bell states. Subsequently, Charlie measures both qubits in the computational basis. For simplicity in this analysis, we treat Charlie's CNOT gate, Hadamard gate, and measurement devices as a single unit, which we refer to as the Bell-measurement unit.

The Bell measurement projects the two-qubit state onto one of the Bell states $|\beta_{ij}\rangle$. The measurement operators are:

$$M_{ij} = |\beta_{ij}\rangle\langle\beta_{ij}|_{a_2 b_1} \otimes I_{a_1 b_2}. \tag{10.26}$$

After the measurement, depending on the result, the system will be in one of the following states:

$$|\Psi_1\rangle_{ij} = M_{ij}|\Psi_0\rangle_{a_2 b_1 a_1 b_2}. \tag{10.27}$$

The result of this equation has the following form:

$$|\Psi_1\rangle_{ij} = |\beta_{ij}\rangle_{a_2 b_1} \otimes |\psi\rangle_{a_1 b_2}. \tag{10.28}$$

After some algebraic manipulation, we obtain $|\psi\rangle_{a_1 b_2}$ as in the following table:

10.4 Entanglement Swapping

Outcome ij	State of a_2 and b_1 ($\|\beta_{ij}\rangle_{a_2 b_1}$)	State of a_1 and b_2 ($\|\psi\rangle_{a_1 b_2}$)
00	$\|\beta_{00}\rangle \equiv \|\Phi^+\rangle = \frac{1}{\sqrt{2}}(\|00\rangle + \|11\rangle)$	$\frac{1}{\sqrt{2}}(\|00\rangle + \|11\rangle)$
01	$\|\beta_{01}\rangle \equiv \|\Psi^+\rangle = \frac{1}{\sqrt{2}}(\|01\rangle + \|10\rangle)$	$\frac{1}{\sqrt{2}}(\|01\rangle + \|10\rangle)$
10	$\|\beta_{10}\rangle \equiv \|\Phi^-\rangle = \frac{1}{\sqrt{2}}(\|00\rangle - \|11\rangle)$	$\frac{1}{\sqrt{2}}(\|00\rangle - \|11\rangle)$
11	$\|\beta_{11}\rangle \equiv \|\Psi^-\rangle = \frac{1}{\sqrt{2}}(\|01\rangle - \|10\rangle)$	$\frac{1}{\sqrt{2}}(\|01\rangle - \|10\rangle)$

Examining the above table, we discover that after the measurement with outcome ij, qubits a_1 and b_2 are also in the Bell state $|\beta_{ij}\rangle$! This relationship, which is the heart of entanglement swapping, is captured by the following equation:

$$|\Psi_1\rangle_{ij} = |\beta_{ij}\rangle_{a_2 b_1} \otimes |\beta_{ij}\rangle_{a_1 b_2}. \tag{10.29}$$

3 Classical Communication

Even though qubits a_1 and b_2 are in the Bell state $|\beta_{ij}\rangle$, the measurement outcome ij appears randomly, limiting the usefulness of the entanglement. Next, Charlie transmits the measurement outcomes ij to Bob through a classical communication channel for him to make corrections to the Bell state.

4 State Reconstruction

To transform the state of qubits a_1 and b_2 from $|\beta_{ij}\rangle$ to $|\Psi_2\rangle = |\beta_{00}\rangle$, Bob applies the necessary correction gates to his qubit b_2 according to the two bits ij from Charlie's measurements: an X gate when $j = 1$ and a Z gate when $i = 1$.

Outcome ij	State of a_1 and b_2	Correction Gates
00	$\|\beta_{00}\rangle \equiv \|\Phi^+\rangle = \frac{1}{\sqrt{2}}(\|00\rangle + \|11\rangle)$	I (None)
01	$\|\beta_{01}\rangle \equiv \|\Psi^+\rangle = \frac{1}{\sqrt{2}}(\|01\rangle + \|10\rangle)$	X
10	$\|\beta_{10}\rangle \equiv \|\Phi^-\rangle = \frac{1}{\sqrt{2}}(\|00\rangle - \|11\rangle)$	Z
11	$\|\beta_{11}\rangle \equiv \|\Psi^-\rangle = \frac{1}{\sqrt{2}}(\|01\rangle - \|10\rangle)$	X & Z

Entanglement Swapping: Core Summary

The re-grouped initial state can be decomposed into the orthonormal Bell basis as follows:

$$\begin{aligned}|\Psi_0\rangle_{a_2 b_1 a_1 b_2} &= \frac{1}{2}(|0000\rangle + |0101\rangle + |1010\rangle + |1111\rangle), \\ &= \frac{1}{2} \sum_{i,j \in \{0,1\}} |\beta_{ij}\rangle_{a_2 b_1} \otimes |\beta_{ij}\rangle_{a_1 b_2}.\end{aligned} \tag{10.30}$$

When Charlie measures qubits a_2 and b_1 in the Bell basis $\{\beta_{ij}\}$, each of the Bell states occurs with equal probability. If he obtains $|\beta_{ij}\rangle$ (where $i,j \in \{0,1\}$), the state of the system collapses to $|\beta_{ij}\rangle \otimes |\beta_{ij}\rangle$, and qubits a_1 and b_2 also take

> on the state $|\beta_{ij}\rangle$. They can then be adjusted based on the measurement results (i, j) to reconstruct the original entangled state $|\beta_{00}\rangle$.

10.4.3 ∗ Further Exploration

The concept of entanglement swapping, similar to quantum teleportation, can be extended beyond the basic protocol outlined in this text. Numerous variations and enhancements focus on the nature and dimensionality of the quantum states used, the number of parties involved, and the types of operations executed on the states. For those interested in deepening their understanding, the following topics are suggested for further exploration:

1. Quantum Repeaters with Entanglement Swapping [30]: Explores the use of entanglement swapping in quantum communication infrastructure. Quantum repeaters help overcome distance limitations in quantum communication by using entanglement swapping to establish entanglement between distant parties.

2. Multi-party Entanglement Swapping [94]: Involves more than two parties, with multiple mediators performing Bell measurements and classical communications. This potentially enables large-scale quantum networks with distributed entanglement.

3. Higher-dimensional Entanglement Swapping [25]: Extends entanglement swapping from qubits to qudits, or higher-dimensional quantum systems. This extension can potentially enhance quantum information capacity and the robustness of entanglement.

4. Entanglement Swapping with Mixed States [63]: While most entanglement swapping schemes use pure entangled states, variations that incorporate mixed states exist. These protocols, though more complex, may be more practical as mixed states are commonly encountered in real quantum systems due to decoherence and noise.

5. Entanglement Swapping in Various Systems [95]: This research explores entanglement swapping in pure and noisy systems, analyzing the relationship between initial and final state entanglement using metrics like concurrence and negativity. The study reveals that the initial state's entanglement can influence the final state's average entanglement, with measurement bases and entanglement intensity playing critical roles.

10.5 Quantum Gate Teleportation

Quantum gate teleportation refers to the process of performing a quantum operation (a quantum gate) on one or more qubits that are not locally accessible, leveraging the properties of entanglement and quantum teleportation.

Quantum gate teleportation can be viewed as entanglement-assisted computation, in contrast to the concept of entanglement-assisted communication found in standard quantum teleportation. The primary aim is to perform a computational operation during the teleportation process itself.

Quantum gate teleportation is particularly beneficial in scenarios where direct manipulation of the qubit could lead to decoherence or when the qubits are spatially

10.5 Quantum Gate Teleportation

separated. It can also enable the realization of distributed quantum computing, where quantum gates can be performed on qubits located in different quantum computers, thus enabling more extensive and powerful quantum networks.

10.5.1 ∗Single-Qubit Gate Teleportation

We will first consider single-qubit gate teleportation, where the goal is to apply a single-qubit unitary operation U to a quantum state $|\psi\rangle$ during teleportation.

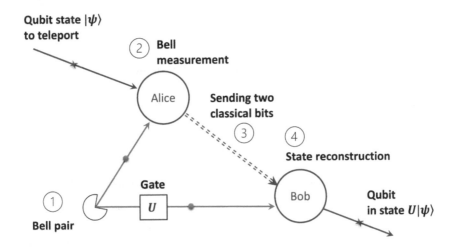

Figure 10.9: Basic Process of Quantum Gate Teleportation

Single-qubit gate teleportation follows a process similar to standard quantum teleportation, encompassing four main steps, as shown in Fig. 10.9. Instead of using the standard Bell state $|\Phi^+\rangle$, Alice and Bob begin with the shared state $(I \otimes U)|\Phi^+\rangle$, in which the unitary operation U is applied to Bob's part of the entangled pair. After Alice sends her measurement results, Bob performs correction operations, much like in standard teleportation, albeit with adjusted correction gates. The final state he obtains is $U|\psi\rangle$ instead of simply $|\psi\rangle$. This modification ensures that the teleportation procedure not only conveys the state $|\psi\rangle$ but also effectively applies the operation U to $|\psi\rangle$, resulting in the final state $U|\psi\rangle$ on Bob's end.

Figure 10.10 presents a quantum circuit for single-qubit gate teleportation that aligns with this procedure. We'll delve into this circuit in detail shortly.

1 Transition from Standard to Gate Teleportation

Let's now explore an intuitive connection between single-qubit gate teleportation and standard quantum teleportation. This exploration offers insight into the discovery process of protocols like gate teleportation. It also lays a foundation for understanding more complex protocols, such as CNOT gate teleportation, which we'll discuss later.

Refer to Fig. 10.11. The segment prior to the dashed visual separator is the standard teleportation circuit from Fig. 10.5. At this circuit's output, the third (bottom) qubit is in state $|\psi\rangle$. Upon application of a unitary gate U, the qubit's state becomes $U|\psi\rangle$. This final state mirrors that of single-qubit gate teleportation, wherein gate U is applied to the teleported $|\psi\rangle$ (as depicted in Fig. 10.10).

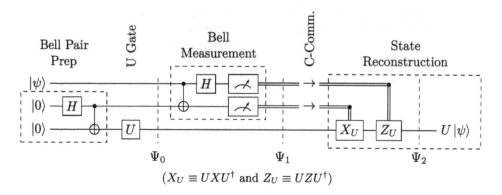

Figure 10.10: Quantum Circuit for Single-Qubit Gate Teleportation

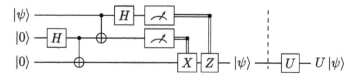

Figure 10.11: Gate U Applied Post-Teleportation

To transform the circuit in Fig. 10.11 into the gate teleportation circuit as shown in Fig. 10.10, we need to shift the gate U to the left of the controlled-Z and controlled-X gates. This transformation is facilitated by the concept of gate sequence equivalence, which is demonstrated in the following table:

Gate Sequence	Equivalent Seq.	Operator Equivalence
—X—U—	—U—X_U—	$UX = UX(U^\dagger U) = (UXU^\dagger)U = X_U U$
—Z—U—	—U—Z_U—	$UZ = UZ(U^\dagger U) = (UZU^\dagger)U = Z_U U$

This concept of gate sequence equivalence clarifies the transformation of the correction gates from X and Z to $X_U \equiv UXU^\dagger$ and $Z_U \equiv UZU^\dagger$ respectively in Fig. 10.10.

 Although quantum circuit diagrams apply gates from left to right, the corresponding operator (matrix) multiplication operates in reverse order, from right to left.

Standard versus Gate Teleportation: Summary

In quantum teleportation, Alice and Bob share an entangled Bell state, $|\Phi^+\rangle$, allowing Alice to teleport an unknown state $|\psi\rangle$ to Bob. This is achieved by Alice measuring her qubits in the Bell basis, resulting in one of four possible outcomes which, on Bob's end, manifest as states $|\psi\rangle$, $X|\psi\rangle$, $Z|\psi\rangle$, or $ZX|\psi\rangle$. Bob corrects his state to $|\psi\rangle$ by applying the corresponding Pauli operator, I, X, Z, or XZ, informed by Alice's classical measurement.

10.5 Quantum Gate Teleportation

> Quantum gate teleportation extends this concept by teleporting a single-qubit gate U. Here, Alice and Bob share $(I \otimes U) |\Phi^+\rangle$ instead. The teleportation procedure leaves Bob with one of four states: $U|\psi\rangle$, $UX|\psi\rangle$, $UZ|\psi\rangle$, or $UZX|\psi\rangle$. Bob then corrects his state to $U|\psi\rangle$ by applying the corresponding operator, I, $UXU^\dagger$, $UZU^\dagger$, or $UXZU^\dagger$, depending on Alice's classical measurement.
>
> Quantum teleportation is equivalent to quantum gate teleportation with $U = I$.

2 ∗ Step-by-Step Analysis

While we've successfully derived the gate teleportation circuit by modifying the standard teleportation circuit, it's worthwhile to conduct a detailed, step-by-step examination of the gate teleportation process. This will serve to reinforce our understanding. Think of single-qubit gate teleportation as an extension of the Bell measurement approach to quantum teleportation, which we discussed in § 10.3.4. The equations that follow will closely mirror those found in the aforementioned section.

1. Initial State

Alice wants to send a quantum state $|\psi\rangle = \alpha|0\rangle + \beta|1\rangle$ to Bob, and have the state transformed by gate U during the process.

Alice and Bob share the state $(I \otimes U)|\Phi^+\rangle$, where U has been applied to Bob's part of the entangled pair. The initial state vector of the system can be written as:

$$|\Psi_0\rangle = \frac{1}{\sqrt{2}}(\alpha|0\rangle + \beta|1\rangle) \otimes (I \otimes U)(|00\rangle + |11\rangle) \quad (10.31\text{a})$$

$$= \frac{1}{\sqrt{2}}(\alpha|0\rangle + \beta|1\rangle) \otimes (|0\rangle U|0\rangle + |1\rangle U|1\rangle) \quad (10.31\text{b})$$

$$= \frac{1}{\sqrt{2}}(\alpha|00\rangle + \beta|10\rangle)U|0\rangle + \frac{1}{\sqrt{2}}(\alpha|01\rangle + \beta|11\rangle)U|1\rangle. \quad (10.31\text{c})$$

Here in the last step, we have singled out the 3rd qubit since we are going to measure the first two qubit in the next step.

2. Bell Measurement

Alice applies a CNOT gate with her qubit as the control and the entangled qubit as the target. Then she applies a Hadamard gate to her qubit. Finally, she measures the two qubits, effectively performing a Bell measurement on them.

The four Bell states form an orthonormal basis for two-qubit states. By performing Bell measurements, Alice projects the two-qubit states onto the Bell states.

The measurement operators, for the overall system of three qubits, are given by:

$$M_{ij} = |\beta_{ij}\rangle\langle\beta_{ij}| \otimes I, \quad (10.32)$$

where $i, j \in \{0, 1\}$, and β_{ij} are the Bell states.

After the measurement, depending on the measurement results, the system will be at one of the following states:

$$|\Psi_1\rangle = M_{ij}|\Psi_0\rangle \tag{10.33a}$$
$$= \frac{1}{\sqrt{2}}(|\beta_{ij}\rangle\langle\beta_{ij}| \otimes I)\big((\alpha|00\rangle + \beta|10\rangle)U|0\rangle + (\alpha|01\rangle + \beta|11\rangle)U|1\rangle\big) \tag{10.33b}$$
$$= \frac{1}{\sqrt{2}}|\beta_{ij}\rangle\langle\beta_{ij}|(\alpha|00\rangle + \beta|10\rangle)U|0\rangle$$
$$+ \frac{1}{\sqrt{2}}|\beta_{ij}\rangle\langle\beta_{ij}|(\alpha|01\rangle + \beta|11\rangle)U|1\rangle \tag{10.33c}$$
$$= |\beta_{ij}\rangle \otimes \begin{cases} \alpha U|0\rangle + \beta U|1\rangle & \text{for } i,j = 0,0 \\ \alpha U|1\rangle + \beta U|0\rangle & \text{for } i,j = 0,1 \\ \alpha U|0\rangle - \beta U|1\rangle & \text{for } i,j = 1,0 \\ \alpha U|1\rangle - \beta U|0\rangle & \text{for } i,j = 1,1 \end{cases} \tag{10.33d}$$
$$= |\beta_{ij}\rangle \otimes \begin{cases} U|\psi\rangle & \text{for } i,j = 0,0 \\ UX|\psi\rangle & \text{for } i,j = 0,1 \\ UZ|\psi\rangle & \text{for } i,j = 1,0 \\ UZX|\psi\rangle & \text{for } i,j = 1,1. \end{cases} \tag{10.33e}$$

3. Classical communication

Alice transmits the measurement outcomes ij, representing two bits of information, to Bob through a classical communication channel.

4. State Reconstruction

Based on the two bits ij resulting from Alice's measurements, Bob will apply the necessary correction gates to his qubit as shown in the following table.

ij	Qubit State	Correction Gates
00	$U|\psi\rangle$	I or None
01	$UX|\psi\rangle$	$UXU^\dagger$
10	$UZ|\psi\rangle$	$UZU^\dagger$
11	$UZX|\psi\rangle$	$UXZU^\dagger$

Consider the case where $ij = 01$. Bob's initial state is $UX|\psi\rangle$, and we aim for it to become $U|\psi\rangle$ after correction. This is achieved by applying $UXU^\dagger$, as $UXU^\dagger UX|\psi\rangle = UXX|\psi\rangle = U|\psi\rangle$, given $X^2 = I$.

After correction, Bob's qubit resides in state $U|\psi\rangle$. This illustrates gate teleportation, where the gate U effect is transferred to Bob's qubit without direct application to $|\psi\rangle$.

Exercise 10.4 In quantum gate teleportation, the U gate is typically applied before the Bell measurement step. However, since the U gate is applied to a different qubit from those involved in the Bell measurement, its application can be delayed until after the measurement. This modified sequence of operations is illustrated in Fig. 10.12. To confirm this modification is valid, repeat the analysis as done above for the regular sequence of operations.

10.5 Quantum Gate Teleportation

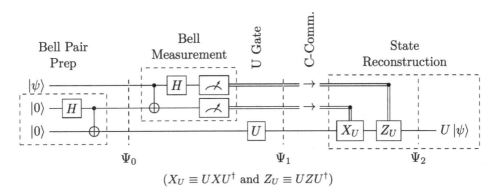

$(X_U \equiv UXU^\dagger \text{ and } Z_U \equiv UZU^\dagger)$

Figure 10.12: Single-Qubit Gate Teleportation with Deferred U Gate

---

Quantum Gate Teleportation: Core Summary

The initial state vector can be decomposed into the orthonormal Bell basis as follows:

$$|\Psi_0\rangle = \frac{1}{\sqrt{2}}(\alpha|0\rangle + \beta|1\rangle) \otimes (I \otimes U)(|00\rangle + |11\rangle),$$

$$= \frac{1}{2}\sum_{i,j\in\{0,1\}} |\beta_{ij}\rangle \otimes U|\psi_{ij}\rangle, \quad (10.34)$$

where $|\beta_{ij}\rangle$ represents the Bell states, and $|\psi_{ij}\rangle$ is defined in Table 10.2 and Eq. 10.22a.

When Alice measures the first two qubits in the Bell basis $\{\beta_{ij}\}$, each of the Bell states occurs with equal probability. If she obtains $|\beta_{ij}\rangle$ (where $i, j \in \{0, 1\}$), the state of the system collapses to $|\beta_{ij}\rangle \otimes U|\psi_{ij}\rangle$, and the third qubit takes on the state $U|\psi_{ij}\rangle$. It can then be adjusted based on Alice's measurement results (i, j) to reconstruct the state $U(\alpha|0\rangle + \beta|1\rangle)$.

---

3 ∗ Teleportation of Single-Qubit Gates in the Clifford Group

You might wonder why we would teleport a gate U if the correction gates $UXU^\dagger$ and $UZU^\dagger$ also contain U. It might seem more straightforward to teleport $|\psi\rangle$ and then directly apply U. However, the process becomes simplified when U belongs to the Clifford group.

By definition, a single-qubit Clifford gate maps any Pauli operator to another Pauli operator under conjugation (up to a phase). This means that if U is in the Clifford group, $UPU^\dagger$ is a Pauli operator for any Pauli operator P.

There are 24 single-qubit Clifford gates, including I, X, Y, Z, H, HX, HY, HZ, S, SX, SY, SZ, and others. Because of their properties under conjugation, Clifford group gates simplify the process of quantum gate teleportation.

Beyond this, the Clifford group also plays a crucial role in error correction within quantum computing. The examples below illustrate how teleportation is streamlined when the gate to be teleported belongs to the Clifford group.

X Gate Teleportation

If we choose $U = X$ as the teleportation gate, we find that the properties of the X gate allow us to simplify the correction gates to basic Pauli matrices. This is due to the transformation properties of the X gate: $UXU^\dagger = XXX^\dagger = X$ and $UZU^\dagger = XZX^\dagger = -Z$. In essence, these correction gates are the same as those used in standard quantum teleportation, as detailed in § 10.3.4.4, albeit with an additional global phase factor.

ij	Qubit State	Correction Gates
00	$X\ket{\psi}$	I or None
01	$XX\ket{\psi}$	X
10	$XZ\ket{\psi}$	$-Z$
11	$XZX\ket{\psi}$	$-XZ$

H Gate Teleportation

The Hadamard gate (H) is one of the single-qubit Clifford gates. With $U = H$, $UXU^\dagger = HXH^\dagger = Z$ and $UZU^\dagger = HZH^\dagger = X$. Therefore, the correction gates become simple Pauli matrices as in the following table.

ij	Qubit State	Correction Gates
00	$H\ket{\psi}$	None
01	$HX\ket{\psi}$	Z
10	$HZ\ket{\psi}$	X
11	$HZX\ket{\psi}$	ZX

S Gate Teleportation

The S gate, also known as the phase gate, is a member of the Clifford group. With $U = S$, $UXU^\dagger = SXS^\dagger = Y$ and $UZU^\dagger = SZS^\dagger = Z$. Therefore, the correction gates become simple Pauli matrices as in the following table.

ij	Qubit State	Correction Gates
00	$S\ket{\psi}$	None
01	$SX\ket{\psi}$	Y
10	$SZ\ket{\psi}$	Z
11	$SZX\ket{\psi}$	YZ

T Gate Teleportation

The T gate, also known as the $\pi/8$ gate, given by $T = \sqrt{S}$, is not a Clifford gate. It doesn't map Pauli matrices to Pauli matrices under conjugation. Therefore, in the case of $U = T$, the correction gates won't simplify to single Pauli gates. Instead, we need to use the full form $TXT^\dagger$, $TZT^\dagger$, and $TXZT^\dagger$. However, these compensations

10.5 Quantum Gate Teleportation

can be simpler to implement than the T gate itself. This property forms the basis for constructing fault-tolerant T gates.

> **Exercise 10.5** Find the correction gates for the teleportation of the Pauli Y gate and Z gate.

10.5.2 ∗ Teleportation of Two-Qubit Gates

The gate teleportation process can be extended to include two-qubit unitary gates, such as the CNOT gate, or a combination of single and two-qubit gates, such as a Bell state generator. In this section, we explore two variations of two-qubit gate teleportation schemes: teleportation of one two-qubit gate and one qubit, which we refer to as 2-1; and teleportation of one two-qubit gate and two qubits, which we refer to as 2-2.

1 2-1 Gate Teleportation

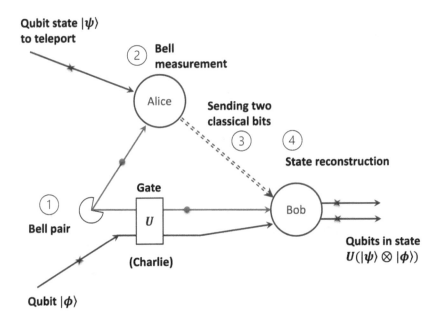

Figure 10.13: Basic Process of 2-1 Gate Teleportation

The 2-1 gate teleportation process is depicted in the diagram shown in Fig. 10.13 and in the quantum circuit in Fig. 10.14.

In this scenario, Alice has a qubit $|\psi\rangle$ that she wants to teleport to Bob. Charlie has a two-qubit gate U and another qubit $|\phi\rangle$. On Bob's end, we aim for him to have $U(|\psi\rangle \otimes |\phi\rangle)$ without directly accessing $|\psi\rangle$ or U.

To achieve this, Charlie prepares a Bell pair with qubits A and B. He sends qubit A to Alice. Alice performs a Bell measurement on $|\psi\rangle$ and qubit A, and sends the measurement results (two classical bits) to Bob via classical communication channel. Charlie also applies U to qubit B and $|\phi\rangle$, and sends both to Bob. Upon receiving the measurement results, Bob applies the appropriate correction gates to the qubits he received from Charlie, and successfully recovers $U(|\psi\rangle \otimes |\phi\rangle)$.

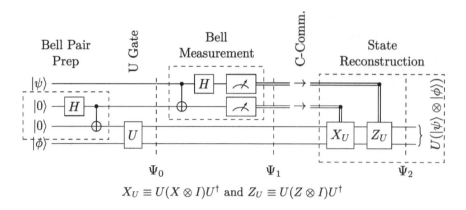

Figure 10.14: Quantum Circuit for 2-1 Gate Teleportation

1. Transition from Standard to Gate Teleportation

Let's now explore an intuitive connection between 2-1 gate teleportation and standard quantum teleportation, depicted in Fig. 10.15.

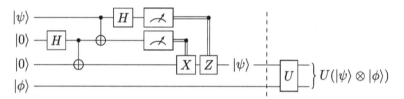

Figure 10.15: Gate U Applied Post-Teleportation

The segment prior to the dashed visual separator is the standard teleportation circuit from Fig. 10.5. At this circuit's output, the third qubit is in state $|\psi\rangle$. Upon application of a unitary gate U on both $|\psi\rangle$ and $|\phi\rangle$, the final state becomes $U(|\psi\rangle \otimes |\phi\rangle)$. This final state mirrors that of 2-1 gate teleportation in Fig. 10.14.

To transform the circuit in Fig. 10.15 into the gate teleportation circuit as shown in Fig. 10.14, we need to shift the gate U to the left of the controlled-Z and controlled-X gates. This transformation is facilitated the following gate sequence equivalence, similar to the single-qubit gate teleportation scenario:

$$U(X \otimes I) = U(X \otimes I)(U^\dagger U) = (U(X \otimes I)U^\dagger)U = X_U U, \quad (10.35a)$$
$$U(Z \otimes I) = U(Z \otimes I)(U^\dagger U) = (U(Z \otimes I)U^\dagger)U = Z_U U, \quad (10.35b)$$

where $X_U \equiv U(X \otimes I)U^\dagger$ and $Z_U \equiv U(Z \otimes I)U^\dagger$. This explains the correction gates X_U and Z_U in Fig. 10.14.

 Although quantum circuit diagrams apply gates from left to right, the corresponding operator multiplication operates in reverse order, from right to left.

10.5 Quantum Gate Teleportation

2-1 Gate Teleportation: Core Summary

The initial state vector can be decomposed into the orthonormal Bell basis as follows:

$$|\Psi_0\rangle = \frac{1}{\sqrt{2}} |\psi\rangle \otimes (I \otimes U)((|00\rangle + |11\rangle) \otimes |\phi\rangle) \quad (10.36\text{a})$$

$$= \frac{1}{\sqrt{2}} (\alpha |0\rangle + \beta |1\rangle) \otimes (I \otimes U)((|00\rangle + |11\rangle) \otimes |\phi\rangle) \quad (10.36\text{b})$$

$$= \frac{1}{2} \sum_{i,j \in \{0,1\}} |\beta_{ij}\rangle \otimes U(|\psi_{ij}\rangle \otimes |\phi\rangle), \quad (10.36\text{c})$$

where $|\beta_{ij}\rangle$ represents the Bell states, and $|\psi_{ij}\rangle$ is defined in Table 10.2 and Eq. 10.22a.

When Alice measures the first two qubits in the Bell basis $\{\beta_{ij}\}$, each of the Bell states occurs with equal probability. If she obtains $|\beta_{ij}\rangle$ (where $i,j \in \{0,1\}$), the state of the system collapses to $|\beta_{ij}\rangle \otimes U(|\psi_{ij}\rangle \otimes |\phi\rangle)$, and the third and fourth qubits take on the state $U(|\psi_{ij}\rangle \otimes |\phi\rangle)$. It can then be adjusted based on Alice's measurement results (i,j) to reconstruct the state $U(|\psi\rangle \otimes |\phi\rangle)$.

2. CNOT Gate Teleportation

You may question the necessity of teleporting a gate U when the correction gates, such as $U(X \otimes I)U^\dagger$, already contain U. It might appear more logical to teleport $|\psi\rangle$ first and then apply U directly. However, the procedure may be simplified when U is a specific two-qubit gate. In this section, we'll consider the CNOT gate as U.

For $U = \text{CNOT}$, it can be demonstrated through matrix multiplication (or through logic operation analysis - see next section) that

$$X_U \equiv U(X \otimes I)U^\dagger = X \otimes X, \quad (10.37\text{a})$$

$$Z_U \equiv U(Z \otimes I)U^\dagger = Z \otimes I. \quad (10.37\text{b})$$

Therefore, for $U = \text{CNOT}$, Fig. 10.14 simplifies to Fig. 10.16, in which the correction gates do not involve CNOT, achieving its true teleportation.

Exercise 10.6 Verify Eq. 10.37 for $U = \text{CNOT}$ through multiplication of the matrices of CNOT, X, and Z.

Exercise 10.7 Following the procedure for the derivation of the CNOT teleportation circuit in Fig. 10.16, design a practical circuit for the teleportation of CNOT with the last qubit (i.e., $|\phi\rangle$) as its control.

2 2-2 Gate Teleportation

Now we examine the teleportation of two qubits, $|\psi\rangle$ and $|\phi\rangle$, along with a two-qubit gate U. This operation is depicted in the diagram shown in Fig. 10.17 and quantum

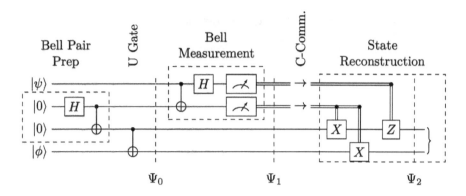

Figure 10.16: Quantum Circuit for 2-1 CNOT Teleportation

circuit in Fig. 10.18. In this scenario, Daniel prepares two Bell pairs: Pair A consists of qubits a_1 and a_2, and Pair B consists of b_1 and b_2. He sends a_1 from Pair A to Alice, and b_2 from Pair B to Bob. He then applies a two-qubit gate U to a_2 and b_1, and sends them to Charlie.

Alice and Bob each possess an additional qubit, c_1 and c_2, respectively, each in states $|\psi\rangle$ and $|\phi\rangle$. Their goal is to teleport these qubits while applying the U gate to them.

To accomplish this, Alice performs a Bell measurement on qubits c_1 and a_1, while Bob does the same for c_2 and b_2. They then transmit their respective measurement results to Charlie via a classical communication channel.

Upon receiving the measurement results, Charlie applies the appropriate correction gates to qubits a_2 and b_1.

At the end of this process, Charlie is left with qubits a_2 and b_1 in the state $U|\psi\rangle|\phi\rangle$. This indicates that the states $|\psi\rangle$ and $|\phi\rangle$ have been teleported with the U gate applied to them. This sequence of operations effectively teleports the application of a U gate between two qubits originating from different locations.

1. Transition from Standard Teleportation

Similar to the illustration of transition from standard teleportation to single-qubit gate teleportation in § 10.5.1.1, let's do the same for two-qubit gate teleportation.

In Fig. 10.19, the segment prior to the dashed visual separator are two standard teleportation circuits in parallel, from Fig. 10.5. At the output, the qubits a_2 and b_1 are in the states $|\psi\rangle$ and $|\phi\rangle$, respectively. Upon application of a U gate, the qubit's state becomes $U|\psi\rangle|\phi\rangle$. This final state mirrors that of intended gate teleportation, wherein a U gate is applied to the teleported $|\psi\rangle$ and $|\phi\rangle$ (as depicted in Fig. 10.18).

To transform the circuit in Fig. 10.19 into the gate teleportation circuit as shown in Fig. 10.18, we need to shift the gate U to the left of the controlled-Z and controlled-X gates. This transformation is facilitated by the concept of gate sequence equivalence, which is demonstrated in the following table:

Here $X_1 \equiv X \otimes I$, $Z_1 \equiv Z \otimes I$, $X_2 \equiv I \otimes X$, and $Z_2 \equiv I \otimes Z$. (For example, X_2 represents the operator of an X gate acting on the second qubit.) This concept of gate

10.5 Quantum Gate Teleportation

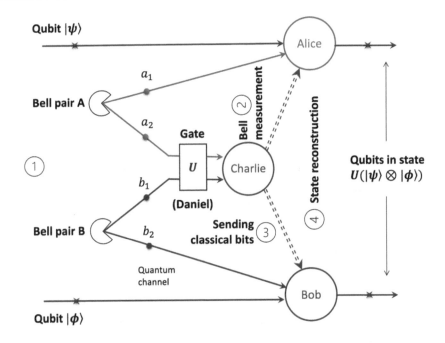

Figure 10.17: Basic Process of 2-2 Gate Teleportation

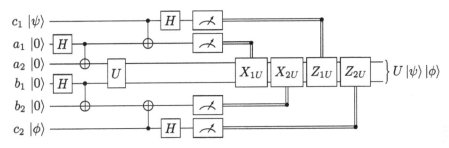

Figure 10.18: Quantum Circuit for 2-2 Gate Teleportation

sequence equivalence underlies the definition of the correction gates $X_{1U} \equiv UX_1U^\dagger$, $Z_{1U} \equiv UZ_1U^\dagger$, $X_{2U} \equiv UX_2U^\dagger$, and $Z_{2U} \equiv UZ_2U^\dagger$ in Fig. 10.18.

2. CNOT Gate Teleportation

For the CNOT gate, the correction gates can be simplified. We can take advantage of the gate sequence equivalence shown in Table 10.6. For a detailed discussion of these sequences, see § 7.4.3. As you'll notice, in this scenario, the U correction gates are simplified to an additional X gate and Z gate, beyond the original four.

With the equivalent gate sequences in table Table 10.6, we have finally arrived at the quantum circuit for CNOT gate teleportation in Fig. 10.20. The two extra gates due to gate sequence equivalence 2 and 3 are highlighted with shading. This is a highly practical circuit as the correction gates no longer involve U directly.

Exercise 10.8 Following the procedure for the derivation of the CNOT teleporta-

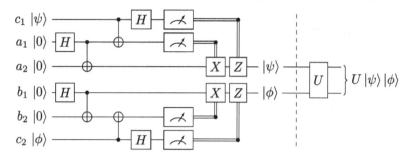

Figure 10.19: Two-Qubit U-Gate Applied Post-Teleportation

Gate Sequence	Equivalent Seq.	Operator Equivalence
X—U	U—X_{1U}	$UX_1 = UX_1(U^\dagger U) = (UX_1 U^\dagger)U = X_{1U}U$
Z—U	U—Z_{1U}	$UZ_1 = UZ_1(U^\dagger U) = (UZ_1 U^\dagger)U = Z_{1U}U$
X—U	U—X_{2U}	$UX_2 = UX_2(U^\dagger U) = (UX_2 U^\dagger)U = X_{2U}U$
Z—U	U—Z_{2U}	$UZ_2 = UZ_2(U^\dagger U) = (UZ_2 U^\dagger)U = Z_{2U}U$

Table 10.5: Gate Sequence Equivalence for Two-Qubit Gate Teleportation

tion circuit in Fig. 10.20, design a practical circuit for the teleportation of the CZ gate.

10.5.3 * Further Exploration

The concept of quantum gate teleportation extends beyond the basic protocol discussed in this text. For those seeking to deepen their understanding, the following topics are suggested for further exploration:

1. Generalized Quantum Gate Teleportation [46]: This work introduces a method for creating a variety of quantum gates by teleporting qubits through specific entangled states. It delves deep into the theoretical framework and outlines potential applications, notably emphasizing a quantum computer model that employs single qubit operations, Bell measurements, and GHZ states.

2. Experimental Teleportation of a Quantum Controlled-NOT Gate [56]: This paper presents an empirical study where a CNOT gate is teleported. The experimental approach and the resulting observations offer valuable insights into the practical implications and challenges of gate teleportation.

3. Optimal Scheme for Teleportation of an n-qubit Quantum State [82]: This research introduces a quantum circuit tailored for teleporting an n-qubit quantum state. The scheme is noted for its optimization in terms of quantum resource utilization.

Index	Gate Sequence	Equivalent Sequence
1	$\|x\rangle$ —•— , $\|y\rangle$ —X—⊕— $\}\|\Psi_1\rangle$	$\|x\rangle$ —•— , $\|y\rangle$ —⊕—X— $\}\|\Psi_2\rangle$
2	$\|x\rangle$ —X—•— , $\|y\rangle$ —⊕— $\}\|\Psi_1\rangle$	$\|x\rangle$ —•—X— , $\|y\rangle$ —⊕—X— $\}\|\Psi_2\rangle$
3	$\|x\rangle$ —•— , $\|y\rangle$ —Z—⊕— $\}\|\Psi_1\rangle$	$\|x\rangle$ —•—Z— , $\|y\rangle$ —⊕—Z— $\}\|\Psi_2\rangle$
4	$\|x\rangle$ —Z—•— , $\|y\rangle$ —⊕— $\}\|\Psi_1\rangle$	$\|x\rangle$ —•—Z— , $\|y\rangle$ —⊕— $\}\|\Psi_2\rangle$

Table 10.6: Gate Sequence Equivalence for CNOT Teleportation

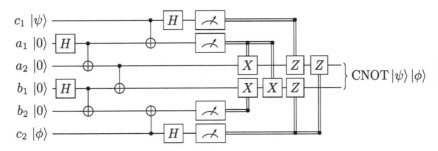

Figure 10.20: Quantum Circuit for CNOT Gate Teleportation

4. **Teleportation of a Quantum Gate Between Two Logical Qubits [36]:** A seminal piece that experimentally demonstrates the teleportation of a CNOT gate between two logical qubits. The research employs real-time adaptive control, making the teleportation deterministic. Additionally, it underscores the significance of error-correctable encoding for maintaining high process fidelity.

5. **Fault-Tolerant Universal Quantum Gate Operations [74]:** This paper provides a detailed account of fault-tolerant universal gate operations on logical qubits within a trapped-ion quantum computer. The research showcases the advantages of the "flag fault tolerance" paradigm and culminates in a demonstration of a fault-tolerant logical T gate. The findings pave the way for further advancements in error-corrected universal quantum computation.

6. **Enhanced Quantum Teleportation with Logical States [60]:** This recent paper proposes a novel approach to quantum teleportation, focusing on the transmission of multi-qubit physical states into error correctable multi-qubit logical states. The work also touches upon an efficient quantum error correction scheme, highlighting its potential in future fault-tolerant quantum technologies.

10.6 E91 Quantum Key Distribution Protocol

In cryptography, encryption entails the transformation of information (plaintext) into a format only accessible by intended recipients through a shared secret key. The security hinges on the secure distribution of this key. Conventionally, transmitting messages through classical channels leaves them vulnerable to undetected eavesdropping. Quantum mechanics offers an alternative, providing the ability to transmit messages while detecting any eavesdropping.

In 1984, Charles H. Bennett and Gilles Brassard introduced the BB84 protocol which utilized four quantum states across two bases. Here, Alice sent particles in one of the four states, and Bob randomly selected bases to measure them. (The BB84 protocol is detailed in § 5.5.)

Building on BB84, Artur Ekert proposed the E91 Quantum Key Distribution (QKD) protocol in 1991. This protocol employs quantum entanglement, local operations, and classical communication channels to securely exchange cryptographic keys, with applications in secure communication and quantum-secured networks. A distinctive feature of the E91 QKD protocol is its intrinsic security, as any attempt to intercept the entangled photons alters their quantum states, leading to detectable anomalies in the shared key. Consequently, Alice and Bob can identify and discard compromised keys due to eavesdropping.

10.6.1 Procedure of the E91 QKD Protocol

The E91 QKD protocol encompasses five primary steps, as depicted in Fig. 10.21. We will explain these steps in detail next.

1 Bell Pair Generation and Distribution

A Bell pair source generates pairs of entangled qubits, typically in the Bell state $|\Psi^-\rangle$. One qubit from each pair is sent to Alice, and the other to Bob.

In the E91 protocol, unlike BB84 where Alice sends qubits to Bob, a central source creates and distributes the entangled qubits. This configuration is considered more practical for real-world implementations, especially over long distances, where a central source such as a satellite can distribute entangled qubits to multiple receivers.

2 Measurement

Upon receiving the qubits, Alice and Bob independently and randomly choose one of three predetermined measurement bases to measure their respective qubits.

The three bases for Alice are:

- a_1 corresponds to measuring in the Z basis (0° on the Bloch sphere).
- a_2 corresponds to measuring in a basis that is 45° rotated from the Z basis on the Bloch sphere, often denoted as $W = \frac{1}{\sqrt{2}}(Z + X)$.
- a_3 corresponds to measuring in the X basis (90° on the Bloch sphere).

For Bob:

- b_1 corresponds to measuring in the W basis (45° on the Bloch sphere).
- b_2 corresponds to measuring in the X basis (90° on the Bloch sphere).
- b_3 corresponds to measuring in a basis that is 135° rotated from the Z basis on the Bloch sphere, often denoted as $V = \frac{1}{\sqrt{2}}(Z - X)$.

10.6 E91 Quantum Key Distribution Protocol

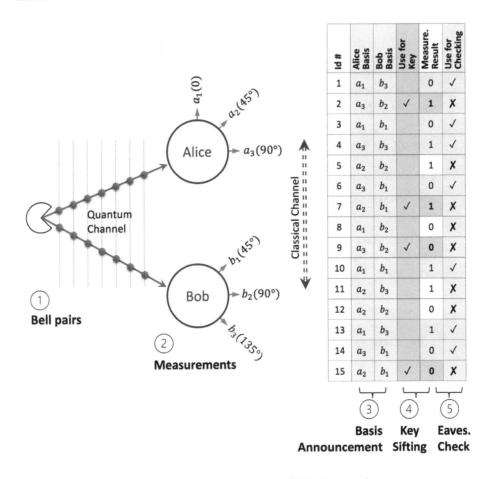

Figure 10.21: Basic E91 QKD Protocol

The BB84 QKD protocol utilizes four measurement settings, two for each participant, while the E91 protocol uses six, with three for each participant. The additional settings in E91 facilitate eavesdropper detection through Bell inequality tests, minimizing the need to reveal key information during the process. This will become clear after we examine the subsequent steps.

 For spin-1/2 systems, the angles on the Bloch sphere equal to the angles of the physical orientation of the Stern-Gerlach apparatus. For photon polarization, however, the physical angles of the polarizers are half the angles on the Bloch Sphere, that is, for Alice a_1 corresponds to $0°$, a_2 $22.5°$, and a_3 $45°$; and for Bob, b_1 $22.5°$, b_2 $45°$, and b_3 $67.5°$.

3   Basis Announcement

After the measurements, Alice and Bob announce the bases they used for each measurement over a public classical channel. They do not disclose their measurement results at this stage.

4 Key Sifting

Using the announced bases, Alice and Bob independently sift their data to retain only the measurement results performed in pairs of compatible bases (a_2, b_1 and a_3, b_2). For these pairs, the measurement results of Alice and Bob are anticorrelated when measuring an entangled state; i.e., if Alice obtains a result of 1, Bob obtains a result of 0, and vice versa (one of them must invert their measurement results).

They retain the results measured with compatible bases as a sifted key. These are indicated in the "Used for Key" column in the table of Fig. 10.21.

5 Eavesdropping Check

Alice and Bob use a subset of their measurement results to perform a Bell's inequality test, designed to detect potential eavesdropping.

The subset includes results from bases a_1, a_3, b_1, and b_3, i.e., excluding the results from a_2 and b_2. These are indicated in the "Used for Checking" column in the table of Fig. 10.21.

The specific choice of angles for the bases a_1, a_3, b_1, and b_3 is designed to maximize the value of the Bell CHSH quantity, which reaches $|S| = 2\sqrt{2}$ for maximally entangled states. Normally, without eavesdropping, the measured photons are in a Bell state and $|S| = 2\sqrt{2}$. However, if an eavesdropper has tampered with the qubits, then $|S| < 2\sqrt{2}$.

By comparing a portion of their measurement results (announced over a classical channel), Alice and Bob can identify any violations of Bell's inequality. If the violations exceed the quantum limit, they conclude that eavesdropping has occurred, and the generated key is unsafe.

6 Key Generation

If the channel is deemed secure following the eavesdropping check, Alice and Bob use the remaining sifted key data from the compatible bases for secure communication. They can also apply error correction and privacy amplification techniques to enhance the security of this key.

10.6.2 Comparison to the BB84 Protocol

Both the E91 and BB84 protocols are Quantum Key Distribution (QKD) protocols that aim to establish a secure key between two parties, typically referred to as Alice and Bob. Despite their shared goal, the two protocols differ significantly in their mechanisms and practical implications. (We have explored the BB84 protocol in § 5.5.)

1 Basis of Security

- The E91 protocol is rooted in quantum entanglement and Bell's inequalities. Alice and Bob each receive one of an entangled pair of qubits. The quantum correlations between measurements on these entangled pairs are exploited to detect the presence of an eavesdropper.

- The BB84 protocol is based on Heisenberg's Uncertainty Principle. An eavesdropper (Eve) cannot measure the quantum states sent by Alice without

10.6 E91 Quantum Key Distribution Protocol

causing a disturbance. This disturbance can be detected by Alice and Bob during the error-checking phase.

2. **Detection of Eavesdropping**

   - In E91, eavesdropping is detected by a decrease in the correlation between Alice's and Bob's measurement outcomes, indicating a violation of Bell's inequalities. This can be observed without any further communication between Alice and Bob beyond sharing their measurement results.

   - In BB84, Alice and Bob detect eavesdropping by comparing a subset of their key bits. If Eve tries to measure the qubits, she introduces errors because of the uncertainty principle, which Alice and Bob can detect.

3. **Number of Bases Used**

   - The E91 protocol uses three bases for encoding and measuring qubits. Alice and Bob randomly select one of the three bases to measure their qubits for each run.

   - The BB84 protocol utilizes two non-commuting bases for encoding and measuring qubits. Alice randomly selects one of these bases to encode each bit of the key, and Bob randomly selects one of the bases to measure each bit.

4. **Practicality for Real-world Applications**

   - The E91 protocol, with its reliance on entangled pairs of qubits, can be technologically challenging to implement, especially in terms of creating, maintaining, and transmitting entangled qubits over long distances. Additionally, the requirement to measure in three different bases may add to the complexity. However, in scenarios where a trusted source can distribute entangled photons, such as satellite-based quantum communication, E91 can have advantages in terms of security and efficiency for long-distance quantum key distribution.

   - The BB84 protocol is often considered more practical for short-distance real-world applications as it requires only the preparation and transmission of single qubits and the capability to measure in two bases. These tasks are relatively easier to achieve with current technology, making BB84 more straightforward for implementation. However, the protocol necessitates a more interactive communication process between Alice and Bob to detect eavesdropping, which can be a limitation in certain scenarios.

10.6.3 ✷ Further Exploration

The concept of the E91 Quantum Key Distribution Protocol, as well as its basis in entanglement and the broader context of quantum cryptography, extends beyond the basic protocol discussed in this text. For those seeking to deepen their understanding of the E91 protocol, its foundations, and its comparisons with other quantum cryptographic schemes, the following are suggested for further exploration:

1. Quantum Cryptography Based on Bell's Theorem [39]: This is the original paper by Artur Ekert, in which he introduced the E91 protocol. It lays the foundation for Quantum Key Distribution based on entanglement and Bell's inequalities.

2. Quantum Key Distribution Network for Multiple Applications [78]: This paper discusses the implementation of a quantum key distribution network where different protocols, including E91, are employed for various applications. It demonstrates the flexibility and real-world utility of the E91 protocol in a network setting.

3. Experimental Demonstration of a Measurement-Device-Independent Quantum Key Distribution [76]: This paper introduces and demonstrates an experimental implementation of measurement-device-independent quantum key distribution, which can be seen as an advancement over the E91 protocol, by removing the need to trust the measurement devices.

4. Quantum Cryptography: Public Key Distribution and Coin Tossing [21]: Though primarily centered around the introduction of the BB84 protocol, this seminal paper by Charles H. Bennett and Gilles Brassard also offers a basis for understanding the context in which the E91 protocol was developed.

5. Quantum Key Distribution in Cryptography: A Survey [13]: This article explores the integration of quantum key distribution (QKD) in cryptographic infrastructures. Highlighting QKD's promise of information-theoretic security, the review underscores its application in renewing symmetric cipher keys and enabling secure key establishment in networks, while addressing inherent challenges and research avenues.

10.7 Summary and Conclusions

Entanglement in Applications

This chapter has taken readers on a comprehensive exploration of the myriad applications of quantum entanglement in quantum computing and quantum communication. Beginning with an overview of the essential role entanglement plays in these domains, we ventured into a systematic analysis of some of the key protocols that harness the power of entanglement.

Our journey began with the elucidation of superdense coding, an elegant method that demonstrated the efficacy of quantum communication. Transitioning from communication to computation, the concept of quantum teleportation showcased the unique capability of transmitting quantum states without the actual physical transfer of qubits, highlighting the distinction between classical and quantum paradigms.

Entanglement swapping, a nuanced yet fundamental process, further underscored the versatility of entanglement in establishing quantum connections over long distances. Such a technique is paramount in the construction of quantum networks, which will potentially revolutionize global communication in the quantum era.

We delved deeper into computation with the discussion of quantum gate teleportation. This approach opens doors to distributed quantum computing, heralding a paradigm where quantum gates can be operationalized across spatially dispersed quantum systems.

Finally, we transitioned back into the realm of quantum communication with the E91 Quantum Key Distribution (QKD) protocol. Rooted in entanglement and the principles of quantum mechanics, this protocol emphasizes the enhanced security quantum systems offer in the face of eavesdropping threats, distinguishing quantum cryptographic techniques from their classical counterparts.

Bridging Theory and Practice

Throughout this chapter, our approach has been two-pronged: we have grounded our discussions in the theoretical constructs of quantum mechanics while constantly alluding to their real-world implications. By doing so, we intend to foster an appreciation for both the abstract beauty of quantum phenomena and their tangible applications in next-generation technologies.

Entanglement, as a quintessential quantum resource, stands at the confluence of many quantum applications. Through the varied techniques and protocols discussed, we aim to have imparted a nuanced understanding of its potential and the avenues it opens up in quantum information science.

Upcoming Topics

Our subsequent chapter promises to immerse readers into the intricate world of quantum algorithms. We will juxtapose quantum algorithms against their classical counterparts, shedding light on the unique advantages and challenges they present. Starting with the pedagogical Deutch Jozsa algorithm, we will then transition into the realm of optimization with algorithms such as CUBO, QAOA, and VQE, which have immediate real-world applications. We will conclude the chapter with a dive into the Quantum Measurement Bomb, an algorithm that underscores the central role of measurements in quantum computing. As we venture into these topics, we will continually build on the foundational principles expounded in the previous chapters, reinforcing and expanding your quantum computing knowledge.

Problem Set 10

10.1 Investigate the following questions related to **superdense coding** and elaborate your answers:

(a) Why is superdense coding considered more efficient than classical transmission if both methods involve two bits?

(b) Does superdense coding enhance transmission security?

(c) How is superdense coding related to quantum telelportation?

(d) Can a single EPR pair be used to transmit more than two bits of classical information?

10.2 Investigate the following questions related to **quantum telelportation** and elaborate your answers:

(a) Can quantum teleportation teleport matter, i.e., move particles from one location to another?

(b) Is quantum teleportation faster than light?

(c) Can quantum teleportation be used to copy quantum states?

(d) After Alice measures the two qubits, are the original qubit pair still entangled?

(e) Can the entangled pair be used twice for quantum teleportation?

(f) What is the relationship between quantum teleportation and the no-cloning theorem?

(g) What is the role of classical communication in quantum teleportation?

10.3 Investigate the following questions related to **entanglement swapping** and elaborate your answers:

(a) How is entanglement swapping related to quantum teleportation?

(b) Is entanglement swapping instantaneous?

(c) Is it possible to swap entanglement more than once?

(d) Why is a mediator required in entanglement swapping?

(e) Can entanglement swapping be used to transmit classical information?

(f) How is entanglement swapping used in a quantum network?

(g) How does the measurement result of the mediator influence the final state of the other two qubits?

(h) Is there a risk of losing entanglement during the swapping process?

(i) Can the mediator control the type of entanglement that is created between the two remaining qubits?

10.4 Investigate the following questions related to **quantum gate teleportation** and elaborate your answers:

(a) How does quantum gate teleportation relate to regular quantum teleportation?

(b) What is the practical use of quantum gate teleportation?

(c) How does quantum gate teleportation impact quantum computing?

(d) How does quantum gate teleportation contribute to quantum cryptography?

(e) What are the limitations of quantum gate teleportation?

(f) Can quantum gate teleportation be achieved without classical communication?

(g) How far apart can two qubits be for successful quantum gate teleportation?

(h) What does it mean to say that the states $|\psi\rangle$ and $|\phi\rangle$ have been teleported with the U gate applied to them?

10.5 Investigate the following questions related to **the E91 QKD protocol** and elaborate your answers:

(a) What advantages does the use of entanglement in E91 provide over the BB84 protocol?

(b) Can the E91 protocol be used for communication beyond key distribution?

(c) How does the E91 protocol ensure that the key has not been compromised?

(d) What are the challenges in implementing the E91 protocol in real-world scenarios?

(e) Can the E91 protocol work with different types of particles, or is it restricted to photons?

(f) How does the E91 protocol deal with noise and loss in a real-world communication channel?

(g) What is the fraction of measurement bits that can be used as sifted keys?

(h) Can the E91 protocol be used in a network with more than two parties?

(i) Is the E91 protocol widely adopted in commercial quantum communication systems?

10.6 Upon reflection, you'll realize that the end result of quantum teleportation is essentially swapping the first and third qubits. In fact, Fig. 10.6 (excluding the final measurements) is equivalent to the swap circuit shown on the right. Your task is to demonstrate this equivalence using gate sequence equivalence developed in §§ 7.4 and 8.5.

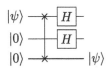

Equivalent Swap Circuit

10.7 Consider the scenario in quantum teleportation where Alice and Bob share the Bell state $|\Psi^-\rangle$ (commonly associated with entangled photon pairs), as opposed to $|\Phi^+\rangle$. The objective is to work through the quantum teleportation protocol given this condition. Specifically, you are tasked with deriving an equation analogous to Eq. 10.23, which concisely encapsulates the principles of quantum teleportation under the premise of the shared $|\Psi^-\rangle$ state.

10.8 Alice has the 2-qubit state $\alpha|00\rangle + \beta|11\rangle$. Demonstrate the process of quantum teleportation by which Alice can send this state to Bob. Specifically, detail the steps she must take to teleport the state using only 2 classical bits of communication and one shared EPR pair. Assume the EPR pair is in the state $\frac{1}{\sqrt{2}}(|00\rangle + |11\rangle)$ and is distributed between Alice (first qubit) and Bob (second qubit) before teleportation begins. Outline how measurement outcomes affect the protocol and how Bob can reconstruct the original state.

10.9 Explore the entanglement swapping protocol assuming that the EPR pairs are in the Bell state $|\Psi^-\rangle$, a state often encountered in entangled photon pairs. Specifically, derive a formulation analogous to Eq. 10.30 that accurately describes the entanglement swapping process when the initial state of the entangled pairs is $|\Psi^-\rangle$.

10.10 Gate teleportation is a sophisticated protocol that extends the concept of quantum teleportation by incorporating quantum gates into the teleportation process. This allows for the execution of quantum gates on a quantum state remotely, without directly interacting with the state itself. In this exercise, consider a scenario where the entangled resource shared between the parties (Alice and Bob) is the Bell state $|\Psi^-\rangle$. Your objective is to work through the gate teleportation protocol given this specific initial entangled state. Derive a formulation equivalent to Eq. 10.34 that succinctly represents the gate teleportation process utilizing the $|\Psi^-\rangle$ state.

10.11 Following the procedure for the derivation of the **2-1** CNOT teleportation circuit in Fig. 10.16, design a practical circuit for the teleportation of the Bell state generator, i.e., a Hadamard gate followed by a CNOT gate on the control qubit.

10.12 Following the procedure for the derivation of the **2-2** CNOT teleportation circuit in Fig. 10.20, design a practical circuit for the teleportation of the Bell analyzer, i.e., CNOT followed by a Hadamard gate on the control qubit.

10.13 ∗ Consider the multipartite Bell states of four qubits as defined in Problem 8.8. These extended Bell states offer a rich framework for advancing quantum communication protocols beyond two-qubit entanglement. Your task is to investigate and extend the following core quantum communication protocols using these states:

(a) Explore how the multipartite Bell states can be utilized to enhance the efficiency of superdense coding. Consider the implications of using four-qubit states for encoding and transmitting more information than traditional two-qubit Bell states.

(b) Extend the concept of quantum teleportation to incorporate four-qubit Bell states.

(c) Extend the concept of entanglement swapping to incorporate four-qubit Bell states.

(d) Investigate the generalization of superdense coding, quantum teleportation, and entanglement swapping using extended Bell states of $2n$ qubits.

In each case, provide a theoretical foundation for your extensions, supported by mathematical formulations where applicable.

IV Quantum Computation & Information

11	Quantum Algorithms: A Sampler	295
12	Quantum Error Correction: A Primer	341
13	Fundamentals of Quantum Information	405

11. Quantum Algorithms: A Sampler

Contents

11.1 The Deutsch-Jozsa Algorithm **297**
11.1.1 Problem Statement 297
11.1.2 Classical vs. Quantum 297
11.1.3 Procedure of the Deutsch-Jozsa Algorithm 298
11.1.4 ✻ Generalized Deutsch-Josza Algorithm 300
11.1.5 ✻ The Phase Oracle 303
11.2 QUBO, VQE, QAOA, and AQC **305**
11.2.1 The Max-Cut Problem 307
11.2.2 ✻ The Ising Model 310
11.2.3 ✻ Variational Quantum Eigensolver (VQE) 311
11.2.4 ✻ Quantum Approximate Optimization Algorithm (QAOA) 314
11.2.5 ✻ Traveling Salesman Problem 317
11.2.6 Adiabatic Quantum Computation and Annealing 319
11.3 Quantum Bomb and Quantum Money **322**
11.3.1 The Quantum Bomb Test Algorithm 322
11.3.2 ✻ Basics of Quantum Money |$\rangle$ 325
11.4 Summary and Conclusions **335**
Problem Set 11 . . . . **336**

An algorithm generally refers to a defined set of instructions or procedures that a computer adheres to in order to solve a particular problem. When these algorithms are designed to run on quantum computers, they are termed quantum algorithms. The objective behind quantum algorithms often centers around executing calculations more efficiently than classical computers, or addressing computations that are practically infeasible for classical computers due to inherent physical limitations (e.g., execution time or memory capacity). In this text, we classify quantum algorithms into three primary categories: Foundational Quantum Algorithms, NISQ Hybrid Algorithms, and Innovative Quantum Algorithmic Concepts.

Quantum Utility Era

As of circa 2023, quantum computing is primarily operating with Noisy Intermediate-Scale Quantum (NISQ) devices. These devices, though limited in scale and afflicted by quantum noise, represent a critical transition phase. They possess computational capabilities that begin to exceed those achievable by classical computer simulations.

In this evolving landscape, we are entering the quantum utility era, a phase where quantum computing starts to yield substantial computational advantages for particular tasks. This marks the onset of quantum computing's practical impact, as it begins to offer tangible solutions to real-world problems, despite the inherent challenges of quantum noise and device limitations.

Evolving Hybrid Algorithms

As quantum computing transitions from the NISQ era to the emerging era of quantum utility, quantum-classical hybrid algorithms are evolving to harness the strengths of both quantum and classical computing paradigms. These algorithms are pivotal for optimizing the performance of both NISQ devices, with their limited qubits and susceptibility to errors, and more advanced utility-scale quantum systems. By strategically integrating quantum computations with classical processes, these hybrid algorithms aim to enhance computational reliability and efficiency, adapting to the evolving capabilities of quantum hardware.

Foundational Quantum Algorithms

Historically, quantum algorithms that showcased the fundamental advantages of quantum over classical computation, such as Shor's and Grover's algorithms, were developed with the assumption of large-scale, fault-tolerant quantum computers for their operation. Such algorithms, which exploit the superposition and entanglement properties of qubits, are termed Foundational Quantum Algorithms. These algorithms serve as pivotal pedagogical tools and epitomize the profound possibilities of quantum computation.

Innovative Quantum Algorithmic Concepts

Beyond the established paradigms of NISQ and foundational algorithms lies a category of innovative quantum algorithmic concepts. These include algorithms and ideas that may have been conceived in the early stages of quantum computing or represent novel approaches in the field. They extend the boundaries of quantum computational research, exploring new methodologies and challenging existing assumptions. Among these are the Quantum Bomb Test Algorithm and Quantum Money, which offer insights into quantum measurements, probabilistic algorithms, and quantum cryptography.

Our Game Plan

In this chapter, we will explore representative quantum algorithms from each of the aforementioned categories:

Deutsch Jozsa Algorithm: A foundational quantum algorithm, it showcases the potential advantages of quantum computation. While its tangible real-world applications may be elusive, we will examine the intuition, working principles, and the generalization of this cornerstone algorithm.

CUBO, QAOA, VQE, AQC: As hallmarks of the NISQ era, these algorithms have proven adept at addressing a myriad of optimization and quantum simulation challenges. Notably, this includes challenges solvable by quantum annealing devices. As a focal point, we will delve into the max-cut problem.

Quantum Measurement Bomb Algorithm and Quantum Money: Representing innovative quantum algorithmic concepts, these examples provide insights into various distinctive aspects such as quantum measurements, probabilistic algorithms, and quantum cryptography.

Through our exploration of these algorithms, we aim to provide readers with a comprehensive understanding of the evolving landscape of quantum computing, tracing its past achievements, present capabilities, and future potential.

11.1 The Deutsch-Jozsa Algorithm

The Deutsch-Jozsa (DJ) algorithm stands as one of the earliest and most elementary instances of a quantum algorithm. While it lacks direct practical applications, it exemplifies the potential advantage of quantum computing over classical computing in specific tasks.

11.1.1 Problem Statement

Consider a binary function $f : \{0,1\}^n \to \{0,1\}$, which takes an n-bit string input and returns a single bit as output. The function is guaranteed to have one of two behaviors:

1. Constant: $f(x)$ is the same for all inputs; that is, either $f(x) = 0$ or $f(x) = 1$ for all x.

2. Balanced: For half of the possible inputs (not necessarily the first half), $f(x) = 0$, and for the other half, $f(x) = 1$.

The algorithm employs an *oracle*, a blackbox mechanism representing the function. One can query the oracle to understand the input-output relationship of the function, but the internal mechanics remain concealed. The objective of the algorithm is to determine with certainty whether the function is constant or balanced across its inputs.

 In a theoretical context, the oracle is conceptualized as the black-box function input to the computation. The algorithm simply assumes its existence without concern for its internal structure. However, for pedagogical completeness, we provide a possible implementation of this oracle in § 11.1.5, which will aid readers in gaining a deeper understanding of the algorithm's operation.

Exercise 11.1 Enumerate all possible constant and balanced functions for $n = 1$ and $n = 2$.

11.1.2 Classical vs. Quantum

For a classical approach, the task of distinguishing whether the function $f(x)$ is constant or balanced necessitates a worst-case scenario of at least $2^{n-1} + 1$ oracle queries.

In contrast, the DJ algorithm showcases the power of quantum computing by achieving this differentiation with just a single oracle query, presenting an exponential speedup in comparison to its classical counterpart. This disparity in performance underscores the profound advantage introduced by quantum parallelism.

11.1.3 Procedure of the Deutsch-Jozsa Algorithm

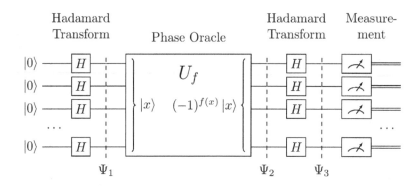

Figure 11.1: Quantum Circuit for the Deutsch-Jozsa Algorithm

As shown in Fig. 11.1, the DJ algorithm involves the following procedure:

1. Initialize all qubits to $|0\rangle$.

2. Apply a Hadamard transform to create a superposition of all possible input states.

3. Query the phase oracle, introducing a phase $(-1)^{f(x)}$ to each state in the superposition.

4. Execute the second Hadamard transform.

5. Measure the qubits. A constant function always produces the state $|000...0\rangle$, while a balanced function results in any state except $|000...0\rangle$.

We will elaborate on the procedure of the DJ algorithm next. While the DJ algorithm might not possess direct practical applications, the strategies and principles elucidated here lay foundation for understanding of more sophisticated quantum algorithms, notably Simon's, Shor's, and Grover's algorithms.

1 Uniform Superposition

Applying a Hadamard gate on each qubit initialized to $|0\rangle$ yields:

$$|\Psi_1\rangle = \bigotimes^n H|0\rangle = \bigotimes^n \frac{1}{\sqrt{2}}(|0\rangle + |1\rangle). \tag{11.1}$$

Expanding the tensor product gives:

$$|\Psi_1\rangle \equiv |\Psi_u\rangle = \frac{1}{\sqrt{2^n}} \sum_{x \in \{0,1\}^n} |x\rangle. \tag{11.2}$$

Here $|\Psi_u\rangle$ is a uniform superposition of the 2^n basis states, from $|000...0\rangle$ to $|111...1\rangle$.

Exercise 11.2 Compute $\bigotimes^n H|1\rangle$. Is this also a uniform superposition of the computational basis?

2 The Phase Oracle

In the context of our quantum circuit, querying the function $f(x)$ necessitates its representation as a unitary transformation, capable of handling superposition states. An innovative realization of this unitary representation is termed the "phase oracle", denoted by U_f. (A detailed account of the oracle's implementation will be provided in § 11.1.5.)

Each potential input string for the function $f(x)$ maps to a distinct basis state, $|x\rangle$, within the n-qubit system. The primary role of the phase oracle is to impart a phase shift of π to a given $|x\rangle$ if its corresponding $f(x) = 1$:

$$U_f|x\rangle = (-1)^{f(x)}|x\rangle, \quad \text{for} \quad x \in \{0,1\}^n. \tag{11.3}$$

After applying the oracle, the state becomes:

$$|\Psi_2\rangle = \frac{1}{\sqrt{2^n}} \sum_{x \in \{0,1\}^n} (-1)^{f(x)}|x\rangle. \tag{11.4}$$

Exercise 11.3 Assume $f(x) = \text{parity}(x)$, and the input state of the oracle is $\frac{1}{\sqrt{2}}(|00...0\rangle + i|11...1\rangle)$. Compute the output state.

3 Hadamard Transform and Measurement

In typical scenarios, one would express the state $|\Psi_3\rangle$ as a result of applying the second Hadamard transform to $|\Psi_2\rangle$. However, within the context of the DJ algorithm, articulating this explicitly is unnecessary. Our primary focus lies in discerning the probability associated with the measurement outcome of $|000...0\rangle$ in the state $|\Psi_3\rangle$.

The component $|000...0\rangle$ in $|\Psi_3\rangle$ originates from the uniform superposition of all basis states in $|\Psi_2\rangle$, denoted as $|\Psi_u\rangle$ in Eq. 11.2.

Therefore, the probability of obtaining the outcome $|000...0\rangle$ hinges on the overlap between states $|\Psi_u\rangle$ and $|\Psi_2\rangle$:

$$P_0 = |\langle\Psi_u|\Psi_2\rangle|^2. \tag{11.5}$$

Upon inspection, we find that if $f(x)$ is a constant function, the state $|\Psi_2\rangle$ retains its uniform superposition, leading to $P_0 = 1$. Conversely, for a balanced function $f(x)$, half the terms in $|\Psi_2\rangle$ carry a negative sign. This results in perfect destructive interference and consequently, $P_0 = 0$. This distinction serves as the central insight behind the DJ algorithm.

11.1.4 ∗ Generalized Deutsch-Josza Algorithm

In the traditional Deutsch-Josza (DJ) algorithm, measurement is performed against the $|000...0\rangle$ basis state to classify functions as either constant or balanced. Extending these measurements to encompass all computational basis states, however, provides deeper insights into the function's periodicity. While this generalization aligns with the principles of the Bernstein-Vazirani algorithm mathematically, it also enhances our understanding of how the DJ algorithm connects to the broader concept of the Fourier transform. This extension not only broadens the scope of problems that can be tackled but also underscores the fundamental relationship between quantum computing algorithms and Fourier analysis.

1 Building Intuition

Each computational basis state can be associated with a unique periodic function. For elucidation, consider the Hadamard transform H_4, illustrated in Table 11.1. Here, the basis state $|0001\rangle$ is tied to a function with a period of 2. Conversely, states $|0010\rangle$ and $|0011\rangle$ indicate functions with a period of 4, with $|0011\rangle$ introducing a phase shift distinct from the square wave of $|0010\rangle$. Similarly, states from $|0100\rangle$ through $|0111\rangle$ denote functions with a period of 8. Furthermore, these functions are orthogonal to each other. In fact, they serve as the basis functions of the Hadamard transform, analogous to the pure harmonics in the Fourier transform.

Row	Basis	Period	Matrix Elements
0	0000	1	1 1 1 1 1 1 1 1 1 1 1 1 1 1 1 1
1	0001	2	1 -1 1 -1 1 -1 1 -1 1 -1 1 -1 1 -1 1 -1
2	0010	4	1 1 -1 -1 1 1 -1 -1 1 1 -1 -1 1 1 -1 -1
3	0011	4	1 -1 -1 1 1 -1 -1 1 1 -1 -1 1 1 -1 -1 1
4	0100	8	1 1 1 1 -1 -1 -1 -1 1 1 1 1 -1 -1 -1 -1
5	0101	8	1 -1 1 -1 -1 1 -1 1 1 -1 1 -1 -1 1 -1 1
6	0110	8	1 1 -1 -1 -1 -1 1 1 1 1 -1 -1 -1 -1 1 1
7	0111	8	1 -1 -1 1 -1 1 1 -1 1 -1 -1 1 -1 1 1 -1
8	1000	16	1 1 1 1 1 1 1 1 -1 -1 -1 -1 -1 -1 -1 -1
9	1001	16	1 -1 1 -1 1 -1 1 -1 -1 1 -1 1 -1 1 -1 1
10	1010	16	1 1 -1 -1 1 1 -1 -1 -1 -1 1 1 -1 -1 1 1
11	1011	16	1 -1 -1 1 1 -1 -1 1 -1 1 1 -1 -1 1 1 -1
12	1100	16	1 1 1 1 -1 -1 -1 -1 -1 -1 -1 -1 1 1 1 1
13	1101	16	1 -1 1 -1 -1 1 -1 1 -1 1 -1 1 1 -1 1 -1
14	1110	16	1 1 -1 -1 -1 -1 1 1 -1 -1 1 1 1 1 -1 -1
15	1111	16	1 -1 -1 1 -1 1 1 -1 -1 1 1 -1 1 -1 -1 1

Table 11.1: Hadamard Transform H_4 and Period Functions

Exercise 11.4 For H_6, determine the period of $f(x)$ corresponding to the basis state $|001100\rangle$. How many periodic occurrences does $f(x)$ exhibit over the 2^6 possible x values?

2 Problem and Solution

Quantum mechanics posits that orthogonal states can be distinctly identified through measurements. Inspired by this observation, we propose a more encompassing version

11.1 The Deutsch-Jozsa Algorithm

of the DJ algorithm:

Given a function in the basis function set of the 2^n-dimensional Hadamard transform, $f : \{0,1\}^n \to \{0,1\}$, mapping an n-bit string to a single-bit outcome, f is certain to display circular periodicity with periods spanning from 1 (for a constant function), through $2, 4, ..., 2^i, ..., 2^n$. Our primary goal is to discern this periodicity.

While a classical method requires at least n function queries, quantum computing leverages the generalized DJ algorithm to accomplish this in merely one query.

The quantum methodology still relies on the original DJ circuit illustrated in Fig. 11.1. Given a measurement outcome of $|\tilde{y}\rangle$, the period is 2^{n-k}, where k denotes the count of leading 0s in $\tilde{y}$'s binary representation.

3 Proof

The Hadamard transform maps the computational basis state $|x\rangle$ into a superposition state of $|y\rangle$:

$$|x\rangle \xrightarrow{H^{\otimes n}} \frac{1}{\sqrt{2^n}} \sum_{y \in \{0,1\}^n} (-1)^{x \cdot y} |y\rangle, \tag{11.6}$$

where $x \cdot y$ represents the binary dot product of x and y, i.e., $x \cdot y = \bigoplus_{i=1}^{n} x_i y_i$, with $x_i, y_i \in \{0,1\}$.

The multi-qubit state across different junctures of the DJ circuit, shown in Fig. 11.1, evolves as:

$$|0\rangle^n \xrightarrow{H^{\otimes n}} |\Psi_1\rangle \equiv |\Psi_u\rangle = \frac{1}{\sqrt{2^n}} \sum_{x \in \{0,1\}^n} |x\rangle \tag{11.7a}$$

$$\xrightarrow{U_f} |\Psi_2\rangle = \frac{1}{\sqrt{2^n}} \sum_{x \in \{0,1\}^n} (-1)^{f(x)} |x\rangle \tag{11.7b}$$

$$\xrightarrow{H^{\otimes n}} |\Psi_3\rangle = \frac{1}{2^n} \sum_{x,y \in \{0,1\}^n} (-1)^{f(x) \oplus (x \cdot y)} |y\rangle. \tag{11.7c}$$

Exercise 11.5 Explain why $(-1)^{f(x)+x \cdot y} = (-1)^{f(x) \oplus (x \cdot y)}$, as utilized in the final transformation of Eq. 11.7.

A key insight of the extended DJ algorithm is that if $f(x)$ is

$$f(x) = x \cdot \tilde{y} \tag{11.8}$$

for a particular $\tilde{y} \in \{0,1\}^n$, then $f(x) \oplus (x \cdot \tilde{y}) = 0$. This results in the corresponding $|\tilde{y}\rangle$ component in $|\Psi_3\rangle$ to have a full amplitude of 1:

$$\frac{1}{2^n} \sum_{x \in \{0,1\}^n} (-1)^{f(x) \oplus x \cdot \tilde{y}} |\tilde{y}\rangle = \frac{1}{2^n} \sum_{x \in \{0,1\}^n} |\tilde{y}\rangle = |\tilde{y}\rangle. \tag{11.9}$$

It follows that $|\Psi_3\rangle$ solely contains $|\tilde{y}\rangle$; all other components nullify due to the normalization condition. Therefore, measuring $|\Psi_3\rangle$ invariably yields $|\tilde{y}\rangle$.

To comprehend why $f(x) = x \cdot \tilde{y}$ symbolizes a circular periodic function, consider $|\tilde{y}\rangle = |00...01...\rangle$ with k initial 0s. The leading digits in $|x\rangle$ corresponding to the leading 0s in $|\tilde{y}\rangle$ do not influence $x \cdot \tilde{y}$, making the periodicity of $f(x)$ equal to 2^{n-k}.

4 Measurement in Alternative Basis

The measurements are not confined solely to computational basis states. Any set of 2^n orthonormal basis states is compatible with the generalized DJ algorithm. However, shifting to an alternative basis might yield functions lacking straightforward properties like apparent periodicity. Moreover, transitioning to an alternative basis necessitates an extra unitary transformation before measurement.

To illustrate, consider grouping the 2^n basis states into 2^{n-1} distinct pairs, which need not be consecutive. Given a pair, say $|i\rangle$ and $|j\rangle$, two orthogonal states can be derived: $\frac{1}{\sqrt{2}}(|i\rangle + |j\rangle)$ and $\frac{1}{\sqrt{2}}(|i\rangle - |j\rangle)$. Together, these states constitute a new orthogonal basis. Measuring against $\frac{1}{\sqrt{2}}(|i\rangle + |j\rangle)$ unequivocally identifies the function $f(x) = x \cdot (i \oplus j)$.

5 Spectral Measurement with the DJ Circuit

To further generalize the DJ algorithm, we leave the nature of the function $f(x)$ unconstrained. Employing repeated measurements on the DJ circuit, as shown in Fig. 11.1, allows us to collect data over numerous iterations. This process enables us to derive an empirical probability distribution for each of the 2^n possible outcomes and construct a length-2^n vector representing the spectral distribution of the function $f(x)$.

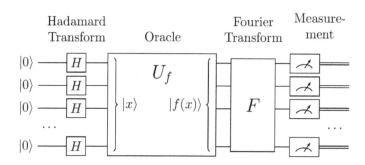

Figure 11.2: Quantum Circuit for Spectral Measurement

Furthermore, while the DJ algorithm traditionally employs the Hadamard transform to facilitate measurements of the spectral distribution in relation to square waves, a more sophisticated approach involves replacing the secondary Hadamard transform with the Fourier transform, denoted as F. This substitution provides a more accurate representation of the spectral density of $f(x)$, as visually depicted in Fig. 11.2. Moreover, we can extend the numerical range by substituting $f(x)$ with a function characterized by a wider output range, moving beyond the binary confines of $\{0, 1\}$.

The theoretical probability distribution, representing the spectral density of $f(x)$, is given by:

11.1 The Deutsch-Jozsa Algorithm

$$P_x(U_f) = |\langle x|FU_f|\Psi_u\rangle|^2, \quad (11.10)$$

where $|\Psi_u\rangle$ is the uniform superposition, as defined in Eq. 11.2, and $x \in \{0,1\}^n$, or equivalently, $x \in [0, 2^n - 1]$ in decimal representation.

DJ Algorithm Revisited

Now, let's analyze how Eq. 11.10 operates in the context of the DJ scenarios, where F is the Hadamard transform H_n.

- Constant functions

  Based on Eq. 11.3, for the constant-0 function, $U_f|x\rangle = |x\rangle$ for any $|x\rangle$, leading to $U_f|\Psi_u\rangle = |\Psi_u\rangle$. Similarly, for the constant-1 function, we have $U_f|\Psi_u\rangle = -|\Psi_u\rangle$. Therefore, for constant functions, $U_f|\Psi_u\rangle = \pm|\Psi_u\rangle$. Since $H_n|\Psi_u\rangle = |0\rangle$, the equation simplifies to:

$$P_x(U_{\text{constant}}) = |\pm\langle x|H_n|\Psi_u\rangle|^2 = |\pm\langle x|0\rangle|^2 = \begin{cases} 1 & \text{for } x = 0, \\ 0 & \text{for } x \neq 0. \end{cases} \quad (11.11)$$

- Balanced functions

  For a balanced function, $U_f|x\rangle = |x\rangle$ for half of the x values, and $U_f|x\rangle = -|x\rangle$ for the other half. Consequently, $U_f|\Psi_u\rangle$ becomes a vector, half of whose components are $\frac{1}{\sqrt{2^n}}$ and the other half $-\frac{1}{\sqrt{2^n}}$. We denote this state as $|\Psi_b\rangle$.

  It's worth noting that $|\Psi_b\rangle$ has no overlap with $|\Psi_u\rangle$, which has all its components equal to $\frac{1}{\sqrt{2^n}}$. Furthermore, since $H_n|0\rangle = |\Psi_u\rangle$, we have:

$$P_x(U_{\text{balanced}}) = \begin{cases} |\langle\Psi_u|\Psi_b\rangle|^2 = 0 & \text{for } x = 0, \\ |\langle x|H_n|\Psi_b\rangle|^2 & \text{for } x \neq 0. \end{cases} \quad (11.12)$$

- Generalized DJ

  In the generalized DJ, or equivalently, the BZ algorithm, where $f(x) = x \cdot \tilde{y}$ (Eq. 11.8), and $U_f|\Psi_u\rangle$ is represented by $|\Psi_3\rangle$ (as defined in Eq. 11.7), Eq. 11.10 can be rewritten as:

$$P_x(U_{\text{gen}}) = \left|\langle x|\frac{1}{2^n}\sum_{z,y\in\{0,1\}^n}(-1)^{(z\cdot\tilde{y})\oplus(z\cdot y)}|y\rangle\right|^2 \quad (11.13a)$$

$$= \left|\frac{1}{2^n}\sum_{z\in\{0,1\}^n}(-1)^{(z\cdot\tilde{y})\oplus(z\cdot x)}\right|^2 \quad (\because \langle x|y\rangle = 0 \text{ unless } y = x) \quad (11.13b)$$

$$= \begin{cases} 1 & \text{for } x = \tilde{y}, \quad (\because (z\cdot x)\oplus(z\cdot x) = 0) \\ 0 & \text{for } x \neq \tilde{y}. \end{cases} \quad (11.13c)$$

11.1.5 *The Phase Oracle

We now focus on the phase oracle, a quantum unitary realization of a specific instance of the binary function $f(x)$. This notion of a phase oracle extends to several quantum algorithms, not just DJ. Two critical facets require elucidation: the quantum realization of $f(x)$, and the phase-kickback mechanism.

1 Unitary Representation of a Function

Harnessing quantum parallelism necessitates that the oracle is constructed as a unitary transformation, capable of handling superposition states. We will exemplify this with a specific instance detailed in the succeeding exercise.

> **Exercise 11.6** It is known that the parity function $f(x) = \text{XOR}(x) = x_1 \oplus x_2 \oplus \cdots \oplus x_n$ is a balanced function (see Problem 11.2).
>
> Now establish that inverting any number of input bits in the above function results in another balanced function, for instance, $f(x) = \overline{x_1} \oplus x_2 \oplus \cdots \oplus x_n$.

Figure 11.3 displays a circuit implementing the function $f(x) = \overline{x_1} \oplus x_2 \oplus x_3 \oplus \overline{x_4}$ across five qubits. The fifth qubit conveys the outcome, $|f(x)\rangle$. The operation $\oplus$ is realized via CNOT gates, while X gates achieve the inversions.

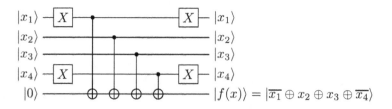

Figure 11.3: An Oracle for a Balanced Function

> **Exercise 11.7** Confirm that the circuit illustrated in Fig. 11.3 outputs $|f(x)\rangle$ on its auxiliary qubit, where $f(x) = \overline{x_1} \oplus x_2 \oplus x_3 \oplus \overline{x_4}$.

2 Phase Kickback

A limitation of the oracle in Fig. 11.3 is its dependence on an auxiliary qubit to relay the function result. For the DJ algorithm, the function result must be reflected as changes in the original qubits. This can be achieved by initializing the auxiliary qubit as $|-\rangle$, typically done with an X gate followed by an H gate, as depicted in Fig. 11.4.

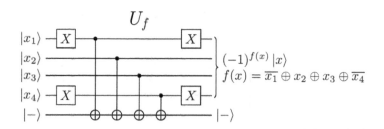

Figure 11.4: A Phase Oracle for a Balanced Function

This adjustment ensures the auxiliary qubit remains $|-\rangle$, but introduces a phase of -1 to the original qubits if $f(x) = 1$, a phenomenon termed as phase kickback.

The root cause behind this behavior stems from the properties of the CNOT gate (see § 7.1.1):

$$\text{CNOT} |0\rangle |-\rangle = |0\rangle |-\rangle, \quad (11.14a)$$
$$\text{CNOT} |1\rangle |-\rangle = - |1\rangle |-\rangle. \quad (11.14b)$$

Consequently, the state of any input, $|x_i\rangle$, transitions to $(-1)^{x_i} |x_i\rangle$.

Following all the CNOT operations, the entire system reaches the state:

$$(-1)^{\overline{x_1}} |\overline{x_1}\rangle (-1)^{x_2} |x_2\rangle (-1)^{x_3} |x_3\rangle (-1)^{\overline{x_4}} |\overline{x_4}\rangle |-\rangle.$$

Subsequent X gates negate the inversion effects on $\overline{x_1}$ and $\overline{x_4}$. This results in:

$$(-1)^{f(x)} |x\rangle |-\rangle.$$

The phase kickback is now complete.

> (i) Though the phase $(-1)^{f(x)}$ may seemingly be aligned with either $|x\rangle$ or $|-\rangle$, its dependence on x restricts the alignment with the former for superposition states. For instance:
>
> $$U_f \frac{1}{\sqrt{2}}(|0000\rangle + |1000\rangle) |-\rangle = \frac{1}{\sqrt{2}}(|0000\rangle - |1000\rangle) |-\rangle.$$

> (i) Given its limited utility within the oracle, the auxiliary qubit, commonly termed an ancilla qubit, can be disregarded outside the oracle.

The phase oracle can also be implemented with CZ gates instead of CNOTs, as showcased in Fig. 11.5.

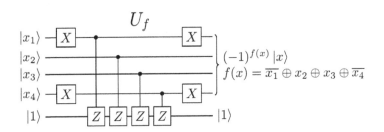

Figure 11.5: A Phase Oracle Using CZ Gates

11.2 QUBO, VQE, QAOA, and AQC

In an era where computational demands continually escalate and classical computing paradigms show limitations, the exploration of novel computational methods has become imperative. As we transition from limited Noisy Intermediate-Scale Quantum (NISQ) devices to the burgeoning era of quantum utility, these quantum systems emerge as a promising avenue. Despite their limitations, NISQ and emerging utility-era quantum computers exhibit potential to outperform classical counterparts in areas like optimization, quantum simulation, and machine learning[23].

Quantum Computing Applications in the NISQ/Unity Era

In the NISQ and emerging quantum utility era, quantum computing, particularly through hybrid quantum-classical algorithms, is poised to benefit real-world applications in several key areas:

- Optimization: Essential for complex problem-solving in finance, logistics, and more.
- Quantum Simulation: Useful in material science, drug discovery, and battery efficiency.
- Machine Learning: Enhancing data processing, pattern recognition, and predictive analytics.

Due to the limited scope of this text, our focus will be on Optimization. Quantum simulation requires an advanced understanding of quantum mechanics, and quantum machine learning necessitates a comprehensive knowledge of classical machine learning.

Central to this pursuit is the Quadratic Unconstrained Binary Optimization (QUBO) problem. As a combinatorial optimization paradigm, QUBO finds applications spanning a broad spectrum from finance and economics to advanced machine learning endeavors. Owing to its NP-hard nature, several classical challenges from theoretical computer science, such as max-cut, graph coloring, and the partition problem, have been mapped into the QUBO framework. Furthermore, this mapping extends to sophisticated machine learning models like support-vector machines, clustering algorithms, and probabilistic graphical models.

NP-hard Problems

Loosely speaking, the term NP-hard designates a class of problems that are particularly challenging to solve in polynomial time. This means that as the size of the problem increases, the time required to find an exact solution grows much more rapidly than a polynomial function of the problem size. These types of problems frequently arise in various real-world applications, such as logistics and cryptography, where finding an optimal solution is crucial yet computationally demanding.

In the quest to solve QUBO, two prominent quantum-classical hybrid algorithms emerge: the Quantum Approximate Optimization Algorithm (QAOA) and the Variational Quantum Eigensolver (VQE). The QAOA leverages quantum mechanics to generate approximate solutions to combinatorial problems by alternating between unitary transformations. On the other hand, VQE aims to find the ground state energy of a given Hamiltonian by using a parameterized quantum circuit and iteratively refining its parameters.

Moreover, given QUBO's intrinsic relationship with Ising models, it is recognized as a key problem class in adiabatic quantum computation (AQC). Herein, solutions are procured through a specialized physical process known as quantum annealing.

To elucidate the core principles and subtleties of these algorithms and com-

11.2 QUBO, VQE, QAOA, and AQC

putational strategies, we will hone in on the Max-Cut problem as our illustrative example.

> **Achieving Quantum Advantages with VQE and QAOA**
>
> VQE (Variational Quantum Eigensolver) and QAOA (Quantum Approximate Optimization Algorithm) exhibit three key traits that make them promising for realizing quantum advantages, particularly on near-term, noisy, intermediate-scale quantum (NISQ) devices: hybrid nature, efficiency, and adaptability.
>
> The hybrid structure of VQE and QAOA enables effective utilization of both classical and quantum resources, making them well-suited for NISQ devices. These algorithms partition the computational workload between a classical optimizer and a quantum processor. This division allows for a fallback to classical techniques for error mitigation and tasks that may not yet be efficiently manageable on current quantum hardware. Such a distributive strategy is particularly beneficial during the transitional phase leading to fully fault-tolerant quantum computing.
>
> Efficiency in VQE and QAOA manifests through their judicious use of quantum resources and their capability to yield good approximations using limited qubits and shallow circuits. Both algorithms employ parameterized quantum circuits with depths usually falling within the coherence times of existing NISQ systems. This minimizes the error probabilities associated with longer computations and maximizes the computational capabilities of current quantum hardware, an essential feature for near-term applications.
>
> The adaptability of these algorithms stems from their iterative mechanisms and parameter fine-tuning capabilities. This not only confers robustness against quantum errors but also offers flexibility in addressing a wide range of optimization and eigenvalue problems. Both VQE and QAOA can be adapted to specific problem characteristics, enabling the inclusion of specialized heuristics or alternative cost functions. This level of adaptability enhances their applicability across various domains, further substantiating their relevance in the context of near-term quantum computing.

A variety of concepts and acronyms are introduced in this section. For easy reference, see Table 11.2.

11.2.1 The Max-Cut Problem

The Max-Cut problem, originated from graph theory, is an examplary QUBO problem. Given an undirected graph with a set of vertices and edges, the objective is to partition the vertices into two disjoint sets (or equivalently, color the vertices with two distinct colors) such that the number of edges connecting these sets (or having differently colored endpoints) is maximized. This scenario is illustrated in Fig. 11.6 comprising five vertices.

A useful analogy is visualizing each vertex as a city and each edge as a road connecting two cities. The Max-Cut problem can then be understood as optimally segregating these cities into two regions to maximize the number of interconnecting roads.

Another analogous representation is the influence maximization challenge in social networks. Envision a system of interconnected individuals influencing one another. Each individual corresponds to a vertex, while their interactions are represented by

Term	Description	Relationships
QUBO	Quadratic Unconstrained Binary Optimization. Problems aiming to minimize or maximize an objective function that comprises quadratic terms in binary variables.	Can be represented by the Ising model. Max-cut and TSP are examples.
Max-Cut	Optimization problem in graph theory.	Can be formulated as a QUBO or Ising Model.
TSP	Traveling Salesman Problem	Can be formulated as a QUBO or Ising Model.
VQA	Variational Quantum Algorithms. A class of algorithms that leverage variational principles combined with ansatzes.	VQE and QAOA are examples.
VQE	Variational Quantum Eigensolver. An algorithm for finding the ground state of a Hamiltonian.	A special case of VQA.
QAOA	Quantum Approximate Optimization Algorithm. Inspired by adiabatic quantum evolution.	A special case of VQA.
Ansatz	Parameterized trial state vector implemented as a quantum circuit.	Component of VQE and QAOA.
Ising Model	Model describing interactions of $\frac{1}{2}$-spins in magnetic fields.	Can represent QUBO problems, including Max-Cut and TSP.
AQC	Adiabatic Quantum Computing. Computation model based on initializing a system in the ground state of a simple Hamiltonian and then slowly evolving the Hamiltonian to one representing the problem.	Includes Quantum Annealing. Inspired QAOA.
Quantum Annealing	Optimization technique using quantum fluctuations to escape local minima. Implemented on dedicated hardware (quantum annealers).	Subset of AQC.

Table 11.2: Relationships Between Terms Related to Quantum Optimization

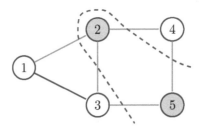

Figure 11.6: Max-Cut Problem Example

edges. This model effectively captures marketing dynamics where the strength of influence between individuals, represented by edge weights, governs their purchasing decisions. For a marketing strategy offering free products to select individuals, the central goal becomes determining the ideal subset of these individuals to optimize revenue.

Being NP-hard, the Max-Cut problem implies that as graph size grows, the computational time to determine the precise solution can expand exponentially, rendering classical methods ineffective for substantial graphs. Nonetheless, due to its significance as a representative for numerous optimization challenges, this problem, despite its deceptive simplicity, underscores the computational intricacies inherent in obtaining optimal solutions, establishing it as a prime candidate for quantum computational techniques such as QAOA and quantum annealing.

1 Problem Formulation

The weighted max-cut problem can be formally defined as follows:

For an undirected graph $G = (V, E)$ comprising n nodes, where V denotes the vertex set and E symbolizes the edge set, edge weights are given by w_{ij} for $(i, j) \in E$. A cut bifurcates the primary set V into two subsets, labelled 0 and 1, respectively. The function to optimize, in this context, is the aggregate weights on edges connecting distinct subsets across the cut. By attributing $x_i = 0$ or $x_i = 1$ to each node i, the objective becomes maximizing the global profit function:

$$\tilde{C}(x) = \sum_{(i,j) \in E} w_{ij}(x_i \oplus x_j), \tag{11.15}$$

where $\oplus$ denotes the XOR function. Specifically, $x_i \oplus x_j = 1$ if $x_i \neq x_j$, meaning that nodes i and j belong to different subsets; the value is 0 otherwise.

In the simplified marketing archetype, w_{ij} characterizes the likelihood that individual j will purchase a product if i has received it as a free sample. Notably, the value of w_{ij} can exceed 1 or even be negative. A value greater than 1 could reflect scenarios where individual j is motivated to buy multiple products, while a negative value could indicate dissuasion. The objective is to maximize the cumulative purchasing probabilities, which equates to revenue optimization. If the anticipated profit probability exceeds the initial sample costs, the strategy is considered profitable.

In an extended model, nodes themselves possess weights, denoted by w_i, mirroring the probability that an individual receiving a free sample repurchases the product. Incorporating this facet, the objective function evolves to:

$$\tilde{C}(x) = \sum_{(i,j)\in E} w_{ij}(x_i \oplus x_j) + \sum_{i\in V} w_i x_i. \qquad (11.16)$$

The objective function for the extended Max-Cut problem (Eq. 11.16) is mathematically equivalent to that of a Quadratic Unconstrained Binary Optimization (QUBO) problem.

Exercise 11.8 Formulate the profit function $\tilde{C}(x)$ for the max-cut problem in Fig. 11.6, assuming $w_{ij} = 1$ and $w_i = 0$.

11.2.2 ∗ The Ising Model

The Ising model, introduced in § 6.5.3, describes a system of spins under the influence of a transverse magnetic field. It provides a bridge between physical systems and computational problems, allowing the use of quantum devices to solve problems of practical interest.

To derive a solution to the weighted max-cut problem via a quantum computer, an initial step is the mapping of this problem to an Ising Hamiltonian. We map the binary variables $x_i, x_j \in \{0, 1\}$ in Eq. 11.16 to $z_i, z_j \in \{-1, 1\}$ (since the eigenvalues of Z are $\{-1, 1\}$):

$$x_i \to \frac{1}{2}(1 - z_i), \quad x_i \oplus x_j \to \frac{1}{2}(1 - z_i z_j). \qquad (11.17)$$

Replacing z_i and z_j with the corresponding Pauli operators Z_i and Z_j, we deduce that:

$$C(Z) = \sum_{(i,j)\in E} \frac{w_{ij}}{2}(I - Z_i Z_j) + \sum_{i\in V} \frac{w_i}{2}(I - Z_i) \qquad (11.18a)$$

$$= -\frac{1}{2} \sum_{(i,j)\in E} w_{ij} Z_i Z_j - \frac{1}{2} \sum_{i\in V} w_i Z_i + \text{const}, \qquad (11.18b)$$

where const is a Z-independent term and can be ignored.

$C(Z)$ is a matrix while $\tilde{C}(x)$ is a number.

Hence, the weighted max-cut problem can be equated to minimizing the Ising Hamiltonian:

$$H = \frac{1}{2} \sum_{(i,j)\in E} w_{ij} Z_i Z_j + \frac{1}{2} \sum_{i\in V} w_i Z_i, \qquad (11.19)$$

which corresponds to identifying the ground state of the spin system portrayed by the Ising model.

11.2 QUBO, VQE, QAOA, and AQC

 In the above notation, $Z_i Z_j$ is a shorthand for $Z_i \otimes Z_j$, acting on qubits i and j.

The eigenvalues $+1$ and -1 of the Z operator signify the two separate sets into which the vertices are partitioned. When the $Z_i Z_j$ product equals -1, this denotes that vertices i and j are placed in different sets, and the corresponding edge is a *cut edge*. Conversely, if $Z_i Z_j$ results in $+1$, this implies that vertices i and j belong to the same set, and the edge is an *uncut edge*.

Exercise 11.9 Express Eq. 11.19 explicitly as an 8×8 matrix, for a system of three nodes on a triangle, assuming $w_{ij} = 1$ for all edges and $w_i = 1$ for all nodes.

Exercise 11.10 Formulate the Hamiltonian H for the max-cut problem in Fig. 11.6, assuming $w_{ij} = 1$ and $w_i = 0$.

The Ising model can subsequently be tackled using a diverse array of quantum algorithms, including the Quantum Approximate Optimization Algorithm (QAOA) and Variational Quantum Eigensolver (VQE), or via quantum annealing apparatuses like the D-Wave systems.

> **Applications of the Ising Model**
>
> The applications of the Ising model are extensive, surpassing just the Max-Cut:
>
> Combinatorial Optimization Problems: Beyond the max-cut problem explored here, challenges such as the traveling salesman problem and other NP-hard problems can be transposed onto an Ising framework.
>
> Machine Learning: The Ising model's reach extends to quantum machine learning, particularly in modeling Boltzmann machines, which are identified as stochastic recurrent neural networks. Herein, the parameters of the Ising model encapsulate the weights and biases of the Boltzmann machine, and the model's equilibrium state offers a probabilistic representation of the data.
>
> Quantum Error Correction: Grounded in the Ising model, topological quantum error correction codes emerge as promising candidates for trustworthy quantum computation. The toric code, a two-dimensional Ising variant with additional constraints, facilitates the comprehension and realization of quantum error correction.
>
> Condensed Matter Physics: The quantum version of the Ising model plays an instrumental role in simulating quantum phase transitions, a subject capturing significant interest within the domain of condensed matter physics. More specifically, the Ising model provides insights into the transitionary phases between distinct states of matter induced by quantum perturbations. This encompasses the exploration of phenomena like ferromagnetism, anti-ferromagnetism, and other unconventional states of matter.

11.2.3 ∗ Variational Quantum Eigensolver (VQE)

The Variational Quantum Eigensolver (VQE) is a hybrid quantum-classical algorithm designed to harness the capabilities of quantum computers. While its initial development targeted quantum chemistry applications—often considered the most practical use-case for Noisy Intermediate-Scale Quantum (NISQ) computers—the algorithm's scope extends beyond this, making it also suitable for solving optimization problems.

At the heart of VQE is the mapping of the problem onto a Hamiltonian. The ground state of this Hamiltonian encodes the solution. The variational principle ensures that for any chosen trial state, $\psi(\theta)$, its energy expectation value will always be greater than or equal to the ground state energy, E_0. Thus, by iteratively optimizing the parameters of the trial state, the VQE approximates the ground state energy and the associated quantum state.

1 The Variational Principle

The variational principle is a fundamental concept in quantum mechanics, formally stated as:

$$\langle \psi(\theta)|H|\psi(\theta)\rangle \geq E_0, \tag{11.20}$$

where E_0 represents the ground state energy.

This principle lays the groundwork for the VQE algorithm. In quantum systems with many qubits, the state space becomes intractably large for exhaustive energy minimization. However, a quantum computer can efficiently prepare trial states and evaluate their corresponding energy expectation values. Following these quantum computations, a classical optimizer adjusts the parameters θ in order to minimize the energy, thus approximating the ground state of the system.

 When maximizing a profit function is the objective, it is equivalent to minimizing the negative of that function.

2 Ansatz

The ansatz in VQE refers to the trial quantum state, $\psi(\theta)$, parameterized by an array of parameters θ. It is an educated guess of the solution, based on certain heuristics or insights. Implemented as a parameterized quantum circuit, the quality of the ansatz is crucial to the accuracy of the VQE. If the chosen ansatz is not expressive enough to approximate the ground state of the problem, the optimization is unlikely to yield accurate results.

3 Measurement

The ansatz is run repeatedly for m times. For each run, the n qubits are measured in the computational basis; the k-run yields n measurement values $z_1^{(k)}, z_2^{(k)}, ..., z_n^{(k)}$, where $z_i^{(k)} = 1$ for measurement outcome $|0\rangle$, and -1 for $|1\rangle$. Essentially, we are measuring each qubit with the Z observable.

Then the expected value of the Hamiltonian is estimated according to Eq. 11.19:

$$\langle \psi|H|\psi\rangle = \frac{1}{2m}\sum_{k=1}^{m}\sum_{i\in V} w_i z_i^{(k)} + \frac{1}{2m}\sum_{k=1}^{m}\sum_{(i,j)\in E} w_{ij} z_i^{(k)} z_j^{(k)}. \tag{11.21}$$

Note the above calculation is only valid for the Ising model, where the Hamiltonian involves Z solely and is hence diagonal in the computational basis. The general case is more complicated, to be discussed at the end of this subsection.

4 Iteration Procedure

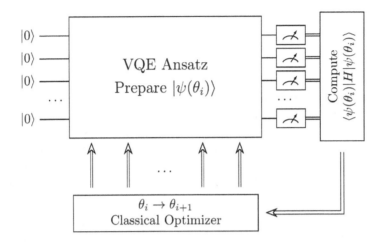

Figure 11.7: VQE Ansatz and Iteration Procedure

The VQE follows an iterative procedure (Fig. 11.7) comprising:

1. Crafting a parameterized quantum circuit that serves as the ansatz, $|\psi(\theta)\rangle$.

2. Measuring the ansatz qubits on a quantum computer and processing the outcomes classically to estimate the expected value $\langle\psi(\theta)|H|\psi(\theta)\rangle$.

3. Employing a classical optimizer to propose adjustments to the ansatz's parameters.

The iterations continue until convergence, or when the change in the expected value is below a predefined threshold.

 The weights $w_{i,j}$ are not directly incorporated into the variational parameters θ; instead, θ is utilized to construct the trial ground state vector within the quantum circuit framework. The weights $w_{i,j}$ are used to compute the expected value of the Hamiltonian on a classical computer. An optimization routine, executed on a classical computer, then uses this calculated expected value to iteratively adjust θ. This process aims to minimize the expected value of the Hamiltonian, thereby accurately approximating the ground state with an optimal θ.

5 ✶ General Case

In the general case, such as when VQE is used in chemistry simulation, computation of the expected value $\langle\psi(\theta)|H|\psi(\theta)\rangle$ is non-trivial. The Hamiltonian must first be decomposed into Pauli strings, P_i:

$$H = \sum_i c_i P_i, \quad P_i \in \{I, X, Y, Z\}^{\otimes n}. \tag{11.22}$$

With the Pauli basis being orthogonal and having a norm of 2, the coefficients c_i can be expressed as:

$$c_i = \frac{1}{2^n} \operatorname{tr}(HP_i). \tag{11.23}$$

Consequently, the expected value of H is:

$$\langle \psi | H | \psi \rangle = \sum_i c_i \langle \psi | P_i | \psi \rangle. \tag{11.24}$$

Sampling the Pauli Operators

When dealing with a Hamiltonian H that comprises not just the Z operator but also X and Y, distinct measurement strategies are employed. Direct measurements in the computational basis suffice for evaluating $\langle \psi | Z | \psi \rangle$. However, to obtain expected values for X and Y, one must rotate the basis prior to measurement, using the Hadamard gate for X and $R_x(-\frac{\pi}{2})$ for Y, as described in § 3.4.5.

In this more general context, calculating the overall expected value $\langle \psi | H | \psi \rangle$ involves measuring each term in the Hamiltonian's Pauli basis expansion (Eq. 11.24). Each term generally necessitates a unique run of the quantum circuit (or ansatz), meaning up to 4^n runs could be required in the worst-case scenario. However, many practical quantum problems feature Hamiltonians with far sparser representations, thereby reducing the number of circuit runs needed to achieve a reliable estimation of the Hamiltonian's expected value.

6 The Ground State

For the Max-Cut problem, since the Hamiltonian is diagonal in the computational basis, its eigenstates will also be computational basis states. Hence, the ground state that minimizes the cost function will correspond to a computational basis state.

In the general case, this may no longer be the case. The ground state may be a superposition state in the computational basis.

11.2.4 ∗ Quantum Approximate Optimization Algorithm (QAOA)

The Quantum Approximate Optimization Algorithm (QAOA) is based on the Variational Quantum Algorithm (VQA) framework. It is specifically designed to solve combinatorial optimization problems such as Max-Cut. The primary objective is to identify the quantum state that minimizes the expected value of a Hamiltonian which encapsulates the cost function of the problem at hand.

While QAOA shares the iterative nature of the Variational Quantum Eigensolver (VQE), it employs a specialized ansatz, as depicted in Fig. 11.8.

1 Structure of the QAOA Ansatz

The QAOA ansatz employs an alternating sequence of unitary transformations, which are derived from two Hamiltonians: the cost Hamiltonian H_C and a mixing Hamiltonian H_B. Specifically, the ansatz consists of repeated blocks of two types of unitary transformations, $U_C(\gamma)$ and $U_B(\beta)$, expressed as:

$$U_B(\beta) = e^{-i\beta H_B}, \quad U_C(\gamma) = e^{-i\gamma H_C}, \tag{11.25}$$

where β and γ are variational parameters optimized through classical means.

11.2 QUBO, VQE, QAOA, and AQC

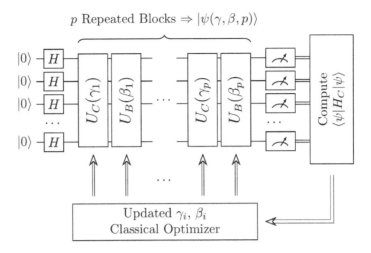

Figure 11.8: QAOA Ansatz and Iterative Procedure

2 Cost Hamiltonian

The cost Hamiltonian, H_C, encodes the objective function of the optimization problem. It assigns an "energy" or "cost" to specific configurations of the system. The goal for combinatorial optimization problems is to find the configuration that minimizes this energy. A representative expression for H_C in the context of Max-Cut is:

$$H_C = \frac{1}{2} \sum_{(i,j) \in E} w_{ij} Z_i Z_j + \frac{1}{2} \sum_{i \in V} w_i Z_i, \tag{11.26}$$

Here, V and E are the sets of vertices and edges in the graph $G = (V, E)$. w_{ij} and w_i are the weights for edges and vertices, and Z_i is the Pauli-Z operator on the i-th qubit.

3 Mixing Hamiltonian

The mixing Hamiltonian H_B facilitates transitions between different cut configurations, thereby aiding the system in exploring the space of potential solutions. A common choice for H_B is:

$$H_B = \sum_i X_i, \tag{11.27}$$

where X_i is the Pauli-X operator on the i-th qubit.

4 Interplay and Iteration

The interplay between H_C and H_B lies at the heart of QAOA's functionality. While H_C encodes problem-specific penalties or rewards, H_B keeps the quantum state in a superposition of various configurations. By iteratively applying these Hamiltonians and fine-tuning the variational parameters, QAOA aims to direct the quantum system toward H_C's ground state. The optimal depth p is generally determined

5 Example Application to Max-Cut

Consider a simple Max-Cut problem depicted in Fig. 11.6. The graph has vertices $V = \{1, 2, 3, 4, 5\}$ and edges $E = \{(1,2), (1,3), (2,3), (2,4), (3,5), (4,5)\}$, with all edge weights set to unity, and node weights to zero.

The solution is obviously a cut shown by the dashed curve: with only one un-cut edge $(1, 3)$, and nodes partitioned into $\{1, 3, 4\}$ and $\{2, 5\}$. But we will walk through the QAOA process as a demonstration of its concepts and procedure.

Its cost Hamiltonian can be expressed using the Ising model as:

$$H_C = \frac{1}{2} \sum_{(i,j) \in E} Z_i Z_j. \tag{11.28}$$

When the product of $Z_i Z_j$ is -1, it means the vertices i and j are in different sets (a cut edge), contributing negatively to the total energy. Therefore, minimizing $\langle H_C \rangle$ corresponds to maximizing the number of cuts, the goal of the Max-Cut problem.

The unitary block $U_C(\gamma)$ is given by:

$$U_C(\gamma) = e^{-i\gamma H_C} = \prod_{(i,j) \in E} e^{-i\frac{\gamma}{2} Z_i Z_j} = \prod_{(i,j) \in E} ZZ_{ij}(\frac{\gamma}{2}), \tag{11.29}$$

where each factor in the product is a $ZZ(\frac{\gamma}{2})$ gate (see § 7.1.5) acting on qubits i and j.

The unitary block $U_B(\beta)$ is given by:

$$U_B(\beta) = e^{-i\beta H_B} = \prod_{i \in V} e^{-i\beta X_i} = \prod_{i \in V} R_{x_i}(\beta), \tag{11.30}$$

where each factor in the product is an $R_x(\beta)$ gate acting on qubit i.

The building block for the QAOA ansatz comprising $U_C(\gamma)$ and $U_B(\beta)$ is shown in Fig. 11.9.

After sufficient number of iterations, the algorithm arrives at a ground state expected value $\langle H_C \rangle = -2$. The resulting ground state is $|01001\rangle$ (or equivalently $|10110\rangle$), yielding a partition of $\{1, 3, 4\}$ and $\{2, 5\}$ as anticipated.

> **Exercise 11.11** Formulate the cost function and the Ising Hamiltonian for the six-node Max-Cut problem shown in the figure below. Assume all the edge weights are 2. Node weights are 1 for even nodes and -1 for odd nodes. Solve the problem to find the optimal cut using QAOA, implemented in Qiskit, Cirq or a similar platform. Provide the final (i.e., most probable) state vector and the measure of the quality of the cut.

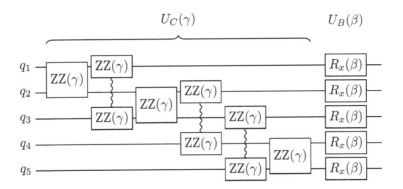

Figure 11.9: Example of a QAOA Ansatz Building Block

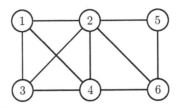

11.2.5 *Traveling Salesman Problem

The Traveling Salesman Problem (TSP) is a seminal problem in the realm of QUBO. It involves finding the shortest route through a list of cities, ensuring that each city is visited exactly once before returning to the starting city. This can be represented as a weighted graph where the cities are vertices and the distances between them serve as weighted edges, as shown in Fig. 11.10. Much like the Max-Cut problem, TSP is NP-hard, meaning that computational time grows substantially with an increase in the number of cities.

Owing to its wide range of applications in fields such as logistics and transportation, the TSP has attracted attention from both classical and quantum optimization techniques. Despite its seeming simplicity, the problem is deeply rooted in combinatorial optimization, underlining the value of quantum approaches for solving complex problems.

1 Problem Formulation

The TSP is mathematically framed as follows: one seeks to minimize the tour distance in a graph $G = (V, E)$, where n nodes represent cities and d_{ij} denotes the distance between nodes i and j. The tour distance for a cycle is defined as:

$$\tilde{L}(x) = \sum_{i=1}^{n} \sum_{j=1, j \neq i}^{n} d_{ij} x_{ij}, \tag{11.31}$$

where $x_{ij} \in \{0, 1\}$ indicates if the tour goes from node i to j, with $x_{ij} = 1$ if so.

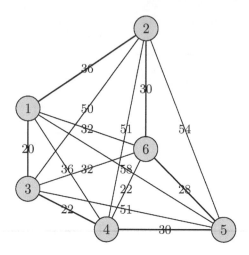

Figure 11.10: Example of the Traveling Salesman Problem

2 Constraints

To ensure that each city is visited exactly once, the following constraints are imposed:

$$S_j \equiv \sum_{i,i\neq j} x_{ij} = 1 \quad \text{for } j = 1, 2, ..., n, \tag{11.32a}$$

$$S_i \equiv \sum_{j,j\neq i} x_{ij} = 1 \quad \text{for } i = 1, 2, ..., n. \tag{11.32b}$$

The first equation states that for each city j, there must be exactly one city i such that $x_{ij} = 1$. This ensures that a path enters each city exactly once.

The second equation states that for each city i, there must be exactly one city j such that $x_{ij} = 1$. This ensures that a path exits each city exactly once.

3 Objective Function

The objective function, which includes both the tour distance and the constraints, is:

$$\tilde{C}(x) = \tilde{L}(x) + A\sum_{j=1}^{n}(1-S_j)^2 + A\sum_{i=1}^{n}(1-S_i)^2, \tag{11.33}$$

with A as a free parameter large enough to enforce the constraints, generally $A > \max(d_{ij})$.

This combined objective function is quandratic in x_{ij}:

$$\tilde{C}(x) = \sum_{i,j,k,l} w_{ijkl} x_{ij} x_{kl} + \sum_{ij} w_{ij} x_{ij}. \tag{11.34}$$

4 Quantum-Classical Hybrid Solution

The TSP can also be solved using hybrid quantum-classical algorithms like the Quantum Approximate Optimization Algorithm (QAOA). Due to the problem's complexity, the above TSP implementation requires n^2 qubits, as opposed to the n qubits needed for Max-Cut. Modern quantum computing platforms often provide libraries to simplify this formulation.

For a more in-depth understanding, the reader is directed to the following references:

- Efficient traveling salesman problem solvers using the Ising model with simulated bifurcation [98].

- Clustering approach for solving traveling salesman problems via Ising model-based solver [37].

> **Exercise 11.12** Work through the six-node TSP example shown in Fig. 11.10. Formulate the cost function and the Ising Hamiltonian, and then solve the problem using QAOA. Use Qiskit, Cirq or a similar platform for implementation. Provide the final (i.e., most probable) state vector and the shortest tour distance. (If your platform does not have the capacity for simulating TSP with six nodes, reduce the number to five or four.)

11.2.6 Adiabatic Quantum Computation and Annealing

Quantum annealing is a specialized approach tailored for optimization problems. We will discuss this further using Max-Cut as an example.

1 From Universal Quantum Computing to Quantum Annealing

In the ever-evolving landscape of quantum computing, two primary paradigms have gained prominence: gate-based universal quantum computing and quantum annealing.

Gate-based universal quantum computers employ a series of quantum gates to control qubits. This model's universality makes it apt for diverse computational problems, from integer factorization to quantum system simulations. Algorithms like Deutsch-Jozsa, VQE, and QAOA, as well as Shor's and Grover's algorithms, belong to this paradigm. Some algorithms are not discussed here due to the scope of this text.

Adiabatic quantum computation (AQC) is another form of quantum computing, as introduced in § 4.4. Here, the problem is mapped to a quantum system whose ground state represents the solution. Starting from an initial, easily-understandable Hamiltonian, the system is slowly evolved to a final Hamiltonian that encodes the solution.

Quantum annealing is a subset of AQC tailored for optimization problems. It avoids the need for precise quantum gate sequences, allowing the system to evolve under its natural quantum mechanics.

This natural evolution makes quantum annealing particularly useful for optimization problems in various domains like machine learning, finance, and logistics. The lack of need for exact gate sequences can ease the construction of larger quantum systems, although some problems may not be well-suited for this approach.

2 Solving Max-Cut with Quantum Annealing

In AQC, we begin with the quantum system in its ground state, corresponding to the lowest energy eigenstate of an initial Hamiltonian H_{initial}. The Hamiltonian is then evolved towards a target H_{target}, which encodes the optimization problem. The evolving Hamiltonian, $H(t)$, is given by:

$$H(t) = (1 - s(t))H_{\text{initial}} + s(t)H_{\text{target}}, \tag{11.35}$$

where $s(t)$ transitions from 0 to 1 during the annealing process. According to the adiabatic theorem, a slow evolution ensures the system remains in its ground state, facilitating a smooth transition from H_{initial} to H_{target} and ultimately arriving at the solution.

A commonly used H_{initial} in quantum annealing the transverse-field Ising Hamiltonian, given by:

$$H_{\text{initial}} = -\Gamma \sum_{i=1}^{n} X_i, \tag{11.36}$$

where Γ represents the strength of the transverse magnetic field, and X_i is the Pauli X operator acting on the i-th qubit.

The corresponding ground state is a uniform superposition of all computational basis states, represented as:

$$|\psi_{\text{initial}}\rangle = \bigotimes_{i=1}^{n} |+\rangle = \sum_{i=0}^{n-1} |i\rangle. \tag{11.37}$$

The target Hamiltonian, H_{target}, is given by:

$$H_{\text{target}} = \frac{1}{2} \sum_{(i,j) \in E} w_{ij} Z_i Z_j + \frac{1}{2} \sum_{i \in V} w_i Z_i, \tag{11.38}$$

where $Z_i Z_j$ represents $Z_i \otimes Z_j$, with Pauli Z acting on qubits i and j. (See Eq. 11.19.)

Once these Hamiltonians are defined, the quantum annealing machine takes over the computation. The annealing process initiates by preparing the quantum system in the ground state of H_{initial}. The system then evolves under the influence of the time-dependent Hamiltonian $H(t)$, which is a mixture of the initial and target Hamiltonians controlled by the parameter $s(t)$.

The annealing schedule, often predefined, controls the evolution of $s(t)$ from 0 to 1. As the system evolves, it naturally explores the energy landscape, guided by quantum fluctuations. The slow change ensures that the system remains near its ground state throughout the process, due to the adiabatic theorem.

At the end of the annealing schedule, the quantum state of the system approximates the ground state of H_{target}. This state is then measured, yielding a classical bit string that represents a near-optimal solution to the encoded optimization problem. Subsequent repetitions of this process can improve the quality of the solution through statistical sampling.

11.2 QUBO, VQE, QAOA, and AQC

It's worth mentioning that while the annealing process is heuristic in nature, it has shown promising results in various optimization problems, offering a complementary approach to gate-based quantum computing.

Exercise 11.13 Solve the six-node Max-Cut problem shown in the figure below using a quantum annealing platfrom such as D-Wave NetworkX. Assume all the edge weights are 2. Node weights are 1 for even nodes and −1 for odd nodes. Provide the final state vector and calculate the value of the cut as the measure of its quality. Compare your solution with the QAOA counterpart.

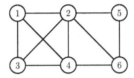

Exercise 11.14 Solve the six-node sraveling salesman problem shown in Fig. 11.10 using a quantum annealing platfrom such as D-Wave NetworkX. Provide the final state vector and the shortest tour distance. Compare your solution with the QAOA counterpart. (If your platform does not have the capacity for simulating TSP with six nodes, reduce the number to five or four.)

3 Conditions for Quantum Speedup and Limitations

In adiabatic quantum computation, and more specifically in quantum annealing, as H_{initial} evolves towards H_{target} (see Eq. 11.35), the energy gap between the ground state and the first excited state of the quantum system varies over time. This gap can become very small, potentially even reaching zero, a point known as a level crossing. The efficiency of the algorithm is closely tied to the behavior of the minimum energy gap, denoted as G.

The required annealing time τ is inversely proportional to the square of G, expressed as $\tau \propto 1/G^2$. Should G decrease exponentially with the size of the problem, or the number of qubits n, in the form $G \propto e^{-\alpha n}$ for some positive constant α, then τ will increase exponentially. This exponential increase in time negates any potential quantum speedup, as the system would then require an impractically long time to evolve.

For quantum annealing to be efficient and to offer a potential quantum speedup, the energy gap G must decrease polynomially with the problem size n. If G scales as $G \propto 1/n^k$ for some constant k, the annealing time τ would then increase only polynomially, remaining manageable and enabling a substantial quantum speedup compared to classical algorithms.

It is important to recognize that the behavior of the energy gap G can significantly vary depending on the specific problem and its representation within the Hamiltonian of the quantum system. While some problems may naturally exhibit a polynomially decreasing gap, others may have an exponentially shrinking gap, rendering them unsuitable for efficient quantum annealing. Notably, the well-known NP-Hard problem, 3SAT, has been shown to present an exponentially decreasing energy gap, thus categorizing it among the less favorable problems for quantum annealing. In addition, for many practical problems, the behavior of their minimum energy gap is unknown beforehand, making it challenging to schedule the annealing process effectively and validate the results.

As of 2023, quantum annealing hardware, such as systems produced by D-Wave, typically operates on stoquastic Hamiltonians. A stoquastic Hamiltonian is one where all the off-diagonal elements in the standard computational basis are real and non-positive, which leads to certain computational simplifications. This limitation constrains the types of problems that quantum annealing can effectively address.

11.3 Quantum Bomb and Quantum Money

In this section, we explore the diversity of quantum algorithms through the study of two distinctive examples: the quantum bomb test algorithm and quantum money. These algorithms, while less commonly cited than DJ (Deutsch-Jozsa) and BV (Bernstein-Vazirani), Simon, or Shor, demonstrate the breadth and versatility of quantum computing. They elucidate a variety of intricate aspects, including quantum measurements, probabilistic computational methods, and post-quantum cryptographic techniques.

11.3.1 The Quantum Bomb Test Algorithm

This algorithm was initially conceptualized in a 1993 thought experiment by A. Elitzur and L. Vaidman[40]. It provides a compelling showcase for the unique principles of quantum measurements, distinguishing them sharply from their classical counterparts. This algorithm is also a cornerstone in the study of interaction-free measurements in quantum mechanics.

1 The Scenario

Consider a situation where you are presented with an assortment of indistinguishable boxes. While some contain quantum bombs, others are empty or 'duds'. The quantum bombs are sensitive to a single photon in horizontal polarization. The task at hand is to ascertain whether a given box contains a bomb without detonating it.

In the classical realm, any attempt to test the box's contents would result in a conundrum: if the box does not explode, you cannot conclusively determine whether it contains a bomb. If it explodes, you obviously know it was a bomb, but the object is destroyed. Essentially, it is a mission impossible in the classical realm.

2 Measuring in the Computational Basis

To cast the problem within the quantum measurement framework, let's represent a dud by an identity gate I. A measurement of an input state $|0\rangle$ will result in the output state $|0\rangle$, and similarly for $|1\rangle$.

A bomb, on the other hand, acts as a measuring device that triggers if it encounters the input state $|1\rangle$. Although not a unitary transformation, such a device can be physically implemented without involving an actual bomb. For instance, a polarization splitter in photonics could execute this conditional action.

3 Measuring in the $\{|+\rangle, |-\rangle\}$ Basis

When considering quantum properties, the scenario changes drastically. By introduc-

11.3 Quantum Bomb and Quantum Money

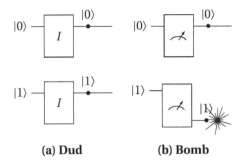

Figure 11.11: Quantum Bomb Tester - Classical Scenario

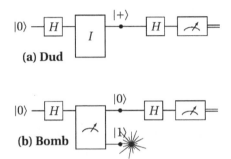

Figure 11.12: Quantum Bomb Tester - Measuring in the $\{|+\rangle, |-\rangle\}$ Basis

ing the bomb to a superposition state and measuring it in a different basis, we open up new possibilities. In particular, we can initialize our circuit with $|0\rangle$ and use two Hadamard gates to sandwich either the dud or the bomb. The initial Hadamard gate transforms $|0\rangle$ into $|+\rangle$. The subsequent Hadamard gate changes the measurement basis to $\{|+\rangle, |-\rangle\}$.

In this setup, the bomb has a 50% probability of exploding and a 50% chance of remaining intact. If it doesn't explode, a subsequent measurement will yield either $|0\rangle$ or $|1\rangle$ with equal probability. For the dud, the output remains as the input, invariably resulting in the measurement outcome $|0\rangle$.

Thus, there exists a net 25% chance of identifying a bomb without causing it to detonate. Even though not perfect, a significant improvement over the classical scenario!

4 The Quantum Bomb Test Algorithm

We now elaborate on an algorithm designed to detect the presence of a quantum bomb with an arbitrarily low probability of detonation. This algorithm is encapsulated in the quantum circuit depicted in Fig. 11.13.

The input state employed for this algorithm is as follows:

$$|\varepsilon\rangle = \cos\varepsilon\,|0\rangle + \sin\varepsilon\,|1\rangle, \tag{11.39}$$

where ε is a small angle defined by $\varepsilon = \frac{\pi}{2N}$, with N representing a large integer.

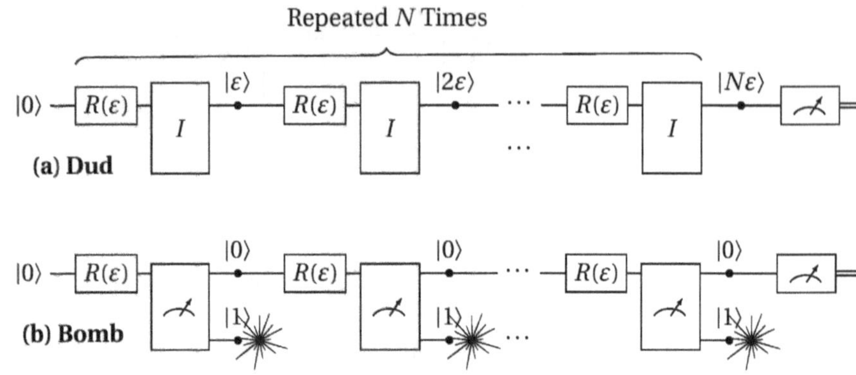

Figure 11.13: The Quantum Bomb Test Algorithm

To achieve this input state, $|\varepsilon\rangle$, we apply the $R(\varepsilon)$ gate to an initial state of $|0\rangle$:

$$R(\varepsilon) = \begin{bmatrix} \cos\varepsilon & -\sin\varepsilon \\ \sin\varepsilon & \cos\varepsilon \end{bmatrix}. \tag{11.40}$$

In the case of a dud, the qubit evolves as $R(\varepsilon)^N |0\rangle = \cos(N\varepsilon)|0\rangle + \sin(N\varepsilon)|1\rangle$ before measurement. The outcome will yield $|1\rangle$ with a probability of $\sin^2(N\varepsilon) = 1$, correctly identifying the dud.

In the case of a bomb, upon measurement, the probability of receiving $|1\rangle$ and thus triggering the bomb is $\sin^2(\varepsilon) \approx \varepsilon^2$, as per the small-angle approximation. If this does not occur, the qubit will collapse to $|0\rangle$, and the circuit will iterate.

After N iterations, the probability of triggering the bomb by that point is approximately $\frac{\pi^2}{4N}$, which can be made arbitrarily low with a large enough N. Conversely, the likelihood of identifying the bomb without triggering it, given by $\cos^{2N}(\varepsilon)$, is approximately $1 - \frac{\pi^2}{4N}$, and can be made arbitrarily close to 1.

> **Key Takeaways**
>
> The Quantum Bomb Test Algorithm elucidates several key concepts:
>
> 1. Similar to certain classical probabilistic algorithms, quantum algorithms can be designed to perform specific tasks with arbitrary precision, involving tradeoffs in resource usage or running time, albeit never achieving absolute certainty.
>
> 2. While quantum algorithms are typically conceptualized as a sequence of unitary gates culminating in measurements—a concept that might be referred to as the unitary computing model—they may also benefit from incorporating intermediate non-unitary *branching* devices.
>
> 3. Interaction-free quantum measurements, allowing for the detection or inference of an object's presence without directly interacting with it, can be conducted with arbitrary precision, showcasing a unique aspect of quantum measurements that has no direct analogue in classical physics.

5 Further Information

For further developments and in-depth analysis of this subject, the reader is referred to the work on quantum measurement detection algorithms by Lugilde Fernandez et al. [68].

11.3.2 ∗ Basics of Quantum Money $|\$\rangle$

Quantum money employs quantum mechanical principles to create coins (or banknotes) that are inherently secure against counterfeiting. This notion was pioneered by Wiesner in 1983 and has since seen extensions and variations by subsequent researchers.

Unlike traditional digital currencies, which derive their security from computationally hard problems, quantum money exploits unique features of quantum mechanics. In particular, the no-cloning theorem establishes that an arbitrary unknown quantum state cannot be perfectly cloned. This quantum feature offers a security paradigm that transcends what is feasible with classical approaches. Nonetheless, the development of publicly verifiable quantum money remains an elusive, open challenge.

Two primary types of quantum money exist: private-key and public-key. Each has its own set of advantages, disadvantages, and appropriate use-cases, with both types leveraging quantum mechanics for enhanced security.

While digital currencies like Bitcoin are well-established in the financial world, quantum money is currently an active area of research, particularly significant in the field of quantum cryptography, exploring new frontiers in secure transactions.

1 Key Concepts

Money State

The money state, denoted as $|\$\rangle$, is a multi-qubit state embodying a quantum coin. It is designed to resist cloning, leveraging the no-cloning theorem. It encodes quantum information critical for the authentication and verification of the value of quantum money.

$$s = \ldots\ldots$$
$$|\$_s\rangle = \bigotimes_{i=1}^{n} |\psi_i\rangle$$

A Quantum Coin $(s, |\$\rangle)$

The number of qubits in $|\$\rangle$, denoted n, is chosen as a tradeoff between security and efficiency, e.g., 256.

Serial Number

The serial number, represented as s, is a classical identifier uniquely associated with each money state. While the serial number is publicly readable, the corresponding quantum state, which constitutes the money, is usually secret and challenging to replicate. In various quantum money protocols, the serial number is employed to query information about the money state and is not directly used in the verification and authentication processes.

Private Key

The private key is confidential information held by the issuing bank (or mint), essential for authenticating or verifying a quantum money state. In private-key schemes, the private key allows for the validation of the money state's authenticity when correlated with its serial number. In public-key schemes, the private key is involved in generating the serial number and money state. In all schemes, the bank rigorously protects the private key to prevent unauthorized access and potential counterfeiting.

Public Key

The public key is disseminated openly, enabling the public to authenticate a quantum money state. Public-key schemes enable decentralized verification without requiring the bank's involvement in each transaction. Despite its public accessibility, the public key is designed not to reveal enough information to counterfeit the money state.

2 Private-Key Quantum Money

Private-key quantum money schemes necessitate the involvement of the issuing bank for authentication, ensuring centralized control. Here, the no-cloning theorem guarantees the integrity and uniqueness of the quantum money state. The basic protocol is outlined below:

1. Money State Generation: The bank generates a serial number s and generates a corresponding private key K_s that remains confidential. Based on the private key, the bank prepares a quantum money state, $|\$_s\rangle$.

2. Issuance to Customer: The bank transfers $|\$_s\rangle$ and its serial number s to a customer.

3. Centralized Verification: The bank alone can validate the quantum money state using the private key K_s, which contains the required measurement angles for measurement.

4. Transaction Confirmation: Upon successful verification, the transaction is approved.

Private-key quantum money offers a more straightforward implementation but introduces scalability issues, as the bank needs to be part of every transaction, similar to a credit card payment. The scheme also shares challenges common to public-key systems, such as the need for quantum memories and error correction.

3 The Wiesner Quantum Money

The Wiesner quantum money, introduced in 1983[90], is a private-key scheme that utilizes a simple yet effective approach for secure quantum currency. In this scheme, each quantum coin, represented by $(s, |\$_s\rangle)$, is associated with a corresponding private key $K_s \in \{0, 1, +, -\}^n$. The subset $\{0, 1\}$ corresponds to the computational basis states $\{|0\rangle, |1\rangle\}$ in $|\$_s\rangle$, and $\{+, -\}$ corresponds to the Hadamard basis states, $\{|+\rangle, |-\rangle\}$. For example, $K_s = (0, +, 1, 0, -, +, \ldots)$ corresponds to $|\$_s\rangle = |0 + 10 - + \ldots\rangle$.

To verify a coin, the bank measures each qubit of $|\$_s\rangle$ in the appropriate basis using K_s, checking for the correct outcome. This process ensures the integrity of the authentic coin's state $|\$_s\rangle$. Conversely, measurements against an incorrect K_s on a counterfeit coin will alter $|\$_s\rangle$ due to state collapse, rendering replication or verification without K_s impossible.

11.3 Quantum Bomb and Quantum Money

In the original scheme, the issuing bank maintained a database of all (s, K_s) pairs. Modern adaptations often derive K_s from s using a secure hash function combined with a secret key (or salt) held by the bank, thereby eliminating the need for such a database. However, this approach introduces potential security concerns associated with the classical hash function, as it is not unconditionally secure.

The Wiesner scheme bears resemblance to the BB84 QKD protocol discussed in § 5.5. In fact, it served as an inspiration for the development of BB84 in 1984.

It is fascinating to examine how quantum money can be forged (which we refer to as attacks) and the possible countermeasures. Through examining these, we can gain significant insight into the issues and protocols of quantum security.

Measure and Replicate Attack

Suppose a counterfeiter measures each qubit of a quantum coin in the computational basis. Since the money state consists of $\{|0\rangle, |1\rangle\}$ with a 50% probability, and $\{|+\rangle, |-\rangle\}$ with a 50% probability, the results would be correct for half of the qubits.

The counterfeiter can then duplicate as many coins as he likes. But what is the chance for his forged coin to pass authentication at the bank? For each qubit, if the correct basis is indeed $\{|0\rangle, |1\rangle\}$, he gets a PASS, which has a probability of $\frac{1}{2}$. If the correct basis happens to be $\{|+\rangle, |-\rangle\}$, measuring a computational basis state in the Hadamard basis has a $\frac{1}{2}$ chance to obtain the authentic result. So the overall passing probability for each qubit is $\frac{1}{2} + \frac{1}{2} \cdot \frac{1}{2} = \frac{3}{4}$. For $n = 256$ qubits, the overall passing probability is $\left(\frac{3}{4}\right)^n$ 10^{-32}. Therefore, the counterfeiter has to replicate approximately 10^{32} coins for this simple attack to be profitable, which is deemed impractical.

A possible countermeasure for the bank is limiting the rate of authentication trials per serial number, say to 1 per second. Note the bank should not simply blacklist a serial number after seeing a bad authentication, because doing so would allow a bad actor to cancel another person's quantum money.

Qubit-by-Qubit Learning Attack

Suppose the bank returns a quantum coin to the customer even if the authentication FAILs. In this scenario, a counterfeiter can devise an attack by learning the qubits one by one [69]. The attacker modifies the state of a single qubit (the i-th) in a legitimate coin to, say, $|0\rangle$, and sends this tampered coin to the bank for authentication. If the correct state should be $|1\rangle$, the coin will be rejected with a probability of 100%. If the correct state is $|+\rangle$ or $|-\rangle$, there is a $\frac{1}{2}$ chance of the coin being rejected. Therefore, the probability for a coin with an incorrect qubit state to be rejected is $\frac{1}{3} + \frac{2}{3} \cdot \frac{1}{2} = \frac{2}{3}$, and $\frac{1}{3}$ for it to not be rejected.

Since the authentication process does not alter the other qubits, which are in the eigenstates of the measurement bases, the counterfeiter can repeat this process. After m trials, the probability for a qubit with the wrong state to be rejected by the bank is at least $1 - \left(\frac{1}{3}\right)^m$. For example, with $m = 10$ trials, the attacker can determine with high probability if the new state of the coin is incorrect. By systematically testing all four states, $|0\rangle$, $|1\rangle$, $|+\rangle$, and $|-\rangle$, the attacker can ascertain the state of the i-th qubit. This process can be repeated to learn the states of all the qubits, requiring a total of mn trials. As the number of trials is linear with n, this strategy represents a highly practical attack.

From this analysis, it is evident that to prevent such attacks, the bank should not return a failed authentication coin to the customer.

The Adaptive Attack

Suppose the bank destroys any coin (or blacklists its serial number) if authentication FAILs, but returns the coin to its customer if authentication PASSes. A clever attack exploiting this situation is derived from the quantum bomb test algorithm (refer to § 11.3.1)[31]. Here, an authentication-FAIL (resulting in the coin's destruction) is analogous to the bomb in the quantum bomb test. The coin undergoes N measurements for each test, but the total probability of triggering an authentication failure and destroying the coin is capped at $\frac{\pi^2}{4N}$, which can be made arbitrarily small for a sufficiently large N.

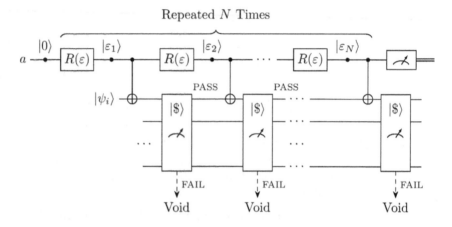

Figure 11.14: Adaptive Attack on Wiesner Quantum Money

The attack scheme, illustrated in Fig. 11.14, assumes the attacker can attach an ancilla qubit a to the money state $|\$\rangle$ and apply a quantum gate (e.g., CNOT) between it and any qubit $|\psi_i\rangle$ in $|\$\rangle$. Before each authentication, the ancilla qubit is rotated by a small angle $\varepsilon = \frac{\pi}{2N}$ and then CNOT'ed with $|\psi_i\rangle$. The state of the ancilla qubit just before the first CNOT gate is:

$$|\varepsilon_1\rangle = \cos\varepsilon\,|0\rangle + \sin\varepsilon\,|1\rangle. \tag{11.41}$$

After the CNOT gate, the combined state of the ancilla and the i-th money qubit is:

$$|\psi_{\text{combined}}\rangle = \text{CNOT}(|\varepsilon_1\rangle \otimes |\psi_i\rangle) \tag{11.42a}$$

$$= \begin{cases} \cos\varepsilon\,|00\rangle + \sin\varepsilon\,|11\rangle & \text{for } |\psi_i\rangle = |0\rangle, \\ \cos\varepsilon\,|01\rangle + \sin\varepsilon\,|10\rangle & \text{for } |\psi_i\rangle = |1\rangle, \\ (\cos\varepsilon\,|0\rangle + \sin\varepsilon\,|1\rangle) \otimes |+\rangle & \text{for } |\psi_i\rangle = |+\rangle, \\ (\cos\varepsilon\,|0\rangle - \sin\varepsilon\,|1\rangle) \otimes |-\rangle & \text{for } |\psi_i\rangle = |-\rangle. \end{cases} \tag{11.42b}$$

Exercise 11.15 Verify Eq. 11.42.

11.3 Quantum Bomb and Quantum Money

The probability of the bank's subsequent authentication measurement yielding a FAIL is:

$$p(\text{FAIL}) = \begin{cases} \sin^2\varepsilon \approx (\frac{\pi}{2N})^2 & \text{for } |\psi_i\rangle = |0\rangle, \\ \sin^2\varepsilon \approx (\frac{\pi}{2N})^2 & \text{for } |\psi_i\rangle = |1\rangle, \\ 0 & \text{for } |\psi_i\rangle = |+\rangle, \\ 0 & \text{for } |\psi_i\rangle = |-\rangle. \end{cases} \quad (11.43)$$

It becomes apparent that $|\psi_i\rangle = |+\rangle$ or $|-\rangle$ corresponds to a dud, while $|\psi_i\rangle = |0\rangle$ or $|1\rangle$ corresponds to a bomb. If the authentication succeeds, which is highly probable, the money state reverts to $|\psi_i\rangle$. For $|\psi_i\rangle = |0\rangle$ or $|1\rangle$, the ancilla and coin qubit were entangled, causing the ancilla to collapse back to $|0\rangle$. For $|\psi_i\rangle = |+\rangle$, the ancilla remains unchanged, while for $|\psi_i\rangle = |-\rangle$, it changes to $\cos\varepsilon|0\rangle - \sin\varepsilon|1\rangle$.

If the coin survives m measurements, the ancilla state becomes:

$$|\varepsilon_m\rangle = \begin{cases} |0\rangle & \text{for } |\psi_i\rangle = |0\rangle, \\ |0\rangle & \text{for } |\psi_i\rangle = |1\rangle, \\ \cos m\varepsilon|0\rangle + \sin m\varepsilon|1\rangle & \text{for } |\psi_i\rangle = |+\rangle, \\ \cos\varepsilon|0\rangle - \sin\varepsilon|1\rangle & \text{for } |\psi_i\rangle = |-\rangle \text{ and odd } m, \\ |0\rangle & \text{for } |\psi_i\rangle = |-\rangle \text{ and even } m. \end{cases} \quad (11.44)$$

After N measurements, assuming N is even and $N\varepsilon = \frac{\pi}{2}$, the ancilla state resolves to:

$$|\varepsilon_N\rangle = \begin{cases} |0\rangle & \text{for } |\psi_i\rangle = |0\rangle, |1\rangle, \text{ or } |-\rangle, \\ |1\rangle & \text{for } |\psi_i\rangle = |+\rangle. \end{cases} \quad (11.45)$$

At this juncture, the attacker measures the ancilla qubit. If the result is $|1\rangle$, it indicates that the i-th money qubit is definitively $|+\rangle$. However, the attacker cannot discern whether the correct basis is $\{|0\rangle, |1\rangle\}$ or $\{|+\rangle, |-\rangle\}$. To determine this, the attacker replaces the CNOT gate with a controlled-negative-NOT gate (refer to § 7.1.3), thus discerning if the qubit is in the $|-\rangle$ state. With a total of $2nN$ trials, the attacker has a high probability, approximately $1 - \frac{n\pi^2}{2N}$, of learning the correct measurement bases without detection. This strategy enables the measurement of each qubit in its correct basis to deduce the entire money state.

This attack demonstrates that Wiesner's money scheme can only be secure if the bank replaces validated coins.

 If an attacker can replicate a quantum coin with a high probability, does this violate the no-cloning theorem? No, because the theorem pertains to the deterministic replication of an arbitrary quantum state.

Exercise 11.16 Work out the attack procedure for the adaptive attack on Wiesner's coin using controlled-Z and controlled-negative-Z gates.

4 Challenge-and-Response Authentication

Private-key quantum money protocols, akin to modern credit card systems, require centralized authentication, imposing significant demands on quantum networking and communication infrastructure. To mitigate these challenges, several alternative protocols have been proposed, including challenge-and-response and public-key systems.

Consider Wiesner's quantum money for illustrating the challenge-and-response authentication process. It initiates with the coin owner transmitting the serial number s to the bank via a *classical*, potentially insecure, channel. The bank then generates a *random* challenge string $C_s \in \{0, 1, +, -\}^n$. This string, formatted like the private key K_s, shares a 50% overlap with it. The owner measures the money state $|\$\rangle$ with respect to C_s and relays the outcomes back to the bank. The bank then authenticates the coin by comparing these measurements with the recorded states of the overlapping qubits.

$$s = \ldots\ldots$$

$$|\$_{s,1}\rangle = \bigotimes_{i=1}^{n} |\psi_i\rangle |1\rangle$$

$$\ldots$$

$$|\$_{s,m}\rangle = \bigotimes_{i=1}^{n} |\psi_i\rangle |m\rangle$$

A Quantum Coin with Multiple $|\$\rangle$

This approach is reminiscent of the key-sifting protocol in the BB84 quantum key distribution (QKD) method (see § 5.5). A significant limitation for quantum money using this approach, however, is the degradation of qubits with each authentication attempt. Specifically, half of the qubits in the coin are compromised during the process. To counter this, quantum coins could be designed to associate multiple identical money states with a single serial number, while the bank maintains records of the invalidated states.

Informationally Secure vs. Computationally Secure

This challenge-and-response authentication protocol, as well as the BB84 and E91 QKD protocols, are informationally secure.

Informationally secure systems guarantee security based on the principles of information theory, ensuring that a cryptographic system remains secure regardless of the computational power available to an adversary. This means that even with unlimited computing resources, an attacker cannot gain any useful information. The one-time pad is a classic example of an informationally secure system.

In contrast, computationally secure systems rely on the assumption of limited computational resources. Their security is based on the difficulty of solving certain mathematical problems, like factoring large numbers, within a practical amount of time or resources. Most modern cryptographic systems, such as RSA and ECC, are computationally secure. Their security could potentially be compromised by future advancements in computing power.

---

5 Public-Key Quantum Money

Basic Protocol

The essence of public-key quantum money lies in creating a money state that can be efficiently produced by the bank (mint) and verified by the merchant, yet is

11.3 Quantum Bomb and Quantum Money

infeasible to duplicate under both quantum and classical computational security frameworks.

This system combines quantum computing with classical cryptography to enable a secure and decentralized protocol. The fundamental steps are:

1. Money State Generation: The bank generates a quantum coin (banknote) with a distinct serial number s, a quantum money state $|\$_s\rangle$, and a corresponding verification protocol P_s. These components must satisfy:

    (a) The quantum state $|\$_s\rangle$ consistently passes the verification test defined by P_s.

    (b) The verification protocol P_s is non-destructive, preserving the state $|\$_s\rangle$ post-verification.

    (c) It is computationally infeasible for a potential counterfeiter, even with access to both $|\$_s\rangle$ and P_s, to replicate the quantum state such that both originals and duplicates pass P_s.

    The bank publishes a registry of valid serial numbers $\{s\}$ along with their associated verification protocols $\{P_s\}$.

2. Issuance to Customer: The bank issues the quantum coin $|\$_s\rangle$, along with its serial number s, to a customer.

3. Public Verification: The authenticity of $|\$_s\rangle$ can be verified by anyone through two criteria: (1) confirming that the serial number s is publicly listed and (2) validating that $|\$_s\rangle$ satisfies the publicly accessible protocol P_s.

4. Transaction Confirmation: The quantum money is accepted by the recipient, and the transaction is deemed complete, if both criteria are successfully met.

While conceptually appealing, the practical implementation of public-key quantum money faces hurdles, such as the need for durable quantum memories and efficient, error-tolerant quantum operations. Furthermore, developing a practical realization of the tuple $(s, |\$_s\rangle, P_s)$ that fulfills all the stated requirements, even under computational security, remains a formidable challenge.

A Basic Implementation Scheme

Implementing public-key quantum money is an intricate task, involving quantum computing, advanced mathematics, and cryptography. We present a simplified, abstract protocol to highlight key concepts and challenges. Each step here is non-trivial:

1. Set Partition: Begin with a vast set of quantum states, such as:

$$B = \{|i\rangle\}, \quad \text{with } i \in [1, 2^n], \quad \text{where } n = 1024. \tag{11.46}$$

Choose a large subset T of B, but significantly smaller than B, for instance,

$$T \subset B, \quad |T| = 2^{\frac{n}{2}}. \tag{11.47}$$

Define a many-to-one function f:

$$f : B \mapsto T, \tag{11.48}$$

such that $f^{-1}(t)$ has a substantial number of preimages for any $t \in T$, say,

$$N \equiv \frac{|B|}{|T|} = 2^{\frac{n}{2}}. \qquad (11.49)$$

2. Oneway Function: Ensure that computing $f(b)$ is feasible in polynomial time, whereas finding $f^{-1}(t)$ is infeasible, even with quantum resources, for large n.

3. Money Generation: Under the aforementioned conditions, the bank generates money states and serial numbers as follows:

   a. Create a uniform superposition of all states in B:

   $$|\Psi_0\rangle = \frac{1}{\sqrt{|B|}} \sum_{b \in B} |b\rangle |0\rangle, \qquad (11.50)$$

   where $|0\rangle$ is an $\frac{n}{2}$-qubit register initialized to zero.

   b. Compute f into the second register:

   $$|\Psi_1\rangle = \frac{1}{\sqrt{|B|}} \sum_{b \in B} |b\rangle |f(b)\rangle. \qquad (11.51)$$

   c. Measure the second register to obtain a value s, collapsing the state to (see § 6.4.3 for a detailed explanation):

   $$|\Psi_2\rangle = \frac{1}{\sqrt{N}} \sum_{\substack{b \in B \\ f(b) = s}} |b\rangle |s\rangle. \qquad (11.52)$$

   The random measurement outcome s from $\{f(b)\}$ can be used as the serial number, with the first register of $|\Psi_2\rangle$ as the money state $|\$_s\rangle$:

   $$|\$_s\rangle = \frac{1}{\sqrt{N}} \sum_{\substack{b \in B \\ f(b) = s}} |b\rangle. \qquad (11.53)$$

   Since $f^{-1}(t)$ is assumed to be computationally hard, $|\$_s\rangle$ is infeasible to compute from s.

4. Public Verification: An authentic $|\$_s\rangle$ is a uniform superposition of all $|b\rangle$'s for which $f(b) = s$. Verification involves two key aspects:

   a. Alignment with the Serial Number: The money state $|\$_s\rangle$ should be such that when f is applied to each $|b\rangle$ in the superposition, the output consistently equals s. This can be implemented as an eigenvalue relationship $\hat{f}(|\$_{s,1}\rangle) = s |\$_{s,1}\rangle$, where $\hat{f}$ is the quantum version of f.

   b. Fidelity of the Superposition: In the scheme presented here, $|\$_s\rangle$ must be a full, uniform superposition of all $|b\rangle$'s such that $f(b) = s$, not just a single state or a different superposition.

11.3 Quantum Bomb and Quantum Money

The bank consolidates this verification method into a verification operator P_s, and publishes its formula along with the serial number s. The verification protocol must be implemented such that replicating the quantum state $|\$_s\rangle$ is computationally infeasible, even with access to both s and P_s.

For instance, knot-invariance based quantum money, as in [41], utilizes Alexander polynomials for the function f. It employs a superposition of all possible Reidemeister moves — elementary transformations preserving knot invariance — to ensure that $|\$_s\rangle$ is a full, uniform superposition.

6 ✷ Current Development

Quantum Money Protocols

A variety of quantum money protocols have been proposed, with many experiencing subsequent cryptographic breaks. Key contributions in this area (as of 2023) include:

- Farhi, Gosset, Hassidim, Lutomirski, Shor, 2009: based on knot invariants [41].
- Aaronson and Christiano, 2012: predicated on subspaces [12] (proven broken).
- Mark Zhandry, 2017: rooted in complexity-theoretic foundations [96] (proven broken).
- Kane, Sharif, Silverberg, 2018, 2021: utilizing quaternion algebra [57].
- Shor, Khesin, Lu, 2021, 2022: founded on lattice cryptography [59] (proven broken).

For a more comprehensive understanding, consult the following works:

- Another round of breaking and making quantum money [67].
- Cryptanalysis of three quantum money schemes [24].

Quantum Lightning

Quantum lightning [96] is a concept related to quantum money, representing an even stronger version of it. In a quantum money scheme, the bank (or mint) issues quantum banknotes that are hard to counterfeit due to their quantum properties. The mint is the only entity that can produce these valid banknotes, and anyone can verify their authenticity.

In contrast, quantum lightning takes this concept further by ensuring that not even the mint can duplicate the banknotes once they are issued. Each quantum banknote in a quantum lightning scheme is unique and cannot be reproduced, not even by the issuer. This feature offers a higher level of security and uniqueness for each banknote.

Quantum lightning is still a theoretical concept and part of ongoing research in quantum cryptography. It presents significant challenges in terms of practical implementation and security proof.

Post-Quantum Cryptography

Cryptosystems fundamentally rely on mathematically challenging problems. Such problems vary widely and lead to a diverse set of cryptosystems, including RSA, discrete logarithm, lattices, multivariate equations, knapsack problems, decoding challenges, among others.

Although most of these schemes are characterized by lengthy public keys, RSA and discrete logarithm based systems have gained popularity primarily due to their shorter key lengths. This advantage is often attributable to the structured mathematics underlying these schemes. Ironically, this very structure renders them vulnerable to quantum attacks, particularly through the application of Shor's algorithm.

In light of this, lattice-based cryptography has emerged as a frontrunner in post-quantum cryptographic research. It offers a balanced combination of security, efficiency, and speed.

For additional resources on this subject, refer to:

- Post-quantum cryptosystems for Internet-of-Things: A survey on lattice-based algorithms [14].
- The science and information organization [42].
- Evaluation and comparison of lattice-based cryptosystems for a secure quantum computing era [77].

Significance of the Field

The advent of quantum computing has cast a dual shadow on the field of cryptography. On one hand, Shor's algorithm poses a significant challenge by threatening the integrity of classical encryption schemes through efficient integer factorization. On the other hand, quantum computing furnishes a suite of novel techniques that bolster encryption and security protocols. This positions quantum computing not merely as a disruptive force but also as an enabling technology. A prime example of such an innovation is quantum money.

Quantum money transcends its role as a singular cryptographic construct and serves as an elemental foundation for various unclonable cryptographic architectures. It underpins more elaborate protocols like quantum copy protection and chainless blockchain systems. Consequently, mastering the subtleties of quantum money serves as a seminal vehicle for a comprehensive understanding of quantum cryptography at large.

The study of quantum money bears significance that extends well beyond the conventional bounds of encryption and security. It functions as a versatile instrument for securing a wide array of digital assets, rendering them virtually induplicable. This places quantum money at the confluence of quantum computing, cryptography, and information theory, and opens up new horizons for secure, efficient, and verifiable transactions in the quantum epoch.

Such multifaceted relevance elevates the subject of quantum money from a point of theoretical interest to a pivotal research area that has the potential to reshape the landscape of secure transactions and information exchange in the future.

11.4 Summary and Conclusions

Foundational Algorithms

The central focus of this chapter lies in offering a panorama of quantum algorithms. We began by delving into the foundational quantum algorithms, exemplified by the Deutsch-Jozsa Algorithm. While its practical applications may be limited, the algorithm serves as a seminal touchstone, elucidating the potential computational advantages of quantum systems over classical ones. We scrutinized its fundamental principles and proposed a generalized version of the algorithm.

NISQ/Utility Hybrid Algorithms

Transitioning from foundational algorithms, we examined the state of the art in quantum computing as of early 2024, spanning from NISQ devices to the nascent era of quantum utility. We focused on Quantum-Classical Hybrid Algorithms, highlighting their role in navigating the limitations of NISQ technologies and the emerging utility-era systems.

Optimization in the NISQ/Utility Era

The chapter then transitioned to explore algorithms particularly relevant to the NISQ era, focusing on solving the Quadratic Unconstrained Binary Optimization (QUBO) problem. We introduced the Variational Quantum Eigensolver (VQE) and the Quantum Approximate Optimization Algorithm (QAOA), both tailored to operate within the constraints of current-day quantum devices. By employing these algorithms, we attempted to solve problems of practical significance, like the Max-Cut problem, thereby demonstrating their utility in real-world applications.

Innovative Quantum Algorithmic Concepts

The chapter concluded with an exploration of innovative quantum algorithms, such as the Quantum Bomb Measurement and Quantum Money. These algorithms not only push computational boundaries but also introduce new aspects like intricate quantum measurements and post-quantum cryptographic techniques.

Pedagogical Aims

Throughout this chapter, our goal has been to provide a comprehensive, albeit not exhaustive, overview of the quantum algorithm landscape. By selecting representative algorithms from different categories, we aimed to illustrate the past milestones, current challenges, and future potentials in the realm of quantum computing.

Upcoming Topics

In the ensuing chapter, we will delve into the critical topic of Error Correction, focusing initially on the mathematical tool of density operators. This chapter will provide a detailed exploration of coherent vs incoherent errors and discuss the famous Shor Codes as a key technique in quantum error correction. The objective is to equip the reader with a foundational understanding of error types and mechanisms to manage them, a subject indispensable for advancing in both theoretical and practical realms of quantum computing.

Problem Set 11

11.1 Referring to the DJ algorithm detailed in § 11.1, for a given n, determine the number of constant functions and the number of balanced functions.

11.2 Assume $f(x) = \text{parity}(x)$, or equivalently, $f(x) = \text{XOR}(x)$. Show that $f(x)$ is a balanced function.

11.3 Validate that the circuit in Fig. 11.5 accurately represents a phase oracle for $f(x) = \overline{x_1} \oplus x_2 \oplus x_3 \oplus \overline{x_4}$.

11.4 The transformations detailed in Eq. 11.7 delineate the foundational mechanics of the DJ algorithm. These mechanics also underpin other quantum algorithms, including those of Simon and Shor.

Examine and elucidate the rationale underlying each transformation step detailed in Eq. 11.7. Your objective is to competently deduce and articulate the derivation of each step, utilizing the circuit representation provided in Fig. 11.1 as your foundational reference.

11.5 The transform below is a quantum operation used in the Deutsch-Jozsa (DJ) and Bernstein-Vazirani (BV) algorithms. It is defined as:

$$U_s |x\rangle |y\rangle = |x\rangle |y \oplus f(x)\rangle,$$

where $f(x) = x \cdot s$ is a bitwise product of x and a secret string s, and $\oplus$ denotes bitwise XOR (or addition modulo 2).

Consider a system with n qubits for x, and an additional qubit for y. The total system size is $n + 1$ qubits.

(a) Write out the matrix form of U_s for a 4-qubit system ($n = 3$).

(b) Verify that U_s is a unitary transform for any n.

(c) Compute the state $U_s |0^{\otimes n}\rangle |0\rangle$ for a given string s.

(d) Compute the state $U_s |0^{\otimes n}\rangle |1\rangle$ for a given string s.

(e) Compute the state $U_s |1^{\otimes n}\rangle |0\rangle$ for a given string s.

(f) Find a general formula for $U_s |x_1 x_2 \cdots x_n\rangle |y\rangle$ where $x_i, y \in \{0, 1\}$.

11.6 The **Number Partitioning Problem** (NPP) poses the following question: Given a set of N positive numbers $S = \{n_1, \ldots, n_N\}$, can this set be partitioned into two disjoint subsets, such that the sums of the numbers in each subset are equal? For instance, is it possible to evenly divide a collection of assets with values $n_1, \ldots, n_N$ between two individuals?

This problem can be modeled using an Ising Hamiltonian expressed as:

$$H = \left(\sum_{i=1}^{N} n_i Z_i\right)^2,$$

where Z_i represents the spin variable associated with the i-th number, taking values of either $+1$ or -1.

A solution to the NPP corresponds to a configuration of the Ising model where $H = 0$. Such a configuration ensures that the sum of the numbers n_i

corresponding to the $+1$ spins is equal to the sum of the numbers for the -1 spins. Consequently, if the ground state energy of the system is $H = 0$, the set S can be partitioned as described in the problem statement.

Your task is to create an array of N random integers and determine whether a fair partition is possible. You are to find solutions using QAOA or VQE, implemented on platforms such as Qiskit or Cirq, or via a quantum annealer. Choose a value of N that is practical for the quantum computing resources available to you, ensuring that $N > 10$ to render the problem significant.

11.7 The **Graph Coloring Problem** (GCP) is a well-known challenge in both computer science and discrete mathematics. It entails the assignment of colors to the vertices of a graph in such a way that adjacent vertices are not colored the same, with the aim often being to minimize the total number of colors used. This problem finds practical applications in a variety of fields, including scheduling, resource allocation, frequency assignment in mobile networks, and register allocation in compiler design.

Formulating as a CUBO Problem

The GCP can be effectively formulated as a Constraint Satisfaction Problem (CSP), which is then convertible into a Combinatorial Unconstrained Binary Optimization (CUBO) problem, as detailed in § 11.2. This conversion involves several steps:

(a) Define Binary Variables: For a graph comprising V vertices, and with the intention to use K colors, a binary variable $x_{i,k}$ is defined for each combination of vertex i and color k. Here, $x_{i,k} = 1$ indicates that vertex i is assigned color k, and $x_{i,k} = 0$ otherwise.

(b) Objective Function: The CUBO formulation for the GCP prioritizes optimization in the absence of direct constraints on the binary variables. The main goal here is to minimize a particular metric or to satisfy certain conditions that are implicitly encoded within these variables. In the case of the GCP, it is crucial to encode constraints ensuring that no two adjacent vertices share the same color.

(c) Incorporate Constraints: Constraints are integrated into the objective function in the form of penalties. The primary constraints are as follows:

- Ensure each vertex is assigned just one color: $\sum_{k=1}^{K} x_{i,k} = 1$ for all i.

- Prevent adjacent vertices from sharing the same color: $x_{i,k} + x_{j,k} \leq 1$ for every edge (i, j) in the graph, applicable across all colors k.

Introducing penalty terms for these constraint violations allows the GCP to be modeled as a CUBO problem, with the objective of minimizing the total penalty to ideally achieve a value of zero when a valid coloring is found. This strategy is akin to the approach employed in solving the Traveling Salesman Problem, as elucidated in § 11.2.5.

Mapping to the Ising Model

The Ising model, as introduced in § 11.2.2, describes ferromagnetism through a mathematical framework but extends its utility to quantum computing. It characterizes systems using discrete spin variables, which can assume states of either $+1$ or -1, with the system's energy defined by the interactions among these spin pairs and any external magnetic fields.

Mapping the CUBO formulation of the GCP onto the Ising model involves the following procedures:

(a) **Variable Transformation:** The binary variables $x_{i,k}$ are transformed into spin variables $Z_{i,k}$, following the relation $x_{i,k} = (Z_{i,k} + 1)/2$. This step converts the binary $\{0,1\}$ representation into spin states $\{-1,+1\}$.

(b) **Energy Function:** Constraints and the objective function from the CUBO model are translated into an energy function within the Ising model. Consequently, each component of the CUBO objective that relies on binary variables is mapped to a corresponding term in the Ising energy function, dependent on spin variables. The penalty terms for constraint violations manifest as interaction terms among spins.

(c) **Minimization Objective:** The ultimate aim is to identify the spin configuration (the Ising model solution) that minimizes the energy function, which correlates with the optimal or a feasible solution for the original GCP.

QAOA, VQE, and quantum annealers, as discussed in §§ 11.2.4 and 11.2.6, are proficient in minimizing such Ising models, thus facilitating the resolution of GCPs on quantum computing platforms. In the simple encoding scheme explained above, referred to as the 'one-hot encoding', for a graph with n vertices and the goal of employing k distinct colors, a total of kn qubits are needed.

Quantum algorithms such as QAOA VQE, and quantum annealing discussed in §§ 11.2.4 and 11.2.6, are effective in minimizing Ising models that represent combinatorial optimization problems like the GCP. The one-hot encoding scheme, as described above, allows quantum computing platforms to address GCP by allocating a unique set of qubits to represent each possible color for every vertex. This approach requires a total of kn qubits for a graph with n vertices to be colored using k distinct colors. More efficient schemes, such binary encoding for the colors, also exist.

Your Tasks

You are tasked with generating a random graph containing n vertices and exploring its colorability with 3 distinct colors. You should formulate this problem according to the guidelines provided and seek solutions employing QAOA or VQE, utilizing platforms such as Qiskit or Cirq, or through a quantum annealer. Select an n that is feasible given the constraints of the system at your disposal, yet ensure $n > 6$ for the exercise to be meaningful.

11.8 Elitzur-Vaidman Bomb-Test Algorithm The following topics build upon the core idea of the Elitzur-Vaidman bomb-tester (detailed in § 11.3.1 and § 11.3.2.3)—gathering information without direct interaction—to push the boundaries of what's possible within quantum mechanics. Researchers continue to explore these concepts not only theoretically but also experimentally, which could lead to new technologies and protocols in quantum information science.

Investigate two of the following topics, and prepare a presentation on each, discussing its concept, principles, applications, and current status.

(a) Quantum Zeno Effect: The Quantum Zeno Effect can be seen as an extension where frequent observations prevent the evolution of a quantum state. It has been proposed that by frequently checking for the presence of the bomb (the system's state), one can effectively freeze its evolution, leading to a form of interaction-free measurement.

(b) Counterfactual Quantum Computation: This application allows for computation to occur without running the computer in the traditional sense. It uses the principles of quantum superposition and interference to infer the result of a computation without actually performing all the steps physically.

(c) Counterfactual Quantum Communication: This method of transmitting information without particles traveling between the sender and receiver exploits quantum entanglement and the principles behind the Elitzur-Vaidman bomb-tester to achieve communication that is effectively "interaction-free."

(d) Chained Quantum Zeno Effect: A series of interaction-free measurements can be chained together to create a more complex system that can perform tasks such as imaging or communication with a lower chance of interaction than the original setup.

(e) Counterfactual Cryptography: Inspired by the bomb-tester, this extension involves secure communication where the presence of an eavesdropper can be detected without any qubits carrying the information directly from the sender to the receiver.

12. Quantum Error Correction: A Primer

Contents

12.1	**Preliminary Concepts**	**342**
12.2	**✷ Mixed States, Density Operators, and CPTP Maps**	**346**
12.2.1	Definitions	346
12.2.2	Interpretation and Properties	348
12.2.3	Probabilities and Expected Values	350
12.2.4	✷ General Single-Qubit States in the Bloch Sphere	352
12.2.5	✷ Unitary Evolution	354
12.2.6	✷ Measurements	356
12.2.7	✷ CPTP Maps and Quantum Channels	358
12.2.8	✷ Partial Trace and Reduced Density Operators	362
12.2.9	✷ Extension of No-Cloning and No-Communication Theorems	367
12.3	**Error Mechanisms in Quantum Computing**	**369**
12.3.1	✷ Coherent Errors	369
12.3.2	✷ Incoherent Errors	371
12.3.3	✷ Other Types of Errors	377
12.3.4	Performance Merit Parameters	377
12.4	**Introduction to Error Correction Codes**	**384**
12.4.1	Three-Qubit Bit-Flip Code	385
12.4.2	Three-Qubit Phase-Flip Code	388
12.4.3	✷ Nine-Qubit Shor Code	389
12.4.4	✷ Parity Measurements, Stabilizers, and Code Distance	394
12.4.5	✷ Further Exploration	396
12.5	**Summary and Conclusions**	**399**
	Problem Set 12	400

In this chapter, we undertake a comprehensive examination of Quantum Error Correction (QEC) in the realm of quantum computing. We commence by discussing the mathematical underpinnings of density operators, before transitioning to an exploration of both coherent and incoherent errors. The chapter culminates with an analysis of Shor Codes, a pioneering approach in quantum error correction. The primary objective is to equip the reader with a thorough understanding of the errors that afflict quantum systems and the methodologies available to amend them. This knowledge is essential for advancing both the theoretical and practical facets of quantum computing.

12.1 Preliminary Concepts

Error correction in quantum computing is a multifaceted topic that necessitates a foundational understanding of certain core concepts.

1 The Noise Problem: An Illustrative Example

To elucidate the impact of noise, consider a quantum computer programmed to generate the Bell state $|\Phi^+\rangle = \frac{1}{\sqrt{2}}(|00\rangle + |11\rangle)$:

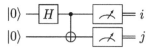

In an ideal setting devoid of noise, measurements would invariably yield $ij = 00$ and 11 with equal probabilities. Contrarily, a real-world, noise-impacted system might also produce outcomes such as 01 and 10, albeit with lower probabilities like 1%. Such deviations manifest due to noise, which disrupts the quantum state and results in aberrant measurement outcomes.

2 Decoherence, Noise, and Error: Definitions and Distinctions

The terms "decoherence," "noise," and "error" are often used interchangeably, yet each has a distinct meaning in quantum computing.

- Decoherence: Decoherence arises from the inevitable interaction between a quantum system and its surrounding environment, leading to a loss of quantum coherence (i.e., superposition). It is often modeled using density operators and is responsible for the transition of quantum states from pure to mixed.

- Noise: Noise encompasses all undesired transformations that quantum states may undergo during computation, stemming from various factors such as decoherence, control errors, and thermal fluctuations. Quantum channels and their associated Kraus operators are often employed to model noise.

- Error: An error in quantum computing refers to the deviation between the actual and expected outputs of a computation. Such errors typically stem from noise and are categorized as either coherent, which are correctable through calibration, or incoherent, which are less predictable and therefore more challenging to manage.

3 From NISQ to Fault Tolerance

The evolution of quantum computing is characterized by distinct phases, each representing a leap forward in capability and application. Currently transitioning from the Noisy Intermediate-Scale Quantum (NISQ) era, the field is moving towards an interim phase known as the "era of quantum utility," eventually leading to the ultimate goal of full fault tolerance.

12.1 Preliminary Concepts

The NISQ Era

Quantum computing is currently marked by limitations in qubit count and coherence times. Amid the ongoing quest for quantum advantage, research is mainly focused on algorithmic strategies that are compatible with these NISQ constraints.

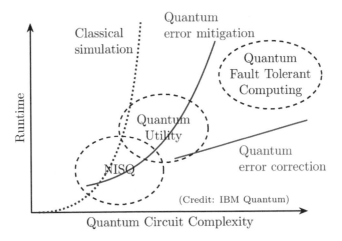

Figure 12.1: From NISQ to Fault Tolerance

The Era of Quantum Utility

As we step into the quantum utility era (circa 2024), quantum computing begins to assert its potential as a transformative technology. This phase is characterized by quantum computers not only solving problems that exceed the capabilities of brute force classical simulations but also surpassing classical systems that employ problem-specific approximation methods. This era is not merely a testament to quantum computing's theoretical prowess but also marks its evolution as an indispensable tool for scientific exploration.

The quantum utility era is distinguished by its ability to deliver reliable and substantial computations for large-scale problems, which were previously deemed infeasible without quantum intervention. With the maturation of quantum hardware, we are now witnessing quantum systems reliably handling simulation tasks involving more than 100 qubits. This milestone is underscored by the emergence of numerous "utility scale experiments," which collectively signify a paradigm shift. These experiments represent not just a leap in quantum computational capacity but also highlight the role of quantum systems in offering novel solutions and perspectives in tackling complex problems. References such as [19, 35, 43, 62, 81, 92, 93] elucidate these advancements.

As we navigate through this era, the expectation is not just the enhancement of computational capabilities but the realization of significant, practical computational advantages. Quantum systems, in this phase, are anticipated to deliver not just solutions but also tangible utility, addressing real-world challenges and enriching the landscape of computational science.

> **Quantum Utility, Advantage, and Supremacy**
>
> Quantum Utility refers to the stage where quantum computers begin to solve complex, real-world problems, offering useful solutions beyond the capabilities of classical computers.
>
> Quantum Advantage is achieved when a quantum computer outperforms classical computers for specific practical tasks, delivering solutions with superior speed, cost-effectiveness, or accuracy.
>
> Quantum Supremacy is the threshold where a quantum computer solves a problem that is infeasible for classical computers within a reasonable time frame. While it's a significant demonstration of the potential of quantum computing, the specific problems solved may not have immediate practical applications or commercial viability.

Fault-Tolerant Quantum Computing

The Accuracy Threshold Theorem is a foundational principle in quantum error correction (QEC), suggesting that Fault-Tolerant Quantum Computing (FTQC) is attainable when the error rate per physical qubit and gate operation is suppressed below a certain critical threshold, denoted by ε_{th}. Achieving error rates below this threshold enables the effective logical error rate to be substantially reduced, potentially to arbitrarily low levels.

The precise threshold value, ε_{th}, is contingent upon the specific error-correcting code employed and the noise model assumed. Present-day quantum devices generally exhibit error rates between 10^{-2} and 10^{-3}. Recent advances in error-correcting codes and error management strategies, like the conversion of general errors into more tractable erasure errors, have fostered optimism that achieving error rates around 10^{-3} could soon render practical error correction viable. Nonetheless, further reduction of the error rate would significantly alleviate the overhead, particularly reducing the ratio of physical qubits required per logical qubit.

4 Strategies for Error Handling

Several strategies are employed to manage errors as we transition from NISQ devices to fully fault-tolerant quantum computing:

- Error Suppression: This approach involves modifying control signals or incorporating additional ones to suppress errors at their source, ensuring that quantum processors produce the intended outcomes more reliably.

- Error Mitigation: Techniques such as Zero Noise Extrapolation (ZNE) and Probabilistic Error Cancellation (PEC) aim to reduce or neutralize the impact of noise on the calculated expected values. While error mitigation has shown promise in systems with the order of 100 qubits, it faces an exponential increase in overhead as the system size grows, rendering it less feasible for very large quantum systems.

- Error Correction: The cornerstone of fault-tolerant quantum computing, error correction involves a set of operations and measurements designed to identify and correct errors without compromising the quantum information. The overhead for error correction scales polylogarithmically with the size of the quantum system, making it a viable strategy for scaling to larger, fault-tolerant quantum computers.

12.1 Preliminary Concepts

- Error Simulation: Simulating errors provides critical insights into the performance of quantum algorithms under realistic, noisy conditions. This strategy supports the refinement of both quantum hardware and error correction protocols, facilitating the development of more robust quantum computing systems.

5 Challenges of Quantum Error Correction

Error correction in quantum computing presents a fundamentally more complex challenge than in classical computing. This arises from a few key properties inherent to quantum mechanics:

- Superposition: Quantum bits (qubits) can exist in a superposition of states, in contrast to classical bits that are strictly binary. This added layer of complexity widens the scope for potential errors, as undesirable changes in the relative probabilities of the states could occur.

- Entanglement: The phenomenon of entanglement means that the state of one qubit can be inexplicably linked with the state of another. Consequently, an error in one qubit can reverberate through its entangled partners, making error detection and correction more intricate.

- No-Cloning Theorem: The no-cloning theorem (see § 5.1.3) states that it is impossible to duplicate an arbitrary unknown quantum state. This principle negates the common classical error correction practice of data copying, rendering it unviable for quantum data.

- Measurement Disturbs Superposition: The act of measuring a qubit disrupts its superposition, collapsing it into one of its basis states (0 or 1). The mere action of checking a qubit for an error can irreversibly alter its state, potentially eliminating the superposition necessary for computation.

6 The Path Forward

As we navigate the evolving landscape of quantum computing, several interrelated challenges emerge as central to advancing the field:

- Quantum Error Correction and Fault Tolerance: These remain foundational to the viability of quantum computing. Current research efforts are aimed at minimizing qubit overhead and developing innovative error-correcting codes that can more efficiently protect quantum information against errors.

- Enhancing Quantum Coherence and Control: Achieving extended coherence times, high-fidelity gate operations, effective decoherence control, and overall clock speed is crucial. These factors are not only vital for successful error correction but also for executing large-scale quantum algorithms. Additionally, specific error types such as coherent, calibration, state preparation, and measurement errors fall outside the purview of traditional QEC and must be addressed through other means to ensure overall system reliability.

- Scalability and System Integration: As we aim to increase the number of qubits, the challenges of scalability and system integration become more pronounced. This pillar focuses on the architectural and connectivity solutions required to manage an expanding quantum system effectively.

- Quantum Software and Algorithm Development: Parallel to hardware advancements, the creation of sophisticated quantum software and algorithms is

12.2 ∗Mixed States, Density Operators, and CPTP Maps

To model a diverse range of quantum operations, including both pure and noisy quantum states, a more comprehensive representation beyond state vectors and unitary gates is needed. Such generalization is particularly useful for capturing the effects of noise, decoherence, or entanglement with an external environment. Density operators, also known as density matrices, and the associated mixed quantum states, serve as this general mathematical framework. This framework serves as an essential tool in quantum information and communication.

12.2.1 Definitions

1 For Pure Quantum States

Until now, we have utilized state vectors, denoted as $|\psi\rangle$, to describe quantum systems. However, these systems can also be characterized using a density operator ρ. For a quantum system isolated from its environment and in a pure state, the density operator is simply the outer product of the state vector $|\psi\rangle$ with itself. Formally, we define it as:

$$\rho = |\psi\rangle\langle\psi|. \tag{12.1}$$

■ **Example 12.1** Consider $|+\rangle = \frac{1}{\sqrt{2}}(|0\rangle + |1\rangle)$. The corresponding density operator is:

$$\rho = \frac{1}{2}\begin{bmatrix} 1 & 1 \\ 1 & 1 \end{bmatrix}. \tag{12.2}$$

■

■ **Example 12.2** For $|\psi\rangle = \alpha|0\rangle + \beta|1\rangle$, the density operator is:

$$\rho = \begin{bmatrix} |\alpha|^2 & \alpha\beta^* \\ \beta\alpha^* & |\beta|^2 \end{bmatrix}. \tag{12.3}$$

■

The state vectors $|\psi\rangle$ and $e^{i\phi}|\psi\rangle$, with ϕ a real number, yield identical density operators, thereby demonstrating that density operators inherently account for the physical irrelevance of the global phase in quantum state vectors.

> **Exercise 12.1** Show that for a pure state, $\rho^2 = \rho$, i.e., ρ is a rank-1 projection operator.

> **Exercise 12.2** Derive the density operator for the Bell state $|\Phi^+\rangle = \frac{1}{\sqrt{2}}(|00\rangle + |11\rangle)$.

2 For Mixed Quantum States

There are instances where the exact state of a quantum system cannot be definitively ascertained. Such a system may exist in any of n pure quantum states $|\psi_i\rangle$, each with a classical probability p_i. In these cases, the density operator provides a convenient representation:

$$\rho = \sum_i p_i |\psi_i\rangle\langle\psi_i|. \tag{12.4}$$

Here, the density operator for the mixed quantum state serves as a weighted average of the density operators for its constituent pure states.

■ **Example 12.3** Suppose a lab prepares an ensemble of quantum states: $|0\rangle$ with a probability of $\frac{3}{7}$ and $|1\rangle$ with a probability of $\frac{4}{7}$. This defines a *mixed state* described by:

$$\rho = \frac{3}{7}|0\rangle\langle 0| + \frac{4}{7}|1\rangle\langle 1|. \tag{12.5}$$

Note that this is completely different from the pure state $\frac{3}{7}|0\rangle + \frac{4}{7}|1\rangle$ (not normalized) or $\frac{3}{5}|0\rangle + \frac{4}{5}|1\rangle$ (normalized). ■

(i) As demonstrated in the following exercise, n can be different from d, the dimension of the Hilbert space of $|\psi_i\rangle$. However, ρ is always a $d \times d$ matrix.

Exercise 12.3 A machine produces the state $|0\rangle$ 25% of the time, the state $|1\rangle$ 25% of the time, and the state $|+\rangle$ 50% of the time. Find the density operator of the qubit.

■ **Example 12.4 — The Uniform Mixed State.** The density operator for a mixed state with equal probabilities (50%) of being in $|0\rangle$ and $|1\rangle$ is given by:

$$\rho = \frac{1}{2}\begin{bmatrix} 1 & 0 \\ 0 & 1 \end{bmatrix}. \tag{12.6}$$

For a d-dimensional qudit, the density operator for the uniform mixed state is

$$\rho = \frac{1}{d} I_{d \times d}. \tag{12.7}$$

Remarkably, the density operator for a mixed state with 50% probability in $|+\rangle$ and 50% probability in $|-\rangle$ yields the same matrix. The same holds true for a mixed state with 50% probability in an arbitrary state $|\psi\rangle$ and 50% in its orthogonal state $|\psi_\perp\rangle$.

Hence, the density operator for the uniform mixed state is basis-invariant. This characteristic can be contrasted with the density operator for $|+\rangle$, as discussed in Example 12.1, which is not basis-invariant. ■

Exercise 12.4 Show that the mixed state with 50% probability in an arbitrary state $|\psi\rangle$ and 50% in its orthogonal state $|\psi_\perp\rangle$ has a dentity operator $\rho = \frac{1}{2}I$.

Density Operator of Density Operators

A density operator can also be expressed as a linear combination of other density operators. For example, consider two density operators ρ_1 and ρ_2, which themselves could be representing mixed states. A new density operator can be formed as:

$$\rho' = q_1\rho_1 + q_2\rho_2, \tag{12.8}$$

where $q_1 + q_2 = 1$ and $q_1, q_2 \geq 0$. In this case, ρ' is a convex combination of ρ_1 and ρ_2. Each of these density operators, ρ_1 and ρ_2, could in turn be a weighted average of density operators for pure states. Then ρ' is also ultimately a weighted average of density operators for pure states.

3 Distinguishing Pure and Mixed States

The above example underscores an important point: although $|+\rangle$ and the uniform mixed state both yield probabilities of 50% for measurements in the computational basis, they can be distinguished in other bases. For instance, $|+\rangle$ can be identified unambiguously in the $\{|+\rangle, |-\rangle\}$ basis, whereas the uniform mixed state cannot. In general, pure states and mixed states can be distinguished in this manner.

Another criterion for distinguishing pure and mixed states is based on the value of the "purity" parameter:

$$0 < \gamma \equiv \operatorname{tr}(\rho^2) \leq 1. \tag{12.9}$$

For a pure state, we have $\rho^2 = \rho$, leading to $\gamma = 1$. On the other hand, for a mixed state, $\gamma < 1$; for a d-dimensional uniform mixed state, γ reaches its minimum at $\frac{1}{d}$.

12.2.2 Interpretation and Properties

1 Diagonal and Off-diagonal Elements

The diagonal elements of a density operator ρ denote the probabilities of the system being observed in each of the basis states after a measurement. Specifically, the element $\rho_{ii} = \langle i|\rho|i\rangle$ gives the probability of detecting the system in the state $|i\rangle$.

The off-diagonal elements, ρ_{ij} for $i \neq j$, encapsulate the coherences between states $|i\rangle$ and $|j\rangle$, characterizing the system's quantum superpositions. These elements may be complex, and both their magnitudes and phases are important, carrying information about the coherence of the superposition states.

The underlying reason for the above properties is demonstrated in Example 12.2 for pure states. This reasoning also applies to mixed states, as they are statistical mixtures of pure states, preserving these diagonal and off-diagonal characteristics in their density operators.

2 Diagonal Form

In its eigenbasis, a density operator ρ becomes diagonal, with eigenvalues representing the probabilities of the system residing in each eigenstate:

12.2 * Mixed States, Density Operators, and CPTP Maps

$$\rho = \sum_i \lambda_i |\lambda_i\rangle\langle\lambda_i|. \tag{12.10}$$

Each eigenstate of ρ is a pure state $|\lambda_i\rangle$. The diagonal form of ρ represents the probability distribution of the system state across these eigenstates.

This diagonal form highlights the system's probabilistic nature:

- In a pure state, ρ has exactly one diagonal element equal to 1, reflecting absolute certainty about the system's quantum state.

- For a mixed state, ρ has multiple non-zero diagonal elements, with each representing the probability of the system being found in a particular eigenstate.

Exercise 12.5 The state of a quantum system is such that it is 50% in the state $|+\rangle$ and 50% in the state $|0\rangle$. Find the diagonal form of its density operator.

3 Basic Properties

Density operators exhibit several key characteristics that confirm their suitability for physical interpretations:

1. Hermiticity: $\rho = \rho^\dagger$. This property guarantees that the eigenvalues of ρ, which are the probabilities in its eigenbasis, are real.

2. Unit Trace: $\text{tr}(\rho) = 1$. This reflects the total probability axiom, as the trace of ρ sums the probabilities for all possible outcomes of a measurement, which must equal one.

3. Positive Semi-Definiteness: For any vector $|\phi\rangle$ in the Hilbert space, $\langle\phi|\rho|\phi\rangle \geq 0$. This ensures that all eigenvalues of ρ are non-negative, thus validating that probabilities derived from ρ are always non-negative.

These properties are foundational, ensuring that density operators provide a consistent and physically plausible framework for quantum states, aligned with the postulates of quantum mechanics. Furthermore, the converse is also true: any operator (or matrix) satisfying these properties can represent a quantum state, be it pure or mixed.

Exercise 12.6 Prove the Hermitian property.

Proof. **Trace One Property**

$$\text{tr}\,\rho = \text{tr}\left(\sum_i p_i |\psi_i\rangle\langle\psi_i|\right) \tag{12.11a}$$

$$= \sum_i p_i \,\text{tr}\,(|\psi_i\rangle\langle\psi_i|) \tag{12.11b}$$

$$= \sum_i p_i = 1. \tag{12.11c}$$

□

Proof. **Positive Semi-definite Property**

Using the definition of ρ, we have:

$$\langle\phi|\rho|\phi\rangle = \langle\phi|\left(\sum_i p_i |\psi_i\rangle\langle\psi_i|\right)|\phi\rangle \quad (12.12a)$$

$$= \sum_i p_i \langle\phi|\psi_i\rangle\langle\psi_i|\phi\rangle \quad (12.12b)$$

$$= \sum_i p_i |\langle\phi|\psi_i\rangle|^2 \quad (12.12c)$$

$$\geq 0, \quad (12.12d)$$

because each term $p_i|\langle\phi|\psi_i\rangle|^2$ is non-negative. □

4 Composition Property

Consider two independent systems, represented by ρ_A and ρ_B. The density operator of the composite system can be expressed as a tensor product of their individual density operators, $\rho_{AB} = \rho_A \otimes \rho_B$. A proof is outlined below:

$$\rho_A \otimes \rho_B = \sum_n \sum_{n'} p_n^A p_{n'}^B \left(|\psi_n^A\rangle\langle\psi_n^A|\right) \otimes \left(|\psi_{n'}^B\rangle\langle\psi_{n'}^B|\right) \quad (12.13a)$$

$$= \sum_n \sum_{n'} p_n^A p_{n'}^B \left(|\psi_n^A\rangle \otimes |\psi_{n'}^B\rangle\right)\left(\langle\psi_n^A| \otimes \langle\psi_{n'}^B|\right) \quad (12.13b)$$

$$= \rho_{AB}. \quad (12.13c)$$

Note that this composition property holds true only for independent systems. In the case of dependent systems, such as entangled systems, the joint density operator ρ_{AB} of a general bipartite composite system may not be expressible as the tensor product of two individual density operators.

12.2.3 Probabilities and Expected Values

The density operator provides a complete description of the quantum state, enabling the determination of outcome probabilities for any system measurement. Observable operators facilitate the extraction of measurable properties from this state, with physical quantities typically represented as statistical averages or expected values of these observables.

1 Measurement Probabilities

The probability of a system described by the density operator ρ being found in any pure state $|\psi\rangle$ is given by the expected value of the projection onto that state:

$$p = \langle\psi|\rho|\psi\rangle. \quad (12.14)$$

This equation is valid for any pure state $|\psi\rangle$, not just the basis states $\{|\psi_i\rangle\}$ in which ρ is expressed. We suggest that the reader verify this as an exercise.

Exercise 12.7 Given the density operator for a state $|\psi\rangle = \alpha|0\rangle + \beta|1\rangle$, represented by:

12.2 * Mixed States, Density Operators, and CPTP Maps

$$\rho = \begin{bmatrix} |\alpha|^2 & \alpha\beta^* \\ \beta\alpha^* & |\beta|^2 \end{bmatrix},$$

compute the probability of measuring the system in the state $|+\rangle$ using Eq. 12.14.

---

* Similarity Between Mixed States

In general, there is no direct probability measure of one mixed state within another, but there are measures of similarity and distinguishability:

Fidelity

The fidelity between two density matrices ρ and ρ' is a common measure of similarity. It is defined as:

$$F(\rho, \rho') = \left(\text{tr} \sqrt{\sqrt{\rho}\rho'\sqrt{\rho}} \right)^2. \tag{12.15}$$

Fidelity equals 1 if and only if $\rho = \rho'$, and it decreases as the states become more distinguishable.

Trace Distance

Another measure is the trace distance, which quantifies the distinguishability of two quantum states. It is given by:

$$D(\rho, \rho') = \frac{1}{2} \text{tr} |\rho - \rho'|, \tag{12.16}$$

where $|A| = \sqrt{A^\dagger A}$. The trace distance ranges from 0 to 1, where 0 means the states are identical and 1 means they are completely distinguishable.

---

2 Expected Values of Observables

For a pure quantum state represented by $|\psi\rangle$, the expected value of an observable A is calculated as:

$$\langle A \rangle = \langle \psi | A | \psi \rangle. \tag{12.17}$$

In the case of mixed quantum states, the situation is more intricate as the system may exist in multiple states $|\psi_n\rangle$ with corresponding classical probabilities p_n. The expected value for a mixed state is essentially the weighted average of the expected values for its constituent pure states:

$$\langle A \rangle = \sum_n p_n \langle \psi_n | A | \psi_n \rangle. \tag{12.18}$$

It is noteworthy that the expected value for an observable in a mixed state can be concisely expressed using the trace operation on the density operator ρ:

$$\langle A \rangle = \text{tr}(A\rho). \tag{12.19}$$

Proof. The following steps show that $\text{tr}(A\rho) = \langle A \rangle$:

$$\mathrm{tr}(A\rho) = \mathrm{tr}\left(A\sum_n p_n |\psi_n\rangle\langle\psi_n|\right) \quad (12.20\mathrm{a})$$

$$= \sum_n p_n \, \mathrm{tr}\left(A|\psi_n\rangle\langle\psi_n|\right) \quad (12.20\mathrm{b})$$

$$= \sum_n p_n \langle\psi_n| A |\psi_n\rangle \quad (12.20\mathrm{c})$$

$$= \langle A \rangle, \quad (12.20\mathrm{d})$$

where we used the linearity of the trace operation in the second line and the cyclic property of the trace in the third line. □

Exercise 12.8 For the mixed state with equal probabilities (50%) of being in $|0\rangle$ and $|1\rangle$, calculate the expected values of the Pauli observables $\langle X \rangle$, $\langle Y \rangle$, and $\langle Z \rangle$. Compare to $\langle X \rangle$, $\langle Y \rangle$, and $\langle Z \rangle$ for the pure state $|+\rangle$.

12.2.4 *General Single-Qubit States in the Bloch Sphere

1 General Single-Qubit States

Since the Pauli matrices $\{I, X, Y, Z\}$ form a complete orthogonal basis in $\mathbb{C}^{2\times 2}$, an arbitrary single-qubit density operator can be decomposed as:

$$\rho = \frac{1}{2}(aI + xX + yY + zZ), \quad (12.21)$$

where the coefficients are given by the inner products of $\{I, X, Y, Z\}$ with ρ (defined as $A \cdot B \equiv \mathrm{tr}(A^\dagger B)$):

$$a = \mathrm{tr}(\rho I), \quad x = \mathrm{tr}(\rho X), \quad y = \mathrm{tr}(\rho Y), \quad z = \mathrm{tr}(\rho Z). \quad (12.22)$$

According to Eq. 12.19, $\mathrm{tr}(\rho A)$ equals to the expectation values of A. Therefore,

$$a = \langle I \rangle = 1, \quad x = \langle X \rangle, \quad y = \langle Y \rangle, \quad z = \langle Z \rangle. \quad (12.23)$$

Thus, we have the following important identity for single-qubit states:

$$\rho = \frac{1}{2}(I + \langle X \rangle X + \langle Y \rangle X + \langle Z \rangle X) \quad (12.24)$$

Exercise 12.9 The state of a quantum system is such that it is 50% in the state $|+\rangle$ and 50% in the state $|0\rangle$. Determine the Pauli representation of this quantum state.

2 Representation in the Bloch Sphere

According to Eq. 12.21, a single qubit state, pure or mixed, is uniquely identified by a vector $\boldsymbol{r} = (x, y, z)$, known as the *polarization vector* of the qubit. For a pure state representing a spin in the direction specified by spherical coordinate angles (θ, ϕ), the expectation values are given by (see § 3.2.3):

12.2 ∗ Mixed States, Density Operators, and CPTP Maps

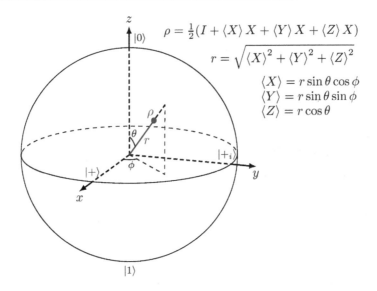

Figure 12.2: Mixed State Inside the Bloch Sphere

$$x = \sin\theta\cos\phi,\ y = \sin\theta\sin\phi,\ z = r\cos\theta. \qquad (12.25)$$

Generalizing this, we can interpret $\boldsymbol{r}$ geometrically as:

$$\boldsymbol{r} = (r\sin\theta\cos\phi,\ r\sin\theta\sin\phi,\ r\cos\theta). \qquad (12.26)$$

where $r^2 \equiv x^2 + y^2 + z^2 \leq 1$.

Therefore, while a pure state corresponds to a point on the Bloch sphere (see § 2.4), a mixed state can be represented as a point inside the Bloch sphere, as illustrated in Fig. 12.2.

Intuitively, the length of the polarization vector (or Bloch vector), r, gives an indication of the "average" behavior of the mixed state. Specifically, r provides a measure of the "purity" of the state, varying between 0 and 1. A length of $r = 1$ corresponds to pure states, while $r < 1$ corresponds to mixed states. The closer r is to zero, the more mixed the state is.

Mathematically, Eq. 12.24, the density operator of an arbitrary qubit state, can be expressed succinctly as:

$$\rho = \frac{1}{2}(I + \boldsymbol{r}\cdot\boldsymbol{\sigma}), \qquad (12.27)$$

and Eq. 12.23, the expectation of $\boldsymbol{\sigma}$ in this state, as:

$$\langle\boldsymbol{\sigma}\rangle = \boldsymbol{r}, \qquad (12.28)$$

where

$$\boldsymbol{\sigma} = (\sigma_1 \equiv X, \sigma_2 \equiv Y, \sigma_3 \equiv Z). \qquad (12.29)$$

Exercise 12.10 Show that for the uniform mixed qubit state (see Example 12.4), the Bloch vector is at the origin of the Bloch sphere. Therefore, the uniform mixed state is the maximally mixed qubit state.

Exercise 12.11 Explain why there is a factor $\frac{1}{2}$ in Eqs. 12.21 and 12.27.

12.2.5 *Unitary Evolution

1 Unitary Transformations

Recall (in Chapter 4) that in a closed quantum system governed by a Hamiltonian H, a pure quantum state $|\psi\rangle$ evolves according to the equation

$$|\psi\rangle \to U |\psi\rangle, \tag{12.30}$$

where the unitary operator U is defined as:

$$U = e^{-iHt/\hbar}. \tag{12.31}$$

Consequently, the density operator ρ for a mixed state transforms as:

$$\rho \to \rho' = \sum_n p_n U |\psi_n\rangle\langle\psi_n| U^\dagger. \tag{12.32}$$

This can be simplified to:

$$\rho \to \rho' = U\rho U^\dagger. \tag{12.33}$$

Exercise 12.12 The state of a quantum system is such that it is 50% in the state $|+\rangle$ and 50% in the state $|0\rangle$. An X gate is applied to the system. Using the above transformation, find the new density operator for the system.

■ **Example 12.5 — Application of Pauli Gates.** Consider the application of an X gate to a general quantum state $\rho = \frac{1}{2}(aI+xX+yY+zZ)$ in the Pauli-decomposition representation. Note that x, y, z coincide with the expected values $\langle X \rangle, \langle Y \rangle, \langle Z \rangle$ (see Eq. 12.24). The Pauli X gate, representing a π rotation around the x-axis, leaves $\langle X \rangle$ unchanged but flips the signs of $\langle Y \rangle$ and $\langle Z \rangle$. Therefore, X transforms (x, y, z) to $(x, -y, -z)$, or,

$$X\frac{1}{2}(aI + xX + yY + zZ)X = \frac{1}{2}(aI + xX - yY - zZ). \tag{12.34}$$

Extending this to a general rotation around the x-axis, $R_x(\theta)$ transforms the state (x, y, z) to $(x, y\cos\theta, z\cos\theta)$. ■

Exercise 12.13 How do the Pauli operators Y and Z, as well as the combined operator ZX, individually transform a quantum state represented by its Bloch vector $r = (x, y, z)$?

2 * The von Neumann Equation

The time evolution of a density operator in a quantum system is governed by the von Neumann equation, which is a generalization of the Schrödinger equation for mixed states. For a pure state $|\psi(t)\rangle$, governed by the Schrödinger equation:

$$i\hbar \frac{\partial}{\partial t}|\psi(t)\rangle = H|\psi(t)\rangle, \tag{12.35}$$

the corresponding von Neumann equation for the density operator ρ is given by:

$$i\hbar \frac{\partial \rho}{\partial t} = [H, \rho], \tag{12.36}$$

derived as follows:

$$\frac{\partial \rho}{\partial t} = \frac{\partial}{\partial t}(|\psi\rangle\langle\psi|) \tag{12.37a}$$

$$= \left(\frac{\partial}{\partial t}|\psi\rangle\right)\langle\psi| + |\psi\rangle\left(\frac{\partial}{\partial t}\langle\psi|\right) \tag{12.37b}$$

$$= \frac{-i}{\hbar}H|\psi\rangle\langle\psi| + \frac{i}{\hbar}|\psi\rangle\langle\psi|H \tag{12.37c}$$

$$= \frac{-i}{\hbar}[H, \rho]. \tag{12.37d}$$

Given the linearity of quantum mechanics and density operators, this derivation extends naturally to mixed states, which are statistical ensembles of pure states, validating the von Neumann equation for mixed states as well.

If the Hamiltonian is time-independent, the von Neumann equation has the solution

$$\rho(t) = e^{-iHt/\hbar}\rho(0)e^{iHt/\hbar}, \tag{12.38}$$

which mirrors the unitary evolution of density operators (Eq. 12.33).

For a single qubit, the Hamiltonian can be decomposed in the Pauli basis, similar to Eq. 12.27:

$$H = \frac{1}{2}(H_0 I + \boldsymbol{H} \cdot \boldsymbol{\sigma}), \tag{12.39}$$

where $\boldsymbol{\sigma}$ is given by Eq. 12.29.

Then the von Neumann equation leads to:

$$\frac{\partial \langle \sigma_n \rangle}{\partial t} = \mathrm{tr}\left(\sigma_n \frac{\partial \rho}{\partial t}\right) = -i\,\mathrm{tr}\left(\sigma_n [H, \rho]\right) = \sum_{jk} \varepsilon_{njk} H_j \langle \sigma_k \rangle, \tag{12.40}$$

where $n, j, k \in \{1, 2, 3\}$, or

$$\frac{\partial \langle \boldsymbol{\sigma} \rangle}{\partial t} = \boldsymbol{H} \times \langle \boldsymbol{\sigma} \rangle. \tag{12.41}$$

12.2.6 *Measurements

Transitioning from the ket representation to density operators offers a more encompassing description of quantum states. This subsection explores how quantum measurement theory (§ 3.4) is adapted to align with the density operator framework. We will discuss the characterization of post-measurement states, the emergence of mixed states from measurements, measurements in an orthonormal basis, and the concept of POVM (Positive Operator-Valued Measure) measurements.

1 Post-Measurement State

The action of a measurement on a quantum system can be captured through operators (see § 3.4.2). For a pure state $|\psi\rangle$ and a measurement operator M_j, the state after measurement is given by:

$$|\psi\rangle \to \frac{M_j |\psi\rangle}{\|M_j |\psi\rangle\|}. \tag{12.42}$$

For a mixed state represented by a density operator ρ, the transformation is as follows:

$$\rho \to \rho' = \frac{\sum_n p_n M_j |\psi_n\rangle\langle\psi_n| M_j^\dagger}{\text{tr}\left[\sum_n p_n M_j |\psi_n\rangle\langle\psi_n| M_j^\dagger\right]}. \tag{12.43}$$

Note that the normalization of a density operator is designed to produce a unity trace. This ensures that the probabilities of all possible outcomes of a measurement sum to 1. Simplifying using the procedure to prove Eq. 12.20, we obtain:

$$\rho \to \rho' = \frac{M_j \rho M_j^\dagger}{p_j}, \tag{12.44}$$

where p_j is the normalization factor or the probability of measuring M_j:

$$p_j = \text{tr}(M_j \rho M_j^\dagger). \tag{12.45}$$

> **Exercise 12.14** The state of a system, initially 50% in the state $|+\rangle$ and 50% in the state $|0\rangle$, is measured in the $\{|0\rangle, |1\rangle\}$ basis, and the outcome $|0\rangle$ is obtained. What is the new density operator of the system?

2 Ensemble of Measurement Outcomes

Instead of considering a single measurement operator M_j, we now extend our focus to a complete set of such operators $\{M_j\}$ (see § 3.4.2) that satisfy the completeness condition

$$\sum_j M_j^\dagger M_j = I. \tag{12.46}$$

The resulting ensemble of measurement outcomes is a mixed state represented by a density operator as follows:

12.2 * Mixed States, Density Operators, and CPTP Maps

$$\rho_M = \sum_j p_j \frac{M_j \rho M_j^\dagger}{p_j}. \tag{12.47}$$

Here, p_j is the probability of measuring M_j given by Eq. 12.45. Apparently, p_j cancels out, yielding the following crucial relationship:

$$\rho_M = \sum_j M_j \rho M_j^\dagger. \tag{12.48}$$

The ensemble of measurement outcomes encapsulates the probabilistic nature of quantum measurements. The density operator ρ_M represents the statistical mixture of all possible states that a quantum system can be in after being subjected to a measurement process. It offers a mechanism to represent the measurement and superposition principles of quantum mechanics statistically, thereby providing a complete picture of the system's possible states after a measurement.

The concept of the ensemble of measurement outcomes is closely related to the phenomenon of measurement-induced decoherence, a process that describes how quantum superpositions are reduced to classical mixtures as a result of measurement. A quantum system can exist in a superposition of states. When a measurement is performed, this superposition appears to collapse to a specific state corresponding to the measurement outcome. However, from a more comprehensive perspective, what happens is that the system becomes entangled with the measurement apparatus, leading to decoherence. The density operator ρ_M describes the state of the system after this decoherence has occurred.

Exercise 12.15 The state of a system, initially $|+\rangle$, is measured in the $\{|0\rangle, |1\rangle\}$ basis. What is the density operator of the post-measurement ensemble?

3 Measurements in an Orthonormal Basis

Now consider the special case where $\{M_j\}$ is the set of projection operators corresponding to an orthonormal basis $\{|\phi_j\rangle\}$:

$$M_j = |\phi_j\rangle\langle\phi_j|. \tag{12.49}$$

Applying the relationship from Eq. 12.48 for these projectors, we obtain the post-measurement mixed state as follows:

$$\rho_M = \sum_j |\phi_j\rangle\langle\phi_j| \rho |\phi_j\rangle\langle\phi_j| \tag{12.50a}$$

$$= \sum_j \langle\phi_j| \rho |\phi_j\rangle |\phi_j\rangle\langle\phi_j| \tag{12.50b}$$

$$= \sum_j p_j |\phi_j\rangle\langle\phi_j|, \tag{12.50c}$$

where

$$p_j = \langle\phi_j| \rho |\phi_j\rangle \tag{12.51}$$

is the probability of obtaining the outcome corresponding to $|\phi_j\rangle$.

State Collapse

The matrix ρ_M is diagonal in the basis $\{|\phi_j\rangle\}$, containing the probabilities p_j as its diagonal elements, while all off-diagonal elements are zero. If ρ was not initially diagonal in this basis, then ρ_M will differ from ρ. This difference is a consequence of the measurement process: a projective measurement collapses the system into a statistical mixture of the states $|\phi_j\rangle$ weighted by their respective probabilities p_j, effectively destroying any initial coherences between different basis states present in ρ. Hence, post-measurement, the system's state is a diagonal mixture of the measurement basis states. The system's final state coincides with the original ρ if and only if ρ was already diagonal in the chosen measurement basis.

4 POVM Measurements

Measurements in quantum mechanics can be more generally described using Positive Operator-Valued Measures (POVMs). A set of POVM operators $\{E_j\}$ satisfies two key conditions: each E_j is positive semidefinite ($E_j \geq 0$), and the sum of all POVM elements equals the identity operator ($\sum_j E_j = I$). These conditions guarantee that the calculated probabilities will sum to one. In this framework, each E_j corresponds to a distinct measurement outcome. The probability p_j of observing outcome j is given by:

$$p_j = \mathrm{tr}(\rho E_j). \tag{12.52}$$

This probability can also be interpreted as the inner product $\langle \rho, E_j \rangle$ between the density operator ρ and the POVM element E_j.

In the traditional formulation of quantum measurements using a set of measurement operators $\{M_j\}$, each POVM element E_j is expressed as $E_j = M_j^\dagger M_j$. However, POVM measurements generalize this concept. The set of POVM operators $\{E_j\}$ can have a different cardinality from the dimension of the quantum state ρ, and the POVM elements are not required to be orthogonal. This broader framework is more aligned with the principles of quantum information theory.

12.2.7 ∗ CPTP Maps and Quantum Channels

1 Completely Positive and Trace-Preserving (CPTP) Maps

Unitary evolution, as given by Eq. 12.33, and the quantum measurement process, described by Eq. 12.48, both transform the density operator of a quantum state. Since density operators must maintain a unit trace and be positive semidefinite, such transformations are referred to as Completely Positive and Trace-Preserving (CPTP) maps. In general, a CPTP map is an operation that transforms a density operator to another which can have a different dimensionality.

Definition

A CPTP map $\mathcal{E}$ is a transformation that satisfies the following properties:

- Completely Positive: A map $\mathcal{E}$ is said to be *positive* if for any density operator ρ, the map $\mathcal{E}(\rho)$ preserves the positive semidefiniteness of ρ, ensuring the physical validity of the transformed state.

  A map $\mathcal{E}$ is defined as *completely positive* if, when extended to any larger system that includes the original system as a subsystem, it preserves the positive

12.2 ∗ Mixed States, Density Operators, and CPTP Maps

semidefiniteness of all states in this larger system. Formally, for any density operator ρ of the combined system (the original system plus an auxiliary system), the extended map $(\mathcal{E} \otimes I)(\rho)$ is positive, where I is the identity map on the auxiliary system.

- Trace-Preserving: For any density operator ρ, $\mathrm{tr}(\mathcal{E}(\rho)) = \mathrm{tr}(\rho)$. This property ensures that the total probability (sum of probabilities of all outcomes) is conserved in the transformation.

- Linearity: $\mathcal{E}(a\rho + b\sigma) = a\mathcal{E}(\rho) + b\mathcal{E}(\sigma)$ for any density operators ρ, σ, and scalars a, b.

Applications

CPTP maps offer a comprehensive framework for describing the dynamics of quantum states, and are a fundamental tool in quantum computation and information theory. They encompass a wide range of physical processes:

- Unitary Evolutions: Unitary operations, such as the action of quantum gates, are reversible and a special case of CPTP maps.

- Quantum Measurements: These maps model the process of quantum measurement, including the decoherence effect, as in Eq. 12.48.

- Simulation of Open Systems. CPTP maps can model open quantum systems where interactions with external environments are significant. They facilitate the inclusion of auxiliary systems in the analysis, which serve to represent environmental effects. In this context, an auxiliary system functions as a supplementary quantum system, considered in conjunction with the primary system to simulate the combined effect of the environment and the system dynamics.

- Reduction of Systems via Partial Trace: CPTP maps can represent the process of tracing out part of a quantum system, effectively reducing it to a subsystem. The partial trace operation ensures that the reduced state remains a valid density operator.

- Other Quantum Operations: These maps include a broad class of quantum operations, such as those involving interactions with an environment. These operations can lead to various effects like quantum noise, dissipation, and more general quantum dynamics beyond purely unitary transformations or simple measurements.

2 Quantum Channels

CPTP maps can be modeled using quantum channels, which are often represented by Kraus decompositions or operator-sum representations. A Kraus decomposition provides an intuitive and computationally convenient method to express a CPTP map by breaking it down into a set of operations on the quantum state. Thus, a quantum channel, describing the evolution of a quantum state under a CPTP map, is represented as:

$$\mathcal{E}(\rho) = \sum_j s_j K_j \rho K_j^\dagger, \qquad (12.53)$$

where the K_j are Kraus operators that describe the quantum channel, and s_j is the probability that the transformation represented by K_j occurs. Here K_j does not have to be a square matrix.

These operators must satisfy the modified completeness relation:

$$\sum_j s_j K_j^\dagger K_j = I, \qquad (12.54)$$

ensuring that the map $\mathcal{E}$ is trace-preserving.

An alternative representation, which has the probability s_j factored into the operator K_j: $\sqrt{s_j} K_j \mapsto K_j$, and thus does not explicitly use the probabilities s_j, is given by:

$$\mathcal{E}(\rho) = \sum_j K_j \rho K_j^\dagger, \qquad (12.55)$$

with the completeness condition:

$$\sum_j K_j^\dagger K_j = I. \qquad (12.56)$$

We have seen that quantum measurement can be represented as a Kraus decomposition in Eq. 12.48. Below are additional examples:

▪ **Example 12.6 — Bit-Flip Noise Channel.** A bit-flip error flips the state of a qubit from $|0\rangle$ to $|1\rangle$, or vice versa, with some specified probability, s. It can be modeled by applying an X gate, randomly with probability s, as illustrated below:

$$\rho \begin{cases} -\!\!-(p=s)-\!\!\boxed{X}\!\!- \\ -\!\!-(p=1-s)-\!\!\boxed{I}\!\!- \end{cases} \mathcal{N}(\rho)$$

The Kraus representation for the associated noise channel is:

$$\mathcal{N}(\rho) = (1-s)I\rho I + sX\rho X = (1-s)\rho + sX\rho X. \qquad (12.57)$$

We will delve into such random noise channels in the next section. ▪

▪ **Example 12.7 — Complete Dephasing Noise Channel.** Consider the general qubit state $|\psi\rangle = \alpha|0\rangle + \beta|1\rangle$ with density operator

$$\rho = |\psi\rangle\langle\psi| = \begin{bmatrix} |\alpha|^2 & \alpha\beta^* \\ \alpha^*\beta & |\beta|^2 \end{bmatrix}. \qquad (12.58)$$

If we measure it in the standard basis, but actually "forget" the outcome, we can see that the resulting state will output $|0\rangle$ with output probability $|\alpha|^2$, and output $|1\rangle$ with probability $|\beta|^2$. Thus, the output is:

$$\mathcal{N}(\rho) = \begin{bmatrix} |\alpha|^2 & 0 \\ 0 & |\beta|^2 \end{bmatrix}. \qquad (12.59)$$

The effect is to zero out all the off-diagonal terms in ρ, thus destroying all coherence. This process is often referred to as complete dephasing. ▪

12.2 * Mixed States, Density Operators, and CPTP Maps

Exercise 12.16 Consider an initial state $|\psi\rangle = \alpha|0\rangle + \beta|1\rangle$. If a bit flip is introduced to the system with a random probability of occurrence s, determine the resulting density operator after this noise channel acts on the state.

3 Stinespring Dilation: Unitary Embedding of a CPTP Map

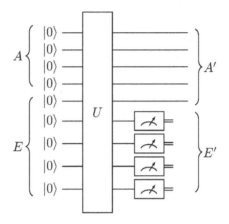

Figure 12.3: Stinespring Dilation: Unitary Embedding of a CPTP Map

Stinespring dilation provides a way to visualize and understand general quantum operations (CPTP maps) as unitary evolutions in a larger space, which includes the system and the environment. Stinespring dilation and Kraus decomposition are mathematically equivalent, yet they provide distinct perspectives on the nature of quantum operations.

Stinespring dilation can be illustrated using the quantum circuit in Fig. 12.3. In this schematic, qubits in group A are considered the input qubits of the system, while those in group E function as ancillary qubits, simulating an environment. The combined system $A + E$ is subject to a unitary transformation U. Following this transformation, the qubits are categorized into groups A' and E', which correspond to the transformed states of A and E, respectively. The qubits in E' are subsequently measured (typically in the computational basis) and then discarded (traced out), while the remaining qubits in A' constitute the output of the CPTP map.

This process is encapsulated by the Stinespring Dilation Theorem, which asserts that any CPTP map can be implemented as a unitary operation on an expanded Hilbert space that includes an ancillary system, followed by tracing out some degrees of freedom—specifically, those associated with the environment. This theorem demonstrates that operations which appear non-unitary, such as decoherence or measurement, can be modeled within the framework of unitary quantum mechanics by accounting for interactions with an ancillary system or external environment.

Formally, the Stinespring Dilation Theorem states that given a CPTP map $\mathcal{E}$ acting on a state ρ in system A, there exists an ancillary system E, a unitary operation U, and a state $|\psi\rangle$ in the combined system such that:

$$\mathcal{E}(\rho) = \text{tr}_E \left(U(\rho \otimes |\psi\rangle\langle\psi|) U^\dagger \right), \tag{12.60}$$

where tr_E denotes the partial trace (discussed in § 12.2.8) over the ancillary system E, effectively capturing the essence of the CPTP map's action on the state ρ.

A common question arises: If unitary operations are inherently reversible, does that imply CPTP maps, including processes like decoherence, are also reversible? The answer is "No." The key lies in the irreversible act of tracing out the environment subsystem E' after the unitary operation. This step, essential for modeling the effect of the environment, discards information, making the process inherently irreversible. Thus, despite the unitary transformation's reversibility in the combined system $A + E$, the practical application of CPTP maps remains irreversible due to the loss of access to the environment's state post-interaction, highlighting a fundamental quantum-to-classical transition.

12.2.8 ∗ Partial Trace and Reduced Density Operators

The concept of the partial trace of density operators in quantum mechanics bears a strong analogy to marginal probability in classical probability theory (see § 13.1.4). Both methodologies enable focusing on a subset within a larger system by summing over and thereby effectively disregarding the other components of the system.

Partial trace serves a complementary role to tensor product. While tensor product is used to construct a composite quantum system from individual states, partial trace allows us to extract and examine the state of a specific subsystem, effectively removing information about other parts of the system.

Partial trace is a fundamental tool for describing parts of larger quantum systems. Its applications range widely, including understanding quantum entanglement, studying the effects of environmental interactions on quantum systems (such as decoherence), and analyzing quantum information processes.

1 Definition

Consider a bipartite composite system represented by a density operator ρ_{AB}. The partial trace of ρ_{AB} over subsystem B is defined as:

$$\rho_A \equiv \text{tr}_B(\rho_{AB}) = \sum_i (I_A \otimes \langle i|_B) \rho_{AB} (I_A \otimes |i\rangle_B), \qquad (12.61)$$

or, in the partial product notation (see Appendix C),

$$\text{tr}_B(\rho_{AB}) = \sum_i \langle i|_B \, \rho_{AB} \, |i\rangle_B \,, \qquad (12.62)$$

where $\{|i\rangle_B\}$ is any orthonormal basis for the Hilbert space of subsystem B. In this case, the result of the partial trace, ρ_A, is also referred to as the reduced density operator of subsystem A.

The reduced density operator ρ_A represents the state of subsystem A after averaging over the degrees of freedom of subsystem B. This operation effectively encompasses all possible outcomes of partial measurements on subsystem B, providing a statistical description of subsystem A irrespective of the specific state or measurement outcome of subsystem B.

Similarly, the partial trace over subsystem A, or the reduced density operator of subsystem B, is defined as:

12.2 * Mixed States, Density Operators, and CPTP Maps

$$\rho_B \equiv \text{tr}_A(\rho_{AB}) = \sum_i \langle i|_A \, \rho_{AB} \, |i\rangle_A, \tag{12.63}$$

where $\{|i\rangle_A\}$ is any orthonormal basis for the Hilbert space of subsystem A.

> **Exercise 12.17** The reduced density operator given in Eq. 12.62 remains the same regardless of the chosen orthonormal basis for the Hilbert space of subsystem B. demonstrate this.

2 Independent Subsystems

In general, $\rho_{AB} \neq \rho_A \otimes \rho_B$. However, if the subsystems are independent, then $\rho_{AB} = \rho_A \otimes \rho_B$, which can be demonstrated as follows. In this special case, Eq. 12.61 simplifies to

$$\text{tr}_B(\rho_{AB}) = \sum_i (I_A \otimes \langle i|_B)(\rho_A \otimes \rho_B)(I_A \otimes |i\rangle_B) \tag{12.64a}$$

$$= \rho_A \otimes \sum_i \langle i|_B \, \rho_B \, |i\rangle_B \tag{12.64b}$$

$$= \rho_A \otimes \text{tr}(\rho_B) \tag{12.64c}$$

$$= \rho_A, \tag{12.64d}$$

as expected, since $\text{tr}(\rho_B) = 1$.

3 Examples

In the following, we will explore partial trace through several examples. We begin with a visual representation of matrix elements in a two-qubit system to establish a foundational understanding. Subsequently, we examine the partial trace in different contexts: a correlated mixed state to demonstrate how classical correlations are handled, and a Bell state to reveal the intriguing nature of quantum entanglement. We then extend our exploration to consider the effects of local operations in entangled systems, which leads us to a deeper understanding of quantum nonlocality.

■ **Example 12.8 — General Two-Qubit System: Matrix Visualization.** The density operator of a generic two-qubit system can be represented as

$$\rho_{AB} = \begin{array}{c} \\ |00\rangle \\ |01\rangle \\ |10\rangle \\ |11\rangle \end{array} \begin{array}{c} \langle 00| \quad \langle 01| \quad \langle 10| \quad \langle 11| \\ \begin{bmatrix} \rho_{0000} & \rho_{0001} & \rho_{0010} & \rho_{0011} \\ \rho_{0100} & \rho_{0101} & \rho_{0110} & \rho_{0111} \\ \rho_{1000} & \rho_{1001} & \rho_{1010} & \rho_{1011} \\ \rho_{1100} & \rho_{1101} & \rho_{1110} & \rho_{1111} \end{bmatrix} \end{array}. \tag{12.65}$$

Here, for clearer visualization, we have used binary indices $00, 01, 10, 11$ (instead of $1, 2, 3, 4$) to label the matrix elements.

To compute ρ_A, the reduced density operator for qubit A, we perform the partial trace of ρ_{AB} over subsystem B. This process can be visualized as combining elements of ρ_{AB} that correspond to the same state of qubit A. For each state of qubit A, we add together the elements of ρ_{AB} where qubit B is in either state $|0\rangle$ or $|1\rangle$:

$$\rho_A = \text{tr}_B(\rho_{AB}) = \begin{bmatrix} \rho_{0000} + \rho_{0101} & \rho_{0010} + \rho_{0111} \\ \rho_{1000} + \rho_{1101} & \rho_{1010} + \rho_{1111} \end{bmatrix}. \tag{12.66}$$

Here, the element $\rho_{0000} + \rho_{0101}$, for example, represents the sum of probabilities where qubit A is in state $|0\rangle$, irrespective of the state of qubit B. It corresponds to $\langle 0|_B (\rho_{0000} |00\rangle\langle 00|) |0\rangle_B + \langle 1|_B (\rho_{0101} |01\rangle\langle 01|) |1\rangle_B$ in Eq. 12.62. This 'partial tracing' effectively removes the detailed information about qubit B, leaving us with the reduced state of qubit A.

Similarly, the reduced density operator for qubit B, ρ_B, can be obtained by summing the matrix elements corresponding to the states of qubit A:

$$\rho_B = \text{tr}_A(\rho_{AB}) = \begin{bmatrix} \rho_{0000} + \rho_{1010} & \rho_{0001} + \rho_{1011} \\ \rho_{0100} + \rho_{1110} & \rho_{0101} + \rho_{1111} \end{bmatrix}. \tag{12.67}$$

∎

Exercise 12.18 Consider a two-qubit product state $|\psi\rangle = |++\rangle$, where $|+\rangle = \frac{1}{\sqrt{2}}(|0\rangle + |1\rangle)$.

(a) Express $|\psi\rangle$ in terms of $|0\rangle$ and $|1\rangle$ and compute the density operator $\rho_{AB} = |++\rangle\langle++|$.

(b) Compute the reduced density operators ρ_A and ρ_B from the partial trace of ρ_{AB}. Verify that $\rho_A = \rho_B = |+\rangle\langle+|$.

(c) Verify that $\rho_{AB} = \rho_A \otimes \rho_B$.

This exercise is designed to show that the reduced density operators derived from partial tracing align with our expectations for simple product states.

∎ **Example 12.9 — Partial Trace of a Correlated Mixed State.** Consider a two-qubit system with the density operator ρ_{AB} given by:

$$\rho_{AB} = \frac{1}{2}|00\rangle\langle 00| + \frac{1}{2}|11\rangle\langle 11|, \tag{12.68}$$

representing a mixed state with classical correlations between the two qubits. It is not an entangled state.

We want to find the reduced density operator for qubit A. The partial trace over qubit B is calculated as:

$$\text{tr}_B(\rho_{AB}) = \langle 0|_B \rho_{AB} |0\rangle_B + \langle 1|_B \rho_{AB} |1\rangle_B. \tag{12.69}$$

Substituting the value of ρ_{AB}, we get:

12.2 ∗ Mixed States, Density Operators, and CPTP Maps

$$\operatorname{tr}_B(\rho_{AB}) = \langle 0|_B \left(\frac{1}{2} |00\rangle\langle 00| + \frac{1}{2} |11\rangle\langle 11| \right) |0\rangle_B \qquad (12.70a)$$

$$+ \langle 1|_B \left(\frac{1}{2} |00\rangle\langle 00| + \frac{1}{2} |11\rangle\langle 11| \right) |1\rangle_B$$

$$= \frac{1}{2} \langle 0|_B |00\rangle\langle 00| |0\rangle_B + \frac{1}{2} \langle 1|_B |11\rangle\langle 11| |1\rangle_B \qquad (12.70b)$$

$$= \frac{1}{2} |0\rangle\langle 0| + \frac{1}{2} |1\rangle\langle 1| \qquad (12.70c)$$

$$= \frac{1}{2} \begin{bmatrix} 1 & 0 \\ 0 & 1 \end{bmatrix}. \qquad (12.70d)$$

Therefore, the reduced density operator for qubit A is:

$$\rho_A = \frac{I}{2}, \qquad (12.71)$$

which indicates that qubit A is in a maximally mixed state. Similarly, $\rho_B = \frac{I}{2}$. ∎

■ **Example 12.10 — Partial Trace of a Bell State.** The Bell state $|\Psi\rangle = \frac{1}{\sqrt{2}}(|00\rangle + |11\rangle)$ has a density operator given by:

$$\rho_{AB} = \frac{1}{2}(|00\rangle + |11\rangle)(\langle 00| + \langle 11|). \qquad (12.72)$$

Expanding the terms yields:

$$\rho_{AB} = \frac{1}{2}(|00\rangle\langle 00| + |00\rangle\langle 11| + |11\rangle\langle 00| + |11\rangle\langle 11|). \qquad (12.73)$$

Let's now calculate the partial trace over the B subsystem, using the computational basis as our orthonormal basis $\{|i\rangle\}$:

$$\rho_A = \operatorname{tr}_B(\rho_{AB}) \qquad (12.74a)$$

$$= \frac{1}{2} \left(\langle 0|_B \rho_{AB} |0\rangle_B + \langle 1|_B \rho_{AB} |1\rangle_B \right) \qquad (12.74b)$$

$$= \frac{1}{2} (|0\rangle\langle 0| + |1\rangle\langle 1|) \qquad (12.74c)$$

$$= \frac{I}{2}. \qquad (12.74d)$$

Similarly, $\rho_B = \frac{I}{2}$.

Apparently, both ρ_A and ρ_B represent a uniform mixed state with 50% probability of $|0\rangle$ and 50% probability of $|1\rangle$, a surprising result considering that the joint state of A and B is a pure, entangled state. This illuminates the non-intuitive properties of quantum entanglement, where subsystems can be mixed even when the global system is in a pure state.

Comparing with Example 12.9, we see that the reduced state of a subsystem can be maximally mixed, regardless of whether the overall system is in a correlated mixed state or an entangled pure state. Thus, in quantum mechanics, the properties of parts of a system do not always straightforwardly reflect the properties of the whole system.

Can we differentiate between this Bell state and the classically correlated mixed state in Example 12.9? The distinction becomes clear with the application of the Bell inequality and the Bell test experiments. For a more comprehensive understanding of these concepts, refer to Chapter 9. ∎

> **Exercise 12.19** Compute ρ_A and ρ_B for the Bell state $|\Psi\rangle = \frac{1}{\sqrt{2}}(|00\rangle + |11\rangle)$, as in Example 12.10, but now using the $\{|+\rangle, |-\rangle\}$ basis, where $|+\rangle = \frac{1}{\sqrt{2}}(|0\rangle + |1\rangle)$ and $|-\rangle = \frac{1}{\sqrt{2}}(|0\rangle - |1\rangle)$.

■ **Example 12.11 — Local Operations in Entangled Systems.** Let's continue from the previous example, and consider the scenario where Alice and Bob are spatially separated and share a Bell pair described by the density operator in Eq. 12.72. Suppose Bob performs a local operation on his qubit, such as a measurement or a state change. We analyze whether Alice can perceive this action through her part of the Bell pair using the concept of the partial trace.

Case 1: Local Unitary Operations

Suppose Bob performs a local unitary operation U_B on his qubit. The state of the system after this operation is given by:

$$\rho'_{AB} = (I_A \otimes U_B)\rho_{AB}(I_A \otimes U_B^\dagger). \tag{12.75}$$

Alice's reduced state, obtained by tracing out Bob's subsystem, is $\rho'_A = \text{tr}_B(\rho'_{AB})$. Since the trace operation is invariant under unitary transformations, $\text{tr}_B(\rho'_{AB}) = \text{tr}_B(\rho_{AB})$. Therefore,

$$\rho'_A = \rho_A = \frac{I}{2}, \tag{12.76}$$

indicating that Alice's reduced state is unaffecte. Although the unitary operation changes the state of Bob's qubit, it does not collapse the quantum state, nor does it change the entanglement properties in a way observable to Alice.

Case 2: Local Measurements

Now, consider Bob performs a local measurement. This collapses the state of the system into one of the eigenstates corresponding to the measurement outcome. However, without further communication, Alice would not know which eigenstate Bob's qubit has collapsed to. So she would experience a mixed state corresponding to the statistical ensemble of both measurement outcomes, given by the density operator in Eq. 12.68. As we have worked out in Example 12.9, Alice's reduced density operator is $\frac{I}{2}$.

Thus, Alice's observable statistics remain consistent with the reduced density operator calculated before Bob's actions. She cannot distinguish whether her mixed state is due to the original entanglement or the collapse caused by Bob's measurement, highlighting the non-observability of Bob's local measurements from Alice's perspective without additional classical communication.

This example illustrates a fundamental aspect of quantum mechanics: local operations on part of an entangled system do not have observable consequences on the distant part of the system, and entanglement itself does not allow for superluminal

12.2 ∗ Mixed States, Density Operators, and CPTP Maps

communication. This underscores the principles of the no-communication theorem in quantum mechanics. ∎

■ **Example 12.12 — GHZ State and the Monogamy of Entanglement.** Consider the GHZ state $|\text{GHZ}\rangle = \frac{1}{\sqrt{2}}(|000\rangle + |111\rangle)$, which extends the concept of entanglement to three qubits, held by Alice, Bob, and Charlie respectively. Unlike pairs of qubits in the Bell state, the three qubits in the GHZ state are mutually entangled; measuring one qubit instantly determines the states of the other two. Its density operator is given by:

$$\rho_{ABC} = \frac{1}{2}(|000\rangle + |111\rangle)(\langle 000| + \langle 111|). \tag{12.77}$$

What happens if we observe just Alice and Bob's qubits without Charlie's? To find out, we perform a partial trace over qubit C, yielding:

$$\rho_{AB} = \frac{1}{2}|00\rangle\langle 00| + \frac{1}{2}|11\rangle\langle 11|. \tag{12.78}$$

Surprisingly, ρ_{AB} resembles the density operator of a classically correlated mixed state, not the entangled Bell state one might expect. This shows that the entanglement in the three-qubit GHZ state is fully manifested only when all three qubits are considered together; with only two, we observe merely classical correlations.

This phenomenon is often referred to as the monogamy of entanglement, highlighting its "all-or-nothing" nature. As a result, the triple cannot share additional entanglement with any external system. Attempting to entangle a fourth qubit with the GHZ state would inevitably weaken its intrinsic entanglement. The principle of monogamy underlies the security of certain quantum cryptography protocols, where any eavesdropping attempt on part of an entangled state disrupts the entanglement in a detectable manner.

Another interesting three-qubit entangled state is the W state, $|\text{W}\rangle = \frac{1}{\sqrt{3}}(|001\rangle + |010\rangle + |100\rangle)$. In the W state, there is partial entanglement between each pair of qubits, but none are maximally entangled. Readers are encouraged to explore this further by performing a partial trace on the density operator of the W state. ∎

12.2.9 ∗ Extension of No-Cloning and No-Communication Theorems

The no-cloning and no-communication theorems, initially introduced within the context of unitary operations in closed quantum systems, are equally applicable to more general quantum operations defined by Completely Positive Trace-Preserving (CPTP) operations. These operations encompass measurements and other non-reversible effects, thereby broadening the applicability of these fundamental theorems. In this subsection, we outline proofs of the no-cloning and no-communication theorems within the framework of CPTP operations.

1 The No-Cloning Theorem: A Proof Outline

The No-Cloning Theorem (see § 5.1.3) asserts that it is impossible to create an identical copy of an arbitrary unknown quantum state. A proof was provided under unitary transformations in § 5.7. The proof can be extended to CPTP maps, which can include measurements and other quantum operations.

Consider an arbitrary quantum state, $|\psi\rangle$. Assume there exists a CPTP map $\mathcal{E}$ that can clone any quantum state, such that $\mathcal{E}(|\psi\rangle\langle\psi|) = |\psi\rangle\langle\psi| \otimes |\psi\rangle\langle\psi|$. The map $\mathcal{E}$ must simultaneously clone $|0\rangle$, $|1\rangle$, and $|+\rangle = \frac{1}{\sqrt{2}}(|0\rangle + |1\rangle)$:

$$\mathcal{E}(|0\rangle\langle 0|) = |0\rangle\langle 0| \otimes |0\rangle\langle 0|, \tag{12.79a}$$
$$\mathcal{E}(|1\rangle\langle 1|) = |1\rangle\langle 1| \otimes |1\rangle\langle 1|, \tag{12.79b}$$
$$\mathcal{E}(|+\rangle\langle +|) = |+\rangle\langle +| \otimes |+\rangle\langle +|. \tag{12.79c}$$

However, if we add the first two equations and multiply the result by $\frac{1}{2}$, we do not get the same result as the third equation. This discrepancy highlights the contradiction and proves that a universal cloning machine cannot exist within the framework of CPTP maps, thus proving the no-cloning theorem.

2 The No-Communication Theorem: A Proof Outline

The No-Communication Theorem (see § 9.7) states that during the measurement of an entangled quantum state, it is fundamentally impossible for one observer to communicate information to another solely through the act of measurement. Here is an outline of the proof of this Theorem, building on the previous example.

Suppose we have a composite quantum system described by a joint density operator ρ_{AB}. Let $\mathcal{E}$ be a completely positive, trace-preserving (CPTP) map that acts only on subsystem B. Physically, $\mathcal{E}$ can represent a quantum operation (such as a measurement or a unitary operation) performed by Bob on his subsystem.

Now, consider the reduced density operator of subsystem A. Before Bob's operation, it is given by $\text{tr}_B(\rho_{AB})$. After Bob's operation, the state of the total system becomes:

$$\rho'_{AB} = (I \otimes \mathcal{E})(\rho_{AB}), \tag{12.80}$$

where I is the identity map acting on subsystem A.

The reduced density operator of subsystem A is then

$$\text{tr}_B(\rho'_{AB}) = \text{tr}_B[(I \otimes \mathcal{E})(\rho_{AB})]. \tag{12.81}$$

A CPTP map $\mathcal{E}$ maintains the trace of a density operator, i.e., $\text{tr}(\mathcal{E}\rho) = \text{tr}(\rho)$. When taking the partial trace over B, we are effectively summing over the degrees of freedom of subsystem B. Since $\mathcal{E}$ is trace-preserving and only acts on B, the overall trace of the entire system over B remains unchanged. Thus, we have

$$\text{tr}_B[(I \otimes \mathcal{E})(\rho_{AB})] = \text{tr}_B(\rho_{AB}). \tag{12.82}$$

Therefore,

$$\text{tr}_B(\rho'_{AB}) = \text{tr}_B(\rho_{AB}). \tag{12.83}$$

This result indicates that the reduced density operator of subsystem A (i.e., Alice's subsystem) remains unchanged regardless of what operation Bob performs on subsystem B. Consequently, Bob cannot send any information to Alice through his actions on his subsystem.

12.3 Error Mechanisms in Quantum Computing

This proof assumes that Alice and Bob's subsystems do not interact during the process. If there is an interaction, information can be transferred, but it cannot occur faster than light as required by the theory of relativity.

12.3 Error Mechanisms in Quantum Computing

In quantum computing, especially in the Noisy Intermediate-Scale Quantum (NISQ) era, errors can be broadly divided into coherent and incoherent errors, which are the focus of the section. These errors manifest differently in the dynamics of the quantum system and, consequently, have different implications for error correction and fault tolerance. This section also introduces merit parameters that serve as quantitative metrics for characterizing the impact and prevalence of these errors.

12.3.1 ✻ Coherent Errors

Coherent errors typically stem from inaccuracies in control parameters such as pulse amplitude, duration, and frequency. They often manifest as unitary transformations that deviate from the ideal transformations. Below, we discuss two main types of coherent errors: gate calibration errors and cross-talk errors.

1 Gate Calibration Errors

Gate calibration errors arise when there is a discrepancy between the ideal and the actual implementations of a quantum gate. These errors are frequently due to inaccuracies in control parameters such as pulse amplitude, frequency, or timing.

To illustrate, consider a simple X-gate applied to a qubit initially in the state $|0\rangle$. In an ideal situation, the gate would transform the qubit into the state $|1\rangle$. Repeated applications of an ideal X-gate would result in the qubit state oscillating between $|0\rangle$ and $|1\rangle$, as shown in Fig. 12.4.

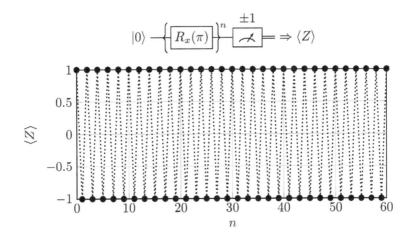

Figure 12.4: Repeated Application of an Ideal X Gate

The X-gate is typically implemented as a π rotation around the x-axis: $X = iR_x(\pi)$ (see §§ 4.5.2 and 5.2.5). However, calibration errors can occur. For instance, the amplitude of the control pulse may be slightly higher than required, resulting in an actual operation $R_x(\pi + \varepsilon)$. When we repeat this off-calibration X-gate, the resulting qubit state will not simply oscillate between $|0\rangle$ and $|1\rangle$, but may evolve into a superposition state. This is illustrated in Fig. 12.5.

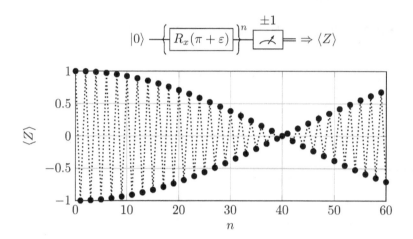

Figure 12.5: Repeated Application of an Off-Calibrated X Gate

Exercise 12.20 Deduce the value of the calibration error ε based on the graph in Fig. 12.5. Hint: $(R_x(\pi + \varepsilon))^n = R_x(n\pi + n\varepsilon)$. When $n\varepsilon$ accumulates to $\frac{\pi}{2}$, the final state is a uniform superpostion, yielding $\langle Z \rangle = 0$.

As demonstrated in the preceding example, the impact of gate calibration errors can be exacerbated in circuits with increasing depth (n). Specifically, as n grows, these errors may accumulate, leading to significant deviations in the final computational results from the expected outcomes.

2 Cross-Talk Errors

Cross-talk errors occur due to unintended interactions between qubits that are either adjacent or otherwise coupled. These interactions can introduce undesired rotations and entanglement, thereby affecting the fidelity of gate operations.

Consider, for instance, an unwanted ZZ coupling governed by the following two-qubit Hamiltonian:

$$H_{ZZ} = -\frac{\hbar\omega}{2}ZZ, \qquad (12.84)$$

where ω represents the coupling strength, and ZZ denotes $Z \otimes Z$.

This Hamiltonian leads to a unitary evolution given by $R_{zz}(\omega t)$, which can be viewed as an unwanted $ZZ(\theta)$ gate with $\theta = \omega t$:

$$R_{zz}(\omega t) = \cos\frac{\omega t}{2}I + i\sin\frac{\omega t}{2}ZZ. \qquad (12.85)$$

Such an interaction results in a phase walk-off in the computational basis. To measure this effect, we sandwich the $R_{zz}(\omega t)$ unitary between Hadamard gates, as

12.3 Error Mechanisms in Quantum Computing

depicted in Fig. 12.6. In the absence of coupling ($\omega = 0$), the measurement on the first qubit yields $\langle ZI \rangle = 1$, corresponding to the expected outcome $|00\rangle$.

However, with coupling, the measurement outcome changes to $\langle ZI \rangle = \cos \omega t$, indicating the emergence of unwanted phase shift that deteriorates over time. (Similarly for $\langle IZ \rangle$ measured on the second qubit.)

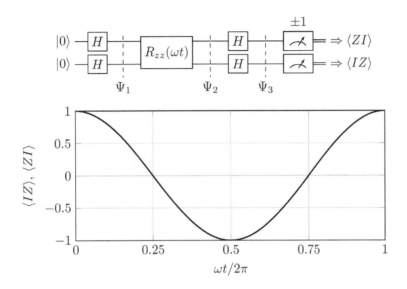

Figure 12.6: Effect of ZZ Cross-Talk over Time

Below is an ouline for the derivation of $\langle ZI \rangle = \cos \omega t$:

$$|\Psi_1\rangle = \text{HH} |00\rangle = |++\rangle, \tag{12.86a}$$

$$|\Psi_2\rangle = R_{zz}(\omega t) |\Psi_1\rangle = \cos \frac{\omega t}{2} |++\rangle + i \sin \frac{\omega t}{2} |--\rangle, \tag{12.86b}$$

$$|\Psi_3\rangle = \text{HH} |\Psi_2\rangle = \cos \frac{\omega t}{2} |00\rangle + i \sin \frac{\omega t}{2} |11\rangle, \tag{12.86c}$$

where we have used the following properties: $H|0\rangle = |+\rangle$, $H|1\rangle = |-\rangle$, $H|+\rangle = |0\rangle$, $H|-\rangle = |1\rangle$, $Z|+\rangle = |-\rangle$, $Z|-\rangle = |+\rangle$.

$$\langle ZI \rangle = \langle \Psi_3 | ZI | \Psi_3 \rangle \tag{12.87a}$$

$$= \cos^2 \frac{\omega t}{2} - \sin^2 \frac{\omega t}{2} = \cos \omega t. \tag{12.87b}$$

Exercise 12.21 Plot $\langle ZZ \rangle$ as function of ωt.

12.3.2 ∗ Incoherent Errors

Incoherent errors are fundamentally stochastic in nature and primarily arise due to uncontrolled interactions between the quantum system and its external environment. Unlike coherent errors, which can be modeled deterministically, incoherent errors are statistically represented using density operators and noise channels.

There are numerous types of incoherent errors that affect both single-qubit and multi-qubit systems. In the following subsections, we will focus on three key examples: bit-flip, phase-flip, and depolarizing errors. We will also outline the noise channel representations for other types of errors. The concepts and formulas presented here are essential for understanding and simulating noise in quantum computing.

> **Pure States, Mixed States, Noise, and Noise Channels**
>
> Here is a recapitulation of some key concepts from § 12.2.
>
> **Pure States:** In quantum mechanics, a pure state is described by a unique ket vector $|\psi\rangle$ in a Hilbert space. For a pure state, the associated density operator ρ is a projection operator given by $\rho = |\psi\rangle\langle\psi|$. A pure state is free from 'quantum noise,' meaning that repeated quantum measurements, under identical conditions, yield the same probabilities for each possible outcome, which in the case of certain observables can produce deterministic results in quantum computing.
>
> **Mixed States:** In contrast, a mixed state is a statistical ensemble of multiple potential quantum states. It is represented by a density operator ρ that is not expressible as $|\psi\rangle\langle\psi|$ for any ket $|\psi\rangle$. A mixed state introduces classical uncertainty, leading to non-deterministic results in quantum computing because the system can be in any of the states of the ensemble with a certain probability.
>
> **Noise and Errors:** In quantum systems, "noise" often refers to processes that convert pure states into mixed states. Such processes include decoherence, dissipation, and interaction with the environment, among others. This random noise manifests as computational errors, which can affect the outcomes of quantum computations.
>
> **Noise Channels and Kraus Representation:** Noise and errors in quantum computing are commonly modeled through quantum channels, which act as trace-preserving transformations on density operators. This is also known as the Kraus representation:
>
> $$\mathcal{N}(\rho) = \sum_j s_j K_j \rho K_j^\dagger,$$
>
> where $\{K_j\}$ are the Kraus operators and $\{s_j\}$ are the corresponding probabilities.
>
> This representation serves as a powerful framework for both understanding and mitigating the impacts of noise in quantum systems.

1 Bit-Flip Error

A bit-flip error inverts the state of a qubit from $|0\rangle$ to $|1\rangle$ or vice versa with a predetermined probability s. This error can be modeled by applying an X gate randomly with probability s. The corresponding noise channel for a bit-flip error is represented as

$$\mathcal{N}_{\text{bit-flip}}(\rho) = (1-s)\rho + sX\rho X. \tag{12.88}$$

As discussed in Example 12.5, an X gate transforms a general quantum state, represented by the Bloch vector (x, y, z), into $(x, -y, -z)$. By combining both terms in $\mathcal{N}_{\text{bit-flip}}(\rho)$, we obtain:

$$\mathcal{N}_{\text{bit-flip}}(x, y, z) = (x, (1-2s)y, (1-2s)z). \tag{12.89}$$

12.3 Error Mechanisms in Quantum Computing

Hence, the influence of bit-flip errors can be visualized as a contraction of the terminal points of the qubit states in the Bloch sphere along the y and z axes by a factor of $1 - 2s$, as depicted in Fig. 12.7.

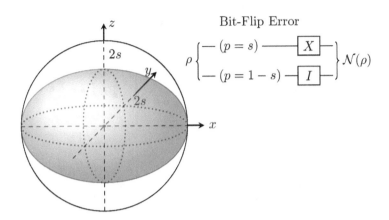

Figure 12.7: Bit-Flip Error in the Bloch Sphere

To elucidate the impact of bit-flip errors further, consider their combined effect with a sequence of X (or $R_x(\pi)$) gates, as presented in Fig. 12.4. The resulting Kraus representation becomes:

$$\mathcal{N}_{\text{X-and-BF}}(\rho) = X\mathcal{N}_{\text{bit-flip}}(\rho)X = (1-s)X\rho X + s\rho, \tag{12.90}$$

or, when expressed in terms of Bloch vectors,

$$\mathcal{N}_{\text{X-and-BF}}(x, y, z) = (x, (2s-1)y, (2s-1)z). \tag{12.91}$$

Upon n repeated applications of this transformation, one obtains:

$$\mathcal{N}^n_{\text{X-and-BF}}(x, y, z) = (x, (2s-1)^n y, (2s-1)^n z). \tag{12.92}$$

Assuming the initial state is $|0\rangle$, corresponding to $(0, 0, 1)$, we have:

$$\mathcal{N}^n_{\text{X-and-BF}}(0, 0, 1) = (0, 0, (2s-1)^n). \tag{12.93}$$

Therefore,

$$\langle Z \rangle = (2s-1)^n = (-1)^n(1-2s)^n. \tag{12.94}$$

The above equation is graphed in Fig. 12.8 with $s = 1\%$. While the unaltered application of the X gate results in an oscillation between the qubit states $|0\rangle$ and $|1\rangle$ (see Fig. 12.4), the presence of bit-flip errors induces a damping effect on the amplitude of this oscillation by the factor $(1-2s)^n$.

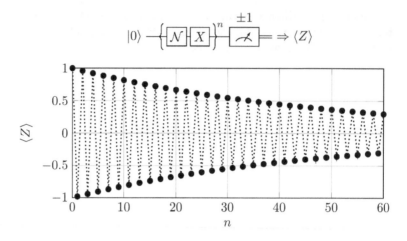

Figure 12.8: Repeated Applications of an X Gate with Bit-Flip Noise

2 Phase-Flip Error

A phase-flip error results in a sign change in the $|1\rangle$ state of a qubit, and is represented by the random application of the Pauli-Z operator with a predetermined probability s. The Kraus representation of the phase-flip error is given by:

$$\mathcal{N}_{\text{phase-flip}}(\rho) = (1-s)\rho + sZ\rho Z. \tag{12.95}$$

The Z gate modifies a general quantum state represented by the Bloch vector (x, y, z) into $(-x, -y, z)$. Integrating both terms in $\mathcal{N}_{\text{phase-flip}}(\rho)$, we find:

$$\mathcal{N}_{\text{phase-flip}}(x, y, z) = ((1-2s)x, (1-2s)y, z). \tag{12.96}$$

Thus, phase-flip errors can be visualized as a contraction of the terminal points of the qubit states along the x and y axes in the Bloch sphere by a factor of $1 - 2s$, as depicted in Fig. 12.9.

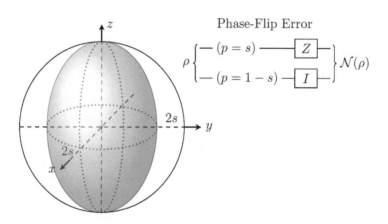

Figure 12.9: Phase-Flip Error in the Bloch Sphere

3 Depolarizing Error

Depolarizing error describes the transition of a qubit from a pure state into a completely mixed state comprising $|0\rangle\langle 0|$ and $|1\rangle\langle 1|$. In this process, the qubit loses both its phase information and the superposition between its basis states. The Kraus representation of depolarizing error is:

$$\mathcal{N}_{\text{dephase}}(\rho) = (1-s)\rho + \frac{s}{3}(X\rho X + Y\rho Y + Z\rho Z), \qquad (12.97)$$

where the parameter s quantifies the noise strength.

This error manifests as a uniform contraction of the terminal points of the qubit states within the Bloch sphere, as illustrated in Fig. 12.10.

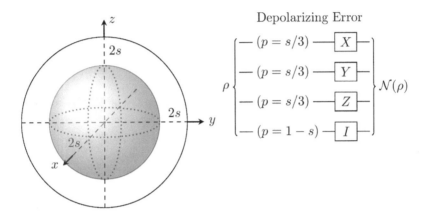

Figure 12.10: Depolarizing Error in the Bloch Sphere

4 Dephasing Error

Dephasing error occurs when the relative phase between the basis states $|0\rangle$ and $|1\rangle$ becomes randomized. The phase-flip error can be considered a special case of dephasing error. Dephasing error can be modeled mathematically by applying a random phase using the $R_z(\theta)$ gate to the qubit state:

$$\mathcal{N}_{\text{dephase}}(\rho) = (1-s)\rho + sR_z(\theta)\rho R_z^\dagger(\theta), \qquad (12.98)$$

where s is the probability of the noise, and θ is a random variable in $[0, \pi)$.

A more abstract definition that does not specify the random process (or equivalently, averaging over all θ values) leading to the dephasing but focuses on the overall effect is:

$$\mathcal{N}_{\text{dephase}}(\rho) = (1-s)\rho + s\,\text{diag}(\rho), \qquad (12.99)$$

where $\text{diag}(\rho)$ represents zeroing out off-diagonal elements of ρ and retaining only the diagonal elements.

The complete dephasing channel in Example 12.7 can be considered as the special case of $s = 1$.

5 Amplitude Damping Error

Amplitude damping error models the irreversible decay of the qubit state $|1\rangle$ to the state $|0\rangle$. For example, in a quantum computer, the state $|1\rangle$ typically has a higher energy than the state $|0\rangle$, prompting the system to relax from $|1\rangle$ to $|0\rangle$. The Kraus representation is as follows:

$$\mathcal{N}_{\text{amp-damping}}(\rho) = K_1 \rho K_1^\dagger + K_2 \rho K_2^\dagger, \quad (12.100)$$

where $K_1 = \begin{bmatrix} 1 & 0 \\ 0 & \sqrt{1-s} \end{bmatrix}$, $K_2 = \begin{bmatrix} 0 & \sqrt{s} \\ 0 & 0 \end{bmatrix}$, and s is the probability of decay.

Note that amplitude damping error is not a simple random occurrence of two unitary transformations; instead, the probabilities are embedded within the Kraus operators.

6 Pauli Channel

This is a general single-qubit noise channel. Pauli X, Y, and Z matrices serve as Kraus operators to model this channel:

$$\mathcal{N}_{\text{Pauli}}(\rho) = (1 - s_x - s_y - s_z)\rho + s_x X \rho X + s_y Y \rho Y + s_z Z \rho Z, \quad (12.101)$$

where s_x, s_y, and s_z are the probabilities of each Pauli error. Restrictions on these probabilities are not imposed unless specified.

7 Two-qubit Bit-Flip Error

This channel describes the bit-flip process in a two-qubit quantum state. It is an extension of the single-qubit bit-flip error. The Kraus representation of this error channel is given by:

$$\mathcal{N}_{\text{bit-flip-2}}(\rho) = (1-s)\rho + \frac{s}{3}(IX\rho IX + XI\rho XI + XX\rho XX), \quad (12.102)$$

where s is the strength of the noise. The last three terms represent the flip of the first qubit, the second qubit, and both, respectively. The factor of $\frac{1}{3}$ in the Kraus representation ensures that each of the bit-flip error terms contributes equally to the overall noise process.

8 Two-qubit Dephasing Error

This channel describes the dephasing process in a two-qubit quantum state. It is an extension of the single-qubit phase-flip error. The Kraus representation is:

$$\mathcal{N}_{\text{dephase-2}}(\rho) = (1-s)\rho + \frac{s}{3}(IZ\rho IZ + ZI\rho ZI + ZZ\rho ZZ), \quad (12.103)$$

where s is the strength of the noise. The factor of $1/3$ in the Kraus representation is conventional and allows for equal contributions from each term.

9 Two-qubit Pauli Channel

This is a general two-qubit noise channel that can model loss of quantum phase, superposition between basis states, and entanglement. The Kraus representation is:

12.3 Error Mechanisms in Quantum Computing

$$\mathcal{N}_{\text{Pauli-2}}(\rho) = \sum_i s_i K_i \rho K_i^\dagger, \tag{12.104}$$

where K_i is one of the sixteen two-qubit Pauli matrices, formed by taking tensor products of single-qubit Pauli matrices, and s_i is the probability of the corresponding K_i.

12.3.3 ∗Other Types of Errors

Apart from coherent and incoherent errors, there are specific errors that occur during the initialization and readout stages of quantum computation. State preparation and measurement (SPAM) errors are often critical as they can set the stage for or amplify subsequent errors during the computation process.

1 State Preparation Errors

Occurring during the initialization phase, state preparation errors arise when an intended initial state like $|0\rangle$ is realized as a different state such as $|1\rangle$. These errors can be either coherent or incoherent.

State preparation errors can be modeled using a density matrix representation. If the intended initial state is $|\psi\rangle$, the actual state can be described as:

$$\rho_{\text{actual}} = (1-\varepsilon)|\psi\rangle\langle\psi| + \varepsilon \rho_{\text{error}}, \tag{12.105}$$

where ε is the error rate and ρ_{error} is the density matrix representing the error state.

2 Readout Errors (Measurement Errors)

Readout errors, also known as measurement errors, manifest during the readout process. A state such as $|0\rangle$ might be mis-measured as $|1\rangle$. Similar to state preparation errors, readout errors can also be either coherent, arising from systematic biases, or incoherent, occurring randomly.

Mathematically, readout errors can be modeled using a confusion matrix C, where the element C_{ij} represents the probability of measuring state $|j\rangle$ when the actual state is $|i\rangle$. For a two-level system, the confusion matrix can be expressed as:

$$C = \begin{bmatrix} 1-\varepsilon_0 & \varepsilon_1 \\ \varepsilon_0 & 1-\varepsilon_1 \end{bmatrix}, \tag{12.106}$$

where ε_0 and ε_1 are the probabilities of incorrectly reading the states $|0\rangle$ and $|1\rangle$, respectively.

12.3.4 Performance Merit Parameters

Merit parameters in quantum computing provide a quantitative framework to assess the capabilities and limitations of quantum systems. In contrast to classical computers, which are relatively homogenous in technology, quantum computers are developed across various and rapidly advancing hardware platforms. Standardized performance metrics are crucial for comprehensively understanding, comparing, and validating the operations of these diverse quantum technologies.

While the number of qubits is frequently cited as a primary comparison metric, implying that more qubits equate to a more potent system able to manage larger or more complex algorithms, this perspective oversimplifies the intricacies of quantum computation. Given the susceptibility of current quantum computing hardware to noise and consequential error rates that can inhibit performance, a combined assessment of the quantity and the quality of qubits yields a more nuanced evaluation of different quantum systems.

These merit parameters encompass temporal, spatial, and computational dimensions from a physical standpoint. Operationally, they are examined at various levels: qubit, circuit, Quantum Processing Unit (QPU), and the overall system. Furthermore, metrics concerning fault-tolerance and scalability are critical. They offer insights into the long-term practicality and effectiveness of quantum computing architectures, particularly those that aim for fault-tolerant operations and scalable designs.

1 Temporal Metrics

Relaxation Time: T_1

The relaxation time T_1 measures the rate at which an excited qubit returns to its ground state due to energy relaxation, also known as amplitude damping. This parameter determines the effective lifetime of the excited state.

An exponential decay model for the relaxation of the excited state can be represented as:

$$\rho(t) = e^{-t/T_1}\rho(0) + (1 - e^{-t/T_1})|0\rangle\langle 0|, \qquad (12.107)$$

where $\rho(t)$ denotes the system's state at time t. As $t \to \infty$, the qubit eventually stabilizes in the state $|0\rangle$.

In quantum error correction, the effects of energy relaxation are typically abstracted and modeled as bit-flip errors (§ 12.3.2.1 and § 12.4.1), capturing the transition from excited to ground states. The exponential decay of the excited state corresponds to a constant rate of occurrence for bit-flip errors over time, which can be modeled by a Poisson distribution.

Coherence Time: T_2

The coherence time T_2 quantifies the duration over which a qubit maintains a coherent superposition before dephasing errors transform it into a mixed state. This factor is critical because quantum computation relies on maintaining the phase coherence of qubits for quantum interference and entanglement.

For a generic pure state $|\psi\rangle = \alpha|0\rangle + \beta|1\rangle$, the density matrix is:

$$\rho = \begin{bmatrix} |\alpha|^2 & \alpha\beta^* \\ \beta\alpha^* & |\beta|^2 \end{bmatrix}. \qquad (12.108)$$

Dephasing can be modeled by an exponential decay of the off-diagonal elements:

$$\rho(t) = \begin{bmatrix} |\alpha|^2 & \alpha\beta^* e^{-t/T_2} \\ \beta\alpha^* e^{-t/T_2} & |\beta|^2 \end{bmatrix}. \qquad (12.109)$$

As $t \to \infty$, the qubit evolves into a mixed state $\rho = |\alpha|^2|0\rangle\langle 0| + |\beta|^2|1\rangle\langle 1|$.

12.3 Error Mechanisms in Quantum Computing

For the purposes of quantum error correction, dephasing effects are commonly represented through phase-flip error channels (§ 12.3.2.2 and § 12.4.2), which model the loss of phase coherence between superposition states.

Qubit Lifetime: T_q

The qubit lifetime T_q is the aggregate period during which a qubit remains in a computationally useful state. Simplistically, T_q can be approximated by $\min(T_1, T_2)$. Typically, T_2 is shorter than T_1, as phase coherence is susceptible to any environmental fluctuations that can perturb the qubit states, even in the absence of energy exchange.

This metric is affected by various elements, such as the quality of qubit fabrication, the efficiency of error correction strategies, and the specific quantum computing architecture employed. Different qubit technologies exhibit a wide range of gate lifetimes and other key temporal parameters. For instance, superconducting qubits often display qubit lefetimes ranging from microseconds to milliseconds. In contrast, systems based on optical qubits, like trapped ions, can sustain coherence for significantly longer durations, spanning from seconds to minutes in some cases.

Gate Time: T_{gate}

The operational speed of a QPU is intrinsically related to the duration of its gate operations. The gate time T_{gate} is the time required to execute a single quantum gate operation.

Coherence Time to Gate Time Ratio

The ratio of coherence time T_q to gate time T_{gate} holds practical significance. If T_{gate} is considerably shorter than T_q, a larger number of successive gate operations can be executed before coherence is lost. Consequently, the achievable depth of quantum circuits on a QPU is generally limited by this ratio.

For instance, while optical systems may exhibit greater coherence times, their control pulse durations are also longer. This means that the ratio of qubit coherence to gate operation times may not necessarily be more advantageous than that of superconducting or semiconducting systems, which operate with pulses on the nanosecond (ns) timescale.

2 Spatial Metrics

Connectivity or Topology

Connectivity, or topology, in quantum computing describes the manner in which qubits within a system are interconnected, facilitating quantum operations between them. This aspect is important in determining how easily multi-qubit gates can be executed.

High connectivity eases the implementation of quantum algorithms by reducing the need for additional SWAP gates, which are otherwise necessary to bring non-adjacent qubits into interaction range. However, increased connectivity can also complicate the physical design of the system and introduce additional error sources, such as cross-talk.

Connectivity types range from nearest-neighbor (where qubits interact only with adjacent ones) to all-to-all (permitting direct interaction between any pair of qubits). The connectivity pattern is closely tied to the quantum computer's architecture and shows significant differences across technologies. For example, solid-state devices like superconducting circuits and semiconducting qubits are typically limited in connectivity due to the constraints of physical wiring or resonators. These systems often have just two or three directly connected neighbors. Entangling distant qubits in such architectures requires additional operations, like SWAP gates or teleportation, to transfer quantum information.

In contrast, trapped ion and optical systems are not constrained by a two-dimensional circuit layout, thereby offering greater flexibility in qubit addressing and entanglement. These technologies can often achieve an all-to-all connectivity, facilitating more direct and versatile qubit interactions for algorithm execution and logic processing. However, it is important to note that the physical movement of ions necessary for these interactions in trapped ion systems can be time-consuming, potentially impacting the overall operation speed of the quantum computer.

Cross-Talk

Cross-talk in quantum computing refers to the inadvertent interactions between qubits that are not intended to be coupled at a particular moment. Such interference typically stems from electromagnetic disturbances, stray couplings, or leakage in control signals.

This phenomenon can inadvertently alter the states of qubits not actively participating in a specific operation, potentially leading to errors and reduced fidelity in quantum computations. To combat cross-talk, strategies such as meticulous design of the physical layout, shielding, implementation of decoupling sequences, and precise calibration routines are utilized. Advanced control methods are also employed for dynamic compensation of cross-talk effects.

As quantum systems scale up, understanding and mitigating cross-talk becomes increasingly vital for developing reliable and fault-tolerant quantum circuits. The challenge lies in achieving high connectivity while simultaneously maintaining qubit coherence and minimizing cross-talk and other error sources in quantum computer design.

3 Computational Metrics

Fidelity

Fidelity quantifies the precision with which a quantum gate or circuit performs in comparison to its ideal theoretical model. A fidelity value of 1 signifies flawless execution, whereas values less than 1 reflect discrepancies from the ideal operation.

The expected output of an ideal quantum gate or circuit is typically a pure state, denoted as $|\psi_{\text{ideal}}\rangle$. However, due to noise and other factors, the actual output might be a mixed state, represented by ρ_{actual}. The fidelity in this scenario is defined as:

$$F = \langle \psi_{\text{ideal}} | \rho_{\text{actual}} | \psi_{\text{ideal}} \rangle . \tag{12.110}$$

This formulation encompasses both the desired unitary operations and the non-unitary effects of noise. For cases where the actual output is also a pure state $|\psi_{\text{actual}}\rangle$, fidelity simplifies to $F = |\langle \psi_{\text{ideal}} | \psi_{\text{actual}} \rangle|^2$.

Fidelity can be assessed at both the gate and circuit levels. Generally, circuit fidelity tends to decrease with increasing circuit depth due to error accumulation, as expressed by:

12.3 Error Mechanisms in Quantum Computing

$$F \approx F_G^D, \qquad (12.111)$$

where F is the overall circuit fidelity, F_G represents the average gate fidelity, and D is the circuit depth.

Notably, the fidelity of two-qubit gates, such as CNOT or ZZ, frequently determines the upper limit of the system's overall fidelity. This is because two-qubit gates generally present more challenges in achieving high precision.

Error Rate

Error rate quantifies the probability of an error occurring during a quantum operation, which could be a single gate or a sequence of gates. Techniques such as randomized benchmarking are often used to estimate this rate.

Fidelity (F) and error rate (ε) are complementary metrics. They are related by:

$$\varepsilon = 1 - F. \qquad (12.112)$$

Circuit Depth

Circuit depth refers to the maximum number of sequential gate operations that can be performed before the system accrues an unacceptable level of errors. As a rough approximation, circuit depth (D) and the gate-level error rate (ε_G) are inversely related:

$$D \approx \frac{1}{\varepsilon_G}, \qquad (12.113)$$

indicating that higher error rates limit the feasible depth of quantum circuits.

4 System-Level Composite Metrics

In the rapidly evolving field of quantum computing, single-number composite metrics offer a holistic approach to evaluate the performance of quantum computing systems. These metrics aim to distill the complex, multi-faceted characteristics of quantum systems into a single, comprehensive figure, facilitating a more straightforward comparison and understanding of quantum processors from an application-level perspective. While beneficial for providing an overarching view of a quantum computer's capabilities, these metrics are continually evolving, reflecting the dynamic nature of quantum technology and its diverse applications. Below are a few representative examples.

Quantum Volume

Quantum Volume (QV), introduced by IBM, serves as a composite metric designed to provide a unified framework for evaluating and comparing quantum processors. It accounts for factors such as the number of qubits, coherence, gate fidelity, and connectivity.

QV is determined by the size of the largest square circuit that can be successfully run on a QPU. The term "square" refers to a circuit where the number of qubits (width) is equal to the number of layers of two-qubit gates (depth). The protocol for determining QV involves running a series of circuits that increase in complexity, both in terms of width and depth, and identifying the largest circuit that the quantum computer can reliably execute. These circuits consist of layers of qubit permutation

(SWAP) and random single and double qubit gates. If the number of qubits in the largest successfully executed square circuit is n, then QV is given by 2^n.

QV, being dependent on both circuit depth and width, inherently relates QPU performance to its topology, error rates, and overall size. As such, QV is increasingly used as a gauge of QPU performance, offering a holistic and comprehensive view through a single value.

CLOPS

While Quantum Volume (QV) provides insight into the potential capabilities of a quantum computer, it does not fully capture the operational speed at a system level. This is particularly relevant for quantum-classical interfacing in full applications, where algorithms require multiple calls to a QPU. The efficiency of the runtime system, which facilitates quantum-classical communication, is crucial for high performance.

CLOPS, or Circuit Layer Operations Per Second, measures the speed and capability of a quantum computer to solve problems of a certain complexity. It defines how many QV circuits a quantum processor can process within a given timeframe, thus providing an operational speed measure in practical scenarios. CLOPS emphasizes the practical throughput of executing quantum algorithms, taking into account the efficiency of quantum-classical communication and the runtime system.

A pertinent metric in this context is "teraquops", the quantum counterpart of "teraops" in classical computation. It signifies the number of quantum operations (quops) a quantum system is capable of performing each second, scaled to a tera-order of magnitude (trillions of operations per second). This metric gauges the raw computational power and speed of quantum systems, complementing the throughput-centric perspective provided by CLOPS.

Algorithmic Qubits

Quantum Volume (QV) is a widely recognized metric, yet there is a growing consensus that it may not fully encompass the nuances of quantum computational power, especially for specific algorithms or practical use cases. Concerns have been raised that QV might not accurately reflect the performance of quantum computers in scenarios that do not require deep circuits or extensive qubit connectivity. Furthermore, QV may not represent the capabilities of various quantum computing platforms, such as quantum annealers or trapped ion systems, as effectively as it does for superconducting qubit systems.

In response, the concept of Algorithmic Qubits (AQ) has been proposed as an alternative metric. Unlike QV, which assesses a QPU's ability to execute increasingly complex quantum circuits, AQ focuses on the "algorithmic strength" of a quantum computer by considering error rates, qubit coherence, gate fidelity, and the overall system architecture in relation to specific quantum algorithms.

AQ provides valuable insights into the practical usability of a quantum processor for executing complex algorithms, offering an application-oriented perspective. It indicates the number of qubits that can be effectively used for algorithmic tasks, factoring in error correction and mitigation techniques necessary for meaningful computational outcomes. AQ thus complements QV, aiding in the selection of the most appropriate quantum processor for specific algorithmic requirements.

5 Fault-Tolerance and Scalability Metrics

Fault-tolerance and scalability metrics are instrumental in providing insights into the viability and efficiency of fault-tolerant quantum computing architectures. By analyzing these metrics, researchers and engineers are equipped to identify optimal architectures and error correction strategies, thereby facilitating the practical realization of robust and high-performance quantum computers.

These metrics primarily encompass an analysis of noise structure, the relationship between physical noise levels and fault-tolerant thresholds (quantified by the Λ factor), the scaling properties of quantum error-correcting codes (QECC), the complexity of QECC decoding, and the computational resource requirements essential for achieving fault tolerance. The specific metrics of interest include:

The Λ Factor

The Λ factor is an essential metric, defined as the ratio of the threshold error rate for a given QECC, $\varepsilon_{\text{threshold}}$, to the actual physical error rate of quantum operations, $\varepsilon_{\text{physical}}$:

$$\Lambda = \frac{\varepsilon_{\text{threshold}}}{\varepsilon_{\text{physical}}}. \tag{12.114}$$

A higher Λ factor is indicative of physical error rates being substantially below the threshold, thereby facilitating effective error correction and extending coherence times within the system.

Application Footprint

The application footprint of a specific quantum computing architecture is a measure that quantifies the resource requirements for a quantum computing task. It is gauged in terms of the number of physical qubits necessary to realize a certain number of logical qubits with a target error rate (e.g., 10^{-12}) or to address a particular computational problem (e.g., factoring a 2048-bit integer).

This metric elucidates the scalability of a quantum computing architecture, underscoring the trade-offs between the overhead of error correction and computational capability. It is invaluable for evaluating the feasibility of employing large-scale quantum algorithms for real-world applications.

Subthreshold Scaling

Subthreshold scaling is a metric that examines the performance dynamics of a QECC as the physical error rate falls below the threshold error rate. This analysis yields insights into the efficiency and dependability of the error correction code, particularly as the system's size and complexity expand. This metric is crucial for assessing the long-term sustainability and performance viability of QECCs.

QECC Decoding Complexity

The complexity of decoding QECCs varies significantly among different types. For instance, surface codes demand minimal additional processing for error correction. In contrast, Low-Density Parity-Check (LDPC) codes necessitate extensive computational resources for real-time decoding, which presents challenges for scalability. This measure of decoding complexity serves as an indicator of the resources required—specifically in terms of time, memory, and classical computational effort—to effectively identify and correct errors in a quantum computing system.

Capability to Suppress a Broad Spectrum of Errors

Some errors, particularly coherent errors such as those encountered in state preparation, gate calibration, and measurement, cannot be effectively corrected by current QECCs. Nevertheless, for a quantum computing system to scale effectively, it is crucial to develop strategies capable of suppressing a comprehensive range of errors that undermine computational accuracy.

12.4 Introduction to Error Correction Codes

Achieving fault tolerance is an essential yet elusive goal for the advancement of quantum computing. Theoretically, full fault tolerance—often considered the Holy Grail in the field—becomes attainable once the actual error rate falls below a certain threshold. However, this comes at the expense of requiring a significant qubit overhead, making effective error correction a primary research focus.

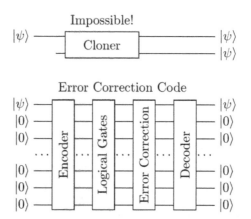

Figure 12.11: Structure of Practical Error Correction Code

Quantum error correction follows a systematic procedure known as an error correction code. As illustrated in Fig. 12.11, this procedure encompasses the encoding of logical qubits into a larger set of physical qubits, the detection and identification of errors, and the application of corrective operations. Subsequently, decoding is performed to retrieve the corrected logical information. The ultimate objective is to approximate a nearly perfect logical qubit by employing a sufficiently large array of physical qubits.

In this section, we introduce the fundamentals of quantum error correction codes through the study of Shor codes, an archetypal example in the field. Shor codes use a blend of quantum entanglement, local operations, and principles from classical error correction to bolster the fidelity of quantum information processing. Beyond its utility in fault-tolerant quantum computing, Shor codes also find applications in secure quantum communications.

We will focus on the comprehensive nine-qubit Shor code. To fully grasp the sophistication behind Shor's code, we start by examining two simpler but foundational

Peter Shor

Source: International Centre for Theoretical Physics

12.4 Introduction to Error Correction Codes

codes: the three-qubit bit-flip code and the three-qubit phase-flip code. Shor's code is a masterful combination of these elementary codes, enabling it to correct both bit-flip and phase-flip errors simultaneously, and thus effectively addressing any one-qubit error.

12.4.1 Three-Qubit Bit-Flip Code

The three-qubit bit-flip code is a quantum error correction code designed to guard against single bit-flip errors. The quantum circuit diagram for this code is depicted in Fig. 12.12.

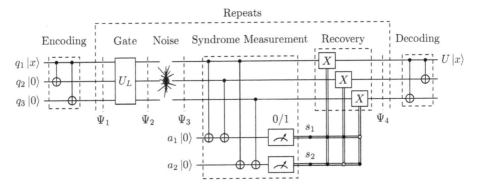

Figure 12.12: Quantum Circuit for the Three-Qubit Bit-Flip Code

---

Computation in Place

In quantum circuits, gates do not exist simultaneously as they might on a classical printed circuit board or integrated circuit. Contrarily, in the quantum realm, neither gates nor wires manifest as palpable physical entities. What's physically implemented are only the qubits and their corresponding external controls. In this context, the wires represent the temporal continuity of the qubit states, and the quantum gates denote specific transformations of these qubit states, directed by the external controls.

For a more information, refer to § 5.3.2.

---

1 Encoding

The encoding process shown in Fig. 12.12 utilizes three data qubits q_1, q_2, and q_3 and two CNOT gates. In this context, $|x\rangle$ represents the basis states $|0\rangle$ and $|1\rangle$. After encoding, it becomes either $|0_L\rangle \equiv |000\rangle$ or $|1_L\rangle \equiv |111\rangle$. A generic single qubit state $|\psi\rangle = \alpha |0\rangle + \beta |1\rangle$ is mapped to a logical qubit state as:

$$|\Psi_1\rangle = |\psi_L\rangle \equiv \alpha |0_L\rangle + \beta |1_L\rangle \equiv \alpha |000\rangle + \beta |111\rangle. \tag{12.115}$$

 It's essential to note that $|\psi_L\rangle \neq |\psi\psi\psi\rangle$. This means that while $|\psi\rangle$ spreads across three qubits via entanglement, it isn't replicated thrice. Making such copies would violate the no-cloning theorem.

2 Logical Gates

When employing an error-correction code, qubit gates, such as the Pauli X and Z, must be translated into logical gates that act on the encoded logical state. These logical gates may affect multiple data qubits, but their operation does not take the system out of the code space, i.e., the space spanned by the logical basis states $\{|000\rangle, |111\rangle\}$. Consider a unitary gate U:

$$U(\alpha |0\rangle + \beta |1\rangle) = \gamma |0\rangle + \delta |1\rangle. \tag{12.116}$$

The logical U gate, given by U_L, transforms the state of the system into:

$$|\Psi_2\rangle = \gamma |000\rangle + \delta |111\rangle. \tag{12.117}$$

Exercise 12.22 Show that $X_L \equiv XXX$ functions as the logical bit-flip operator for the three-qubit bit-flip code.

This task exemplifies the concept of *transversal gates*, which involve applying physical gates individually to each qubit of a quantum error-correcting code word, without necessitating multi-qubit operations that could potentially propagate errors.

3 Noise Channel

The gate operation of U_L can be a long, complex process during which errors might occur. Bit-flip errors can be modeled as random applications of the X gates on data qubits. For instance, applying X to q_2 (or equivalently, IXI to all three qubits) flips the bit in q_2. This noise alters the state of the data qubits to:

$$|\Psi_3\rangle = \gamma |010\rangle + \delta |101\rangle. \tag{12.118}$$

4 Syndrome Measurement

The syndrome measurement circuit identifies any modifications to the qubits resulting from noise. It employs two ancilla qubits. The first, a_1, checks the parity of data qubits q_1 and q_2 with two CNOT gates. If q_1 and q_2 share the same state, either $|0\rangle$ or $|1\rangle$, the dual CNOT gates negate each other, and the measurement produces $s_1 = 0$. Thus, in the absence of noise, the state $\gamma |000\rangle + \delta |111\rangle$ invariably results in $s_1 = 0$, regardless of γ and δ values.

If a bit flip in q_2 has occured, q_1 and q_2 now have contrasting states, and the measurement on a_1 returns $s_1 = 1$, highlighting the core principle of parity measurement which discerns the parity of two data qubits without perturbing their states.

In a similar vein, the second ancilla, a_2, measures the parity of data qubits q_1 and q_3, generating a classical bit s_2. This parity information, represented by the ancilla outputs $\{s_1, s_2\}$, is termed the error syndrome. Table 12.1 summarizes the possible error syndromes, corresponding errors, and their corrective measures.

12.4 Introduction to Error Correction Codes

$s_{1,2}$	Error	Correction Gates
00	No error	None
01	Bit flip on q_3	X on q_3
10	Bit flip on q_2	X on q_2
11	Bit flip on q_1	X on q_1

Table 12.1: Error Syndrome and Correction for the Three-qubit Bit-flip Code

5 Error Correction and State Recovery

Upon detecting the error syndrome, the error can be rectified by applying an X gate to the relevant qubit as specified by Table 12.1. This action flips the aberrant qubit back to its noise-free state. Now the system state is recovered:

$$|\Psi_4\rangle = |\Psi_2\rangle = \gamma|000\rangle + \delta|111\rangle. \tag{12.119}$$

Utilizing the deferred measurement principle, as detailed in § 7.4.5, the classically controlled X gates can be substituted with CCNOT gates.

6 Sequential Gate Operations

A quantum computation task is typically carried out with a sequence of gates. As errors accumulates above a certain threshold, error detection and state recovery are executed. This is represented as repeating the gate-noise-syndrome-recovery block in Fig. 12.12.

7 Decoding

The decoder's role is to transform the logical qubit state $\gamma|000\rangle + \delta|111\rangle$ back to $\gamma|0\rangle + \delta|1\rangle$, which is $U|\psi\rangle$. This is achieved by applying two CNOT gates in the inverse sequence of the encoder.

8 Undetected Errors

It is important to address certain limitations of the Shor three-qubit bit-flip code. Referring to Table 12.1, we have made the assumption that only a single qubit experiences a bit-flip error. However, the syndrome 01 can be a result of bit-flip errors on both q_1 and q_2, corresponding to the error operator XXI. Similarly, the syndrome 11 can emerge from the error operator IXX, and the syndrome 00 might be due to XXX. Such double or triple bit-flip errors, although theoretically possible, are less frequent. Their probability is of the order $O(\varepsilon^2)$ or lower, making them less likely when compared to single bit-flip errors, especially when the single qubit error rate ε is sufficiently low.

Moreover, the Shor three-qubit bit-flip code is specifically designed to counteract bit-flip errors. Consequently, errors involving phase flips, represented by IIZ, IZI, or ZII, are not detected. Each of these transforms the state $\gamma|000\rangle + \delta|111\rangle$ to $\gamma|000\rangle - \delta|111\rangle$, with the parity unaffected. This limitation underscores the necessity of the phase-flip code.

Exercise 12.23 Reflect on the potential benefits of incorporating a third ancilla to measure the parity of q_2 and q_3. Could it enhance the capability to detect more types of errors?

Exercise 12.24 Investigate if the Shor three-qubit bit-flip code can detect and correct errors that results from random applications of Y gates. Aim to construct a table akin to Table 12.1 and an error correction code analogous to Fig. 12.12.

12.4.2 Three-Qubit Phase-Flip Code

The Shor phase-flip code is similar to its bit-flip counterpart, as illustrated in Fig. 12.13. The application of a Hadamard rotation transforms bit-flip operations to correct phase flips.

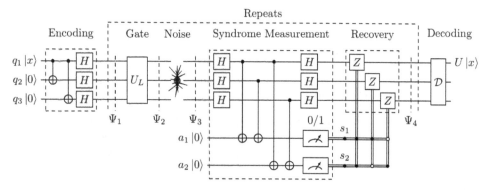

Figure 12.13: Quantum Circuit for the Three-Qubit Phase-Flip Code

The Three-Qubit Phase-Flip Code maps the initial qubit state $|\psi\rangle = \alpha|0\rangle + \beta|1\rangle$ to a three-qubit state as follows:

$$|\Psi_1\rangle = |\psi\rangle_L \equiv \alpha|+_L\rangle + \beta|-_L\rangle \equiv \alpha|+++\rangle + \beta|---\rangle, \quad (12.120)$$

where $|+\rangle = \frac{1}{\sqrt{2}}(|0\rangle + |1\rangle)$ and $|-\rangle = \frac{1}{\sqrt{2}}(|0\rangle - |1\rangle)$.

After the application of the logical gate U_L, the system state becomes:

$$|\Psi_2\rangle = \gamma|+++\rangle + \delta|---\rangle. \quad (12.121)$$

Exercise 12.25 Show that for the three-qubit phase-flip code, ZZZ functions as the logical bit-flip operator X_L.

A phase-flip error can be modeled as random applications of Z gates on data qubits, since $Z|+\rangle = |-\rangle$ and $Z|-\rangle = |+\rangle$. For instance, applying Z to q_1 (equivalent to ZII) flips the phase in q_1. This noise alters the state of the data qubits to:

$$|\Psi_3\rangle = \gamma|-++\rangle + \delta|+--\rangle. \quad (12.122)$$

12.4 Introduction to Error Correction Codes

Using the Hadamard gates, we can translate $\gamma\left|-++\right\rangle + \delta\left|+--\right\rangle$ into $\gamma\left|100\right\rangle + \delta\left|011\right\rangle$ and revert it. This process enables the CNOT gates to detect any noise-induced modifications to the qubits. For instance, in our scenario, the error manifests as a syndrome 11.

Table 12.2 summarizes the possible error syndromes, corresponding errors, and their corrective measures.

$s_{1,2}$	Error	Correction Gates
00	No error	None
01	Phase flip on q_3	Z on q_3
10	Phase flip on q_2	Z on q_2
11	Phase flip on q_1	Z on q_1

Table 12.2: Error Syndrome and Correction for the Three-Qubit Phase-Flip Code

The error can be rectified by applying a Z gate to the affected qubit (in our scenario, q_1), restoring the system state to:

$$\left|\Psi_4\right\rangle = \left|\Psi_2\right\rangle = \gamma\left|+++\right\rangle + \delta\left|---\right\rangle. \tag{12.123}$$

The decoder circuit, represented by $\mathcal{D}$ in Fig. 12.13, mirrors the encoder circuit. It decodes $\left|\Psi_4\right\rangle$ back to $\gamma\left|0\right\rangle + \delta\left|1\right\rangle$, equivalent to $U\left|\psi\right\rangle$.

Similar to the three-qubit bit-flip code, the three-qubit phase-flip Code is not equipped to detect double or triple phase-flip errors, nor any bit-flip errors.

> **Exercise 12.26** Modify the circuit in Fig. 12.13 for the three-qubit phase-flip code to employ classically controlled X gates instead of classically controlled Z gates for error correction.

12.4.3 ∗ Nine-Qubit Shor Code

The nine-qubit Shor code, introduced by Peter Shor in 1995, represents one of the pioneering quantum error-correcting codes. This code is devised by concatenating the three-qubit bit-flip and three-qubit phase-flip codes, as demonstrated in Fig. 12.14. This construction allows for the correction of any single-qubit errors, which will be elaborated upon subsequently.

1 Encoding

The encoding process integrates both the phase-flip and bit-flip codes, encoding the logical qubits of the phase-flip code using the bit-flip code.

The initial CNOT and Hadamard gates perform phase-flip code encoding on the qubits q_1, q_4, and q_7. Following this first-level encoding, the state of the entire nine-qubit system evolves as:

$$\left|0\right\rangle \mapsto \frac{1}{\sqrt{8}}(\left|0\right\rangle + \left|1\right\rangle)\left|00\right\rangle\,(\left|0\right\rangle + \left|1\right\rangle)\left|00\right\rangle\,(\left|0\right\rangle + \left|1\right\rangle)\left|00\right\rangle, \tag{12.124a}$$

$$\left|1\right\rangle \mapsto \frac{1}{\sqrt{8}}(\left|0\right\rangle - \left|1\right\rangle)\left|00\right\rangle\,(\left|0\right\rangle - \left|1\right\rangle)\left|00\right\rangle\,(\left|0\right\rangle - \left|1\right\rangle)\left|00\right\rangle. \tag{12.124b}$$

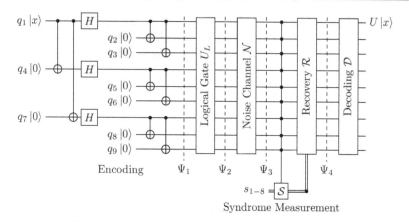

Figure 12.14: Quantum Circuit for the Nine-Qubit Shor Code

The CNOT gates that follow execute bit-flip code encoding on qubits q_1 through q_9. This operation transitions the basis states to their corresponding logical states:

$$|0\rangle \mapsto |0_L\rangle = \frac{1}{\sqrt{8}}(|000\rangle + |111\rangle)(|000\rangle + |111\rangle)(|000\rangle + |111\rangle), \quad (12.125a)$$

$$|1\rangle \mapsto |1_L\rangle = \frac{1}{\sqrt{8}}(|000\rangle - |111\rangle)(|000\rangle - |111\rangle)(|000\rangle - |111\rangle). \quad (12.125b)$$

A quantum state generally represented as $|\psi\rangle = \alpha|0\rangle + \beta|1\rangle$ is encoded into:

$$|\Psi_1\rangle = |\psi_L\rangle = \alpha|0_L\rangle + \beta|1_L\rangle. \quad (12.126)$$

Exercise 12.27 Elucidate the reason that the encoding procedure, as presented in Fig. 12.14, does not directly result in $|0\rangle \mapsto |+++++++++\rangle$ and $|1\rangle \mapsto |---------\rangle$, but rather adheres to the transformation depicted in Eq. 12.125. Reference: § 7.3.3.

2 Logical Gates

When employing an error-correction code, qubit gates must be translated into the corresponding logical gates (i.e., logical operations) that act on the encoded logical states. Consider a unitary gate U:

$$U(\alpha|0\rangle + \beta|1\rangle) = \gamma|0\rangle + \delta|1\rangle. \quad (12.127)$$

The logical U gate, denoted by U_L, is to be compatible with the error correction code, meaning it transforms a logical basis state to another logical basis state or a superposition of such states. After applying U_L, the system state is given by:

$$|\Psi_2\rangle = \gamma|0_L\rangle + \delta|1_L\rangle \quad (12.128a)$$
$$= \frac{\gamma}{\sqrt{8}}(|000\rangle + |111\rangle)(|000\rangle + |111\rangle)(|000\rangle + |111\rangle)$$
$$+ \frac{\delta}{\sqrt{8}}(|000\rangle - |111\rangle)(|000\rangle - |111\rangle)(|000\rangle - |111\rangle). \quad (12.128b)$$

12.4 Introduction to Error Correction Codes

Designing logical gates poses significant challenges. The following exercises offer an introductory exploration. The Clifford group, which includes Pauli gates (X, Y, Z), the Hadamard gate (H), the phase gate (S), and the CNOT gate, can be efficiently implemented with fault tolerance in various quantum error-correcting codes. For more discussion, refer to § 7.2.2.

Exercise 12.28 The operator $Z_1 Z_4 Z_7$ interchanges the amplitudes of the logical qubit states, effectively acting as a logical X gate (X_L). Demonstrate how applying $Z_1 Z_4 Z_7$ to the state $\alpha |0_L\rangle + \beta |1_L\rangle$ results in $\alpha |1_L\rangle + \beta |0_L\rangle$.

Similarly, a logical Z gate (Z_L) modifies the phase between the logical states. Show that applying $Z_L = X_1 X_2 X_3$ to $\alpha |0_L\rangle + \beta |1_L\rangle$ transforms it into $\alpha |0_L\rangle - \beta |1_L\rangle$, thus serving as a logical Z gate.

The nine-qubit Shor code supports multiple implementations of both X_L and Z_L. Identify two additional operators for each that fulfill the logical X and Z functions, respectively.

3 Error Syndrome Measurement and Correction

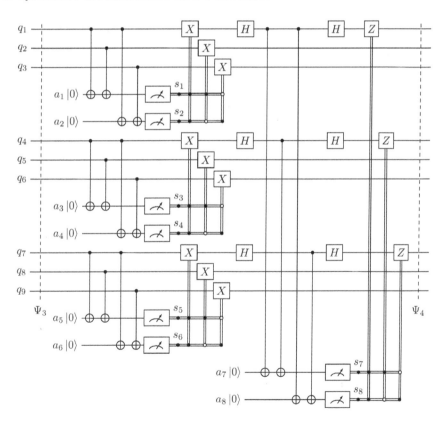

Figure 12.15: Error Syndrome Measurement and Correction for the Shor Code

As shown in Fig. 12.15, the nine-qubit Shor code utilizes three blocks of the three-qubit bit-flip code. Each block can detect and correct a bit-flip error within its group of three qubits. Similarly, for phase errors, the Shor code employs the three-qubit phase-flip code on three encoded qubits, one from each of the three

4 Bit-Flip Error

When a bit-flip error occurs, for instance, on q_5, the system state becomes:

$$|\Psi_3\rangle = \frac{\gamma}{\sqrt{8}}(\cdots)\,(|0\underline{1}0\rangle + |1\underline{0}1\rangle)\,(\cdots) + \frac{\delta}{\sqrt{8}}(\cdots)\,(|0\underline{1}0\rangle - |1\underline{0}1\rangle)\,(\cdots). \quad (12.129)$$

For clarity, in the preceding equation and in subsequent discussions, the changed bit is highlighted, and the unchanged blocks are represented using $(\cdots)$.

This bit-flip error is detected by the second block of the bit-flip code, yielding the error syndrome $s_3 s_4 = 10$. Consequently, the corrective X gate is applied to q_5, returning the system state to $|\Psi_4\rangle = |\Psi_2\rangle$.

5 Phase-Flip Error

When a phase-flip error takes place, for instance, on q_5, the system state evolves to:

$$|\Psi_3\rangle = \frac{\gamma}{\sqrt{8}}(\cdots)\,(|000\rangle - |111\rangle)\,(\cdots) + \frac{\delta}{\sqrt{8}}(\cdots)\,(|000\rangle + |111\rangle)\,(\cdots). \quad (12.130)$$

This phase-flip error is identified by the phase-flip code, generating the error syndrome $s_7 s_8 = 10$. As a result, the corrective Z gate is applied to q_4, bringing the system state back to $|\Psi_4\rangle = |\Psi_2\rangle$.

Note that the Shor code cannot distinguish which of the qubits q_4, q_5, or q_6 has experienced a phase-flip. This inability to differentiate is referred to as error degeneracy, a phenomenon where different errors yield the same syndrome. However, this lack of specificity in pinpointing the exact qubit does not hinder the ability to correct the error. The Shor code can effectively correct the phase flip without needing to know which of the qubits was affected.

6 General Single-Qubit Error

Consider a single-qubit error acting on q_5 such that:

$$|0\rangle \to a\,|0\rangle + b\,|1\rangle, \quad (12.131a)$$
$$|1\rangle \to c\,|0\rangle + d\,|1\rangle. \quad (12.131b)$$

The resulting system state can be represented as:

$$\begin{aligned}|\Psi_3\rangle =\ & \frac{\delta}{\sqrt{8}}(\cdots)\,(a\,|000\rangle + b\,|0\underline{1}0\rangle + c\,|1\underline{0}1\rangle + d\,|111\rangle)\,(\cdots) \\ & + \frac{\gamma}{\sqrt{8}}(\cdots)\,(a\,|000\rangle + b\,|0\underline{1}0\rangle - c\,|1\underline{0}1\rangle - d\,|111\rangle)\,(\cdots).\end{aligned} \quad (12.132)$$

To categorize bit-flip and phase-flip errors, we introduce the following relationships: $p + q = a$, $p - q = d$, $u + v = b$, and $u - v = c$. The terms are then rearranged as outlined in Table 12.3.

12.4 Introduction to Error Correction Codes

Term in $\|\Psi_3\rangle$	$s_{3,4,7,8}$	Error
$\frac{p}{\sqrt{8}}(\cdots)\,\bigl(\delta(\|000\rangle+\|111\rangle)+\gamma(\|000\rangle-\|111\rangle)\bigr)\,(\cdots)$	0000	No error
$\frac{q}{\sqrt{8}}(\cdots)\,\bigl(\delta(\|000\rangle-\|111\rangle)+\gamma(\|000\rangle\pm\|111\rangle)\bigr)\,(\cdots)$	0010	Phase flip
$\frac{u}{\sqrt{8}}(\cdots)\,\bigl(\delta(\|0\underline{1}0\rangle+\|1\underline{0}1\rangle)+\gamma(\|0\underline{1}0\rangle-\|1\underline{0}1\rangle)\bigr)\,(\cdots)$	1000	Bit flip
$\frac{v}{\sqrt{8}}(\cdots)\,\bigl(\delta(\|0\underline{1}0\rangle-\|1\underline{0}1\rangle)+\gamma(\|0\underline{1}0\rangle\pm\|1\underline{0}1\rangle)\bigr)\,(\cdots)$	1010	Bit & phase flip

Table 12.3: Error Syndrome for the Nine-Qubit Shor Code

As delineated in Table 12.3, a general single-qubit error can manifest in one of four scenarios:

1. No error.
2. Bit flip, corrected using an X gate.
3. Phase flip, corrected using a Z gate.
4. Both bit and phase flip, corrected using both X and Z gates.

Exercise 12.29 Construct an error syndrome and correction table similar to Table 12.3, assuming a general single-qubit error has occurred in qubit q_7, instead of q_5. For each scenario, identify the appropriate correction gate(s) in Fig. 12.15.

A pertinent question arises: how is it possible for a general error characterized by continuous coefficients a, b, c, and d to be effectively corrected using only four discrete scenarios? The answer is rooted in the fundamental properties of quantum measurements. When one conducts bit- and phase-flip parity-check measurements on the system, it collapses the system's state to one wherein either a bit or phase flip has transpired, or it hasn't, contingent on the outcome of the measurement. This inherent quantum behavior equips us with the capability to address a spectrum of errors merely by executing bit- and phase-flip checks.

Alternatively, we can analyze nine-qubit Shor code using a Pauli noise channel (see § 12.3.2), which is expressed as:

$$\mathcal{N}_{\text{Pauli}}(\rho) = (1 - s_x - s_y - s_z)I\rho I + s_x X\rho X + s_y Y\rho Y + s_z Z\rho Z, \quad (12.133)$$

where s_x, s_y, and s_z are the probabilities of each Pauli error. We find that scenario (1) above corresponds to I in the Pauli channel, scenario (2) to X, (3) to Z, and (4) to Y.

7 Decoding

The decoder's role is to transform the logical qubit state $\gamma\|0_L\rangle + \delta\|1_L\rangle$ back to $\gamma\|0\rangle + \delta\|1\rangle$, which is $U\|\psi\rangle$. This is achieved by applying the CNOT gates and H gates in the reverse order of the encoder.

8 Sequential vs. Oneshot Operation

A quantum computation task is typically carried out with a sequence of gates. As errors accumulate above a certain threshold, error detection and state recovery are executed. The version of the Shor code illustrated in Fig. 12.14 is capable of this purpose by repeating the blocks between $|\Psi_2\rangle$ and $|\Psi_4\rangle$.

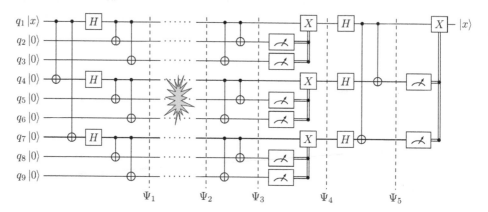

Figure 12.16: Nine-Qubit Shor Code Simplified for Oneshot Operation

The Shor code can also be used for quantum communication in which quantum states are transmitted subject to noise. In this case, the repetition of the middle blocks is often unnecessary, and the circuit can be simplified, as illustrated in Fig. 12.16. In particular, the error correction process can be postponed until after the decoding, which eliminates the need for ancillary qubits.

9 Limitations

We have seen that the Shor code can correct an arbitrary error on a single qubit. When considering simultaneous multi-qubit errors, the situation becomes more complex. Nevertheless, these scenarios are less frequent. Their probability is on the order of $O(\varepsilon^2)$, making them less probable compared to single bit-flip errors, especially when the single qubit error rate ε is sufficiently small.

While repetition codes, such as the Shor code, elucidate the principle of error correction, they are often too inefficient for practical application. This has motivated the search for more efficient codes. A notable example is the seven-qubit Steane code. To delve deeper into such error correction codes, additional mathematical tools, like the stabilizer formalism, are required. These will be introduced in the subsequent subsection.

> **Exercise 12.30** For the simplified version of the nine-qubit Shor code depicted in Fig. 12.14, derive the system state at different stages ($|\Psi_1\rangle$ through $|\Psi_5\rangle$) for each type of single-qubit error: bit-flip, phase-flip, and general errors.

12.4.4 ∗Parity Measurements, Stabilizers, and Code Distance

Fundamental to the study of quantum error-correction codes are the inter-related concepts of parity measurements, stabilizer codes, and code distance.

1 Parity Measurements

Parity measurements constitute a type of joint measurement on multiple qubits, as discussed in § 6.2.2. For a two-qubit system, $P_Z \equiv Z \otimes Z$ can distinguish

12.4 Introduction to Error Correction Codes

between the even parity states $|\psi_{\text{even}}\rangle \equiv \alpha|00\rangle + \beta|11\rangle$ and odd parity states $|\psi_{\text{odd}}\rangle \equiv \gamma|01\rangle + \delta|10\rangle$. Similarly, $P_X \equiv X \otimes X$ can be used to measure phase parity.

In the discussion of the Shor code in preceding sections, we have employed parity measurements. Specifically, P_Z is used for the error syndrome measurement associated with bit flips, while P_X facilitates the error syndrome measurement related to phase flips.

The stabilizer formalism discussed next can be viewed as a generalization of the concept of parity measurements.

2 The Stabilizer Formalism

The stabilizer formalism is a framework that enables the efficient description and manipulation of quantum states and operations pertinent to quantum error correction. Rather than representing these entities in Hilbert space, the formalism treats them in the operator picture as elements of a stabilizer group. This framework proves essential in analyzing advanced error correction codes.

Stabilizers are a set of operators that commute with each other and stabilize a given quantum state, meaning they leave the state invariant under their action. Formally, if $|\psi\rangle$ is a quantum state and S is an operator belonging to the stabilizer group, then $S|\psi\rangle = |\psi\rangle$.

The stabilizer formalism is useful because it allows for a compact representation of certain quantum error-correcting codes. For instance, in the three-qubit bit-flip code, the encoded logical state has the form $|\psi\rangle_L = \alpha|000\rangle + \beta|111\rangle$. The set of these error-free states, referred to as the code space, are described by the associated stabilizers ZZI, IZZ, and ZIZ. All states within the code space are simultaneous $+1$ eigenstates of these stabilizers.

Stabilizer place constraints on the permissible states of the code space. When an error affects a qubit, it alters the eigenvalues of the stabilizers. Measuring these stabilizers allows one to diagnose which qubit is in error and subsequently correct it, without disturbing the error-free logical states, which are eigenstates of the stabilizers.

For example, the error syndrome detection for the three-qubit bit-flip code shown in Fig. 12.12 employs two parity measurements: one between the first and second qubits, and another between the first and third qubits. These parity measurements correspond to measuring the stabilizers ZZI and ZIZ, respectively. Using only two out of the three stabilizers is sufficient because the product of any two stabilizers yields the third.

For the nine-qubit Shor code, the stabilizer group consists of operators that act on nine qubits, including:

$$ZZIIIIIII, \quad ZIZIIIIII, \quad IIIZZIIII, \quad IIIZIZIII,$$
$$IIIIIIZZI, \quad IIIIIIZIZ, \quad XXXXXXIII, \quad IIIXXXXXX.$$

The above operators form the minimal set of independent stabilizers that uniquely define the code space, often referred to as stabilizer generators. Every state in the code space is a $+1$ eigenstate of each of these stabilizers. These stabilizer generators are specifically the operators that are measured in the error syndrome detection circuit, as illustrated in Fig. 12.15. Their measurements make it possible to identify and correct both bit-flip (X) and phase-flip (Z) errors without disturbing the encoded quantum information.

> **Exercise 12.31** Explain why these are not stabilizers for the nine-qubit Shor code: $ZIIZIIIII$, $ZZZIIIIII$, $XXXIIIIII$, and $ZZZZZZIII$.

> **Exercise 12.32** Explain why these are stabilizers but not included as generators in the table above for the nine-qubit Shor code: $IZZIIIIII$, and $XXXIIIXXX$.

3 Code Distance

The code distance, often denoted as d, is a metric in quantum error correction that signifies the minimum number of qubit errors required to transform one valid encoded state (codeword) into another without detection by the code's syndrome measurements.

The three-qubit bit-flip code can correct a single bit-flip error, such as XII. It can also detect a double bit-flip errors like XXI, even though cannot correct it due to syndrome ambiguity with the single bit-flip error IIX. Therefore, the code distance for bit-flip errors is $d = 3$. On the other hand, the code is not designed to detect or correct phase-flip errors, rendering its code distance for phase-flip errors as $d = 1$.

In contrast, the nine-qubit Shor code is designed to correct for one arbitrary error on any of the nine qubits. It would take at least three such arbitrary errors to go undetected. Therefore, it has a code distance of $d = 3$ for both bit-flip and phase-flip errors.

12.4.5 ∗ Further Exploration

The field of quantum error correction (QEC) is critical in advancing the development of large-scale, fault-tolerant quantum computers. Given that real-world quantum devices invariably encounter various types of noise and errors, a robust understanding and effective mitigation of these imperfections are fundamental. For those interested in delving deeper into error correction strategies and noise reduction techniques in quantum computing, the following topics and references are recommended for extended exploration:

1 Advancements in Efficient QEC

The development of quantum error-correcting codes has been a cornerstone in the advancement of quantum computing. The first such code, known as Shor's 9-qubit code, laid the foundation for subsequent innovations. Following this, Andrew Steane introduced a more efficient code, known as Steane's code, which could detect and correct any single-qubit error using only 7 physical qubits to encode 1 logical qubit[84]. Further progress was made by Raymond Laflamme and colleagues, who developed codes that required as few as 5 qubits[64].

Surface codes represent another significant class of quantum error-correcting codes. Characterized as topological quantum error-correcting codes, they are implemented on a two-dimensional lattice of qubits[34]. The error correction in surface codes leverages the lattice's topology. Although surface codes generally necessitate a larger

12.4 Introduction to Error Correction Codes

number of qubits than the likes of Steane's or Laflamme's codes, they are highly valued for their high fault tolerance and adaptability to two-dimensional grid layouts with nearest-neighbor qubit interactions.

Recent advancements in quantum error correction have seen the emergence of new codes, notably within the Low-Density Parity Check (LDPC) family, that promise a tenfold increase in efficiency compared to traditional methods[29]. These innovative LDPC codes achieve effective error correction with a reduced qubit requirement, marking a pivotal step towards the realization of scalable and practical quantum computing. This evolution reflects the ongoing progress in quantum error correction, moving from the original Shor's 9-qubit code to increasingly resource-efficient solutions.

An emerging strategy in quantum error correction is the concept of erasure conversion. This method transforms various quantum errors into erasure errors, whose locations are known, thereby simplifying their correction. Recognized for its potential to reduce error correction complexity and overhead, erasure conversion is seen as a key strategy for enhancing quantum computing efficiency. This approach could significantly streamline the error correction process, marking a crucial step towards scalable and practical quantum computing[49, 91].

2 General References on QEC

1. Quality, Speed, and Scale: Key Attributes of Near-Term Quantum Computers [88]: This paper emphasizes the three primary attributes to evaluate the performance of near-term quantum computers, guiding the assessment of quantum progress.

2. Foundations of Quantum Error Correction [34]: A comprehensive review tailored for computer scientists unfamiliar with quantum physics. It discusses the foundational principles of QEC, practicality of QEC codes, and challenges for NISQ computers.

3 Fault Tolerance Threshold

1. Comparison of Quantum Error-Correction Threshold for Exact and Approximate Errors [51]: This study uses classical simulations of stabilizer circuits to assess quantum code thresholds. It integrates Clifford and Pauli operators for improved noise approximation, focusing on the accuracy of Steane code and the Pauli twirling approximation.

4 Error Mitigation and Simulation

1. How to Simulate Quantum Measurement Without Computing Marginals [28]: A novel approach to quantum simulation that bypasses the need to compute marginals during quantum measurement.

2. Error Mitigation for Universal Gates on Encoded Qubits [73]: This work delves into advanced techniques for error mitigation when using universal gates on encoded qubits.

3. Mitigating Coherent Noise Using Pauli Conjugation [33]: A proposed method to mitigate coherent noise in quantum computations leveraging the Pauli conjugation technique.

4. Mitigating Depolarizing Noise on Quantum Computers [87]: This paper offers insights into noise-estimation circuits designed to counteract depolarizing noise, ensuring enhanced computational fidelity.

5 Holonomic and Topological Quantum Computation

Holonomic quantum computation (HQC) and topological quantum computation (TQC) are advanced approaches to quantum computing. Both leverage specific non-local properties of quantum systems for quantum operations, offering inherent resistance to certain types of errors.

The underlying principle of HQC involves exploiting the global properties (or geometry) of quantum systems, rather than their local properties (such as the quantum state). In HQC, computation is realized by cyclically moving the system within its parameter space (for instance, on the surface of the Bloch sphere). During this movement, the system acquires geometric phases (or Berry's phases) corresponding to the solid angles enclosed by its trajectory. As a result of relying on global properties, HQC displays a commendable resilience against local errors.

On the other hand, TQC derives its computational capability from the braiding of anyons. These anyons are exotic quasiparticles present in two-dimensional systems, distinguished from both fermions and bosons (see § 6.3.4). The topological nature of the quantum state, upon which the computation is built, ensures that local disturbances have minimal impact on the results. Such topological protection provides a potential pathway for fault-tolerant quantum computation.

As of early 2024, while HQC principles have seen experimental validation across several qubit systems, TQC remains more elusive. Although TQC boasts a robust theoretical foundation, its experimental realization presents significant challenges. Notably, the controlled observation and braiding of anyons remain at the forefront of quantum research.

For further exploration, readers are encouraged to consult the following references:

1. Enhancing Error Resilience Using Geometric Phase-Based Quantum Gates [83]: An introductory paper explaining the basics of geomtric phase and HQC, exploring the maximal error resilience achievable using quantum gates designed with geometric phase shifts, and their compatibility with other quantum error-correction techniques.

2. Berry's Phase in *Introduction to Quantum Mechanics* [3]: This book section delves into the concept of Berry's phase (geometric phase) and its relation to cyclic adiabatic evolution. The Berry phase is determined solely by the path in the parameter space, irrespective of the traversal rate along that path. The book exemplifies this using the physical system of an electron undergoing cyclotron motion, providing a comprehensive explanation in the dedicated section on Berry's phase.

3. Quantum Algorithm Design Exploiting Geometric and Holonomic Gates [97]: A recent paper offering an overview on the use of intrinsic geometric properties in quantum-mechanical state spaces for the realization of quantum logic gates. It also investigates the development of novel quantum algorithms that can take advantage of the error resilience and unique properties of geometric and holonomic quantum gates.

4. **Harnessing Non-Abelian Phases for Robust Quantum Information Processing** [85]: A comprehensive overview of leveraging non-Abelian quantum phases in topological quantum computation, emphasizing their nonlocal nature that renders quantum information resilient against environmental interactions and protocol imperfections. The review further highlights various solid-state systems, such as those hosting Majorana fermions and non-Abelian quantum Hall states, as potential platforms for implementing these concepts.

5. **Non-Abelian Anyons: A Cornerstone of Fault-Tolerant Quantum Computation** [70]: An in-depth exploration of the significance of non-Abelian anyons in the realm of topological quantum computation. The article discusses the unique braiding statistics of these anyons and how they enable fault-tolerant quantum gate operations. Emphasis is placed on the inherent resilience imparted by the nonlocal encoding of quasiparticle states, detailing potential architectures and experiments for realizing a topological quantum computer.

12.5 Summary and Conclusions

Essentials of Quantum Error Correction

In this chapter, our emphasis was on elucidating the nuances of quantum error correction, a domain that has emerged as vital for the progression and practicality of quantum computing. Given the inherent sensitivity of quantum systems to noise and their potential to lose coherence, understanding errors, their causes, and methods to rectify them is paramount.

We began our journey by emphasizing the importance of preliminary concepts, shedding light on the challenges posed by noise and decoherence. By drawing distinctions between decoherence, noise, and errors, we underscored the unique features and manifestations of each, leading to a clearer understanding of the quantum noise problem.

Density Operators and Mixed States

Transitioning from state vectors, we introduced the mathematical framework of density operators, a powerful tool to capture the subtleties of mixed quantum states. The relevance of density operators in modeling diverse quantum operations, especially in the context of noise and decoherence, was highlighted.

We navigated through their fundamental properties, delving into the computation of expected values, the representation of qubit states, and their interpretation on the Bloch sphere. This not only expanded our understanding of quantum states but also provided a lens to appreciate the richness of quantum mechanics.

Error Mechanisms and Error Correction

Our discourse then veered towards the core focus: the diverse error mechanisms in quantum computing. We identified and elaborated on coherent errors, characterized by their systematic nature, and incoherent errors, which arise due to stochastic interactions with the environment. Through examples such as bit-flip, phase-flip, and depolarizing errors, we accentuated the varied manifestations of these errors and the challenges they pose.

To combat these challenges, we embarked on an exploration of quantum error correction codes. Here, the emphasis was on introducing and dissecting the principles and techniques that serve as our arsenal against quantum errors. We delved into the Shor codes, highlighting their significance in the realm of quantum error correction, and underscored the importance of parity measurements, stabilizers, and code distance in this context.

Bridging Theory and Practice

The overarching narrative of this chapter bridges the theoretical underpinnings of quantum mechanics with the pragmatic challenges faced in quantum computing. By intertwining the mathematical formalism of density operators with real-world error mechanisms, we have endeavored to prepare the reader for both theoretical research and hands-on experimentation in quantum error correction.

The journey through this chapter offers a holistic perspective on the challenges and solutions in the domain of quantum error correction, laying the groundwork for the subsequent discussions and further explorations in this dynamic field.

Upcoming Topics

The upcoming chapter provides an exploration of quantum information theory. It begins by contrasting quantum probability with classical probability, highlighting unique quantum aspects like entanglement. The chapter then discusses quantum entropy and information, comparing these with classical concepts. Finally, it delves into advanced topics like the quantum data processing inequality and the Holevo theorem, emphasizing their theoretical and practical importance in quantum communications.

Problem Set 12

12.1 Derive the density operator $\rho_\perp$ for the qubit state orthogonal to $|\psi\rangle$: $|\psi_\perp\rangle = \beta^* |0\rangle - \alpha^* |1\rangle$. Show that ρ and $\rho_\perp$ are orthogonal to each other, i.e., $\text{tr}(\rho^\dagger \rho_\perp) = 0$. In fact, $\rho^\dagger \rho_\perp = 0$.

12.2 Derive the density operator for the general qubit state $|\psi\rangle = \cos\frac{\theta}{2}|0\rangle + \sin\frac{\theta}{2}e^{i\phi}|1\rangle$.

12.3 Alice and Bob share pairs of qubits, with half of the pairs in the Bell state $\frac{1}{\sqrt{2}}(|00\rangle - |11\rangle)$ and the remaining half in the state $\rho_A \otimes \rho_B$, where $\rho_A = \frac{1}{4}\begin{bmatrix} 2 & 1 \\ 1 & 2 \end{bmatrix}$ and $\rho_B = |0\rangle\langle 0|$.

Determine the density operator for the composite system of Alice and Bob's shared qubits.

12.4 Represent the following quantum states on (or in) the Bloch sphere:

(a) 20% in $|0\rangle$ and 80% in $|1\rangle$

(b) 50% in $|+\rangle$ and 50% in $|0\rangle$

(c) 25% of $|0\rangle$, 25% of $|1\rangle$, and 50% of $|+\rangle$

(d) 50% in $|+\rangle$ and 50% in $|-\rangle$

12.5 The state of a quantum system is such that it is 50% in the state $|+\rangle$ and 50% in the state $|0\rangle$. A Hadamard gate is applied to the system. Find the new density operator for the system.

12.6 How does the Hadamard gate H transform a quantum state represented by its Bloch vector $\vec{r} = (x, y, z)$?

12.7 Given the initial state $|\psi\rangle = \alpha|0\rangle + \beta|1\rangle$, what is the density operator of the ensemble of measurement outcomes upon measurements in the $\{|0\rangle, |1\rangle\}$ basis?

12.8 The state of a system, initially 50% in the state $|+\rangle$ and 50% in the state $|0\rangle$, is measured in the $\{|0\rangle, |1\rangle\}$ basis. What is the density operator of the post-measurement ensemble?

12.9 Consider the GHZ state, which is defined for three qubits as $|\text{GHZ}\rangle = \frac{1}{\sqrt{2}}(|000\rangle + |111\rangle)$.

   (a) Calculate the reduced density operator of the first two qubits after the third qubit has been traced out (i.e., its degrees of freedom have been disregarded).

   (b) Calculate the reduced density operator of the third qubit after the first two qubits are traced out.

12.10 Consider the three-qubit W state, defined as $|W\rangle = \frac{1}{\sqrt{3}}(|001\rangle + |010\rangle + |100\rangle)$, known for its characteristic partial entanglement among qubits without any pair being maximally entangled. Your task involves:

   (a) Constructing the density operator ρ_W for the W state.

   (b) Calculating the reduced density matrices for each pair of qubits by performing the appropriate partial traces on ρ_W.

   (c) Confirming the partial entanglement between each pair of qubits through analysis of their reduced density matrices.

12.11 Investigate the two-qubit cross-talk error caused by unintended ZX coupling, which is represented by the Hamiltonian $H_{ZX} = -\frac{\hbar\omega}{2}(Z \otimes X)$.

   (a) The undesired ZZ coupling is known to result in a phase walk-off in the computational basis, as discussed in § 12.3.1. Describe the specific type of error that arises from unintended ZX coupling.

   (b) Derive the equations that quantify the cross-talk errors due to ZX coupling in a manner analogous to the errors described by Eqs. 12.86 and 12.87.

   Hint: Consult § 6.5.4 for adiscussion on ZX coupling and its implications in the context of CNOT gate implementation.

12.12 Consider a qubit initially in the state $|\psi\rangle = \alpha|0\rangle + \beta|1\rangle$. Investigate the impact of the following incoherent errors on the qubit, expressing the resultant state as a density operator and visualizing the changes on the Bloch sphere. Detailed discussions of these errors can be found in § 12.3.2.

   (a) Bit-flip error

   (b) Phase-flip error

(c) Depolarizing error

(d) Dephasing error

(e) Amplitude damping error

(f) Pauli channel error

12.13 Consider a two-qubit system initially in the Bell state $|\Phi^+\rangle = \frac{1}{\sqrt{2}}(|00\rangle + |11\rangle)$. Analyze the state of the system expressed as a density operator after it undergoes the following incoherent errors. For detailed discussions on these errors, refer to § 12.3.2.

(a) Two-qubit bit-flip error

(b) Two-qubit dephasing error

(c) Two-qubit Pauli channel

12.14 Unlike classical computers, which largely rely on a homogeneous technology base, quantum computers are being developed using a variety of hardware platforms, each advancing rapidly. For this exercise, select three distinct quantum computing hardware platforms (e.g., superconducting qubits, trapped ion qubits, neutral atom qubits, photonic systems, etc.). For each selected platform, conduct a detailed investigation into the following:

- The specific aspects and applications where it excels.

- The challenges or limitations it faces.

Quantify these strengths and weaknesses using the performance metric parameters outlined in § 12.3.4, such as coherence time, qubit connectivity, gate fidelity, error rates, quantum volume, and scalability, among others.

Prepare a presentation on your findings, ensuring to:

- Clearly define each hardware platform and its operational principles.

- Discuss and quantify the performance metrics for each platform, providing a comparative analysis.

- Highlight the implications of these metrics for future developments and applications of quantum computing.

Consider utilizing reputable scientific journals, technology reports, and official documentation from research institutions as sources for your research to ensure accuracy and relevancy.

This exercise aims to deepen your understanding of the diverse landscape of quantum computing technologies and their potential impact on the field's evolution.

12.15 Construct an alternative version of the Shor three-qubit bit-flip code, as detailed in Fig. 12.12. In this variant, modify the pairs of qubits used for parity measurement. Instead of employing the qubit pairs (q_1, q_2) and (q_1, q_3), use the pairs (q_1, q_2) and (q_2, q_3).

12.16 In Fig. 12.13, the error syndrome measurement for the three-qubit phase-flip code utilizes two sets of Hadamard gates on the data qubits. Design a circuit

Problem Set 12

that performs the same syndrome measurement function, but without using these Hadamard gates.

Hint: Refer to the equivalent gate sequences discussed in § 7.4.

12.17 A logical Pauli Y gate, denoted as Y_L, transforms $\alpha\,|0_L\rangle + \beta\,|1_L\rangle$ into $-i\beta\,|0_L\rangle + i\alpha\,|1_L\rangle$. Design four distinct versions of the logical Y_L gate for the nine-qubit Shor code.

Hint: Recall that $Y = iXZ$. Refer to Exercise 12.28 for examples of X_L and Z_L implementations.

12.18 The Shor code shown in Fig. 12.14 begins with a phase-flip block followed by three bit-flip blocks. Investigate the consequences of starting with a bit-flip block and then following with three phase-flip blocks.

13. Fundamentals of Quantum Information

Contents

13.1	**Quantum Probability Essentials**	**408**
13.1.1	Distinctive Features of Quantum Probability	408
13.1.2	Sample Space, Events, and Probability Distribution	409
13.1.3	Random Variables and Expected Values	411
13.1.4	Multiple Random Variables	413
13.1.5	Random Processes	421
13.1.6	∗ Example: One-Dimensional Quantum Walk	422
13.2	**Quantum Entropy and Information**	**426**
13.2.1	Classical Shannon Entropy	427
13.2.2	∗ Quantum von Neumann Entropy	429
13.2.3	∗ Classical Joint Entropy and Mutual Information	430
13.2.4	∗ Quantum Joint Entropy and Mutual Information	434
13.3	∗ **Core Theorems in Quantum Information**	**437**
13.3.1	∗ The Data Processing Inequality	438
13.3.2	∗ The Holevo Bound and Channel Capacity	440
13.3.3	∗ Subadditivity and Strong Subadditivity	443
13.4	**Further Exploration**	**444**
13.5	**Summary and Conclusions**	**444**
	Problem Set 13	**445**

The field of information theory underwent a significant shift with the introduction of quantum mechanics. The principles that govern the world of atoms and photons provided a new perspective for reimagining the foundation of information. This chapter begins an explorative journey from the familiar territory of classical information to the mysterious and less intuitive realm of quantum information.

Classical information theory, rooted in the seminal work of Claude Shannon, laid the foundation for understanding the transmission, processing, and encoding of information within a probabilistic framework. Central to this theory is the concept of Shannon entropy, which quantifies the uncertainty or information content in a classical system. Classical systems, governed by deterministic laws, utilize probabilities to address the incomplete knowledge about the system. Despite their deterministic nature, probabilistic methods in classical information theory are crucial for managing uncertainties inherent in information transmission and processing.

In contrast, quantum information theory is deeply intertwined with the inherent probabilistic nature of quantum mechanics, a fundamental departure from classical theories. In quantum mechanics, probabilities do not merely represent a lack of knowledge but are an intrinsic aspect of quantum states. This intrinsic probabilistic nature is encapsulated by the Born rule, which governs the probabilities of various outcomes when measurements are made on quantum states. Consequently, quantum information theory offers a richer framework for encoding and processing information, taking into account the phenomena of superposition and entanglement. Analogous to Shannon entropy in classical systems, von Neumann entropy extends these concepts to the quantum realm, accommodating the unique properties of quantum states and their non-classical correlations.

The significance of quantum probability and entropy transcends the mere extension of classical concepts into the quantum domain. They are the cornerstone for understanding quantum coherence and correlations, which have no classical counterpart. Quantum entropy not only measures information but also captures the degree of entanglement—a unique resource for quantum computing and communication. The interplay between quantum probability, entropy, and information fosters new potential for advancements that include the enhanced efficiency of quantum algorithms, the establishment of secure communication channels, and the development of quantum error correction techniques.

In exploring the fundamentals of quantum information, we lay the groundwork that will enable researchers and practitioners to actively engage in and further the quantum revolution in computation, communication, and measurement. This foundational knowledge is essential for those who aspire to drive the forthcoming era of technological innovations in quantum information science.

> **Prerequisite Review: Mixed States and Density Operators**
>
> Mixed states and density operators are detailed in § 12.2. Here is a brief summary of the concepts and formulas essential for this chapter.
>
> - A mixed quantum state is a statistical ensemble of multiple potential pure states. It encapsulates both quantum uncertainty and classical uncertainty. A mixed qudit state with probability p_i of $|\psi_i\rangle \in \mathbb{C}^d$, $i = 1, \ldots, n$, can be described by a density operator:
>
> $$\rho = \sum_{i=1}^{n} p_i |\psi_i\rangle\langle\psi_i|. \qquad (13.1)$$
>
> - Properties of density operators:
>   - Hermitian: $\rho^\dagger = \rho$
>   - Positive Semidefinite: $\langle\psi|\rho|\psi\rangle \geq 0$ for all $|\psi\rangle \in \mathbb{C}^d$
>   - Unit Trace: $\text{tr}(\rho) = \sum_{i=1}^{d} \rho_{ii} = 1$
> - A density operator has a diagonal form in its eigenbasis, corresponding to the canonical states of the quantum system:

$$\rho = \sum_{i=1}^{d} \lambda_i |\lambda_i\rangle\langle\lambda_i|, \qquad (13.2)$$

with $\lambda_i \geq 0$ and $\sum_i \lambda_i = 1$.

- Applying a unitary transformation $U \in \mathbb{C}^{d \times d}$ transforms ρ as

$$\rho \to U\rho U^\dagger. \qquad (13.3)$$

- Measuring ρ in orthonormal basis $\{|\phi_i\rangle\}$ yields outcome $|\phi_i\rangle$ with probability:

$$p_i = \langle\phi_i|\rho|\phi_i\rangle = \mathrm{tr}(\rho|\phi_i\rangle\langle\phi_i|). \qquad (13.4)$$

Following the measurement, the ensemble of measurement outcomes is given by

$$\rho_M = \sum_{i=1}^{d} p_i |\phi_i\rangle\langle\phi_i|, \qquad (13.5)$$

which is diagonal in the measurement basis. This represents the equivalence of state collapse in the context of mixed states.

In the more general POVM measurement framework, a measurement outcome corresponds to a POVM operator E_i, and the probability p_i of observing outcome i is given by:

$$p_i = \mathrm{tr}(\rho E_i). \qquad (13.6)$$

- For a composite system consisting of two independent subsystems, A described by ρ_A and B by ρ_B, the combined density operator is given by

$$\rho_{AB} = \rho_A \otimes \rho_B. \qquad (13.7)$$

- Partial trace of a density operator provides a way to focus on a subset of a larger system by summing over (and effectively ignoring) the rest of the system. The partial trace over subsystem A, or the reduced density operator of subsystem B, is given by

$$\rho_A = \mathrm{tr}_B(\rho_{AB}) = \sum_i \langle i|_B\, \rho_{AB}\, |i\rangle_B, \qquad (13.8)$$

where $\{|i\rangle_B\}$ is any orthonormal basis for the Hilbert space of subsystem B.

Density Operator: Pillar of Quantum Probability and Information

The density operator plays key roles in quantum probability and information, as a versatile and foundational construct in quantum mechanics, underpinning concepts such as von Neumann entropy and quantum information.

- The density operator ρ, being Hermitian, can be diagonalized to reveal a probability distribution over its constituent eigenstates. Each eigenvalue λ_i of ρ corresponds to the probability of finding the system in the associated eigenstate $|\lambda_i\rangle$. This structure allows ρ to represent both pure and mixed states in a unified framework.

- In quantum communication, the density operator can represent mixed states arising from noise and environmental interactions, essential for quantum error correction and channel capacity.

  When a state ρ is measured in an orthonormal basis $\{|\phi_i\rangle\}$, the resulting post-measurement ensemble is captured by a new density operator ρ_M. Diagonal elements of ρ_M, given by $\langle\phi_i|\rho|\phi_i\rangle$, form a probability distribution over the measurement basis states, reflecting the outcome probabilities. (See discussion around Eq. 12.48.)

- Quantum coherence, encapsulated in the off-diagonal elements of ρ in a given basis, is fundamental to quantum computation and information processing. It represents the superposition of states and is a key resource for quantum algorithms, enabling quantum parallelism and interference.

- Purity and entropy, critical in quantum thermodynamics and information theory, are derived from the density operator's properties. Purity, indicating the degree of mixedness of a state, is given by $\text{tr}(\rho^2)$, while von Neumann entropy, measuring the informational content, is $-\text{tr}(\rho\log\rho)$.

13.1 Quantum Probability Essentials

Quantum information science is deeply rooted in the principles of probability theory. In this section, we examine the fundamental components of probability theory as they apply to quantum information science. Beginning with a discussion of each concept within the classical framework, we then establish parallels with their quantum mechanical counterparts. This comparative analysis enhances our understanding of the inherently probabilistic nature of quantum systems.

We will use the following convention for symbols unless stated otherwise:

Symbol	Meaning
X, Y	Random variables
x, y	Values of random variables X and Y, respectively
A, B, C	Subsystems or random events
p	Probability or probability distribution
$p(x)$	Probability that X takes the value x
$p(X)$	Probability distribution of X
$H(p), H(X)$	Shannon entropy of a prob. distribution p or random variable X
$S(\rho), S(X)$	von Neumann entropy of density operator ρ or random var X
$E[X]$	classical expected value (statistical average) of X
$\langle H \rangle$	quantum expected value (statistical average) of observable H
$\sigma_x, \sigma_y, \sigma_z$	Pauli operators

13.1.1 Distinctive Features of Quantum Probability

While quantum probability encompasses the general structure of classical probability, it introduces several key features that set it apart, fundamentally altering our understanding of probabilistic phenomena.

13.1 Quantum Probability Essentials

1. **Complex-Valued Probability Amplitudes**

   In classical probability, probabilities are real numbers between 0 and 1. Quantum mechanics, however, uses complex-valued probability amplitudes. These amplitudes, when squared, give the probabilities of various outcomes. The use of complex numbers allows for the phenomenon of quantum superposition, where a quantum state can simultaneously exist in multiple states. This superposition is a cornerstone of quantum mechanics, leading to interference effects that have no counterpart in classical probability.

2. **Entanglement**

   Entanglement introduces correlations between quantum systems that are non-existent in classical systems. When two particles are entangled, their quantum states become so interlinked that the state of one particle correlates with the state of the other, regardless of the distance separating them. This leads to joint probability distributions that cannot be decomposed into independent probabilities of each entangled component. This nonlocal property of entangled systems profoundly affects the interpretation and calculation of probabilities in quantum systems.

3. **Non-commutativity of Quantum Observables**

   Another fundamental aspect of quantum probability is the non-commutative nature of quantum observables. When two operators do not commute, the precise sequence of measurements affects the observed probabilities. This is exemplified by the Heisenberg uncertainty principle (see § 1.6), which states that certain pairs of observables, like rectilinear and circular polarizations, cannot be simultaneously measured with arbitrary precision. The non-commutativity of observables can lead to complex probability distributions and correlations that cannot be replicated in classical probability theory.

13.1.2 Sample Space, Events, and Probability Distribution

1. **Classical Probability**

   Random Experiments

   In classical probability, a random experiment or trial is defined as a process with a definitive set of possible outcomes, which cannot be precisely predicted in advance. Such an experiment can be replicated under consistent conditions, with each repetition being independent of the others.

   Sample Space

   The sample space, denoted $\Omega = \{s_1, s_2, \ldots, s_n\}$, encompasses all conceivable outcomes of an experiment, with each s_i representing a unique outcome. This space can be finite or countably infinite for discrete cases and must fulfill two criteria:

   1. Exhaustiveness: It should encompass all potential outcomes.

   2. Mutual Exclusivity: No two outcomes should simultaneously occur in a single trial.

   Events

   An event, denoted as E, is any subset of the sample space, $E \subseteq \Omega$. An event E occurs if the outcome of the random experiment is contained within E. An event may be an elementary event (consisting of a single outcome) or a compound event (comprising multiple outcomes).

■ **Example 13.1** The roll of a six-sided die constitutes a random experiment, with the sample space represented by $\Omega = \{1, 2, 3, 4, 5, 6\}$.

The event of rolling an even number on the die is $E = \{2, 4, 6\}$. The event of rolling an odd number on the die is $O = \{1, 3, 5\}$. E and O are exhaustive and mutually exclusive. ■

Probability Distribution

A probability distribution assigns a probability to each experiment outcome. These assignments adhere to:

1. Non-Negativity: Probabilities are always non-negative.
2. Normalization: The sum of all outcome probabilities equals 1.
3. Additivity: The probability of the union of two distinct events is the sum of their individual probabilities.

■ **Example 13.2** For a fair die, the probability distribution over Ω is uniform:

$$p_i \equiv p(\{i\}) = \frac{1}{6} \quad \text{for each } i \in \Omega. \tag{13.9}$$

Thus, the probability of event E (rolling an even number) is:

$$p(E) = p(\{2\}) + p(\{4\}) + p(\{6\}) = \frac{1}{6} + \frac{1}{6} + \frac{1}{6} = \frac{1}{2}. \tag{13.10}$$

■

2 Quantum Generalization

Sample Space: Spectrum of an Observable

In quantum mechanics, the classical concept of a sample space is reinterpreted as the spectrum of an observable, represented by a Hermitian operator H. The eigenvalues $\{\lambda_i\}$ of this operator represent the possible results we can obtain when measuring the observable. Each eigenvalue has a corresponding eigenvector (or a subspace when there's degeneracy), which represents the quantum state the system will be in if we obtain that specific measurement outcome (see § 3.4).

More generally, a quantum sample space can be represented by an orthonormal basis in the Hilbert space of the system, where measurements are conducted.

Events: Quantum Measurements

In the quantum realm, events are paralleled by measurements. For an observable H, distinct eigenvalues correspond to orthogonal eigenstates $\{|\lambda_i\rangle\}$, analogous to classical mutually exclusive events. These eigenstates form a complete basis set in the Hilbert space, akin to a complete set of events.

More generally, quantum measurements can be described by POVM (Positive Operator-Valued Measure) operators $\{E_i\}$. While some scenarios allow E_i to correspond to projective measurements like $|\lambda_i\rangle\langle\lambda_i|$, associated with the eigenstates of observables, POVMs offer a more generalized framework and can represent a

13.1 Quantum Probability Essentials

broader set of outcomes. For example, two POVM operators could represent the aggregate events of rolling an even or odd number on a die.

Probability Distribution: The Born Rule

The Born rule dictates the probabilities of measurement outcomes in quantum mechanics, forming the basis for quantum probability distribution. The probabilistic behavior of a quantum system is fully characterized by its state vector $|\psi\rangle$ for pure states or density operator ρ for mixed states. The state vector $|\psi\rangle$ is associated with the inherent quantum randomness of measurements, while ρ accounts for both quantum and classical randomness due to statistical mixtures of states.

The probability p_i of observing a particular outcome λ_i of an observable is given by:

$$p_i = \begin{cases} |\langle \lambda_i | \psi \rangle|^2 = \langle \psi | \lambda_i \rangle \langle \lambda_i | \psi \rangle & \text{for a pure state,} \\ \langle \lambda_i | \rho | \lambda_i \rangle = \text{tr}(\rho |\lambda_i\rangle\langle\lambda_i|) & \text{for a mixed state.} \end{cases} \quad (13.11)$$

For POVM measurements, outcome probabilities for each E_i are:

$$p_i = \begin{cases} \langle \psi | E_i | \psi \rangle & \text{for a pure state,} \\ \text{tr}(\rho E_i) & \text{for a mixed state.} \end{cases} \quad (13.12)$$

13.1.3 Random Variables and Expected Values

1 Classical Probability

A random variable is a function that assigns a numerical value (or other mathematical constructs) to each outcome in the sample space of a random experiment. For example, a random variable X representing the result of rolling a die conventionally assumes values in the set $\{1, 2, 3, 4, 5, 6\}$, but it may also assume values in some other set, e.g., $\{-3, -2, -1, 1, 2, 3\}$. The assignment may even be many-to-one, e.g., $\{o, e, o, e, o, e\}$.

As introduced above, a probability distribution assigns a probability to each outcome in a sample space. While a probability distribution originates from the concept of a sample space, it can also be defined in the context of random variables. In the latter case, a probability distribution describes how probabilities are distributed over the different possible values that the random variable can take. In the above example of rolling a die, the probability distribution can be expressed as $p(X = x) = \frac{1}{6}$ for $x = 1, 2, \ldots, 6$.

When a random variable employs a many-to-one assignment to map outcomes in the sample space to numerical values, it effectively aggregates those outcomes under the assigned values. This aggregation may result in a probability distribution for the random variable that differs from the original distribution of outcomes in the sample space.

For example, let X represent whether the outcome of rolling a die is even or odd. It takes the value e for outcomes $\{2, 4, 6\}$ and o for $\{1, 3, 5\}$. Now, $p(X = e) = p(X = o) = \frac{1}{2}$.

Events can also be described in terms of random variables. For example,

$E = \{X \leq 3\}$ is an event representing the outcome of rolling a die taking on a value at most 3.

The expected value (or expectation value) quantifies the statistical average of a random variable, weighted by its probability distribution, indicating the central tendency of the distribution. For instance, the expected value of a fair die's roll is

$$E[X] = \sum_{x=1}^{6} x \cdot p(X = x) = 3.5. \tag{13.13}$$

2 Quantum Generalization

An observable in quantum mechanics can be considered a generalization of a classical random variable because it describes the possible outcomes of a measurement (a random process) and their associated probabilities. However, unlike classical random variables, which directly map outcomes to real numbers, the outcomes of measuring an observable are determined by the eigenvalues of its operator, and the probability of each outcome is given by the state of the system before measurement.

Much like in the classical case, the expected value of an observable tells us about the average result we'd see if we measured it many times, given the quantum state of the system. The expected value of an observable H is given by:

$$\langle H \rangle = \begin{cases} \langle \psi | H | \psi \rangle & \text{for a pure state,} \\ \text{tr}(\rho H) & \text{for a mixed state.} \end{cases} \tag{13.14}$$

The variance of an observable H, which is equal to the square of the standard deviation and quantifies the spread of measurement outcomes (see § 3.4.6), is calculated as:

$$(\Delta H)^2 = \langle H^2 \rangle - \langle H \rangle^2 = \begin{cases} \langle \psi | H^2 | \psi \rangle - \langle \psi | H | \psi \rangle^2 & \text{for a pure state,} \\ \text{tr}(\rho H^2) - (\text{tr}(\rho H))^2 & \text{for a mixed state.} \end{cases} \tag{13.15}$$

Exercise 13.1 Consider a two-qubit system where 50% of the system is in the uniform mixed state described by the density operator $\frac{1}{4}I$, with I being the 4×4 identity operator, and 50% of the system is in the Bell state $\frac{1}{\sqrt{2}}(|00\rangle + |11\rangle)$. The system is measured with the observable $H = Z \otimes Z$, where $Z \equiv \sigma_Z$ denotes the Pauli-Z operator. Calculate:

(a) The density operator of the system, ρ.

(b) The eigenvalues and eigenstates of the observable H.

(c) The probability of obtaining each eigenvalue of H as the measurement outcome.

(d) The expected value of the measurement, $\langle H \rangle$.

(e) The variance of the measurement outcomes, $(\Delta H)^2$.

13.1 Quantum Probability Essentials

> **Key Takeaways**
>
> In quantum probability, quantum observables fulfill the role of random variables. The density operator encapsulates both classical and quantum randomness of a system. Measurements correspond to probabilistic events and are made with respect to the eigenbasis of an observable, serving as the sample space. The probability distribution is governed by the Born rule.

13.1.4 Multiple Random Variables

1 Classical Probability

Analyzing multiple random variables involves understanding their interactions and the consequent impact on their probabilities. Although we focus on two random variables, for simplicity, these principles extend to more variables.

Probabilities of Two Random Variables

Consider two random variables, X and Y, defined over the same sample space Ω. We examine specific events associated with these variables, represented by the outcomes x for X and y for Y. The probabilities linked to these events are outlined as follows:

- Joint Probability: The probability of x and y occurring together is denoted as $p(X = x, Y = y)$, $p(x \cap y)$, or simply $p(x, y)$, where $p(x, y) = p(y, x)$. The joint probability distribution, denoted as $p(X, Y)$, represents $p(x, y)$ for all possible values of x and y.

- Conditional Probability: This is the probability of event x given event y, denoted as $p(x|y)$. Note that $p(x|y)$ may not equal $p(y|x)$. We also extend this concept to the conditional probability distributions, $p(X|Y)$ and $p(Y|X)$.

- Marginal Probability: The probability of an event associated with one random variable, irrespective of the other variable, is calculated by summing over all outcomes of the other variable:

$$p(x) = \sum_{y \in \Omega} p(x, Y = y). \tag{13.16}$$

This process, known as "marginalization," can also be expressed for the probability distribution:

$$p(X) = \sum_{y \in \Omega} p(X, Y = y). \tag{13.17}$$

- Probability of the Union: The probability that at least one of the events x or y occurs is denoted as $p(x \cup y)$.

Relationships Between Two Events

- Independent Events: The occurrence of one event does not affect the other. For example, the outcomes of two rolls of a die are independent. The probabilities for independent events satisfy:

$$p(x, y) = p(x) \cdot p(y), \quad p(x|y) = p(x). \tag{13.18}$$

- **Dependent Events:** One event influences the occurrence of the other, such as drawing cards from a deck without replacement. Their probabilities are related by the chain rule:

$$p(x,y) = p(y,x) = p(x|y) \cdot p(y) = p(y|x) \cdot p(x). \tag{13.19}$$

This leads to Bayes' Theorem:

$$p(x|y) = \frac{p(y|x) \cdot p(x)}{p(y)}. \tag{13.20}$$

- **Mutually Exclusive Events:** These events cannot occur simultaneously, implying:

$$p(x,y) = 0, \quad p(x \cup y) = p(x) + p(y). \tag{13.21}$$

- **Overlapping Events:** These events can occur together ($p(x \cap y) \neq 0$) and satisfy the inclusion-exclusion principle:

$$p(x \cup y) = p(x) + p(y) - p(x,y). \tag{13.22}$$

This is visually represented in a Venn diagram as shown in Fig. 13.1.

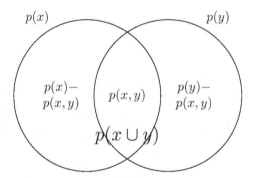

Figure 13.1: Venn Diagram for Probability Relationship

Normalization Conditions

The sum of the probabilities for all possible outcomes of a random variable X equals 1, indicating that some outcome in the sample space will occur with certainty. This normalization condition extends to joint and conditional probabilities:

$$\sum_{x \in \Omega} p(x) = 1, \tag{13.23a}$$

$$\sum_{x \in \Omega} \sum_{y \in \Omega} p(x,y) = 1, \tag{13.23b}$$

$$\sum_{x \in \Omega} p(x|y) = 1. \tag{13.23c}$$

The normalization for conditional probability (Eq. 13.23c) is valid for any given $y \in \Omega$ where $p(y) > 0$.

13.1 Quantum Probability Essentials

■ **Example 13.3 — Classical Joint Distribution and Marginal Probability.** Consider two random variables, X and Y, each taking values in $\Omega = \{1, 2, \ldots, 6\}$ for fair dice.

Given two events: x is rolling a 1 on die one, and y is rolling a 3 on die two. The joint probability $p(x, y)$, the probability of both x and y occurring together, is given by the product of the probabilities of x and y, as the dice rolls are independent:

$$p(x, y) = p(x) \cdot p(y) = \frac{1}{6} \cdot \frac{1}{6} = \frac{1}{36}. \tag{13.24}$$

The probability of the union of events x and y, $p(x \cup y)$, representing either event occurring, is given by:

$$p(x \cup y) = p(x) + p(y) - p(x, y) = \frac{1}{6} + \frac{1}{6} - \frac{1}{36} = \frac{11}{36}. \tag{13.25}$$

Now, we impose the condition that $X + Y$ is even. In this case, X and Y are no longer independent. The joint probability distribution $p(X, Y)$ becomes uniform over the subset of $\Omega \times \Omega$ where the sum of the dice is even, including pairs such as $\{(1, 1), (1, 3), (2, 2), (2, 4), \ldots\}$, each with a probability of $\frac{1}{18}$. However, $p(X, Y)$ is not uniform over $\Omega \times \Omega$, as $p(X, Y) = 0$ if $X + Y$ is odd.

If Alice "holds" X and Bob "holds" Y, using the marginalization formula for Alice's marginal distribution

$$p(X) = \sum_{y \in \Omega} p(X, Y = y), \tag{13.26}$$

we discover that $p(X)$ is still uniform over Ω, each outcome having a probability of $\frac{1}{6}$. Similarly, for Bob's $p(Y)$.

This shows that if X and Y are dependent, the marginal distributions can be uniform over Ω, despite their joint distribution $p(X, Y)$ not being uniform over $\Omega \times \Omega$. We will see a similar phenomenon with a quantum EPR pair in Example 13.4. ■

2 Quantum Generalization

The concepts of classical multi-variable probability are extended to the quantum realm, introducing additional complexities. This complexity arises primarily from the non-commutative nature of quantum observables, where, unlike their classical counterparts, the order of measurement affects the outcome. Furthermore, quantum states of composite systems often exhibit entanglement, a phenomenon where the state of each component cannot be described independently of the others. Additionally, measurements on these systems can be either local, affecting individual subsystems, or joint, involving multiple subsystems simultaneously (see § 6.2.2). We will use examples to build connections with classical probabilities and illustrate these complexities.

Exercise 13.2 Prerequisite Review: To understand quantum probability and information concerning multi-qubit states, readers are highly recommended to review Exercise 6.6, replicated below for convenience.

Consider three qubits (A, B, and C) in the state

$$|\psi\rangle = \sum_{i,j,k\in\{0,1\}} c_{ijk}|ijk\rangle.$$

(a) Find the probability of measuring the third qubit (C) with an outcome $|0\rangle$.

(b) After qubit C is measured and collapsed to $|0\rangle$, what is the joint state of qubits A and B? What is the state of the three-qubit system?

(c) Suppose a Bell measurement is performed on qubits A and B with an outcome $|\Phi^+\rangle = \frac{1}{\sqrt{2}}(|00\rangle + |11\rangle)$. What is the probability of this outcome? What is the post-measurement state of the three-qubit system?

Commuting vs. Non-Commuting Observables

A real-valued quantum random variable corresponds to an observable X, which is a Hermitian operator. The eigenvalues of X represent its possible outcomes.

As detailed in § 1.6, if two observables X and Y belonging to the same quantum system do not commute, i.e., $[X,Y] \neq 0$, there is generally no state in which both X and Y have definite values simultaneously; they do not share an eigenvector. In practical terms, two such variables are mutually uncertain and are not simultaneously measurable.

For instance, consider the non-commuting Pauli operators σ_X and σ_Z for a single qubit in state $|+\rangle = \frac{1}{\sqrt{2}}(|0\rangle + |1\rangle)$. The measurement of one observable affects the system in a way that the subsequent measurement of the other becomes completely uncertain. (See exercise below.)

Thus, in quantum mechanics, the concept of joint and conditional probabilities for non-commuting observables does not apply in the same way it does for commuting observables or in classical probability theory.

Exercise 13.3 Consider the non-commuting Pauli operators σ_X and σ_Z for a single qubit in state $|+\rangle = \frac{1}{\sqrt{2}}(|0\rangle + |1\rangle)$. Compare two measurement sequences: measuring σ_X and then σ_Z on the qubit, and measuring σ_Z and then σ_X. Describe how the first measurement affects the state of the system and thereby the outcome of the second measurement. Discuss the meaning of $\langle\sigma_X\sigma_Z\rangle$ in this context.

Exercise 13.4 Consider two commuting Hermitian operators U and V on a single-qubit system: $U = V = \sigma_Z$, where σ_Z is the Pauli-Z operator. Verify if $\langle UV\rangle = \langle U\rangle\langle V\rangle$ for the following states:

(a) $|\psi\rangle = |0\rangle$.

(b) $|\psi\rangle = |+\rangle = \frac{1}{\sqrt{2}}(|0\rangle + |1\rangle)$.

(c) $\rho = \frac{1}{2}I$.

Marginal Probability Through Partial Trace

The reduced density operator calculated through the partial trace, discussed extensively in § 12.2.8, is the quantum counterpart of classical marginal probability. Both methodologies enable us to focus on a component within a composite system

13.1 Quantum Probability Essentials

by summing over and thereby effectively disregarding the other components of the system.

■ **Example 13.4 — Marginal Probabilities in an EPR Pair.** Let's re-examine the archetypal case where Alice and Bob share an EPR pair, $|\Phi^+\rangle = \frac{1}{\sqrt{2}}(|00\rangle + |11\rangle)$. (See Example 12.10 for more details.)

To calculate the marginal probability, we start with the density operator for $|\Phi^+\rangle$:

$$\rho_{AB} = |\Phi^+\rangle\langle\Phi^+|. \tag{13.27}$$

The reduced density operator for Alice, obtained by tracing ρ_{AB} over B, captures the marginal probability of her qubit:

$$\rho_A = \text{tr}_B(\rho_{AB}) = \frac{1}{2}\begin{bmatrix} 1 & 0 \\ 0 & 1 \end{bmatrix}. \tag{13.28}$$

This is a maximally mixed state, with a uniform probability distribution over $|0\rangle$ and $|1\rangle$, in fact in all measurement bases.

Bob's reduced density operator is identical to Alice's, also showing a uniform probability distribution over $|0\rangle$ and $|1\rangle$. ■

Probabilities in a Product State

Let's now examine local measurements of a composite state of two qubit A and B. In this case, two observables, X and Y, pertaining to qubit A and B, respectively, commute regardless of their commutation properties within their respective systems. This is because the operations act on different Hilbert spaces and do not interfere with each other. (See § 7.4.1.)

If the state is a product state, measurements on one qubit are independent of measurements on the other, similar to the classical independent events.

■ **Example 13.5 — Joint and Conditional Probabilities in a Product State.** Consider two qubits A and B in the product state $|\psi\rangle = |++\rangle$, where $|+\rangle = \frac{1}{\sqrt{2}}(|0\rangle + |1\rangle)$.

(1) <u>Measuring in Computational Basis</u>

Suppose we measure each qubit in the computational basis. That is, we have two random variables $X = \sigma_{ZA}$ on qubit A and $Y = \sigma_{ZB}$ on qubit B, where σ_Z is the Pauli-Z operator.

(a) Probability of measuring qubit A in state $|0\rangle$ and qubit B in $|1\rangle$:

$$p(|0\rangle, |1\rangle) \equiv p(|01\rangle) = |\langle 01|++\rangle|^2 = \tfrac{1}{4}.$$

This probability can also be expressed in terms of the eigenvalues as $p(1, -1)$.

(b) Conditional probability of measuring qubit A in $|0\rangle$ given qubit B is in $|1\rangle$:

$$p(|0\rangle_A \,|\, |1\rangle_B) = p(|0\rangle_A) = \tfrac{1}{2}, \text{ due to independence.}$$

(c) Joint probability distribution of X and Y:

$p(X,Y) = p(X) \cdot p(Y) = \frac{1}{2} \cdot \frac{1}{2} = \frac{1}{4}$ for each outcome in $\{|00\rangle, |01\rangle, |10\rangle, |11\rangle\}$.

(d) Conditional probability distribution of X given Y:

$p(X|Y) = p(X) = \frac{1}{2}$, for each outcome in $\{|0\rangle, |1\rangle\}$.

(e) Probability of either qubit A being in $|0\rangle$ or qubit B being in $|1\rangle$:

$p(|0\rangle \cup |1\rangle) = \frac{3}{4}$, corresponding to the three outcomes $\{|00\rangle, |01\rangle, |11\rangle\}$.

(f) Expected values (where $X = \sigma_{ZA}$ and $Y = \sigma_{ZB}$):

$\langle X \rangle = p(|0\rangle)(1) + p(|1\rangle)(-1) = 0$.

$\langle XY \rangle = p(|00\rangle)(1 \cdot 1) + p(|01\rangle)(1 \cdot (-1)) + p(|10\rangle)((-1) \cdot 1) + p(|11\rangle)(1 \cdot 1) = 0$.

$\langle X+Y \rangle = p(|00\rangle)(1+1) + p(|01\rangle)(1-1) + p(|10\rangle)(-1+1) + p(|11\rangle)(-1-1) = 0$.

Since measuring an observable X on system A and an observable Y on system B are independent events, $p(X,Y) = p(X) \cdot p(Y)$, $p(X|Y) = p(X)$, and $p(X \cup Y) = p(X) + p(Y) - p(X,Y)$.

(2) Measuring in $\{|+\rangle, |-\rangle\}$ Basis

Now we measure each qubit in the $\{|+\rangle, |-\rangle\}$ basis. The corresponding random variables are $X = \sigma_{XA}$ on qubit A and $Y = \sigma_{XB}$ on qubit B, where σ_X is the Pauli-X operator.

(a) Probability of measuring qubit A in state $|+\rangle$ and qubit B in $|-\rangle$:

$p(|+\rangle, |-\rangle) = |\langle +-|++\rangle|^2 = 0$.

(b) Conditional probability of measuring qubit A in $|+\rangle$ given qubit B is in $|-\rangle$:

$p(|+\rangle_A | |-\rangle_B)$ is undefined since qubit B cannot be in $|-\rangle$.

(c) Joint probability distribution of X and Y:

$p(X,Y) = p(X) \cdot p(Y) = 1$ for outcome $|++\rangle$ and 0 for $\{|+-\rangle, |-+\rangle, |--\rangle\}$.

(d) Conditional probability distribution of X given Y:

$p(X|Y)$ is defined only if Y is $|+\rangle$, in which case, $p(X|Y) = 1$ for X-outcome of $|+\rangle$, and 0 for $|-\rangle$.

(e) Probability of either qubit A being in $|+\rangle$ or qubit B being in $|-\rangle$:

$p(|+\rangle \cup |-\rangle) = 1$, because qubit A is definitively in $|+\rangle$.

(f) Expected values (where $X = \sigma_{XA}$ and $Y = \sigma_{XB}$):

$\langle X \rangle = p(|+\rangle)(1) + p(|-\rangle)(-1) = 1$.

$\langle XY \rangle = p(|++\rangle)(1 \cdot 1) + \ldots = 1$.

$\langle X+Y \rangle = p(|++\rangle)(1+1) + \ldots = 2$.

 This example demonstrates that quantum probabilities depend on the basis (or observable) of measurement, and that conditional probabilities may be undefined for specific outcomes.

13.1 Quantum Probability Essentials

(3) Calculating Probabilities from Density Operators

In straightforward cases like this example, probabilities can be directly computed. For more complex scenarios, however, we often rely on density operators to determine these probabilities. To built intuition, let's re-calculate some of the above probabilities and expected values using density operators.

The density operator of the composite system is given by $\rho_{AB} = |++\rangle\langle++|$. The density operator of qubit A is $\rho_A = \text{tr}_B(\rho_{AB}) = |+\rangle\langle+|$. Similarly, for ρ_B. Since this is a product state, $\rho_{AB} = \rho_A \otimes \rho_B$.

(a) $p(|0\rangle, |1\rangle) = \langle 01|\rho_{AB}|01\rangle = \frac{1}{4}$.

(b) $p(|+\rangle, |+\rangle) = \langle++|\rho_{AB}|++\rangle = 1$.

(c) $p(|+\rangle, |-\rangle) = \langle+-|\rho_{AB}|+-\rangle = 0$.

(d) $p(|-\rangle_B) = \langle-|\rho_B|-\rangle = 0$. Hence, $p(|+\rangle_A \mid |-\rangle_B) = \dfrac{p(|+\rangle, |-\rangle)}{p(|-\rangle_B)}$ is undefined.

(e) Expected values:

$\langle \sigma_{ZA} \rangle = \text{tr}(\sigma_Z \rho_A) = 0$.

$\langle \sigma_{ZA}\sigma_{ZB} \rangle = \text{tr}(\sigma_{ZA}\sigma_{ZB}\rho_{AB}) = 0.$ $(\sigma_{ZA}\sigma_{ZB} \equiv \sigma_{ZA} \otimes \sigma_{ZB})$

$\langle \sigma_{ZA} + \sigma_{ZB} \rangle = \text{tr}((\sigma_{ZA}I_B + I_A\sigma_{ZB})\rho_{AB}) = 0$.

$\langle \sigma_{XA} \rangle = \text{tr}(\sigma_X \rho_A) = 1$.

$\langle \sigma_{XA}\sigma_{XB} \rangle = \text{tr}(\sigma_{XA}\sigma_{XB}\rho_{AB}) = 1$.

$\langle \sigma_{XA} + \sigma_{XB} \rangle = \text{tr}((\sigma_{XA}I_B + I_A\sigma_{XB})\rho_{AB}) = 2$.

∎

Probabilities in Entangled Systems

Unlike the product state in Example 13.5, the entanglement embodied in states leads to outcomes that are intertwined in a way that is not possible in classical physics. We will use an example to illustrate how entanglement leads to inseparability of the joint state and to probabilities that exhibit non-classical correlations.

■ **Example 13.6 — Joint and Conditional Probabilities in a Bell State.** Consider two qubits A and B in the Bell state $|\Phi^+\rangle = \frac{1}{\sqrt{2}}(|00\rangle + |11\rangle)$. The joint and reduced density operators are examined in Example 13.4.

(1) Local Measurements in Computational Basis

Local measurements of each qubit in the computational basis are equivalent to observing the outcomes of the Pauli-Z operator applied to each qubit. We define two random variables $X = \sigma_{ZA}$ for qubit A and $Y = \sigma_{ZB}$ for qubit B.

(a) Probability of measuring both qubits in state $|0\rangle$:

$p(|0\rangle, |0\rangle) = |\langle 00|\Phi^+\rangle|^2 = \frac{1}{2}$.

(b) Joint probability distribution of X and Y:

$p(X, Y) = [\frac{1}{2}, 0, 0, \frac{1}{2}]$ for outcomes $[|00\rangle, |01\rangle, |10\rangle, |11\rangle]$.

Note that $p(X,Y) \neq p(X)p(Y)$. This demonstrates that the joint state of an entangled system is inseparable into individual states, leading to probabilities that cannot be factored as in classical independent events.

(c) Conditional probabilities given qubit B is in state $|1\rangle$:

First, $p(|1\rangle_B) = \langle 1|\rho_B|1\rangle = \frac{1}{2}$. Then,

$$p(|0\rangle_A \,|\, |1\rangle_B) = \frac{p(|01\rangle)}{p(|1\rangle_B)} = 0,$$

$$p(|1\rangle_A \,|\, |1\rangle_B) = \frac{p(|11\rangle)}{p(|1\rangle_B)} = 1.$$

Without the condition of qubit B in $|1\rangle$, we would have $p(|0\rangle_A) = p(|1\rangle_A) = \frac{1}{2}$.

 These conditional probabilities demonstrate correlation. The measurement outcome of one qubit affects the probabilities of outcomes for the other qubit.

(d) Expected values:

$\langle X \rangle = p(|0\rangle)(1) + p(|1\rangle)(-1) = 0$. Or equivalently, $\langle \sigma_{ZA} \rangle = \text{tr}(\sigma_Z \rho_A) = 0$.

$\langle XY \rangle = p(|00\rangle)(1 \cdot 1) + p(|01\rangle)(1 \cdot (-1)) + p(|10\rangle)((-1) \cdot 1) + p(|11\rangle)(1 \cdot 1) = 1$. Or equivalently, $\langle \sigma_{ZA}\sigma_{ZB} \rangle = \text{tr}(\sigma_{ZA}\sigma_{ZB}\rho_{AB}) = 1$. Note that, in this case, $\langle XY \rangle \neq \langle X \rangle \langle Y \rangle$, indicating correlation.

$\langle X + Y \rangle = p(|00\rangle)(1+1) + p(|01\rangle)(1-1) + p(|10\rangle)(-1+1) + p(|11\rangle)(-1-1) = 0$.

(2) <u>Joint Measurements in the Bell Basis</u>

The Bell basis, $\{|\Phi^+\rangle, |\Phi^-\rangle, |\Psi^+\rangle, |\Psi^-\rangle\}$, is an orthonormal basis for two-qubit states (see § 8.2). Measuring both qubits jointly in this basis yields a probability distribution of 1 for $|\Phi^+\rangle$ and 0 for all other basis states. ∎

Exercise 13.5 Consider a two-qubit system composed of 50% in the uniform mixed state $\frac{1}{4}I$ and 50% in the Bell state $\frac{1}{\sqrt{2}}(|00\rangle + |11\rangle)$. We perform local measurements of each qubit in the computational basis which is equivalent to observing the outcomes of the Pauli-Z operator (σ_Z) applied to each qubit. We define two random variables $X = \sigma_{ZA}$ for qubit A and $Y = \sigma_{ZB}$ for qubit B. Calculate:

(a) The density operator of the system ρ_{AB}.

(b) The reduced density operators ρ_A and ρ_B.

(c) The joint probability of measuring both qubits in state $|0\rangle$: $p(|0\rangle, |0\rangle)$.

(d) The conditional probabilities given qubit B is in state $|1\rangle$: $p(|0\rangle_A \,|\, |1\rangle_B)$ and $p(|1\rangle_A \,|\, |1\rangle_B)$.

(e) The expectation values $\langle X \rangle$, $\langle Y \rangle$, $\langle XY \rangle$, and $\langle X+Y \rangle$.

13.1 Quantum Probability Essentials

> **Key Takeaways**
>
> In the quantum realm, multiple random variables are represented by observables, whose non-commutative nature influences their measurements. Entanglement in composite systems introduces complex dependencies between variables, making quantum probabilities differ fundamentally from classical counterparts. The concept of marginal probability is extended through reduced density operators, enabling focus on specific subsystems. The nature of quantum probability is further nuanced by the distinction between product and entangled states, with the latter showing non-classical correlations.

13.1.5 Random Processes

A random or stochastic process is essentially a collection of random variables, each indexed by time or space, that describe the evolution of systems in a stochastic, or probabilistic, manner. This concept is central in understanding systems that evolve unpredictably over time or space.

1 Classical Stochastic Processes

Classical stochastic processes are mathematical models used to describe classical systems that evolve over time in a way that is at least partially random. Here are some common examples:

- **Random Walk:** This is a fundamental stochastic process where an object moves step by step, with each step being random. In its simplest form, at each time step, the object moves either one unit up or one unit down with certain probabilities. Random walks are used to model various phenomena, including stock market fluctuations and particle movements in liquids.

- **Markov Chains:** A Markov chain is a stochastic process where the probability of moving to the next state depends only on the current state and not on the sequence of events that preceded it. Markov chains are used in a variety of fields, including economics, game theory, and biology.

    Random walk is a special case of Markov chain where each step is taken randomly and independently, typically with equal probability, in a specific state space like a lattice or a grid.

- **Queueing Models:** Used extensively in operations research, these models study the behavior of queues (or lines). They help in understanding and predicting queue lengths and waiting times, important in designing and managing facilities like call centers, hospitals, and manufacturing plants.

2 Quantum Stochastic Processes

Quantum stochastic processes extend these concepts into the quantum domain, featuring unique quantum mechanical principles like superposition and entanglement. For example:

- **Quantum Walk:** Analogous to classical random walk but in the quantum realm, quantum walk exhibits features such as superposition and entanglement, leading to behavior distinct from their classical counterparts.

- **Quantum Markov Processes:** Quantum Markov Processes, as the quantum counterparts of classical Markov processes, characterize the evolution of quantum states by extending the classical notion of memorylessness to quantum systems. In these processes, the progression of a quantum state is depicted through a sequence of Completely Positive Trace-Preserving (CPTP) maps. These maps ensure each transition from one state to another is independent of prior states or transitions, thereby maintaining the Markov property. Furthermore, CPTP maps can incorporate effects such as measurements, noise, and decoherence, important in accurately representing quantum systems.

- **Quantum Queueing Models:** Emerging in the context of quantum communication and computing, these models apply quantum mechanical principles to queueing theory. Quantum Queueing Models can be used for the efficient management of quantum information processes, such as the transmission of qubits in quantum networks and the execution of operations in quantum computing. They deal with challenges unique to quantum systems, like the no-cloning theorem and quantum entanglement, offering new ways to optimize quantum resource management and information processing.

3 Simulation Techniques in Quantum Stochastic Processes

Simulations help us understand quantum stochastic processes by offering tools to model and study complex quantum behaviors where analytical solutions are often intractable. Among the main methods employed are:

- **Quantum Monte Carlo (QMC) Simulations:** QMC methods use stochastic sampling to solve quantum mechanical problems, particularly valuable in studying many-body systems. These simulations help calculate properties like ground state energies and molecular structures.

  In QMC, random samples are used to estimate the properties of a quantum system, effectively turning a complex quantum problem into a statistical one. Techniques like Variational Monte Carlo (VMC) and Diffusion Monte Carlo (DMC) employ different strategies for handling the probabilistic aspects of quantum mechanics.

- **Time-Dependent Schrödinger Equation:** This fundamental equation in quantum mechanics describes how the quantum state of a physical system changes over time. While QMC employs random sampling, the Schrödinger equation models the probabilistic evolution of quantum states over time.

- **Quantum Operations Sequence:** This approach models quantum stochastic processes through a sequence of quantum operations, including measurements and unitary transformations.

13.1.6 ∗ Example: One-Dimensional Quantum Walk

In this subsection, we delve into the one-dimensional quantum walk as a representative example of quantum probability and quantum stochastic processes. We will contrast its characteristics with those of classical random walks, highlighting the distinctive features of quantum mechanics that come into play in the quantum variant.

13.1 Quantum Probability Essentials

1 Classical 1D Random Walk

The one-dimensional random walk is a simple yet instructive model. It can be visualized as a walk along a line, with each step being of the same length. Starting at the origin, a fair (unbiased) coin is flipped before each step. A heads (H) results in a step forward, while tails (T) leads to a step backward.

The problem of interest is determining the probability of being at a specific point n after N steps, given that the starting point is 0. We denote this probability as $f_N(n)$.

Since the walker must be at some position after N steps, the sum of these probabilities for all possible positions n must equal 1, i.e., $\sum_n f_N(n) = 1$.

The pattern in these probabilities becomes clearer when factoring out $\frac{1}{2^N}$:

n	-5	-4	-3	-2	-1	0	1	2	3	4	5
$f_0(n)$						1					
$2 f_1(n)$					1		1				
$2^2 f_2(n)$				1		2		1			
$2^3 f_3(n)$			1		3		3		1		
$2^4 f_4(n)$		1		4		6		4		1	
$2^5 f_5(n)$	1		5		10		10		5		1

To better understand these probabilities, one can enumerate the sequences of coin flips leading to each position. For instance, in a three-step walk ($N = 3$), the sequence HHH will end at $n = 3$, while HHT, HTH, and THH will end at $n = 1$. The negative positions correspond to the reverse combinations of H and T. With $2^3 = 8$ possible three-step walks, the probabilities for the different positions are: $f_3(-3) = 1/8$, $f_3(-1) = 3/8$, $f_3(1) = 3/8$, $f_3(3) = 1/8$.

These probabilities correspond to the coefficients in Pascal's Triangle, and they are the same as those in the binomial expansion of $(a+b)^N$. Therefore, the probability function is given by:

$$f_N(n) = \frac{1}{2^N} \binom{N}{\frac{N+n}{2}} = \frac{N!}{2^N \left(\frac{N+n}{2}\right)! \left(\frac{N-n}{2}\right)!}, \tag{13.29}$$

where n is an integer such that $-N \leq n \leq N$ and $N + n$ is even.

2 Quantum Simulation of Classical 1D Random Walk

We can simulate a classical 1D random walk using a quantum circuit, as illustrated in Fig. 13.2.

The first qubit represents the state of a fair coin, denoted as $|\text{coin}\rangle$. The second register of d qubits records the position of the walker, denoted as $|\text{pos}\rangle$. This register is initialized to $|100\ldots0\rangle$, representing position $n = 0$. The offset ensures that $n = 0$ is centered in the set $\{0,1\}^d$. We choose a sufficiently large d such that $2^d > 2N$.

The state of the $d + 1$ qubits is given by

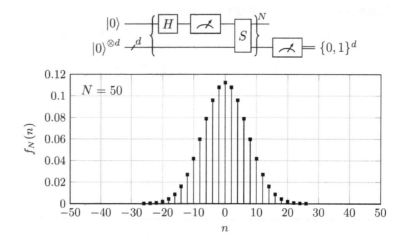

Figure 13.2: Quantum Simulation of 1D Random Walk

$$|\Psi\rangle = |\text{coin}\rangle \otimes |\text{pos}\rangle. \tag{13.30}$$

A quantum random number generator (QRNG, see § 5.4) simulates the coin flip. A single-bit QRNG is implemented as a qubit initialized in the state $|0\rangle$ or $|1\rangle$, rotated by a Hadamard gate, followed by a measurement in the computational basis. After each measurement, the qubit state collapses to $|0\rangle$ or $|1\rangle$, so the QRND is ready for use again.

We construct a quantum gate S, acting on the $d+1$ qubits, such that if $|\text{coin}\rangle = |0\rangle$ (representing Heads), n is incremented; if $|\text{coin}\rangle = |1\rangle$ (representing Tails), n is decremented:

$$S = |0\rangle\langle 0| \otimes \sum_n |n+1\rangle\langle n| + |1\rangle\langle 1| \otimes \sum_n |n-1\rangle\langle n|. \tag{13.31}$$

Exercise 13.6 Verify that the shift operator S, given in Eq. 13.31, is unitary.

We repeat the QRNG and S operations N times and measure the last d qubits. The probability of finding the d-qubit state corresponding to position n yields the function $f_N(n)$:

$$f_N(n) = |\langle n|\text{pos}\rangle|^2. \tag{13.32}$$

The results of a simulation with $N = 50$ and $d = 7$ are graphed in the lower part of Fig. 13.2. The graph mirrors Eq. 13.29, a binomial probability distribution centered around $n = 0$ with a spread of $\sqrt{N}$.

3 1D Quantum Walk

Finally, we are ready to explore the 1D quantum walk. Unlike a classical random walk, a quantum walk is fundamentally deterministic in its evolution, with probabilistic outcomes emerging from quantum measurement. For this reason, we omit "random" in its name.

13.1 Quantum Probability Essentials

Consider a quantum particle as the walker, initially at the origin on a number line. In each step of a quantum walk, the particle enters a superposition of moving one step left and one step right. After multiple steps, the particle does not merely occupy a single position; rather, it has a probability amplitude for every position it could have reached. These amplitudes from different steps interfere, leading to a probability distribution distinct from classical random walks. Notably, some off-center positions may exhibit higher probabilities than the center position $n = 0$ due to constructive interference.

The quantum circuit for simulating this phenomenon is shown in Fig. 13.3. The key difference from the classical walk simulation (Fig. 13.2) is that we no longer measure the QRNG after each step. This lack of measurement allows the system state to maintain superposition and demonstrate interference effects.

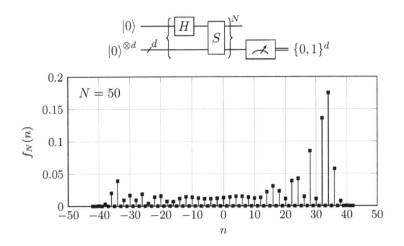

Figure 13.3: Quantum 1D Walk

A simulation with $N = 50$ and $d = 7$ produces a result graphed in the lower part of Fig. 13.3. Unlike the bell-shaped binomial distribution of the classical walk, the quantum walk yields a distinctly shaped distribution, exhibiting higher probabilities for positions further from the start point due to quantum interference.

To elucidate this behavior, we demonstrate the evolution of the system state under Eq. 13.31 in Table 13.1, started with the coin state $|0\rangle$. Noticeably, by the third step, interference shifts the probability distribution, making it asymmetrical. This asymmetry arises because the terms $-|1,-1\rangle + |1,-1\rangle$ cancel each other, eliminating their probability contribution at $x = -1$. Conversely, the terms $|0,1\rangle + |0,1\rangle$ combine to form $2|0,1\rangle$, contributing a probability of $\frac{4}{8}$.

It is important to note that the initial coin state significantly influences the shape of the $f_N(n)$ curve. This initial state can be a general state represented as $\alpha|0\rangle + \beta|1\rangle$.

Exercise 13.7 Construct a table showing the evolution of system states for a 1D quantum walk over the first three steps, following the dynamics described by Eq. 13.31. Assume the initial state of the coin is $|1\rangle$. Your table should be similar to Table 13.1, detailing the coin and walker states, as well as the probability

Steps	Coin & Walker State	Probability
0	$\|0,0\rangle$	$[1]$
1	$\frac{1}{\sqrt{2}}(\|0,1\rangle + \|1,-1\rangle)$	$[\frac{1}{2}, 0, \frac{1}{2}]$
2	$\frac{1}{2}(\|0,2\rangle + \|1,0\rangle + \|0,0\rangle - \|1,-2\rangle)$	$[\frac{1}{4}, 0, \frac{1}{2}, 0, \frac{1}{4}]$
3	$\frac{1}{\sqrt{8}}(\|0,3\rangle + \|1,1\rangle + \|0,1\rangle - \|1,-1\rangle$ $\quad + \|0,1\rangle + \|1,-1\rangle - \|0,-1\rangle + \|1,-3\rangle)$ $= \frac{1}{\sqrt{8}}(\|0,3\rangle + \|1,1\rangle + 2\|0,1\rangle - \|0,-1\rangle + \|1,-3\rangle)$	$[\frac{1}{8}, 0, \frac{1}{8}, 0, \frac{5}{8}, 0, \frac{1}{8}]$

Table 13.1: Evolution of Coin & Walker States in a 1D Quantum Walk

distribution at each step.

4 Applications

Although the one-dimensional quantum walk has been introduced primarily as an illustrative example of quantum probability and quantum stochastic processes, the generalized concept of quantum walks is a rich and complex topic with significant applications in quantum computing and quantum information theory. These applications exploit the unique properties of quantum walks, including superposition and interference. The following are some key areas where quantum walks are applied:

- Quantum Algorithms: Quantum walks are a powerful tool for designing quantum algorithms. They can explore computational spaces more efficiently than their classical counterparts due to quantum superposition and interference. Algorithms based on quantum walks have been shown to offer speedups for certain problems. For example, the quantum walk-based search algorithm can search an unsorted database quadratically faster than any possible classical algorithm.

- Quantum Simulation: Quantum walks can simulate various quantum systems, particularly in studying transport phenomena in quantum systems. They are instrumental in simulating the behavior of electrons in materials, which helps in understanding and designing new quantum materials and devices.

- Quantum Computing Models: The concept of quantum walks underpins certain models of quantum computation, like the Continuous-Time Quantum Walk (CTQW) model. These models provide alternative frameworks to the standard gate-based quantum computing model and are more natural for certain types of problems and physical implementations.

- Quantum Communications: In the realm of quantum communication, quantum walks are used to model and analyze quantum networks, potentially leading to more efficient protocols for quantum information transfer. Quantum walks also have applications in developing new cryptographic protocols.

13.2 Quantum Entropy and Information

The concept of entropy, a cornerstone in information theory, measures the uncertainty

13.2 Quantum Entropy and Information

or information content associated with a random variable. In the quantum regime, this notion extends to encapsulate the uncertainty in quantum states.

In information theory, "information" is defined in a way that is quite different from its everyday usage. It's not about the message, meaning, or knowledge, but rather about the uncertainty or unpredictability of a message source. For instance, if a message is very predictable, it contains less information than a highly unpredictable one. This is quantified using the concept of entropy, usually represented in bits. The higher the entropy, the more information a message carries.

13.2.1 Classical Shannon Entropy

Consider a classical random variable X with a sample space Ω. Each outcome x in Ω occurs according to a probability distribution $p(X)$. The information content (or self-information, to distinguish with joint information) $I(x)$ of a particular outcome x is defined as:

$$I(x) = -\log p(x). \tag{13.33}$$

The logarithmic base determines the unit of information; for base 2, the unit is bits.

The log function in the definition ensures that information is additive. Consider two independent events x and y with probabilities $p(x)$ and $p(y)$, respectively. The joint probability of both events occurring is the product $p(x)p(y)$. This relationship leads to the total information content for both events being $I(x, y) = -\log(p(x)p(y)) = I(x) + I(y)$.

The Shannon entropy is defined as the expected value (statistical average) of the information content of a random variable over all possible outcomes:

$$H(X) = E[I(x)] = -\sum_{x \in \Omega} p(x) \log p(x). \tag{13.34}$$

Thus, in information theory, "information" is quantified as the reduction in uncertainty about a random variable and is measured by the change in entropy, when a message is received.

■ **Example 13.7 — Roll of a Six-sided Fair Die.** Consider the roll of a six-sided fair die, with $\Omega = \{1, 2, 3, 4, 5, 6\}$ and a uniform probability distribution for each outcome. The Shannon entropy in this case is $H(X) = -\sum_{x=1}^{6} \frac{1}{6} \log \frac{1}{6} = \log 6 \approx 2.58$ bits. This value represents the average uncertainty or information content associated with the outcome of a roll. In information theory, it indicates that, on average, about 2.58 bits is needed to optimally encode the outcome of a die roll in a large series of rolls. ■

In the context of information theory and probability simulation, we often use the phrase *"draw from probability distribution p"*. For example, consider a simple probability distribution p over $\Omega = \{a, b\}$, where $p(a) = 0.6$, $p(b) = 0.4$. A "draw from p" means randomly selecting either a or b, with a 60% chance of selecting a

and a 40% chance for b in any single draw.

■ **Example 13.8 — Simulating a Draw from a Distribution.** Suppose we have a probability distribution p over a finite set of outcomes. In a computational simulation, $H(p)$ represents the least number of truly random bits (coin flips) needed on average to generate a draw from p. This is especially accurate when simulating a large number of draws, as the average number of bits used converges to $H(p)$.

For instance, consider $p = \left(\frac{1}{2}, \frac{1}{4}, \frac{1}{4}\right)$ for the sample space $\{a, b, c\}$. To generate a draw from p using minimum number of coin flips, if we get Heads (with a probability of $\frac{1}{2}$), we select a, done; if we get Tails, we flip a second coin, and select b or c according to Heads or Tails (each with overall probability of $\frac{1}{4}$). Thus, on average, it takes $1 \cdot \frac{1}{2} + 2 \cdot \frac{1}{4} = 1.5$ coin flips to generate a draw. This is consistent with the Shannon entropy calculation, $H(p) = 1.5$ bits. ∎

■ **Example 13.9 — Storing a Draw with Optimal Compression.** For a probability distribution p, the average number of bits required to store a draw from p, using the best possible compression scheme, is also given by $H(p)$. This stems from the fact that Shannon entropy defines the limit of lossless compression. For each draw from p, one can store the sequence of random bits used to generate that particular outcome. Over many such draws, the average length of these stored bit sequences will approximate $H(p)$.

As an example, consider again $p = \left(\frac{1}{2}, \frac{1}{4}, \frac{1}{4}\right)$ for the sample space $\{a, b, c\}$. To encode a string drawn from this distribution, we use the shortest code 0 for the most frequent letter a, and then 10 for b, 11 for c. On average, we need 1.5 bits per letter. This is consistent with the Shannon entropy calculation, $H(p) = 1.5$ bits. ∎

Exercise 13.8 Calculate the entropy in bits for each of the following random variables, representing pixel values in an image:

(a) Assume we have an image where each pixel's color is represented by a unique combination of Red, Green, and Blue channels, each of which can take any integer value from 0 to 255 with uniform probability. Calculate the entropy of a single pixel.

(b) Now we remove the red color in the image, i.e., the Red channel is always zero, and only the Green and Blue channels vary from 0 to 255 with uniform probability. Calculate the entropy of a single pixel under this condition.

(c) Finally, assume the color space of the image is limited: the Red channel varies from 0 to 63, the Green channel from 0 to 127, and the Blue channel from 64 to 255, all with uniform probability. Calculate the entropy of a single pixel for this scenario.

Note: Assume each color channel is independent of the others.

Basic Properties

Shannon entroy has a number of important properties:

1. $0 \leq H(p) \leq \log(d)$, where d is the dimension of the sample space.

2. $H(p) = 0$ if $p = 1$ for some event and 0 for the rest, representing a deterministic process.

3. $H(p) = \log(d)$ if p is uniform, $\left(\frac{1}{d}, \ldots, \frac{1}{d}\right)$, representing maximum randomness.

13.2.2 ∗Quantum von Neumann Entropy

In quantum probability theory, observables are a non-commutative generalization of random variables, with their set of eigenvalues playing the role of the set of possible outcomes. A given density operator ρ generalizes the role of a probability distribution by completely encapsulating the probabilistic state of a quantum system, allowing us to compute statistical quantities such as the expected value of an observable A:

$$\langle A \rangle = \text{tr}(\rho A) = \langle \rho, A \rangle. \tag{13.35}$$

To compute the von Neumann entropy, which serves as a measure of uncertainty in quantum systems, analogous to Shannon entropy in classical probability, we consider $A = -\log \rho$ as an observable. The von Neumann entropy $S(\rho)$ of a quantum system described by a density operator ρ is defined as:

$$S(\rho) = -\langle \log \rho \rangle = -\text{tr}(\rho \log \rho). \tag{13.36}$$

This expression can be simplified using the spectral decomposition of ρ. In the eigenbasis of ρ, where ρ is diagonal, the entropy takes the form:

$$S(\rho) = -\sum_i \lambda_i \log \lambda_i, \tag{13.37}$$

where each eigenvalue λ_i represents the probability of the quantum system being in the corresponding eigenstate. This result is invariant under basis changes, thanks to the basis-independence of the trace operation.

The von Neumann entropy quantifies the amount of information present in a quantum system. It has a number of properties similar to those of Shannon entroy:

1. $0 \leq S(\rho) \leq \log(d)$, where d is the dimension of ρ.

2. $S(\rho) = 0$ if some term in $\{\lambda_i\}$ is 1 and the rest 0, representing a pure state.

3. $S(\rho) = \log(d)$ if $\{\lambda_i\}$ is uniform, representing a maximumly mixed state.

■ **Example 13.10 — Pure States.** For the pure state $|\psi\rangle = \alpha |0\rangle + \beta |1\rangle$, the density operator is:

$$\rho = \begin{bmatrix} |\alpha|^2 & \alpha\beta^* \\ \beta\alpha^* & |\beta|^2 \end{bmatrix}.$$

The eigenvalues of ρ are 0 and 1. Therefore, the von Neumann entropy for ρ is:

$$S(\rho) = 0 \log 0 + 1 \log 1 = 0. \tag{13.38}$$

Here we take $0 \log 0 = 0$ in the limiting sense: $\lim_{x \to 0^+} x \log x = 0$.

In fact, the entropy for any pure state of any dimension, including Bell states, is 0, reflecting the absence of uncertainty or information entropy in a pure state. This is because $\rho = |\psi\rangle\langle\psi|$ of a pure state $|\psi\rangle$ is a rank-1 matrix, i.e., with only 1 linearly independent column vector or row vector. For such a matrix, one of the eigenvalues is always 1 (due to the normalization of quantum states), and all others are 0. ∎

■ **Example 13.11 — The Uniform Mixed State.** For a d-dimensional qudit, the density operator for the uniform mixed state is:

$$\rho = \frac{1}{d} I_{d \times d}.$$

This density operator is already diagonal. Its von Neumann entropy is:

$$S(\rho) = -\sum_{i=1}^{d} \frac{1}{d} \log \frac{1}{d} = \log d. \tag{13.39}$$

∎

Exercise 13.9 A machine randomly produces qubits in a mixed state, where the state is $|0\rangle$ with probability p and $|1\rangle$ with probability $1-p$ for $0 \leq p \leq 1$. Define the density matrix ρ of this mixed state. Compute the von Neumann entropy $S(\rho)$ of the qubits. Determine the value of p that maximizes $S(\rho)$, and find the maximum value of $S(\rho)$.

13.2.3 * Classical Joint Entropy and Mutual Information

Building on our understanding of Shannon entropy, this subsection delves into information measures involving multiple random variables. We introduce and examine the concept of mutual information, which plays a key role in understanding the interdependencies and shared information between random variables. We will focus on the case of two random variables, with the principles being extendable to more complex scenarios involving additional variables. These measures not only broaden our perspective on information theory but also lay the groundwork for their quantum counterparts.

1 Joint Entropy

In classical information theory, given two random variables X and Y, the joint entropy, $H(X,Y)$, is defined in terms of the Shannon entropy of the joint probability distribution $p(x,y)$ (see § 13.1.4):

$$H(X,Y) = -\sum_{x \in X} \sum_{y \in Y} p(x,y) \log p(x,y), \tag{13.40}$$

which quantifies the total uncertainty in the joint system of X and Y, encompassing all combinations of random events x and y.

Since $p(x,y) = p(y,x)$, it follows that $H(X,Y)$ is symmetric, i.e., $H(X,Y) = H(Y,X)$.

If X and Y are independent, then $p(x,y) = p(x)p(y)$, resulting in:

$$H(X,Y) = H(X) + H(Y). \tag{13.41}$$

13.2 Quantum Entropy and Information

2 Conditional Entropy

The conditional entropy of Y given X is defined in terms of the conditional probability distribution $p(y|x)$. For a specific outcome $X = x$, the conditional entropy is determined using the Shannon formula:

$$H(Y|x) = -\sum_{y \in Y} p(y|x) \log p(y|x). \tag{13.42}$$

The full conditional entropy of Y given X is defined as the expected value of $H(Y|x)$ over all possible outcomes of X:

$$H(Y|X) = \sum_{x \in X} p(x) H(Y|x) \tag{13.43a}$$

$$= -\sum_{x \in X} \sum_{y \in Y} p(x) p(y|x) \log p(y|x) \tag{13.43b}$$

$$= -\sum_{x \in X} \sum_{y \in Y} p(y, x) \log p(y|x), \tag{13.43c}$$

which represents the average uncertainty in Y given knowledge of X, averaged over all possible outcomes of X.

Similarly, the conditional entropy of X given Y is:

$$H(X|Y) = -\sum_{y \in Y} \sum_{x \in X} p(x, y) \log p(x|y). \tag{13.44}$$

3 Relationships

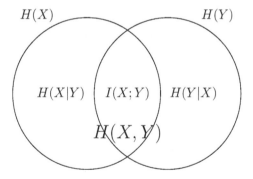

Figure 13.4: Venn Diagram for Shannon Entropy Relationships

Due to the chain rule in probability (Eq. 13.19):

$$p(x, y) = p(y, x) = p(x|y)p(y) = p(y|x)p(x), \tag{13.45}$$

joint and conditional entropies satisfy the identity:

$$H(X, Y) = H(Y, X) = H(Y|X) + H(X) = H(X|Y) + H(Y). \tag{13.46}$$

Or, equivalently:

$$H(Y|X) = H(Y,X) - H(X), \qquad (13.47a)$$
$$H(X|Y) = H(X,Y) - H(Y). \qquad (13.47b)$$

These identities illustrate that the conditional entropy measures the additional uncertainty about one variable when the other is known. These relationships can be illustrated using a Venn diagram as in Fig. 13.4.

In Fig. 13.4, the exclusive region of $H(X)$ relative to $H(Y)$ is represented by $H(X) - I(X;Y)$, equal to $H(X|Y)$.

Conversely, in Fig. 13.1, the exclusive part of $p(x)$ relative to $p(y)$ is shown as $p(x) - p(x,y)$, which differs from $p(x|y)$.

Readers are encouraged to explore the reasons behind this distinction.

4 Mutual Information

Mutual information is defined as a measure of the amount of information that one random variable contains about another. Mathematically, it is given by:

$$I(X;Y) = H(X) + H(Y) - H(X,Y). \qquad (13.48)$$

In this notation, the semicolon (instead of a comma) is used to emphasize that mutual information is a measure of the dependence between the two variables, rather than a joint distribution of them.

The mutual information $I(X;Y)$ quantifies the shared information between two random variables, X and Y. It represents the reduction in uncertainty about one variable given knowledge of the other. Specifically, if X and Y are independent, $I(X;Y) = 0$, indicating that knowledge of X provides no information about Y, and vice versa. It has been established that $I(X;Y) \geq 0$ for all cases, which implies that gaining knowledge about one variable cannot decrease our information about another; such knowledge can only maintain or increase the amount of shared information.

In the Venn diagram in Fig. 13.4, $I(X;Y)$ is the intersection between the two circles representing $H(X)$ and $H(Y)$, while $H(X,Y)$ is their union. Notably, both $I(X;Y)$ and $H(X,Y)$ are symmetric: $I(X;Y) = I(Y;X)$ and $H(X,Y) = H(Y,X)$.

Exercise 13.10 Verify the following equivalent expressions for mutual information:

$$I(X;Y) = H(X) + H(Y) - H(X,Y) \qquad (13.49a)$$
$$= H(X) - H(X|Y) \qquad (13.49b)$$
$$= H(Y) - H(Y|X) \qquad (13.49c)$$
$$= \sum_{x,y} p(x,y) \log \frac{p(x,y)}{p(x)p(y)}. \qquad (13.49d)$$

Mutual information also exhibits additivity for independent variables under certain conditions. For instance, if the joint distribution of two pairs of random variables (X,Y) and (W,Z) can be factorized as $p(X,Y,W,Z) = p(X,Y)p(W,Z)$, indicating that (X,Y) is independent of (W,Z), then the mutual information between these two sets of variables is additive: $I(X,W;Y,Z) = I(X;Y) + I(W;Z)$.

13.2 Quantum Entropy and Information

- **Example 13.12 — Perfectly Correlated Systems.** Suppose that X is a random variable whose entropy $H(X)$ is 8 bits, and that $Y(X)$ is a deterministic function that takes on a different value for each value of X.

  (a) The entropy of Y, $H(Y)$, is also 8 bits because Y is a one-to-one function of X.

  (b) The conditional entropy of Y given X, $H(Y|X) = 0$, since X uniquely determines Y.

  (c) The conditional entropy of X given Y, $H(X|Y) = 0$, because Y uniquely determines X due to the function being one-to-one.

  (d) The joint entropy $H(X,Y)$ is $H(X) + H(Y|X) = 8$ bits, which is just the entropy of X since Y provides no additional uncertainty.

  (e) The mutual information $I(X;Y) = H(X) + H(Y) - H(X,Y) = 8$, consistent with the fact that knowing X provides full information about Y and vice versa.

  (f) Thus, for two perfectly correlated systems,
  $I(X;Y) = H(X,Y) = H(X) = H(Y)$, and
  $H(X|Y) = H(Y|X) = 0$.

- **Example 13.13 — Partially Correlated Systems.** Similar to Example 13.12, X is a random variable whose entropy $H(X)$ is 8 bits. Now, $Y(X)$ is a deterministic function but is no longer invertible; that is, different values of X may correspond to the same value of $Y(X)$.

  (a) Since now different values of X may correspond to the same value of $Y(X)$, the new distribution of Y has lost entropy because it has less variability and therefore $H(Y) < 8$ bits.

  (b) The conditional entropy of Y given X, is still $H(Y|X) = 0$ bits, since X still uniquely determines Y.

  (c) Now, knowledge of Y no longer uniquely determines X, and so the conditional entropy $H(X|Y)$ is no longer zero because there is some uncertainty about X even after knowing Y: $H(X|Y) > 0$.

  (d) The joint entropy $H(X,Y)$ is still $H(X) + H(Y|X) = 8$ bits, which is just the entropy of X since Y provides no additional uncertainty. Even though $H(X|Y) > 0$ and $H(Y) < 8$, we still have $H(Y) + H(X|Y) = 8$.

  (e) The mutual information $I(X;Y) = H(X) + H(Y) - H(X,Y) = H(Y) < H(X)$, measuring the amount of information shared between X and Y which is now smaller than 8 bits.

  (f) Thus, for two partially correlated systems,
  $H(X), H(Y) \leq H(X,Y) \leq H(X) + H(Y)$,
  $0 \leq H(X|Y) \leq H(X)$, $0 \leq H(Y|X) \leq H(Y)$, and
  $0 \leq I(X;Y) \leq H(X), H(Y)$.

- **Example 13.14 — Independent Systems.** Similar to Example 13.12, X is a random variable whose entropy $H(X)$ is 8 bits. But now $Y(X)$ is completely random, i.e., given a value of $X = x$, Y takes on a random value y from a probability distribution with the same entropy as X.

(a) We are given $H(Y) = H(X) = 8$ bits, even though Y and X are now independent.

(b) The conditional entropy of Y given X, is $H(Y|X) = 8$ bits, since X does not influence Y.

(c) Similarly, the conditional entropy of X given Y, is $H(X|Y) = 8$ bits as well.

(d) The joint entropy $H(X,Y)$ is $H(X) + H(Y|X) = H(Y) + H(X|Y) = 16$ bits.

(e) The mutual information $I(X;Y) = H(X) + H(Y) - H(X,Y) = 0$, signifying that no information is shared between X and Y.

(f) Thus, for two independent systems,
$H(X,Y) = H(X) + H(Y)$,
$H(X|Y) = H(X)$, $H(Y|X) = H(Y)$, and
$I(X;Y) = 0$.

■

Exercise 13.11 Draw the corresponding Venn diragram for each of the Examples 13.12 to 13.14.

13.2.4 ∗ Quantum Joint Entropy and Mutual Information

Having established a solid foundation in both the von Neumann entropy and classical information measures, we now turn our attention to quantum information measures involving two random variables. This generalization provides a deeper insight into how quantum mechanics redefines informational relationships and interdependencies.

Given a bipartite quantum system AB described by a density operator ρ_{AB} which encodes the probabilities of the system, the pair of random variables X and Y are replaced by ρ_{AB}. The two subsystems are described by the reduced density operators $\rho_A = \text{tr}_B(\rho_{AB})$ and $\rho_B = \text{tr}_A(\rho_{AB})$. (See § 12.2.8 for details.)

1 Joint and Conditional Entropies

The standard definitions of joint and conditional von Neumann entropies adopt the form of the classical relations, with $S(\rho_{AB})$ in place of the classical joint entropy, $S(\rho_A)$ substituted for $H(X)$, and $S(\rho_B)$ substituted for $H(Y)$. Thus,

$$S(A,B) \equiv S(\rho_{AB}) = -\text{tr}(\rho_{AB} \log \rho_{AB}), \qquad (13.50a)$$
$$S(A) \equiv S(\rho_A) = -\text{tr}(\rho_A \log \rho_A), \qquad (13.50b)$$
$$S(B) \equiv S(\rho_B) = -\text{tr}(\rho_B \log \rho_B). \qquad (13.50c)$$

Since conditional density operators are rarely used, the conditional von Neumann entropies are defined directly according to the classical entropy relationship Eq. 13.47:

$$S(A|B) = S(\rho_{AB}) - S(\rho_B), \qquad (13.51a)$$
$$S(B|A) = S(\rho_{AB}) - S(\rho_A). \qquad (13.51b)$$

2 Mutual Information

Mutual information is defined similarly to the classical case, but with quantum entropies. For two quantum systems A and B, the mutual information $I(A;B)$ is given by:

$$I(A;B) = S(\rho_A) + S(\rho_B) - S(\rho_{AB}). \qquad (13.52)$$

The terms $S(\rho_A)$ and $S(\rho_B)$ represent the individual uncertainties of systems A and B, respectively, while $S(\rho_{AB})$ represents the uncertainty of the joint system. The mutual information $I(A;B)$ quantifies how much the knowledge of one system (say A) reduces the uncertainty about the other system (B), thus signifying the total amount of correlation between two quantum systems, reflecting the interconnectedness of their quantum states.

Notably, since ρ_A and ρ_B are derived from ρ_{AB} through partial trace, the joint and conditional entropies, as well as the mutual information, are all encoded in ρ_{AB}.

3 Relationships

The relationships among quantum entropies and information can be depicted using a Venn diagram, as shown in Fig. 13.5. However, it's important to note that such diagrams cannot capture the phenomenon of negative quantum conditional entropy, which occurs in entangled states.

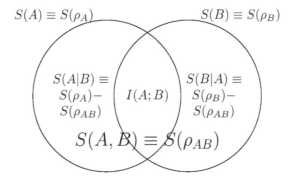

Note: Quantum conditional entropy can be negative in the case of entangled states, which is not captured by the diagram. Refer to the text for details.

Figure 13.5: Venn Diagram for von Neumann Entropy Relationships

In the case where the systems A and B are independent, $\rho_{AB} = \rho_A \otimes \rho_B$, $S(\rho_{AB}) = S(\rho_A) + S(\rho_B)$, and $I(A;B) = 0$.

4 Negative Conditional Entropy

A surprising aspect of quantum conditional entropy is that it can be negative for entangled states, a phenomenon that has no counterpart in classical information theory. This is exemplified in Example 13.15.

In classical systems, conditional entropy quantifies the reduction in uncertainty about one part of the system when another part is known, and it is always nonnegative. This reflects that knowledge about a system can only decrease or leave unchanged the uncertainty associated with its parts. Conversely, quantum systems exhibit a unique property where conditional entropy can become negative. This occurs due to quantum entanglement, indicating that the interconnections between subsystems are so profound that knowing the state of the whole system actually implies a greater certainty (less uncertainty) than the sum of what is known about the individual parts. Such negative conditional entropy is exclusive to quantum mechanics, underscoring the fundamentally non-classical behavior introduced by entanglement.

However, even though quantum conditional entropy can be negative, quantum mutual information is still always non-negative: $I(A; B) \geq 0$. (A proof of this inequality is beyond the scope of this text.)

■ **Example 13.15 — Conditional Entropy of an EPR Pair.** Consider the archetypal case where Alice and Bob share an EPR pair, $|\Phi^+\rangle = \frac{1}{\sqrt{2}}(|00\rangle + |11\rangle)$. Its joint density operator is $\rho_{AB} = |\Phi^+\rangle\langle\Phi^+|$.

As derived in Example 13.4, the reduced density operator for both Alice and Bob is $\rho_A = \rho_B = \text{tr}_B(\rho_{AB}) = \text{tr}_A(\rho_{AB}) = I/2$, representing maximally mixed states.

Consequently, the entropies are $S(\rho_{AB}) = 0$ for the pure joint state, and $S(\rho_A) = S(\rho_B) = 1$ for the maximally mixed reduced states. This leads to a surprising result in the conditional entropies:

$$S(A|B) = S(\rho_{AB}) - S(\rho_B) = -1,$$
$$S(B|A) = S(\rho_{AB}) - S(\rho_A) = -1.$$

The negative values of $S(A|B)$ and $S(B|A)$ signify the presence of entanglement and the departure from classical probabilistic correlations.

The mutual information is

$$I(A; B) = S(\rho_A) + S(\rho_B) - S(\rho_{AB}) = 2,$$

signifying that the total amount of entanglement-based correlation is equivalent to 2 bits of information.

This is in contrast to classical systems, where the mutual information between two perfectly correlated random variables X and Y would be $I(X;Y) = H(X) = H(Y)$. The factor of 2 in $I(A;B) = 2S(\rho_A)$ for Bell states indicates that quantum correlations from entanglement can be stronger than any classical correlation.

In general, if ρ_{AB} represents a pure state with $S(\rho_{AB}) = 0$, then it holds that $S(\rho_A) = S(\rho_B)$. For a product state, where ρ_{AB} is separable into ρ_A and ρ_B, both are pure states and hence $S(\rho_A) = S(\rho_B) = 0$. Conversely, for a maximally entangled state like a Bell state, $S(\rho_A) = S(\rho_B)$ attains its maximum value of 1. Thus, $S(\rho_A) = S(\rho_B)$ is often referred to as the *entanglement entropy* for a pure state, quantifying the degree of entanglement. ■

Exercise 13.12 Consider a two-qubit mixed state which is composed of an EPR pair $|\Phi^+\rangle = \frac{1}{\sqrt{2}}(|00\rangle + |11\rangle)$ with 50% probability, and another entangled state $|\Psi^+\rangle = \frac{1}{\sqrt{2}}(|01\rangle + |10\rangle)$ with 50% probability.

Compute the joint and conditional von Neumann entropies for this mixed state.

> **Key Takeaways**
>
> In quantum information theory, von Neumann entropy fulfills the roles of the classical Shannon entropy. Joint entropy in bipartite quantum systems, derived from the joint density operator ρ_{AB}, quantifies the total uncertainty associated with the combined state of the two systems. Mutual information, derived from the joint and reduced density operators of these systems, measures the amount of information shared between them, reflecting the degree of their quantum correlations. Notably, quantum conditional entropy can be negative in entangled systems, which underscores the non-classical nature of quantum entanglement, a distinctive feature absent in classical information theory.

13.3 ✶ Core Theorems in Quantum Information

In the preceding chapters, we have explored two theorems fundamental to the field of quantum computation and information theory. The no-cloning theorem, as outlined in § 5.1.3, asserts that it is impossible to create an identical copy of an arbitrary unknown quantum state. This theorem highlights a fundamental restriction on quantum information imposed by quantum mechanics. In addition, the no-communication theorem (§ 9.7) establishes that superluminal communication, or faster-than-light transmission of information, is not feasible, despite the presence of non-local correlations inherent in quantum entanglement.

This section ventures into more advanced territories of quantum communications. We will explore two significant theorems that further shape our understanding of quantum information—the quantum data processing inequality and the Holevo theorem—as well as the property of strong subadditivity of von Neumann entropy. Our focus will be on elucidating the background and implications of these theorems and principles, offering insights into their roles in the broader landscape of quantum information science. While the mathematical underpinnings of these concepts are profound, we aim to emphasize their conceptual significance and practical implications, thereby making these advanced topics accessible to a broader audience.

> **Prerequisite Review: Quantum Channels**
>
> This topic is detailed in § 12.2.7. Below is a brief summary of the relevant concepts.
>
> A Completely Positive Trace-Preserving (CPTP) map is an operation that transforms one density operator into another while preserving trace and positivity. It includes both unitary and non-unitary transformations, is applicable to closed and open quantum systems, and allows for changes in dimensionality.
>
> CPTP maps are often modeled using quantum channels, also known as Kraus decompositions. A quantum channel can be expressed as:
>
> $$\mathcal{E}(\rho) = \sum_j K_j \rho K_j^\dagger, \qquad (13.53)$$

> where K_j are the Kraus operators that describe the quantum channel. These operators embody the transformations the system undergoes, and the sum of their effects ensures the preservation of the trace and positivity of the density operator. The specific dynamics of the channel determine the form and number of the Kraus operators.

13.3.1 ∗ The Data Processing Inequality

The Data Processing Inequality (DPI) is a fundamental concept in information theory, both classical and quantum. It dictates how information is transformed and preserved within systems undergoing various processes. This subsection explores the DPI, focusing first on its quantum formulation.

1 The Quantum Data Processing Inequality

The von Neumann entropy is invariant under any unitary transformation U:

$$S(U\rho U^\dagger) = S(\rho). \tag{13.54}$$

This invariance signifies that entropy remains constant through reversible information processing in closed quantum systems.

For more general transformations, quantum systems undergo completely positive, trace-preserving (CPTP) maps, also known as quantum channels in quantum information theory. A CPTP map $\mathcal{E}$ transforms a density operator ρ into another density operator $\mathcal{E}(\rho)$, preserving the positivity and total probability. (See § 12.2.7 for more details.)

The von Neumann entropy does not decrease under CPTP maps [66]:

$$S(\mathcal{E}(\rho)) \geq S(\rho). \tag{13.55}$$

This relationship (which encompasses Eq. 13.54) is known as the quantum Data Processing Inequality (DPI). The proof of this inequality is mathematically sophisticated, involving advanced concepts in quantum theory, and falls outside the purview of this discussion. For a detailed exploration, readers are encouraged to consult the references provided at the end of this chapter.

■ **Example 13.16 — Measuring a Bell State.** Suppose we have a Bell state $|\Phi^+\rangle = \frac{1}{\sqrt{2}}(|00\rangle + |11\rangle)$ with zero entropy. Upon measuring one qubit in the computational basis, the state collapses to a mixed state, either $|00\rangle$ or $|11\rangle$ with equal probability, resulting in an entropy of 1. ■

■ **Example 13.17 — Bit-Flip Noise.** Consider a qubit state $|\psi\rangle = |0\rangle$, which intially has zero entropy:

$$\rho = |0\rangle\langle 0|, \quad S(\rho) = 0. \tag{13.56}$$

A bit-flip noise channel $\mathcal{E}$ acts on each qubit with probability p, flipping $|0\rangle$ to $|1\rangle$ and vice versa:

$$\mathcal{E}(\rho) = (1-p)\rho + p\sigma_X \rho \sigma_X, \tag{13.57}$$

13.3 ∗ Core Theorems in Quantum Information

where σ_X is the Pauli-X (bit-flip) operator. The resulting state $\mathcal{E}(\rho)$ is a mixed state with increased entropy:

$$\mathcal{E}(\rho) = (1-p)|0\rangle\langle 0| + p|1\rangle\langle 1|, \tag{13.58a}$$

$$S(\mathcal{E}(\rho)) = -[(1-p)\log(1-p) + p\log(p)]. \tag{13.58b}$$

$S(\mathcal{E}(\rho))$ has a maximum value of 1 at $p = 0.5$. ∎

Key Takeaways

The quantum DPI indicates that the entropy of a quantum system cannot decrease under the application of CPTP operations. These operations encompass quantum noise, decoherence, and other non-unitary processes commonly encountered in open quantum systems, as well as unitary operations in closed quantum systems.

---

2 Implications

The quantum DPI is fundamental in understanding the behavior of quantum systems, especially in the context of quantum information theory and quantum thermodynamics.

The increase in von Neumann entropy under CPTP maps reflects several fundamental aspects of quantum systems. This includes their tendency to lose coherence and approach an equilibrium characterized by maximally mixed states, indicative of maximal entropy and reduced quantum information. Additionally, it highlights the irreversibility of certain quantum processes, the inherent uncertainties in quantum states, and the dynamics of quantum entanglement, particularly in multipartite systems where increased local entropy can signal entanglement. From an information-theoretic perspective, this entropy increase also represents a degradation in the system capacity to retain and process quantum information.

Quantum thermodynamics is an interdisciplinary field that merges the principles of quantum mechanics with those of thermodynamics. It extends traditional thermodynamics to the quantum realm, exploring how quantum properties like superposition and entanglement influence and are influenced by thermodynamic processes. Key areas of interest include understanding the thermodynamics of small systems at quantum scales, energy exchanges in quantum systems, the thermodynamic cost of quantum information processes, and the second law of thermodynamics in quantum systems. This field provides important insights for developing quantum technologies like quantum computers, quantum batteries, and quantum heat engines.

In this context, quantum DPI is often compared to the second law of thermodynamics because both describe a non-decreasing property of the system entropy. In classical thermodynamics, the second law states that the total entropy of a closed system does not decrease over time, reflecting the natural tendency towards disorder or equilibrium in isolated systems. However, while the second law is a statement about closed classical systems, the quantum DPI often applies to open quantum systems undergoing CPTP operations.

3 Classical DPI Versus Quantum DPI

In classical information theory, the behavior of entropy under different types of transformations diverges from quantum systems (refer to Examples 13.12 to 13.14):

1. When using deterministic and invertible functions g, the entropy $H(X)$ remains unchanged, i.e., $H(g(X)) = H(X)$. In this scenario, each outcome of X uniquely corresponds to an outcome of $g(X)$ and vice versa, preserving the level of uncertainty in the system.

2. With deterministic but non-invertible functions g, entropy may decrease ($H(g(X)) \leq H(X)$). This reduction occurs because non-invertible functions can map multiple outcomes of X into a single outcome of $g(X)$, thereby diminishing the uncertainty and, consequently, the entropy.

3. In the case of stochastic functions g, we observe an increase in entropy ($H(g(X)) \geq H(X)$) due to the additional randomness introduced by g, which heightens the system's uncertainty.

Thus, the classical counterpart of the quantum DPI, namely $H(g(X)) \geq H(X)$, is not universally applicable. Deterministic but non-invertible functions in classical systems can actually decrease the entropy through many-to-one mappings. In contrast, general quantum transformations are described by CPTP maps, which inherently ensure that the von Neumann entropy does not decrease.

13.3.2 ✻ The Holevo Bound and Channel Capacity

This subsection explores a foundational theorem (the HSW theorem) that defines the capabilities of quantum channels in information transmission. It focuses on Holevo's theorem and its implications for channel capacity, providing key insights into the potential and limitations of quantum communication networks.

1 Background

Imagine Alice wishes to send information, encoded in a classical random variable X, to Bob. Using n classical bits, she can transmit a maximum of n bits of information. But what happens if she employs qubits instead? Unlike a classical bit, which represents two discrete states, a pure qubit state can represent any point on the Bloch sphere. For instance, Alice could encode a random n-bit string $s \in \{0,1\}^n$ into a single qubit state as:

$$|\psi\rangle_x = \cos\frac{\pi x}{N}|0\rangle + \sin\frac{\pi x}{N}|1\rangle, \tag{13.59}$$

where x is the decimal representation of the bit string and $N = 2^n$.

When Alice sends a qubit to Bob, either directly or through quantum teleportation, it is natural to question whether Bob can fully recover x (or the n-bit string s). Unfortunately, the answer is negative. To derive classical information from the qubit, Bob must perform a measurement. Upon measuring the qubit, Bob will find it in state $|0\rangle$ with probability $p_0 = \cos^2\left(\frac{\pi x}{N}\right)$ and in state $|1\rangle$ with probability $1 - p_0$. To accurately determine $\frac{\pi x}{N}$, and thus the information encoded by Alice, Bob would need to measure a large number of identically prepared qubits. However, the quantum no-cloning theorem prohibits creating exact copies of an unknown quantum state, which precludes Bob from generating the necessary multiple qubits for precise measurement. Alice could send Bob multiple copies, but the very act would significantly reduce the information transmission efficiency. Consequently, Alice's capacity to encode a wealth of information into a single qubit state does not translate into Bob's ability to retrieve it, primarily due to the probabilistic nature of quantum measurements and the constraints of quantum cloning.

The Holevo Bound addresses this issue by quantifying the maximum efficiency of information transmission using qubits. Surprisingly, the Holevo Bound dictates that an n-qubit stream can convey at most n bits of classical information. This bound confirms that irrespective of the quantum information encoded (as represented by the states of the qubits), an observer is limited to extracting at most n bits of classical information from the measurement of these n qubits.

The optimal efficiency of classical information transmission using a qubit is achieved only when the information is encoded using a set of orthogonal states. Given that a qubit has only two orthogonal states, it can transmit at most one bit of classical information per measurement. However, when encoding information with non-orthogonal states, or in scenarios involving quantum correlations, the amount of accessible classical information per qubit may be less than 1 bit.

Thus, the Holevo Bound demonstrates that despite a qubit's ability to exist in a continuum of states due to superposition, it cannot convey more than 1 bit of classical information to an observer performing any permissible quantum measurement.

In superdense coding (see § 10.2), it is shown that a qubit can transmit two bits of information, achievable through the pre-sharing of an entangled pair. Nevertheless, this still equates to two bits of classical information being transmitted using two qubits.

This result, known as the Holevo-Schumacher-Westmoreland (HSW) theorem, sets an upper limit on the amount of classical information transmissible via a quantum system.

2 The HSW Theorem

The Holevo-Schumacher-Westmoreland (HSW) theorem is anchored in the scenario of a quantum communication channel, wherein a sender (Alice) aims to transmit classical information to a receiver (Bob) via quantum states. In this setup, X denotes the classical information Alice seeks to convey, encoded within quantum states $\{\rho_x\}$, while Y encapsulates the classical information Bob decodes from his measurements of these states. The mutual information, $I(X;Y)$, quantifies the reduction in uncertainty about X (Alice's message) afforded by knowledge of Y (Bob's measurement outcomes), representing the *accessible information*.

Preceding the HSW theorem, the Holevo Bound addresses a related yet simpler query: given a collection of quantum states $\{\rho_x\}$, each with a probability $p(x)$, what is the maximal quantity of classical information that can be obtained from measurements? The Holevo Bound posits that the accessible information, $I(X;Y)$, is constrained by:

$$I(X;Y) \leq S\left(\sum_x p(x)\rho_x\right) - \sum_x p(x)S(\rho_x), \qquad (13.60)$$

where $S(\rho)$ signifies the von Neumann entropy of the state ρ. This limitation is met when the quantum states $\{\rho_x\}$ are mutually orthogonal, such as $\{|0\rangle\langle 0|, |1\rangle\langle 1|\}$.

The HSW theorem extends the Holevo Bound to establish the limits of information transmission through quantum channels. As depicted in Fig. 13.6, the HSW theorem builds on this by considering:

$$(\text{Alice}) \; X : p(x) \to \sum p(x)\rho_x \;\;—\;\boxed{\mathcal{E}}\;—\;\; (\text{Bob}) \sum p(x)\mathcal{E}(\rho_x) \to Y$$

Figure 13.6: Schematic of the Quantum Information Transmission Process

- Alice encoding classical information from a random variable X with a probability distribution $p(x)$ into a set of quantum states $\{\rho_x\}$.
- These quantum states are transmitted to Bob via a quantum channel described by a Completely Positive Trace-Preserving (CPTP) map $\mathcal{E}$.
- Bob performs measurements to decode the information, which results in a new random variable Y containing the classical information he extracts.

The HSW theorem posits that the mutual information $I(X;Y)$ is bounded by the Holevo channel capacity χ, formulated as:

$$\chi = \max_{\{p(x),\rho_x\}} \left(S(\sum_x p(x)\mathcal{E}(\rho_x)) - \sum_x p(x) S(\mathcal{E}(\rho_x)) \right), \tag{13.61}$$

where the maximization runs over all ensembles of states $\{\rho_x\}$ and their probability distribution $p(x)$.

This theorem sets an upper bound on the classical capacity of a quantum channel, illustrating that quantum mechanics, despite allowing for elaborate state representations, limits the classical information extractable from a quantum system. It also emphasizes the role of quantum entropy in determining the capacity of quantum channels.

The HSW theorem is a cornerstone in the field of quantum information theory. Its implications are significant, particularly in quantum cryptography and the creation of efficient quantum communication protocols.

Implications of Holevo's Bound on Quantum Computing

A system of n qubits can encode quantum information in a composite quantum state represented by 2^n complex numbers. This expansive encoding capacity underlies the remarkable parallel computational power of quantum computing. However, transitioning from quantum computations to retrievable classical results necessitates measurements, introducing inherent constraints. Holevo's bound limits the amount of information that can be extracted to n classical bits.

Quantum algorithms, such as Deutsch-Jozsa or Shor's, ingeniously exploit quantum interference to efficiently condense the outcome of quantum computations into these n bits. This condensation process ensures that critical information from the quantum computation is compacted into a form suitable for output measurements. These algorithms showcase the delicate balance between maximizing quantum processing capabilities and navigating the fundamental constraints of information extraction during the quantum-to-classical transition.

The derivation of the HSW theorem is complex, rooted in the depths of quantum information theory, and extends beyond the introductory nature of this text. Those

13.3 ∗ Core Theorems in Quantum Information

seeking a comprehensive understanding are advised to refer to the literature listed at the end of this chapter.

3 Revisiting the Holevo Bound

Having explored the theoretical underpinnings of the HSW theorem, let's revisit our initial discussion about the Holevo Bound. To understand why a quantum channel cannot transmit more than 1 bit of classical information, we analyze the Holevo channel capacity, as expressed in Eq. 13.61. The first term represents the entropy of the average state of the ensemble after passing the channel. For a single qubit, this is at most 1 bit when the state is maximally mixed. The second term is the weighted average of the entropies of the individual quantum states. When each state $\mathcal{E}(\rho_x)$ is pure, its entropy is 0, and consequently, the second term sums to 0.

Therefore, the maximal value of χ is achieved when the average state of the ensemble is maximally mixed, yielding an entropy of 1 bit, and simultaneously, each individual state is pure, contributing no entropy. This scenario results in $\chi_{\max} = 1$ bit.

13.3.3 ∗ Subadditivity and Strong Subadditivity

The subadditivity and strong subadditivity properties of von Neumann entropy mathematically encapsulate how entropy behaves in quantum operations, forming a crucial underpinning for various key theorems. Notably, these properties are instrumental in the proofs of the quantum data processing inequality (DPI) and the Holevo-Schumacher-Westmoreland (HSW) theorem, both of which are central to our understanding of quantum communication and information processing.

1 Subadditivity (SA)

For any composite quantum system AB, subadditivity asserts that the total entropy of the system does not exceed the sum of the entropies of its constituent subsystems. Mathematically, this is expressed as:

$$S(\rho_{AB}) \leq S(\rho_A) + S(\rho_B), \tag{13.62}$$

where ρ_{AB} is the density operator of the composite system AB, and ρ_A and ρ_B are the reduced density operators of the subsystems A and B, respectively.

This principle of subadditivity indicates that the total entropy of combined quantum systems is constrained by the sum of the individual entropies of each system.

A related inequality that sets the lower bound of $S(\rho_{AB})$ is the Araki-Lieb inequality:

$$S(\rho_{AB}) \geq |S(\rho_A) - S(\rho_B)| \tag{13.63}$$

An important implication of subadditivity is that in a quantum system, the mutual information—a measure of the total correlations between subsystems A and B—is always non-negative:

$$I(A; B) = S(\rho_A) + S(\rho_B) - S(\rho_{AB}) \geq 0. \tag{13.64}$$

2 Strong Subadditivity (SSA)

Strong Subadditivity (SSA) of von Neumann entropy further generalizes the concept of subadditivity to tripartite quantum systems, encompassing three subsystems A, B, and C. The SSA is mathematically expressed as:

$$S(\rho_{ABC}) + S(\rho_B) \leq S(\rho_{AB}) + S(\rho_{BC}), \qquad (13.65)$$

where ρ_{ABC} is the density operator for the combined system and ρ_{AB}, ρ_{BC} are the density operators for two of the bipartite subsystems.

This inequality encapsulates a more intricate interplay of entropic relationships in a composite system. It implies that the total entropy of the entire system ABC plus the entropy of one of its parts (B) is no greater than the sum of the entropies of two overlapping subsystems (AB and BC).

While the proof of the classical version of SSA is relatively straightforward, proving the quantum SSA presents significant challenges. This increased difficulty arises from the non-commutative nature of the density operators that describe quantum subsystems. Despite these challenges, several distinct proofs have been developed to establish SSA in quantum systems.

SSA is a powerful tool in quantum information, providing a deeper understanding of entropy dynamics in complex quantum systems and forming a theoretical basis for key theorems in quantum communication and information processing.

13.4 Further Exploration

For further exploration of topics in quantum information, the following resources are recommended:

- Michael A Nielsen and Isaac L Chuang. *Quantum Computation and Quantum Information: 10th Anniversary Edition*. Cambridge University Press, 2010. ISBN: 978-1-107-00217-3.

- John Watrous. *The Theory of Quantum Information*. Cambridge University Press, 2018. ISBN: 978-1-107-18056-7.

- Mark M. Wilde. *Quantum Information Theory*. Cambridge University Press, 2017. ISBN: 978-1-107-17616-4.

These references provide comprehensive insights into quantum computation, information theory, and the intricate dynamics of quantum stochastic processes.

13.5 Summary and Conclusions

Chapter Overview

This chapter on quantum information delves into three core areas: quantum probability, quantum entropy and information, and advanced theorems. It starts by contrasting quantum probability with classical probability, highlighting unique quantum features like entanglement. The discussion then shifts to quantum entropy, comparing it with classical Shannon entropy, and examines quantum mutual information. Finally, the chapter explores advanced concepts like the quantum data processing inequality, the Holevo theorem, and subadditivity in quantum systems, emphasizing their importance in quantum communication and theoretical underpinnings.

Quantum Probability Essentials

The first section delves into the essentials of quantum probability, highlighting its distinctive features compared to classical probability. Key concepts such as complex-valued probability amplitudes, entanglement, and non-commutativity of quantum observables are explored. The section provides a foundational understanding of how quantum probability resembles and differs from classical probability. This exploration lays the groundwork for understanding quantum entropy and information.

Quantum Entropy and Information

Next, the chapter shifts focus to quantum entropy and information. It provides a comprehensive understanding of classical Shannon entropy and its quantum counterpart, von Neumann entropy, highlighting their roles in quantifying uncertainty in classical and quantum systems. Additionally, the concepts of classical joint entropy, mutual information, and their quantum analogs are explored. Through these discussions, the section offers insights into how quantum mechanics reshapes the understanding of informational relationships and interdependencies, demonstrating the unique nature of quantum information.

Core Theorems in Quantum Information

The final part of the chapter delves into core theorems and concepts central to quantum information theory. It begins with an exploration of the quantum data processing inequality, which posits that the entropy of a quantum system cannot decrease under the application of CPTP operations. This principle underlines the non-decreasing nature of entropy in quantum processes, reflecting phenomena such as the loss of coherence, the irreversibility of certain quantum operations, and the constraints on information extraction from quantum systems.

The discussion then moves to the Holevo theorem, which sets fundamental limits on the amount of classical information that can be transmitted using quantum states through quantum channels. This theorem is essential in understanding the boundaries of quantum communication.

Additionally, this section covers the properties of subadditivity and strong subadditivity of von Neumann entropy. These mathematical formulations are useful in analyzing the entropic characteristics of composite quantum systems.

This part of the chapter weaves together these advanced concepts, providing insights into their theoretical importance and practical implications in the broader landscape of quantum information science.

Problem Set 13

13.1 (Review of classical probability and information.) Consider random variables,

X and Y, each taking values in $\Omega = \{1, 2, \ldots, 6\}$, representing the outcomes of rolling fair dice. We impose the condition that the sum $X + Y$ is odd. Your tasks are:

(a) Find the probability distributions of X, Y, (X, Y), and $X + Y$.

(b) Compute the joint entropy $H(X, Y)$.

(c) Compute the individual entropies $H(X)$ and $H(Y)$.

(d) Compute the conditional entropies $H(Y|X)$ and $H(X|Y)$.

(e) Describe what each of these quantities represents in the context of this problem.

(f) Verify Eq. 13.46 using your computations.

13.2 For classical Shannon entropy, we have the property that $H(X) \leq H(X, Y)$, reflecting the intuition that the uncertainty of a single variable X cannot exceed that of the joint system X and Y. This exercise explores whether a similar property holds in the quantum realm with von Neumann entropy, specifically, whether $S(\rho_A) \leq S(\rho_{AB})$ for the following cases:

(a) A Bell state $|\Phi\rangle = \frac{1}{\sqrt{2}}(|00\rangle + |11\rangle)$.

(b) A classically correlated state $\rho_{AB} = \frac{1}{2}(|00\rangle\langle 00| + |11\rangle\langle 11|)$.

13.3 Coherent information is defined as $I(A\rangle B) = S(\rho_B) - S(\rho_{AB})$. It is known to be a measure of quantum correlations. Positivity of coherent information indicates that quantum correlations are present. Use the following cases to demonstrate this.

(a) A Bell state $|\Phi\rangle = \frac{1}{\sqrt{2}}(|00\rangle + |11\rangle)$.

(b) A product state $|\psi\rangle = \frac{1}{\sqrt{2}}(|0\rangle + |1\rangle) \otimes (|0\rangle + |1\rangle)$.

13.4 Consider a mixed state described by the density operator ρ. Since ρ is Hermitian, it can also be regarded as an observable. Set $H = \rho$ in Eq. 13.14 for expected value and Eq. 13.15 for the variance. Simplify these equations and discuss their physical meanings, particularly in the context of the purity of the state.

13.5 Consider a two-qubit (A and B) system composed of 50% in the uniform mixed state $\frac{1}{4}I$ and 50% in the Bell state $\frac{1}{\sqrt{2}}(|00\rangle + |11\rangle)$. We define two random variables $U = \sigma_X \otimes I$ and $V = I \otimes \sigma_Z$, where σ_X and σ_Z are Pauli operators. They represent measuring qubit A in the $\{|+\rangle, |-\rangle\}$ basis, and qubit B in the computational basis.

(a) Calculate the density operator of the two-qubit system ρ_{AB}.

(b) Verify that U and V commute.

(c) Find all the common eigenstates for U and V, e.g., $|+0\rangle \equiv |+\rangle \otimes |0\rangle$.

(d) Calculate the probability distribution on set of common eigenstates for U and V.

(e) Calculate the conditional probabilities given qubit B is in state $|1\rangle$: $p(|+\rangle_A | |1\rangle_B)$ and $p(|-\rangle_A | |1\rangle_B)$.

Problem Set 13

(f) Calculate the expectation values $\langle U \rangle$, $\langle V \rangle$, $\langle UV \rangle$, and $\langle U+V \rangle$.

(g) Calculate the von Neumann entropy of the two-qubit system.

(h) Calculate the reduced density operators ρ_A and ρ_B.

(i) Calculate the joint entropy $S(\rho_{AB})$, and the entropies of the subsystems $S(\rho_A)$ and $S(\rho_B)$.

(j) Calculate the mutual information $I(A; B)$.

13.6 Consider two quantum systems A and B, each in a maximally mixed state of dimension d:

$$\rho_A = \rho_B = \frac{1}{d} \sum_{i=1}^{d} |i\rangle\langle i|.$$

The joint system is described by the density operator:

$$\rho_{AB} = \frac{1}{d} \sum_{i=1}^{d} |i\rangle\langle i| \otimes |i\rangle\langle i|.$$

Prove that the entropies $S(\rho_A)$ and $S(\rho_B)$ are both $\log d$, and that the joint entropy $S(\rho_{AB})$ and mutual information $I(A; B)$ are also $\log d$, indicating perfect classical correlation in the joint system.

13.7 Consider a mixed state represented by the density operator $\rho = p|0\rangle\langle 0| + (1-p)|1\rangle\langle 1|$, where the state is $|0\rangle$ with probability p and $|1\rangle$ with probability $1-p$ for $0 \leq p \leq 1$.

Apply the following CPTP map representing a depolarizing channel:

$$\mathcal{E}(\rho) = (1-s)\rho + \frac{s}{2}I,$$

where s is the probability of depolarization, and I is the identity operator.

Your tasks are:

(a) Calculate the von Neumann entropy of the initial state ρ.

(b) Apply the depolarizing channel $\mathcal{E}$ to ρ and find the resulting state.

(c) Calculate the von Neumann entropy of the state after applying the map.

(d) Compare the entropies before and after applying $\mathcal{E}$ to illustrate that entropy does not decrease.

13.8 ✽ Prove the Data Processing Inequality (DPI) from the Strong Subadditivity (SSA) property.

The data processing inequality in quantum information theory asserts that the entropy S of a quantum state ρ cannot decrease under the action of a Completely Positive Trace-Preserving (CPTP) map $\mathcal{E}$. Formally, this is expressed as:

$$S(\mathcal{E}(\rho)) \geq S(\rho).$$

To prove the data processing inequality using the strong subadditivity of quantum entropy,

$$S(\rho_{ABC}) + S(\rho_B) \leq S(\rho_{AB}) + S(\rho_{BC}),$$

follow these steps:

a. Consider a tripartite system described by the state ρ_{ABC}, with ρ_A being the subsystem of interest and ρ_{BC} an environment subsystem used in the Stinespring dilation for a CPTP map $\mathcal{E}$ (see § 12.2.7.1). Here, $\rho_{ABC} = |\psi\rangle\langle\psi|_{ABC}$ represents a unitary dilation on an expanded system, and $\rho_A = \text{tr}_{BC}(\rho_{ABC})$ is the initial state of subsystem A.

b. Apply the CPTP map $\mathcal{E}$ to the subsystem ρ_C of ρ_{ABC}, resulting in a new state $\rho'_{ABC} = (I_A \otimes I_B \otimes \mathcal{E}_C)(\rho_{ABC})$.

c. Trace out the environment subsystem (BC) in ρ'_{ABC} to obtain $\mathcal{E}(\rho_A) = \rho'_A$, where ρ'_A is the state of subsystem A after the application of the CPTP map, illustrating the Stinespring dilation theorem in action.

d. Utilize the strong subadditivity property and the non-increasing nature of entropy under partial trace to connect $S(\rho_{ABC})$ and $S(\rho_{BC})$ with $S(\rho_A)$ and $S(\mathcal{E}(\rho_A))$.

e. Apply strong subadditivity to demonstrate that $S(\mathcal{E}(\rho_A)) \geq S(\rho_A)$, thereby establishing the data processing inequality.

13.9 ✳ Prove the Holevo Bound using Strong Subadditivity.

The Holevo Bound is a fundamental result in quantum information theory that establishes an upper limit on the amount of classical information that can be transmitted securely using quantum states. Specifically, it limits the accessible information that can be extracted from a quantum ensemble.

Given an ensemble of quantum states $\{\rho_x\}$ with probabilities $\{p(x)\}$, the Holevo Bound states that the mutual information between the classical variable X representing the choice of quantum state and the quantum variable Y representing the measurement outcomes is bounded by:

$$\chi(\{p(x), \rho_x\}) = S\left(\sum_x p(x)\rho_x\right) - \sum_x p(x)S(\rho_x),$$

where χ is the Holevo quantity, $S(\cdot)$ denotes the von Neumann entropy, and $\sum_x p(x)\rho_x$ is the average state of the ensemble.

To prove the Holevo Bound using the Strong Subadditivity (SSA) property, follow these steps:

a. Begin by considering a quantum ensemble $\mathcal{X} = \{p(x), \rho_x\}$, where each quantum state ρ_x is prepared with probability $p(x)$. Construct a tripartite system ρ_{ABC}, where A represents the classical variable X encoded in a quantum state, B is the quantum system prepared in the state ρ_x corresponding to X, and C is an ancillary system introduced to purify the state of AB.

b. Express the average state $\rho = \sum_x p(x)\rho_x$ and consider the purification ρ_{ABC} such that $\rho_{AB} = \sum_x p(x) |x\rangle_A \langle x| \otimes \rho_x$ and $\rho = \text{tr}_{A,C}(\rho_{ABC})$.

c. Apply the strong subadditivity property, $S(\rho_{ABC}) + S(\rho_B) \leq S(\rho_{AB}) + S(\rho_{BC})$, to the constructed quantum ensemble and purification, taking into account that the purification ρ_{ABC} makes $S(\rho_{ABC}) = 0$ and $S(\rho_{AB})$ represents the entropy of the classical-quantum state.

d. Use the definition of the Holevo quantity $\chi(\mathcal{X})$ and relate it to the entropies in the strong subadditivity inequality. Specifically, show how $\chi(\mathcal{X})$ can be expressed in terms of $S(\rho_{AB})$ and $S(\rho_B)$, considering that $S(\rho_B) = S(\rho)$ due to the purification process.

e. Conclude that the Holevo quantity $\chi(\mathcal{X})$, representing the accessible information from the quantum ensemble, is bounded by the difference $S(\rho) - \sum_x p(x) S(\rho_x)$, thereby proving the Holevo Bound.

V Supporting Materials

Essential Mathematics: Quick References . 453

Bibliography 467

List of Figures 477

List of Tables 479

Index 481

Journey Forward 490

Essential Mathematics: Quick References

A Complex Numbers

Basic Relations

Imaginary unit: $i \equiv \sqrt{-1}$. $i^2 = -1$, $i^3 = -i$, $i^4 = 1$, $i^5 = i$, ...

	Cartesian Form	Exponential Form
	$z = x + iy$	$z = re^{i\theta}$
Conjugate	$z^* = x - iy$	$z^* = re^{-i\theta}$
Modulus	$\|z\| = \sqrt{zz^*} = \sqrt{x^2 + y^2}$	$\|z\| = r$
Conversion	$x = r\cos\theta$	$r = \sqrt{x^2 + y^2}$
	$y = r\sin\theta$	$\theta = \operatorname{arctan2}(y, x)$

Basic Operations

Given
$z_1 = x_1 + iy_1 = r_1 e^{i\theta_1}$, $z_2 = x_2 + iy_2 = r_2 e^{i\theta_2}$:

$(z_1 \cdot z_2)^* = z_1^* \cdot z_2^*$ $|z_1 \cdot z_2| = |z_1| \cdot |z_2|$

$z_1 \cdot z_2 = r_1 e^{i\theta_1} \cdot r_2 e^{i\theta_2} = r_1 r_2 e^{i(\theta_1 + \theta_2)}$

$\dfrac{z_1}{z_2} = \dfrac{r_1 e^{i\theta_1}}{r_2 e^{i\theta_2}} = \dfrac{r_1}{r_2} e^{i(\theta_1 - \theta_2)}$

$z_1 \cdot z_2 = x_1 x_2 - y_1 y_2 + i(x_1 y_2 + x_2 y_1)$

$\dfrac{z_1}{z_2} = \dfrac{z_1 z_2^*}{z_2 z_2^*} = \dfrac{x_1 x_2 + y_1 y_2 - i(x_1 y_2 - x_2 y_1)}{x_2^2 + y_2^2}$

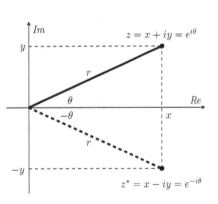

Useful Formulas

Euler's formula: $e^{i\theta} = \cos\theta + i\sin\theta$

De Moivre's formula: $(\cos\theta + i\sin\theta)^n = \cos(n\theta) + i\sin(n\theta)$

Roots of unity ($\omega^n = 1$): $\omega_k = e^{ik\frac{2\pi}{n}}$, where $k = 0, 1, \ldots, n-1$. $\displaystyle\sum_{k=0}^{n-1} \omega_k = 0$.

This content is adapted from *Mathematical Foundations of Quantum Computing* by James M. Yu et al., used with permission.

B Trigonometry

Definitions and Basic Properties

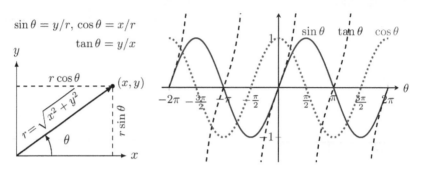

	$\sin\theta$	$\cos\theta$	$\tan\theta$	$\csc\theta$	$\sec\theta$	$\cot\theta$
Definition	y/r	x/r	y/x	r/y	r/x	x/y
Period	2π	2π	π	2π	2π	π
Range	$[-1,1]$	$[-1,1]$	$(-\infty,\infty)$	$(-\infty,-1]\cup[1,\infty)$		$(-\infty,\infty)$
Zeros	$n\pi$	$(n+\tfrac{1}{2})\pi$	$n\pi$			$(n+\tfrac{1}{2})\pi$
Poles			$(n+\tfrac{1}{2})\pi$	$n\pi$	$(n+\tfrac{1}{2})\pi$	$n\pi$
Inv Range	$[-\tfrac{\pi}{2},\tfrac{\pi}{2}]$	$[0,\pi]$	$(-\tfrac{\pi}{2},\tfrac{\pi}{2})$	$[-\tfrac{\pi}{2},\tfrac{\pi}{2}]$	$[0,\pi]$	$(-\tfrac{\pi}{2},\tfrac{\pi}{2})$

Note: n is an integer. $\sin\theta$ is also denoted as $\sin(\theta)$. The inverse of $x = \sin\theta$, $\theta = \arcsin x$, is also written as $\theta = \sin^{-1} x$. The extended inverse arcsin2(y,x), or asin2(y,x), extends the range to $[-\pi,\pi]$ by considering the signs of x and y. Similarly, for other trig functions.

Special Values

θ	0	$\tfrac{\pi}{6}, 30°$	$\tfrac{\pi}{4}, 45°$	$\tfrac{\pi}{3}, 60°$	$\tfrac{\pi}{2}, 90°$	$\tfrac{3\pi}{4}, 135°$	$\pi, 180°$	$\tfrac{5\pi}{4}, 225°$
$\sin\theta$	0	$\tfrac{1}{2}$	$\tfrac{\sqrt{2}}{2}$	$\tfrac{\sqrt{3}}{2}$	1	$\tfrac{\sqrt{2}}{2}$	0	$-\tfrac{\sqrt{2}}{2}$
$\cos\theta$	1	$\tfrac{\sqrt{3}}{2}$	$\tfrac{\sqrt{2}}{2}$	$\tfrac{1}{2}$	0	$-\tfrac{\sqrt{2}}{2}$	-1	$-\tfrac{\sqrt{2}}{2}$
$\tan\theta$	0	$\tfrac{\sqrt{3}}{3}$	1	$\sqrt{3}$	∞	-1	0	1

Interrelations

$$\tan\theta = \frac{\sin\theta}{\cos\theta} \qquad \cot\theta = \frac{1}{\tan\theta} \qquad \sec\theta = \frac{1}{\cos\theta} \qquad \csc\theta = \frac{1}{\sin\theta}$$

Cofunction Formulas

$$\sin\left(\tfrac{\pi}{2} - \theta\right) = \cos\theta \qquad \cos\left(\tfrac{\pi}{2} - \theta\right) = \sin\theta \qquad \cot\left(\tfrac{\pi}{2} - \theta\right) = \tan\theta$$

Pythagorean Identities

$$\sin^2\theta + \cos^2\theta = 1 \qquad \tan^2\theta + 1 = \sec^2\theta \qquad 1 + \cot^2\theta = \csc^2\theta$$

B Trigonometry

Symmetry Properties

$\sin(-\theta) = -\sin\theta \qquad \sin(\pi - \theta) = \sin\theta \qquad \sin(\pi + \theta) = -\sin\theta$

$\cos(-\theta) = \cos\theta \qquad \cos(\pi - \theta) = -\cos\theta \qquad \cos(\pi + \theta) = -\cos\theta$

$\tan(-\theta) = -\tan\theta \qquad \tan(\pi - \theta) = -\tan\theta \qquad \tan(\pi + \theta) = \tan\theta$

Double Angle Formulas

$\sin 2\theta = 2\sin\theta\cos\theta$

$\cos 2\theta = \cos^2\theta - \sin^2\theta = 2\cos^2\theta - 1 = 1 - 2\sin^2\theta$

$\tan 2\theta = \dfrac{2\tan\theta}{1 - \tan^2\theta}$

Half Angle Formulas

$\sin^2\dfrac{\theta}{2} = \dfrac{1 - \cos\theta}{2} \qquad \cos^2\dfrac{\theta}{2} = \dfrac{1 + \cos\theta}{2} \qquad \tan\dfrac{\theta}{2} = \dfrac{\sin\theta}{1 + \cos\theta} = \dfrac{1 - \cos\theta}{\sin\theta}$

Sum and Difference Formulas

$\sin(\alpha \pm \beta) = \sin\alpha\cos\beta \pm \cos\alpha\sin\beta \qquad \cos(\alpha \pm \beta) = \cos\alpha\cos\beta \mp \sin\alpha\sin\beta$

$\tan(\alpha \pm \beta) = \dfrac{\tan\alpha \pm \tan\beta}{1 \mp \tan\alpha\tan\beta}$

Product-to-Sum Formulas

$2\sin\alpha\sin\beta = \cos(\alpha - \beta) - \cos(\alpha + \beta) \qquad 2\cos\alpha\cos\beta = \cos(\alpha - \beta) + \cos(\alpha + \beta)$

$2\sin\alpha\cos\beta = \sin(\alpha + \beta) + \sin(\alpha - \beta) \qquad 2\cos\alpha\sin\beta = \sin(\alpha + \beta) - \sin(\alpha - \beta)$

Sum-to-Product Formulas

$\sin\alpha + \sin\beta = 2\sin\dfrac{\alpha + \beta}{2}\cos\dfrac{\alpha - \beta}{2} \qquad \sin\alpha - \sin\beta = 2\cos\dfrac{\alpha + \beta}{2}\sin\dfrac{\alpha - \beta}{2}$

$\cos\alpha + \cos\beta = 2\cos\dfrac{\alpha + \beta}{2}\cos\dfrac{\alpha - \beta}{2} \qquad \cos\alpha - \cos\beta = -2\sin\dfrac{\alpha + \beta}{2}\sin\dfrac{\alpha - \beta}{2}$

Laws of Sines and Cosines

$\dfrac{\sin A}{a} = \dfrac{\sin B}{b} = \dfrac{\sin C}{c} \qquad a^2 = b^2 + c^2 - 2bc\cos A$

Spherical Coordinate System

Useful Taylor Expansions

$e^x = \displaystyle\sum_{n=0}^{\infty} \dfrac{x^n}{n!} = 1 + x + \dfrac{x^2}{2!} + \dfrac{x^3}{3!} + \cdots$

$\sin x = \displaystyle\sum_{n=0}^{\infty} (-1)^n \dfrac{x^{2n+1}}{(2n+1)!} = x - \dfrac{x^3}{3!} + \dfrac{x^5}{5!} - \cdots$

$\cos x = \displaystyle\sum_{n=0}^{\infty} (-1)^n \dfrac{x^{2n}}{(2n)!} = 1 - \dfrac{x^2}{2!} + \dfrac{x^4}{4!} - \cdots$

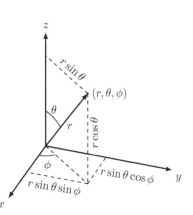

C   Linear Algebra for QCI

Arrays in 0D, 1D, 2D, ...

0D: scalar, c, in $\mathbb{C}$

1D: vector, $|a\rangle = \begin{bmatrix} a_1 \\ a_2 \\ \vdots \\ a_n \end{bmatrix}$, in $\mathbb{C}^n$

2D: matrix, $A = \begin{bmatrix} a_{11} & a_{12} & \cdots \\ a_{21} & a_{22} & \ddots \\ \vdots & & \end{bmatrix}$, in $\mathbb{C}^{n \times n}$

3D, 4D, ...: tensor, in $\mathbb{C}^{n \times n \times \cdots \times n}$ (k times), or $\mathbb{C}^{n^k}$

Matrix Definitions

A^*: complex conjugate of A

A^T: transpose of A

$A^\dagger$: Hermitian conjugate or adjoint of A, $A^\dagger = (A^T)^*$

Hermitian (self-adjoint) matrix: $H^\dagger = H$

Unitary matrix: $U^\dagger = U^{-1}$ or $U^\dagger U = I$

Normal matrix: $AA^\dagger = A^\dagger A$ (includes Hermitian matrix and unitary matrix)

Trace: $\operatorname{tr} A \equiv \sum_i a_{ii}$

Determinant: $\det A = |A|$

Commutator: $[A, B] \equiv AB - BA$

Anti-commutator: $\{A, B\} \equiv AB + BA$

Direct Sum: $A \oplus B$ (block diagonal matrix of A and B)

Ket, Bra, and Braket

Ket: $|v\rangle \equiv \vec{v} = \begin{bmatrix} v_1 \\ v_2 \\ v_3 \end{bmatrix}$

Bra: $\langle v| \equiv |v\rangle^\dagger = \begin{bmatrix} v_1^* & v_2^* & v_2^* \end{bmatrix}$

Braket (inner product):

C Linear Algebra for QCI

$$\langle u|v\rangle \equiv \langle u| \cdot |v\rangle = \begin{bmatrix} u_1^* & u_2^* & u_3^* \end{bmatrix} \begin{bmatrix} v_1 \\ v_2 \\ v_3 \end{bmatrix} = u_1^* v_1 + u_2^* v_2 + u_3^* v_3$$

$$\langle u|v\rangle = \langle v|u\rangle^*$$

Norm: $\|v\| = \sqrt{\langle v|v\rangle} = \sqrt{v_1^* v_1 + v_2^* v_2 + v_3^* v_3}$

Normalization: $|v\rangle \to |\hat{v}\rangle = \dfrac{|v\rangle}{\|v\|}$ so that $\|\hat{v}\| = 1$

Vector Matrix Product

$$A|v\rangle = \begin{bmatrix} a_{11} & a_{12} & a_{13} \\ a_{21} & a_{22} & a_{23} \\ a_{31} & a_{32} & a_{33} \end{bmatrix} \begin{bmatrix} v_1 \\ v_2 \\ v_3 \end{bmatrix}$$

$$(A|v\rangle)^\dagger \equiv \langle Av| = \langle v| A^\dagger$$

$$\langle u|A|v\rangle = \begin{bmatrix} u_1^* & u_2^* & u_3^* \end{bmatrix} \begin{bmatrix} a_{11} & a_{12} & a_{13} \\ a_{21} & a_{22} & a_{23} \\ a_{31} & a_{32} & a_{33} \end{bmatrix} \begin{bmatrix} v_1 \\ v_2 \\ v_3 \end{bmatrix}$$

$$\langle u|A|v\rangle = \langle u|Av\rangle = \langle uA|v\rangle = \langle A^\dagger u|v\rangle$$

Eigenvalues and Eigenvectors

$H|\phi_i\rangle = \lambda_i |\phi_i\rangle$

A normal matrix allows for spectral decomposition. Its eigenvalues are complex, whereas those of Hermitian matrices are real, and anti-Hermitian matrices imaginary. The eigenvectors of these matrices form complete, orthonormal bases.

Orthonormal property: $\langle \phi_i | \phi_j \rangle = \delta_{ij}$

Completeness: $\sum_i |\phi_i\rangle\langle\phi_i| = I$

Spectral decomposition of H: $H = \sum_i \lambda_i |\phi_i\rangle\langle\phi_i|$

Statistics

Observables in QM (A and B) are Hermitian matrices.

Expected value: $\langle A \rangle_\psi = \langle \psi | A | \psi \rangle$

Standard deviation: $\Delta A_\psi = \sqrt{\left\langle \left(A - \langle A \rangle_\psi \right)^2 \right\rangle_\psi} = \sqrt{\langle A^2 \rangle_\psi - \langle A \rangle_\psi^2}$

Cauchy-Schwarz inequality: $|\langle u|v\rangle|^2 \leq \langle u|u\rangle \langle v|v\rangle$

Uncertainty theorem: $\Delta A_\psi \cdot \Delta B_\psi \geq \tfrac{1}{2} |\langle AB - BA \rangle_\psi|$

Outer Product

$$|v\rangle\langle u| = \begin{bmatrix} v_1 \\ v_2 \\ v_3 \end{bmatrix} \begin{bmatrix} u_1^* & u_2^* & u_3^* \end{bmatrix} = \begin{bmatrix} v_1 u_1^* & v_1 u_2^* & v_1 u_3^* \\ v_2 u_1^* & v_2 u_2^* & v_2 u_3^* \\ v_3 u_1^* & v_3 u_2^* & v_3 u_3^* \end{bmatrix}$$

$$|0\rangle\langle 0| = \begin{bmatrix} 1 & 0 \\ 0 & 0 \end{bmatrix}, \quad |1\rangle\langle 1| = \begin{bmatrix} 0 & 0 \\ 0 & 1 \end{bmatrix}, \quad |0\rangle\langle 1| = \begin{bmatrix} 0 & 1 \\ 0 & 0 \end{bmatrix}, \quad |1\rangle\langle 0| = \begin{bmatrix} 0 & 0 \\ 1 & 0 \end{bmatrix}$$

$$\begin{bmatrix} a_{00} & a_{01} \\ a_{10} & a_{11} \end{bmatrix} = a_{00} |0\rangle\langle 0| + a_{01} |0\rangle\langle 1| + a_{10} |1\rangle\langle 0| + a_{11} |1\rangle\langle 1|$$

Projection

Projector onto $|u\rangle$: $P_u = |u\rangle\langle u|$

Projection of $|v\rangle$ onto $|u\rangle$: $P_u |v\rangle = |u\rangle \langle u|v\rangle = \langle u|v\rangle |u\rangle$

Projectors are idempotent: $P_u^2 = P_u$

Unitary Matrices

U is unitary: $U^\dagger U = I$ or $U^\dagger = U^{-1}$

The columns (or rows) of a unitary matrix form an orthonormal basis:

$$(U |i\rangle)^\dagger (U |j\rangle) = \langle i|U^\dagger U|j\rangle = \langle i|j\rangle = \delta_{ij}$$

Unitary transformation preserves inner product:

$$\langle U\phi|U\psi\rangle = (U |\phi\rangle)^\dagger (U |\psi\rangle) = \langle \phi|U^\dagger U|\psi\rangle = \langle \phi|\psi\rangle$$

$\{|\phi_i\rangle\}$ are orthonormal and $|\lambda_i| = 1$ $\Leftrightarrow$ $\sum_i \lambda_i |\phi_i\rangle\langle \phi_i|$ is unitary.

If U and V are unitary, so is $U \otimes V$.

$|\det U| = 1$

Standard (or Computational) Basis

Example: $|0\rangle = \begin{bmatrix} 1 \\ 0 \\ 0 \end{bmatrix}, \quad |1\rangle = \begin{bmatrix} 0 \\ 1 \\ 0 \end{bmatrix}, \quad |2\rangle = \begin{bmatrix} 0 \\ 0 \\ 1 \end{bmatrix}$

Orthonormality and completeness: $\langle i|j\rangle = \delta_{ij}, \quad \sum_i |i\rangle\langle i| = I$

Vector decomposition: $|v\rangle = \sum_i v_i |i\rangle, \quad v_i = \langle i|v\rangle$

Matrix decomposition: $A = \sum_{i,j} a_{ij} |i\rangle\langle j|, \quad a_{ij} = \langle i|A|j\rangle, \quad A |i\rangle = \sum_j a_{ij} |j\rangle$

C Linear Algebra for QCI

Change of Basis

Let $\{|b_i\rangle\}$ and $\{|b_i'\rangle\}$ be complete and orthonormal bases:

$$\langle b_i|b_j\rangle = \langle b_i'|b_j'\rangle = \delta_{ij}, \quad \sum_i |b_i\rangle\langle b_i| = \sum_i |b_i'\rangle\langle b_i'| = I$$

Change of basis from $\{|b_i\rangle\}$ to $\{|b_i'\rangle\}$ via U: $|b_i'\rangle = U|b_i\rangle$

where U is unitary and is given by

$$U = \sum_i |b_i'\rangle\langle b_i| = \sum_{i,j} \langle b_i|b_j'\rangle |b_i'\rangle\langle b_j'| = \begin{bmatrix} \langle b_1|b_1'\rangle & \langle b_1|b_2'\rangle & \cdots \\ \langle b_2|b_1'\rangle & \langle b_2|b_2'\rangle & \\ \vdots & & \ddots \end{bmatrix}$$

Vectors and Matrices under Change of Basis

$|v'\rangle = U^\dagger |v\rangle \qquad A' = U^\dagger A U$

$\langle u'|v'\rangle = \langle u|v\rangle \qquad \langle u'|A'|v'\rangle = \langle u|A|v\rangle$

$H|\psi_i\rangle = \lambda_i|\psi_i\rangle \to H'|\psi_i'\rangle = \lambda_i|\psi_i'\rangle \qquad \operatorname{tr} A' = \operatorname{tr} A$

Change from Standard Basis to Eigenvector Basis

Given H is Hermitian and $H|\phi_i\rangle = \lambda_i|\phi_i\rangle$, to change basis from $\{|i\rangle\}$ to $\{|\phi_i\rangle\}$:

$$|\phi_i\rangle = U|i\rangle, \quad U = \sum_i |\phi_i\rangle\langle i|$$

H is diagonal in $\{|\phi_i\rangle\}$: $H = \sum_i \lambda_i |\phi_i\rangle\langle \phi_i|$

$U^\dagger H U$ is diagonal in $\{|i\rangle\}$: $U^\dagger H U = \sum_i \lambda_i |i\rangle\langle i|$

Tensor Product

$|u\rangle \otimes |v\rangle \equiv |u\rangle|v\rangle \equiv |uv\rangle$

$$|uv\rangle = \begin{bmatrix} u_1 \\ u_2 \end{bmatrix} \otimes \begin{bmatrix} v_1 \\ v_2 \end{bmatrix} = \begin{bmatrix} u_1|v\rangle \\ u_2|v\rangle \end{bmatrix} = \begin{bmatrix} u_1 v_1 \\ u_1 v_2 \\ u_2 v_1 \\ u_2 v_2 \end{bmatrix}$$

$A \otimes B =$

$$\begin{bmatrix} a_{11} & a_{12} \\ a_{21} & a_{22} \end{bmatrix} \otimes \begin{bmatrix} b_{11} & b_{12} \\ b_{21} & b_{22} \end{bmatrix} = \begin{bmatrix} a_{11}B & a_{12}B \\ a_{21}B & a_{22}B \end{bmatrix} = \begin{bmatrix} a_{11}b_{11} & a_{11}b_{12} & a_{12}b_{11} & a_{12}b_{11} \\ a_{11}b_{21} & a_{11}b_{22} & a_{12}b_{21} & a_{12}b_{21} \\ a_{21}b_{11} & a_{21}b_{12} & a_{22}b_{11} & a_{22}b_{11} \\ a_{21}b_{21} & a_{21}b_{22} & a_{22}b_{21} & a_{22}b_{21} \end{bmatrix}$$

In general, $(A \otimes B) \neq (B \otimes A)$

$(A \otimes B)^\dagger = A^\dagger \otimes B^\dagger \qquad (A \otimes B)^{-1} = A^{-1} \otimes B^{-1}$

$(A \otimes B)(C \otimes D) = AC \otimes BD$

$(A \otimes B)(|u\rangle \otimes |v\rangle) \equiv (A \otimes B)|uv\rangle = |Au\rangle \otimes |Bv\rangle$

$A \otimes B = (A \otimes I)(I \otimes B)$

$(A_1 \otimes B_1)(A_2 \otimes B_2) \cdots (A_n \otimes B_n) = (A_1 A_2 \cdots A_n) \otimes (B_1 B_2 \cdots B_n)$

$(A_1 \otimes A_2 \cdots \otimes A_n)(B_1 \otimes B_2 \cdots \otimes B_n) = (A_1 B_1) \otimes (A_2 B_2) \cdots \otimes (A_n B_n)$

Tensor product and inner product: $\langle \phi_1 \otimes \phi_2 | \psi_1 \otimes \psi_2 \rangle = \langle \phi_1 | \psi_1 \rangle \langle \phi_2 | \psi_2 \rangle$

Tensor product and outer product:

$(|i\rangle_1 \otimes |j\rangle_2)(\langle k|_1 \otimes \langle l|_2) = |i\rangle_1 \langle k|_1 \otimes |j\rangle_2 \langle l|_2$

Shorthand notation: $|ij\rangle\langle kl| = |i\rangle\langle k| \otimes |j\rangle\langle l|$

Exponent notation for $\otimes$: $|x\rangle^{\otimes n} \equiv \bigotimes_{i=1}^{n} |x\rangle \equiv |xx \cdots x\rangle$

Basis expansion in linear space $\mathbb{C}^{2^n}$ (i.e., general state vector of an n-qubit system):

$$|\psi\rangle = \sum_{\substack{x_i \in \{0,1\} \\ i \in \{1,2,\cdots,n\}}} c_{x_1 x_2 \cdots x_n} |x_1 x_2 \cdots x_n\rangle \equiv \sum_{k=0}^{2^n-1} c_k |k\rangle$$

Functions of Matrices

If A is Hermitian with real eigenvalues λ_i and eigenvectors $|\phi_i\rangle$, then

$$f(A) = \sum_i f(\lambda_i) |\phi_i\rangle\langle\phi_i|$$

$$e^A = \sum_i e^{\lambda_i} |\phi_i\rangle\langle\phi_i|$$

$$\log A = \sum_i \log \lambda_i |\phi_i\rangle\langle\phi_i| \text{ (for positive } A\text{)}$$

With orthonormal basis $\{|\phi_i\rangle\}$ and $m \in \mathbb{Z}^+$,

$$(|\phi_i\rangle\langle\phi_i|)^m = |\phi_i\rangle\langle\phi_i|, \quad \left(\sum_i \lambda_i |\phi_i\rangle\langle\phi_i|\right)^m = \sum_i \lambda_i^m |\phi_i\rangle\langle\phi_i|$$

Alternative definition: $\exp(A) \equiv e^A = \sum_{n=0}^{\infty} \frac{1}{n!} A^n$

If A is normal, then e^A is also normal, and the eigenvalues of e^A are the exponentials of the eigenvalues of A, with the same eigenvectors.

C Linear Algebra for QCI

If H is Hermitian, then e^{iH} is unitary.

If U is unitary, $e^{UAU^\dagger} = Ue^A U^\dagger$

Generalized Euler formula: If γ is real and $A^2 = I$, then $e^{i\gamma A} = \cos\gamma I + i\sin\gamma A$

In general, $e^{A+B} \neq e^A e^B$, unless A and B commute.

BCH formula: $e^{A+B} = e^A e^B e^{-\frac{1}{2}[A,B]+\cdots}$

In general, $e^{A \otimes B} \neq e^A \otimes e^B$, but $e^{A \otimes I + I \otimes B} = e^A \otimes e^B$.

Partial Product Notations

$$U_A |x_1 x_2\rangle \equiv U|x_1\rangle \otimes |x_2\rangle \qquad U_B |x_1 x_2\rangle \equiv |x_1\rangle \otimes U|x_2\rangle$$

$$\langle a^{(1)}|x_1 x_2\rangle \equiv \langle a|x_1\rangle |x_2\rangle \qquad \langle a^{(2)}|x_1 x_2\rangle \equiv \langle a|x_2\rangle |x_1\rangle$$

$$\langle a^{(j)}|x_1 x_2 \cdots x_j \cdots x_n\rangle \equiv \langle a|x_j\rangle |x_1 x_2 \cdots x_{j-1} x_{j+1} \cdots x_n\rangle$$
$$\equiv (I_{j-1} \otimes \langle a| \otimes I_{n-j}) |x_1 x_2 \cdots x_j \cdots x_n\rangle$$

$$\langle a_1^{(1)} a_2^{(2)} \cdots a_m^{(m)}|x_1 x_2 \cdots x_m \cdots x_n\rangle \equiv \langle a_1 a_2 \cdots a_m|x_1 x_2 \cdots x_m\rangle |x_{m+1} x_{m+2} \cdots x_n\rangle$$
$$\equiv (\langle a_1 a_2 \cdots a_m| \otimes I_{n-m}) |x_1 x_2 \cdots x_m \cdots x_n\rangle$$

Given $|\psi\rangle = \displaystyle\sum_{x_1, x_2, \cdots, x_n \in \{0,1\}} c_{x_1 x_2 \cdots x_n} |x_1 x_2 \cdots x_n\rangle$,

$$\langle a_1^{(1)} a_2^{(2)} \cdots a_m^{(m)}|\psi\rangle = \sum_{x_{m+1}, \cdots, x_n \in \{0,1\}} c_{a_1 a_2 \cdots a_m x_{m+1} \cdots x_n} |x_{m+1} \cdots x_n\rangle$$

Trace

$$\operatorname{tr} A \equiv \operatorname{tr}(A) \equiv \sum_{i=1}^{n} a_{ii}$$

$\operatorname{tr}(A^T) = \operatorname{tr}(A) \qquad \operatorname{tr}(A^\dagger) = \operatorname{tr}(A^*) = (\operatorname{tr} A)^*$
$\operatorname{tr}(cA) = c \operatorname{tr} A \qquad \operatorname{tr}(A+B) = \operatorname{tr} A + \operatorname{tr} B$
$\operatorname{tr}(AB) = \operatorname{tr}(BA) \qquad$ (Note in general, $AB \neq BA$, and $\operatorname{tr}(AB) \neq \operatorname{tr}(A)\operatorname{tr}(B)$)
$\operatorname{tr}([A,B]) = 0$

For Pauli matrices: $\operatorname{tr}\sigma_j = 0, \qquad \operatorname{tr}(\sigma_j \sigma_k) = 2\delta_{jk}$

Similarity invariance: $\operatorname{tr}(PAP^{-1}) = \operatorname{tr} A$

As sum of eigenvalues: $\operatorname{tr}(A) = \sum_{i=1}^{n} \lambda_i, \quad \operatorname{tr}(A^k) = \sum_{i=1}^{n} \lambda_i^k$

Tensor product property: $\operatorname{tr}(A \otimes B) = \operatorname{tr}(A)\operatorname{tr}(B)$

Outer product property: $\langle v|u \rangle = \operatorname{tr}(|u\rangle\langle v|), \quad \langle v|A|u \rangle = \operatorname{tr}(A|u\rangle\langle v|)$

Cyclic product property: $\operatorname{tr}(ABC) = \operatorname{tr}(BCA) = \operatorname{tr}(CAB) \quad (\neq \operatorname{tr}(BAC))$

For orthonormal basis $\{|\phi_i\rangle\}$, $\operatorname{tr}\left(\sum_{i=1}^{n} |\phi_i\rangle\langle\phi_i|\right) = n, \quad \operatorname{tr} A = \sum_{i=1}^{n} \langle\phi_i|A|\phi_i\rangle$

A useful identity: $\langle\phi_i|A|\phi_i\rangle = \operatorname{tr}(\langle\phi_i|A|\phi_i\rangle) = \operatorname{tr}(A|\phi_i\rangle\langle\phi_i|)$

Cauchy-Schwarz inequality: $0 \leq [\operatorname{tr}(AB)]^2 \leq \operatorname{tr}(A^2)\operatorname{tr}(B^2) \leq [\operatorname{tr}(A)]^2 [\operatorname{tr}(B)]^2$

For normal matrix: $\operatorname{tr}(A) = \log\left(\det(e^A)\right)$

Determinant

$\det(A^T) = \det(A) \quad \det(A^\dagger) = \det(A^*) = \det(A)^*$

$\det(A^{-1}) = \det(A)^{-1}$

$\det(AB) = \det(A)\det(B)$

$\det(A \otimes B) = (\det A)^n (\det B)^m$ (where m is the dimension of A, and n of B.)

For unitary matrix: $|\det(U)| = 1$

For normal matrix: $\det(A) = \prod_i \lambda_i \quad \det(e^A) = e^{\operatorname{tr} A}$

Vector Space of Matrices

Inner product of two matrices:

$$\langle A, B \rangle \equiv A \cdot B \equiv \sum_{i=1}^{m}\sum_{j=1}^{n} a_{ij}^* b_{ij} = \operatorname{tr}(A^\dagger B) = \operatorname{tr}(BA^\dagger)$$

Decomposition of $A \in \mathbb{C}^{2^n \times 2^n}$ in Pauli basis:

$P_i \in \{I, X, Y, Z\}^{\otimes n}, \quad P_i \cdot P_j = 2^n \delta_{ij}, \quad \text{with } i,j = 1, 2, ..., 4^n$

$A = \sum_{i=1}^{4^n} a_i P_i, \quad a_i = \frac{1}{2^n} P_i \cdot A \equiv \frac{1}{2^n}\operatorname{tr}(AP_i^\dagger) = \frac{1}{2^n}\operatorname{tr}(AP_i)$

This content is adapted from *Mathematical Foundations of Quantum Computing* by James M. Yu et al., used with permission.

D Pauli Matrices

Definitions

$$\sigma_0 = I = |0\rangle\langle 0| + |1\rangle\langle 1| \quad = \begin{bmatrix} 1 & 0 \\ 0 & 1 \end{bmatrix}$$

$$\sigma_1 = \sigma_x = X = |0\rangle\langle 1| + |1\rangle\langle 0| \quad = \begin{bmatrix} 0 & 1 \\ 1 & 0 \end{bmatrix}$$

$$\sigma_2 = \sigma_y = Y = |0\rangle\langle 1|(-i) + |1\rangle\langle 0|i \quad = \begin{bmatrix} 0 & -i \\ i & 0 \end{bmatrix}$$

$$\sigma_3 = \sigma_z = Z = |0\rangle\langle 0| - |1\rangle\langle 1| \quad = \begin{bmatrix} 1 & 0 \\ 0 & -1 \end{bmatrix}$$

Eigenvalues and Vectors

Pauli Matrix	Eigenvalue	Eigenvector
$\sigma_z \equiv Z$	1	$|0\rangle$
	-1	$|1\rangle$
$\sigma_x \equiv X$	1	$|+\rangle = \frac{1}{\sqrt{2}}(|0\rangle + |1\rangle)$
	-1	$|-\rangle = \frac{1}{\sqrt{2}}(|0\rangle - |1\rangle)$
$\sigma_y \equiv Y$	1	$|+_i\rangle = \frac{1}{\sqrt{2}}(|0\rangle + i|1\rangle)$
	-1	$|-_i\rangle = \frac{1}{\sqrt{2}}(|0\rangle - i|1\rangle)$

Properties

For $i, j, k \in \{1, 2, 3\}$,

Unitary and Hermitian: $\sigma_j^2 = \sigma_0, \quad \sigma_j^\dagger = \sigma_j$

Commutation relation: $[\sigma_j, \sigma_k] = 2i\varepsilon_{jkl}\sigma_l$, or $\sigma_1 \sigma_2 = \sigma_3$, etc.

Anti-commutation relation: $\{\sigma_j, \sigma_k\} = 2\delta_{jk}\sigma_0$, or $\sigma_j \sigma_k = -\sigma_k \sigma_j$ for $j \neq k$

Orthogonality: $\sigma_j \cdot \sigma_k \equiv \operatorname{tr}(\sigma_j^\dagger \sigma_k) = \operatorname{tr}(\sigma_j \sigma_k) = 2\delta_{jk}$

In addition, $\operatorname{tr} \sigma_j = 0, \quad \det \sigma_j = -1, \quad \sigma_j \sigma_k = i\varepsilon_{jkl}\sigma_l + \delta_{jk}\sigma_0$

Basis for 2×2 Matrices

Any matrix $A \in \mathbb{C}^{2 \times 2}$ can be expanded over the set $\{\sigma_0, \sigma_1, \sigma_2, \sigma_3\}$:

$$A = \sum_{j=0}^{3} a_j \sigma_j, \text{ where } a_0 = \tfrac{1}{2}\operatorname{tr} A, \text{ and } a_j = \tfrac{1}{2}\operatorname{tr}(\sigma_j A).$$

Representing Spins in Any Direction

For any real unit vector $\hat{u} = (u_x, u_y, u_z)$ where $u_x^2 + u_y^2 + u_z^2 = 1$,

define $\sigma_u = \hat{u} \cdot \sigma = u_x \sigma_x + u_y \sigma_y + u_z \sigma_z$.

For $\hat{u} = (\sin\theta \cos\phi,\ \sin\theta \sin\phi,\ \cos\theta)$,

$$\sigma_u = \sin\theta \cos\phi\, \sigma_x + \sin\theta \sin\phi\, \sigma_y + \cos\theta\, \sigma_z = \begin{bmatrix} \cos\theta & \sin\theta e^{-i\phi} \\ \sin\theta e^{i\phi} & -\cos\theta \end{bmatrix}$$

Eigenvectors:

$$|+_u\rangle = \cos\frac{\theta}{2}|0\rangle + \sin\frac{\theta}{2}e^{i\phi}|1\rangle = \begin{bmatrix} \cos\frac{\theta}{2} \\ \sin\frac{\theta}{2}e^{i\phi} \end{bmatrix}$$

$$|-_u\rangle = -\sin\frac{\theta}{2}|0\rangle + \cos\frac{\theta}{2}e^{i\phi}|1\rangle = \begin{bmatrix} -\sin\frac{\theta}{2} \\ \cos\frac{\theta}{2}e^{i\phi} \end{bmatrix}$$

Expected values:

$$\langle +_u|\sigma_x|+_u\rangle = \sin\theta \cos\phi$$
$$\langle +_u|\sigma_y|+_u\rangle = \sin\theta \sin\phi$$
$$\langle +_u|\sigma_z|+_u\rangle = \cos\theta$$

Exponentiation

For any real γ, $e^{i\gamma\sigma_j} = \cos\gamma\, \sigma_0 + i\sin\gamma\, \sigma_j$

Also, $e^{i\gamma\sigma_u} = \cos\gamma\, \sigma_0 + i\sin\gamma\, \sigma_u$

Rotation Operators

$$R_j(\gamma) \equiv e^{-i\frac{\gamma}{2}\sigma_j} = \sigma_0 \cos\frac{\gamma}{2} - i\sigma_j \sin\frac{\gamma}{2}$$

Explicitly,

$$R_1(\gamma) \equiv R_x(\gamma) = \begin{bmatrix} \cos\frac{\gamma}{2} & -i\sin\frac{\gamma}{2} \\ -i\sin\frac{\gamma}{2} & \cos\frac{\gamma}{2} \end{bmatrix}$$

$$R_2(\gamma) \equiv R_y(\gamma) = \begin{bmatrix} \cos\frac{\gamma}{2} & -\sin\frac{\gamma}{2} \\ \sin\frac{\gamma}{2} & \cos\frac{\gamma}{2} \end{bmatrix}$$

$$R_3(\gamma) \equiv R_z(\gamma) = \begin{bmatrix} e^{-i\frac{\gamma}{2}} & 0 \\ 0 & e^{i\frac{\gamma}{2}} \end{bmatrix}$$

Rotation about axis $\hat{u} = (n_x, n_y, n_z)$:

$$R_u(\gamma) = e^{-i\frac{\gamma}{2}\sigma_u} = \sigma_0 \cos\frac{\gamma}{2} - i\sigma_u \sin\frac{\gamma}{2}$$

Rotation from $|0\rangle$ to $|+_u\rangle$ and $|-_u\rangle$:

D Pauli Matrices

$$|+_u\rangle = R_z(\phi)R_y(\theta)|0\rangle$$
$$|-_u\rangle = R_z(\phi)R_y(-\theta)|0\rangle$$

Other Forms

$$\sigma_+ \equiv \tfrac{1}{2}(\sigma_1 + i\sigma_2) = \begin{bmatrix} 0 & 1 \\ 0 & 0 \end{bmatrix} \qquad \sigma_- \equiv \tfrac{1}{2}(\sigma_1 - i\sigma_2) = \begin{bmatrix} 0 & 0 \\ 1 & 0 \end{bmatrix}$$

$$\sigma_h \equiv \tfrac{1}{2}(\sigma_0 + \sigma_3) = \begin{bmatrix} 1 & 0 \\ 0 & 0 \end{bmatrix} \qquad \sigma_v \equiv \tfrac{1}{2}(\sigma_0 - \sigma_3) = \begin{bmatrix} 0 & 0 \\ 0 & 1 \end{bmatrix}$$

Gate Relations

$$XY = -YX = iZ$$
$$YZ = -ZY = iX$$
$$ZX = -XZ = iY$$

$$XXX = X, \quad XYX = -Y, \quad XZX = -Z,$$
$$YXY = -X, \quad YYY = Y, \quad YZY = -Z,$$
$$ZXZ = -X, \quad ZYZ = -Y, \quad ZZZ = Z$$

$$HXH = Z, \text{ where } H = \frac{1}{\sqrt{2}}\begin{bmatrix} 1 & 1 \\ 1 & -1 \end{bmatrix}$$

$$HYH = -Y$$
$$HZH = X$$

$$SXS^\dagger = Y, \text{ where } S = \begin{bmatrix} 1 & 0 \\ 0 & i \end{bmatrix}$$

$$SYS^\dagger = -X$$
$$SZS^\dagger = Z$$

Pauli Strings

Definition: $P = A_1 \otimes \cdots \otimes A_n$ with $A_i \in \{I, X, Y, Z\}$, or $P \in \{I, X, Y, Z\}^{\otimes n}$

Orthogonality: $P_i \cdot P_j \equiv \mathrm{tr}(P_i^\dagger P_j) = 2^n \delta_{ij}$, with $i, j = 1, 2, ..., 4^n$.

Two Pauli strings P_i and P_j commute if they have an even number of qubits where their corresponding Pauli matrices are different and anticommute (e.g., one is X and the other is Y or Z, but not I).

This content is adapted from *Mathematical Foundations of Quantum Computing* by James M. Yu et al., used with permission.

Bibliography

Books

[1] Claude Cohen-Tannoudji, Bernard Diu, and Franck Laloë. *Quantum Mechanics, Vol 1: Basic Concepts, Tools, and Applications*. Wiley, 2019. ISBN: 978-3-527-34553-3 (cited on page 36).

[2] Richard P. Feynman, Robert B. Leighton, and Matthew Sands. *Quantum Mechanics (Feynman Lectures on Physics, Volume 3)*. Basic Books, Oct. 4, 2011. ISBN: 978-0465025015 (cited on page 37).

[3] David J. Griffiths and Darrell F. Schroeter. *Introduction to Quantum Mechanics*. English. 3rd edition. Cambridge University Press, Aug. 2018, page 508. ISBN: 978-1107189638 (cited on page 398).

[4] Kurt Jacobs. *Quantum Measurement Theory and its Applications*. English. 1st. Cambridge University Press, 2014, page 554. ISBN: 978-1-107-02548-6 (cited on page 80).

[5] Jonathan Katz and Yehuda Lindell. *Introduction to Modern Cryptography*. CRC Press, 2014. ISBN: 978-1-4665-7027-6 (cited on page 134).

[6] Junichiro Kono. *Quantum Mechanics for Tomorrow's Engineers: New Edition*. English. New edition. Cambridge University Press, Sept. 29, 2022, page 350. ISBN: 978-1108842587 (cited on page 36).

[7] Michael A Nielsen and Isaac L Chuang. *Quantum Computation and Quantum Information: 10th Anniversary Edition*. Cambridge University Press, 2010. ISBN: 978-1-107-00217-3 (cited on pages 186, 219, 265, 444).

[8] J. J. Sakurai and Jim Napolitano. *Modern Quantum Mechanics*. Cambridge University Press, 2017. ISBN: 978-1-108-47322-4 (cited on pages 37, 89, 102).

[9] Gilbert Strang. *Introduction to Linear Algebra*. English. 6th edition. Wellesley-Cambridge Press, Apr. 2023, page 440. ISBN: 978-1733146678 (cited on page 19).

[10] John Watrous. *The Theory of Quantum Information*. Cambridge University Press, 2018. ISBN: 978-1-107-18056-7 (cited on page 444).

[11] Mark M. Wilde. *Quantum Information Theory*. Cambridge University Press, 2017. ISBN: 978-1-107-17616-4 (cited on pages 80, 444).

Articles

[12] Scott Aaronson and Paul Christiano. Quantum Money from Hidden Subspaces. In: (2012) (cited on page 333).

[13] R. Alléaume et al. Using quantum key distribution for cryptographic purposes: A survey. In: Theoretical Computer Science 560 (2014), pages 62–81. ISSN: 0304-3975. DOI: 10.1016/j.tcs.2014.09.018 (cited on pages 134, 288).

[14] Rameez Asif. Post-Quantum Cryptosystems for Internet-of-Things: A Survey on Lattice-Based Algorithms. In: IoT 2.1 (2021), pages 71–91. ISSN: 2624-831X. DOI: 10.3390/iot2010005 (cited on pages 129, 134, 334).

[15] Alain Aspect, Jean Dalibard, and G'erard Roger. Experimental realization of Einstein-Podolsky-Rosen-Bohm gedankenexperiment: A new violation of Bell's inequalities. In: Physical Review Letters 49.2 (1982), page 91. DOI: 10.1103/PhysRevLett.49.91 (cited on page 246).

[16] Alain Aspect, Jean Dalibard, and Gérard Roger. Experimental test of Bell's inequalities using time-varying analyzers. In: Physical Review Letters 49.25 (1982), pages 1804–1807. DOI: 10.1103/PhysRevLett.49.1804 (cited on page 237).

[17] Alain Aspect, Philippe Grangier, and Gérard Roger. Experimental tests of realistic local theories via Bell's theorem. In: Physical Review Letters 47.7 (1981), pages 460–463. DOI: 10.1103/PhysRevLett.47.460 (cited on page 237).

[18] Adriano Barenco et al. Elementary gates for quantum computation. In: Physical Review A 52.5 (1995), page 3457. DOI: 10.1103/PhysRevA.52.3457 (cited on page 186).

[19] Elisa Bäumer et al. Efficient Long-Range Entanglement using Dynamic Circuits. In: (2023) (cited on page 343).

[20] John S Bell. On the Einstein Podolsky Rosen Paradox. In: Physics Physique Fizika 1 (3 Nov. 1964), pages 195–200. DOI: 10.1103/PhysicsPhysiqueFizika.1.195 (cited on page 246).

[21] Charles H Bennett and Gilles Brassard. Quantum Cryptography: Public Key Distribution and Coin Tossing. In: Proceedings of IEEE International Conference on Computers, Systems and Signal Processing (1984), pages 175–179. DOI: 10.1016/j.tcs.2014.05.025 (cited on pages 134, 288).

[22] Charles H. Bennett. Quantum Cryptography Using Any Two Nonorthogonal States. In: Physical Review Letters 68.21 (1992), page 3121. DOI: 10.1103/PhysRevLett.68.3121 (cited on page 134).

[23] Kishor Bharti et al. Noisy intermediate-scale quantum algorithms. In: Reviews of Modern Physics 94.1 (Feb. 2022). DOI: 10.1103/revmodphys.94.015004 (cited on page 306).

[24] A. Bilyk, J. Doliskani, and Z. Gong. Cryptanalysis of Three Quantum Money Schemes. In: Quantum Information Processing 22 (2023), page 177. DOI: 10.1007/s11128-023-03919-0 (cited on page 333).

[25] Jan Bouda and Vladimír Buzek. Entanglement swapping between multi-qudit systems. In: Journal of Physics A: Mathematical and General 34.20 (May 2001), page 4301. DOI: 10.1088/0305-4470/34/20/304 (cited on page 270).

[26] Dik Bouwmeester et al. Experimental quantum teleportation. In: Nature 390.6660 (Dec. 1997), pages 575–579. DOI: 10.1038/37539 (cited on page 265).

[27] Dik Bouwmeester et al. Observation of three-photon Greenberger-Horne-Zeilinger entanglement. In: Physical review letters 82.7 (1999), page 1345. DOI: 10.1103/PhysRevLett.82.1345 (cited on page 219).

[28] Sergey Bravyi, David Gosset, and Yinchen Liu. How to Simulate Quantum Measurement without Computing Marginals. In: Phys. Rev. Lett. 128 (22 June 2022), page 220503. DOI: 10.1103/PhysRevLett.128.220503 (cited on page 397).

[29] Sergey Bravyi et al. High-threshold and low-overhead fault-tolerant quantum memory. In: (2023) (cited on page 397).

[30] H.-J. Briegel et al. Quantum Repeaters: The Role of Imperfect Local Operations in Quantum Communication. In: Phys. Rev. Lett. 81 (26 Dec. 1998), pages 5932–5935. DOI: 10.1103/PhysRevLett.81.5932 (cited on page 270).

[31] Aharon Brodutch et al. An adaptive attack on Wiesner's quantum money. In: (2016) (cited on page 328).

[32] D. Bruß et al. Distributed Quantum Dense Coding. In: Phys. Rev. Lett. 93 (21 Nov. 2004), page 210501. DOI: 10.1103/PhysRevLett.93.210501 (cited on page 256).

[33] Zhenyu Cai, Xiaosi Xu, and Simon C. Benjamin. Mitigating coherent noise using Pauli conjugation. In: npj Quantum Information 6.1 (2020), page 17. ISSN: 2056-6387. DOI: 10.1038/s41534-019-0233-0 (cited on page 397).

[34] Avimita Chatterjee, Koustubh Phalak, and Swaroop Ghosh. Quantum Error Correction For Dummies. In: (2023) (cited on pages 396, 397).

[35] Edward H. Chen et al. Realizing the Nishimori transition across the error threshold for constant-depth quantum circuits. In: (2023) (cited on page 343).

[36] Kevin S. Chou et al. Deterministic teleportation of a quantum gate between two logical qubits. In: Nature 561.7723 (2018), pages 368–373. ISSN: 1476-4687. DOI: 10.1038/s41586-018-0470-y (cited on page 283).

[37] A. Dan et al. Clustering approach for solving traveling salesman problems via Ising model based solver. In: (2020), pages 1–6. DOI: 10.1109/DAC18072.2020.9218695 (cited on page 319).

[38] Albert Einstein, Boris Podolsky, and Nathan Rosen. Can Quantum-Mechanical Description of Physical Reality Be Considered Complete? In: Physical Review 47.10 (1935), page 777. DOI: 10.1103/PhysRev.47.777 (cited on page 246).

[39] Artur K Ekert. Quantum cryptography based on Bell's theorem. In: Physical Review Letters 67.6 (1991), page 661. DOI: 10.1103/PhysRevLett.67.661 (cited on pages 134, 287).

[40] Avshalom C. Elitzur and Lev Vaidman. Quantum mechanical interaction-free measurements. In: Foundations of Physics 23.7 (July 1993), pages 987–997. DOI: 10.1007/bf00736012 (cited on page 322).

[41] Edward Farhi et al. Quantum money from knots. In: (2010) (cited on page 333).

[42] Arman Rasoodl Faridi et al. Blockchain in the Quantum World. In: International Journal of Advanced Computer Science and Applications 13.1 (2022). DOI: 10.14569/ijacsa.2022.0130167 (cited on page 334).

[43] Roland C. Farrell et al. Scalable Circuits for Preparing Ground States on Digital Quantum Computers: The Schwinger Model Vacuum on 100 Qubits. In: (2023) (cited on page 343).

[44] A. Furusawa et al. Unconditional Quantum Teleportation. In: Science 282.5389 (1998), pages 706–709. DOI: 10.1126/science.282.5389.706 (cited on page 265).

[45] Marissa Giustina et al. A significant-loophole-free test of Bell's theorem with entangled photons. In: Physical Review Letters 115.25 (2015), page 250401. DOI: 10.1103/PhysRevLett.115.250401 (cited on page 238).

[46] Daniel Gottesman and Isaac L. Chuang. Demonstrating the viability of universal quantum computation using teleportation and single-qubit operations. In: Nature 402.6760 (Nov. 1999), pages 390–393. DOI: 10.1038/46503 (cited on page 282).

[47] Trent M. Graham et al. Superdense teleportation using hyperentangled photons. In: Nature Communications 6.1 (May 2015), page 7185. ISSN: 2041-1723. DOI: 10.1038/ncomms8185 (cited on page 265).

[48] Daniel M Greenberger, Michael A Horne, and Anton Zeilinger. Going beyond Bell's theorem. In: (1989), pages 69–72. DOI: 10.1007/978-94-017-0849-4_10 (cited on page 219).

[49] Shouzhen Gu, Alex Retzker, and Aleksander Kubica. Fault-tolerant quantum architectures based on erasure qubits. In: (2023) (cited on page 397).

[50] Yu Guo et al. Advances in Quantum Dense Coding. In: Advanced Quantum Technologies 2.5-6 (2019), page 1900011. DOI: 10.1002/qute.201900011 (cited on page 256).

[51] Mauricio Gutiérrez and Kenneth R. Brown. Comparison of a quantum error-correction threshold for exact and approximate errors. In: Physical Review A 91.2 (Feb. 2015). DOI: 10.1103/physreva.91.022335 (cited on page 397).

[52] Laszlo Gyongyosi and Sandor Imre. A Survey on quantum computing technology. In: Computer Science Review 31 (2019), pages 51–71. ISSN: 1574-0137. DOI: 10.1016/j.cosrev.2018.11.002 (cited on page 60).

[53] Johannes Handsteiner et al. Cosmic Bell Test: Measurement Settings from Milky Way Stars. In: Physical Review Letters 118.6 (2017). DOI: 10.1103/PhysRevLett.118.060401 (cited on page 247).

[54] B Hensen et al. Loophole-free Bell inequality violation using electron spins separated by 1.3 kilometres. In: Nature 526.7575 (2015), pages 682–686. DOI: 10.1038/nature15759 (cited on pages 238, 246).

[55] Ryszard Horodecki et al. Quantum entanglement. In: Reviews of Modern Physics 81.2 (2009), pages 865–942. DOI: 10.1103/RevModPhys.81.865 (cited on page 246).

[56] Yun-Feng Huang et al. Experimental Teleportation of a Quantum Controlled-NOT Gate. In: Phys. Rev. Lett. 93 (24 Dec. 2004), page 240501. DOI: 10.1103/PhysRevLett.93.240501 (cited on page 282).

[57] Daniel M. Kane, Shahed Sharif, and Alice Silverberg. Quantum Money from Quaternion Algebras. In: (2022) (cited on page 333).

[58] Daniel Keith et al. Ramped measurement technique for robust high-fidelity spin qubit readout. In: Science Advances 8.36 (2022), eabq0455. DOI: 10.1126/sciadv.abq0455 (cited on page 55).

[59] Andrey Boris Khesin, Jonathan Z. Lu, and Peter W. Shor. Publicly verifiable quantum money from random lattices. In: (2022) (cited on page 333).

[60] Dai-Gyoung Kim et al. Enhanced quantum teleportation using multi-qubit logical states. In: Results in Physics 50 (2023), page 106565. ISSN: 2211-3797. DOI: 10.1016/j.rinp.2023.106565 (cited on page 283).

[61] Sangbae Kim and Byoung S. Ham. Observations of the delayed-choice quantum eraser using coherent photons. In: Scientific Reports 13.1 (June 16, 2023), page 9758. ISSN: 2045-2322. DOI: 10.1038/s41598-023-36590-7 (cited on page 31).

[62] Youngseok Kim et al. Evidence for the utility of quantum computing before fault tolerance. In: Nature 618.7965 (June 2023), pages 500–505. ISSN: 1476-4687. DOI: 10.1038/s41586-023-06096-3 (cited on page 343).

[63] Brian T. Kirby et al. Entanglement swapping of two arbitrarily degraded entangled states. In: Phys. Rev. A 94 (1 July 2016), page 012336. DOI: 10.1103/PhysRevA.94.012336 (cited on page 270).

[64] Raymond Laflamme et al. Perfect Quantum Error Correction Code. In: (1996) (cited on page 396).

[65] Jan-Åke Larsson. Loopholes in Bell inequality tests of local realism. In: Journal of Physics A: Mathematical and Theoretical 47.42 (2014). DOI: 10.1088/1751-8113/47/42/424003 (cited on page 246).

[66] Göran Lindblad. Completely positive maps and entropy inequalities. In: Communications in Mathematical Physics 40.2 (1975), pages 147–151. DOI: 10.1007/BF01609396 (cited on page 438).

[67] J. Liu, H. Montgomery, and M. Zhandry. Another Round of Breaking and Making Quantum Money. In: Lecture Notes in Computer Science 14004 (2023). Edited by C. Hazay and M. Stam. DOI: 10.1007/978-3-031-30545-0_21 (cited on page 333).

[68] G. Lugilde Fernández, E.F. Combarro, and I.F. Rúa. Quantum measurement detection algorithms. In: Quantum Information Process 21.274 (2022). DOI: 10.1007/s11128-022-03614-6 (cited on page 325).

[69] Andrew Lutomirski. An online attack against Wiesner's quantum money. In: (2010) (cited on page 327).

[70] Chetan Nayak et al. Non-Abelian anyons and topological quantum computation. In: Rev. Mod. Phys. 80 (3 Sept. 2008), pages 1083–1159. DOI: 10.1103/RevModPhys.80.1083 (cited on page 399).

[71] A. K. Pati, P. Parashar, and P. Agrawal. Probabilistic superdense coding. In: Phys. Rev. A 72 (1 July 2005), page 012329. DOI: 10.1103/PhysRevA.72.012329 (cited on page 256).

[72] Asher Peres. All the Bell Inequalities. In: (1999) (cited on page 246).

[73] Christophe Piveteau et al. Error Mitigation for Universal Gates on Encoded Qubits. In: Phys. Rev. Lett. 127 (20 Nov. 2021), page 200505. DOI: 10.1103/PhysRevLett.127.200505 (cited on page 397).

[74] Lukas Postler et al. Demonstration of fault-tolerant universal quantum gate operations. In: Nature 605.7911 (2022), pages 675–680. ISSN: 1476-4687. DOI: 10.1038/s41586-022-04721-1 (cited on page 283).

[75] Gustavo Rigolin. Quantum teleportation of an arbitrary two-qubit state and its relation to multipartite entanglement. In: Physical Review A 71.3 (Mar. 2005). DOI: 10.1103/physreva.71.032303 (cited on page 265).

[76] A. Rubenok et al. Real-world two-photon interference and proof-of-principle quantum key distribution immune to detector attacks. In: Physical Review Letters 111.13 (2013), page 130501. DOI: 10.1103/PhysRevLett.111.130501 (cited on page 288).

[77] Maria E. Sabani et al. Evaluation and Comparison of Lattice-Based Cryptosystems for a Secure Quantum Computing Era. In: Electronics 12.12 (2023). ISSN: 2079-9292. DOI: 10.3390/electronics12122643 (cited on page 334).

[78] M. Sasaki et al. Field test of quantum key distribution in the Tokyo QKD Network. In: Optics Express 19.11 (2011), pages 10387–10409. DOI: 10.1364/OE.19.010387 (cited on page 288).

[79] Valerio Scarani et al. Quantum Cryptography Protocols Robust Against Photon Number Splitting Attacks. In: Physical Review Letters 92.5 (2004), page 057901. DOI: 10.1103/PhysRevLett.92.057901 (cited on page 134).

[80] Lynden K Shalm et al. A strong loophole-free test of local realism. In: Physical Review Letters 115.25 (2015), page 250402. DOI: 10.1103/PhysRevLett.115.250402 (cited on page 238).

[81] Oles Shtanko et al. Uncovering Local Integrability in Quantum Many-Body Dynamics. In: (2023) (cited on page 343).

[82] Mitali Sisodia et al. Design and experimental realization of an optimal scheme for teleportation of an n-qubit quantum state. In: Quantum Information Processing 16.12 (2017), page 292. ISSN: 1573-1332. DOI: 10.1007/s11128-017-1744-2 (cited on page 282).

[83] Erik Sjöqvist. A new phase in quantum computation. In: Physics 1 (Nov. 2008), page id. 35. DOI: 10.1103/Physics.1.35 (cited on page 398).

[84] A. M. Steane. Simple quantum error-correcting codes. In: Physical Review A 54.6 (1996), page 4741. DOI: 10.1103/physreva.54.4741 (cited on page 396).

[85] Ady Stern and Netanel H. Lindner. Topological Quantum Computation—From Basic Concepts to First Experiments. In: Science 339.6124 (2013), pages 1179–1184. DOI: 10.1126/science.1231473 (cited on page 399).

[86] Damien Stucki et al. Continuous high speed coherent one-way quantum key distribution. In: Optics Express 17.16 (July 2009), page 13326. DOI: 10.1364/oe.17.013326 (cited on page 134).

[87] Miroslav Urbanek et al. Mitigating Depolarizing Noise on Quantum Computers with Noise-Estimation Circuits. In: Phys. Rev. Lett. 127 (27 Dec. 2021), page 270502. DOI: 10.1103/PhysRevLett.127.270502 (cited on page 398).

[88] Andrew Wack et al. Quality, Speed, and Scale: three key attributes to measure the performance of near-term quantum computers. In: (2021) (cited on page 397).

[89] Gregor Weihs et al. Violation of Bell's inequality under strict Einstein locality conditions. In: Physical Review Letters 81.23 (1998), pages 5039–5043. DOI: 10.1103/PhysRevLett.81.5039 (cited on page 237).

[90] Stephen Wiesner. Conjugate Coding. In: SIGACT News 15.1 (Jan. 1983), pages 78–88. ISSN: 0163-5700. DOI: 10.1145/1008908.1008920 (cited on page 326).

[91] Yue Wu et al. Erasure conversion for fault-tolerant quantum computing in alkaline earth Rydberg atom arrays. In: Nature Communications 13.1 (Aug. 2022). ISSN: 2041-1723. DOI: 10.1038/s41467-022-32094-6 (cited on page 397).

[92] Toshiki Yasuda et al. Quantum reservoir computing with repeated measurements on superconducting devices. In: (2023) (cited on page 343).

[93] Hongye Yu, Yusheng Zhao, and Tzu-Chieh Wei. Simulating large-size quantum spin chains on cloud-based superconducting quantum computers. In: Physical Review Research 5.1 (Mar. 2023). ISSN: 2643-1564. DOI: 10.1103/physrevresearch.5.013183 (cited on page 343).

[94] M. Żukowski et al. 'Event-ready-detectors' Bell experiment via entanglement swapping. In: Phys. Rev. Lett. 71 (26 Dec. 1993), pages 4287–4290. DOI: 10.1103/PhysRevLett.71.4287 (cited on page 270).

[95] Sultan M. Zangi et al. Entanglement Swapping and Swapped Entanglement. In: Entropy 25.3 (Feb. 2023), page 415. ISSN: 1099-4300. DOI: 10.3390/e25030415 (cited on page 270).

[96] Mark Zhandry. Quantum Lightning Never Strikes the Same State Twice. In: (2017) (cited on page 333).

[97] Jiang Zhang et al. Geometric and holonomic quantum computation. In: Physics Reports 1027 (2023), pages 1–53. ISSN: 0370-1573. DOI: 10.1016/j.physrep.2023.07.004 (cited on page 398).

[98] Tingting Zhang and Jie Han. Efficient Traveling Salesman Problem Solvers Using the Ising Model with Simulated Bifurcation. In: DATE '22 (2022), pages 548–551. DOI: 10.23919/DATE54114.2022.9774576 (cited on page 319).

List of Figures

1.1 Role of Quantum Mechanics in Quantum Computing 5
1.2 Light Polarization in Classical and Quantum Realms 8
1.3 A Light Polarization Experiment 10
1.4 Relationship between H/V and D/A Polarizations 13
1.5 Elliptical Polarization 16
1.6 Illustration of the Quantum Observable Postulate 19
1.7 Measuring Photons Through a Polarizer 26
1.8 Repeated Identical Measurements 27
1.9 Measuring Photons Through Crossed Polarizers 28
1.10 Measuring Photons: Malus' Law 30
1.11 Mach-Zehnder Interferometer 32
1.12 Double Slit Interference Experiment 33
1.13 Summary of the Observable and Measurement Postulates 39

2.1 1/2 Spin 45
2.2 Spherical Coordinate System 48
2.3 Key Points on the Bloch Sphere 50
2.4 Pauli X, Y, and Z as Rotations on the Bloch Sphere 52
2.5 Stern-Gerlach Experiment 53
2.6 Cascaded Stern-Gerlach Experiments 54

3.1 Visualization of qubit states on the Bloch Sphere 64
3.2 Basis Rotation $\{|0\rangle, |1\rangle\} \to \{|+\rangle, |-\rangle\}$ 67
3.3 Illustration of Basis Rotation $\{|b_i\rangle\} \to \{|b_i'\rangle\}$ 68
3.4 Measurements in $\{|0\rangle, |1\rangle\}$ and $\{|+\rangle, |-\rangle\}$ Bases 75
3.5 Plot of ΔZ as a function of $|\alpha|^2$ 78

4.1 Example of a Practical Multi-Level System 93
4.2 Larmor Precession on the Bloch Sphere 97
4.3 Plot of the probabilities $P_0(t)$ and $P_1(t)$ as functions of time. 101

5.1 Examples of Classical Irreversible Gates 114
5.2 R_x, R_y, and R_z Gates as Rotations on the Bloch Sphere 120
5.3 Demonstration of Non-Commutative Nature of 3D Rotations 121
5.4 The Unified Rotation Gate (R_u) on the Bloch Sphere 122
5.5 Examples of Quantum Circuits 124
5.6 Quantum Random Number Generator (QRNG) 127
5.7 Fundamental BB84 QKD Protocol 130
5.8 BB84 QKD Implementation with Qubits 133
5.9 Fair Coin Gate - Setup and Rules 135

5.10	Quantum Coin Gate - Setup and Rules	135
5.11	Strategy in Which Alice Always Wins	136
5.12	Strategy Shift: Bob Beats Alice	137
5.13	Strategy Revised: Alice Wins Again	137
5.14	Illustration of the No-Cloning Theorem	138
6.1	Local Measurement Examples	151
6.2	Local Measurement in $\{\lvert+\rangle,\lvert-\rangle\}$ Basis	153
6.3	Measurement on a Group of Qubits	164
6.4	A Partial Measurement Paradigm for Quantum Algorithms	164
6.5	Multi-Qubit Measurement in Alternate Basis	167
6.6	Illustration of Ising Interaction	170
7.1	Examples of Classically Controlled Gates	183
7.2	Parallel Gate on Unentangled State	184
7.3	Gate on Part of an Entangled State	184
7.4	Expansion of the Boolean Representation of CNOT	190
7.5	Example of Deferred Measurements	197
7.6	Deferred Measurement Principle	198
8.1	Bell State Creation using H and CNOT Gates	212
8.2	Bell State Creation from Initial State $\lvert 00\rangle$	212
8.3	Bell State Creation Using Output Phase Shift	213
8.4	Photon Pair Creation with SPDC	213
8.5	Boolean Representation of Bell States	214
8.6	Bell Measurement	215
8.7	Bell Measurement with Post-Measurement State Restored	216
8.8	Converting $\lvert\beta_{00}\rangle \equiv \lvert\Phi^+\rangle$ to Other Bell States	216
8.9	Converting Other Bell States to $\lvert\beta_{00}\rangle \equiv \lvert\Phi^+\rangle$	217
8.10	Additional Conversions Related to $\lvert\beta_{11}\rangle \equiv \lvert\Psi^-\rangle$	217
9.1	Classical Correlation versus Quantum Entanglement	225
9.2	Settings for Violating the Bell-CHSH Inequality	234
9.3	Basic Experimental Setup for Bell Test	237
9.4	Illustration for Problem 9.5	249
10.1	Bell State Generator and Analyzer	253
10.2	Basic Process of Superdense Coding	255
10.3	Basic Quantum Circuit for Superdense Coding	255
10.4	Basic Process of Quantum Teleportation	258
10.5	Fundamental Quantum Circuit for Teleportation	258
10.6	Quantum Circuit for Teleportation with Deferred Measurement	259
10.7	Basic Process of Entanglement Swapping	267
10.8	Fundamental Quantum Circuit for Entanglement Swapping	267
10.9	Basic Process of Quantum Gate Teleportation	271
10.10	Quantum Circuit for Single-Qubit Gate Teleportation	272
10.11	Gate U Applied Post-Teleportation	272
10.12	Single-Qubit Gate Teleportation with Deferred U Gate	275

10.13	Basic Process of 2-1 Gate Teleportation	277
10.14	Quantum Circuit for 2-1 Gate Teleportation	278
10.15	Gate U Applied Post-Teleportation	278
10.16	Quantum Circuit for 2-1 CNOT Teleportation	280
10.17	Basic Process of 2-2 Gate Teleportation	281
10.18	Quantum Circuit for 2-2 Gate Teleportation	281
10.19	Two-Qubit U-Gate Applied Post-Teleportation	282
10.20	Quantum Circuit for CNOT Gate Teleportation	283
10.21	Basic E91 QKD Protocol	285
11.1	Quantum Circuit for the Deutsch-Jozsa Algorithm	298
11.2	Quantum Circuit for Spectral Measurement	302
11.3	An Oracle for a Balanced Function	304
11.4	A Phase Oracle for a Balanced Function	304
11.5	A Phase Oracle Using CZ Gates	305
11.6	Max-Cut Problem Example	309
11.7	VQE Ansatz and Iteration Procedure	313
11.8	QAOA Ansatz and Iterative Procedure	315
11.9	Example of a QAOA Ansatz Building Block	317
11.10	Example of the Traveling Salesman Problem	318
11.11	Quantum Bomb Tester - Classical Scenario	323
11.12	Quantum Bomb Tester - Measuring in the $\{|+\rangle, |-\rangle\}$ Basis	323
11.13	The Quantum Bomb Test Algorithm	324
11.14	Adaptive Attack on Wiesner Quantum Money	328
12.1	From NISQ to Fault Tolerance	343
12.2	Mixed State Inside the Bloch Sphere	353
12.3	Stinespring Dilation: Unitary Embedding of a CPTP Map	361
12.4	Repeated Application of an Ideal X Gate	369
12.5	Repeated Application of an Off-Calibrated X Gate	370
12.6	Effect of ZZ Cross-Talk over Time	371
12.7	Bit-Flip Error in the Bloch Sphere	373
12.8	Repeated Applications of an X Gate with Bit-Flip Noise	374
12.9	Phase-Flip Error in the Bloch Sphere	374
12.10	Depolarizing Error in the Bloch Sphere	375
12.11	Structure of Practical Error Correction Code	384
12.12	Quantum Circuit for the Three-Qubit Bit-Flip Code	385
12.13	Quantum Circuit for the Three-Qubit Phase-Flip Code	388
12.14	Quantum Circuit for the Nine-Qubit Shor Code	390
12.15	Error Syndrome Measurement and Correction for the Shor Code	391
12.16	Nine-Qubit Shor Code Simplified for Oneshot Operation	394
13.1	Venn Diagram for Probability Relationship	414
13.2	Quantum Simulation of 1D Random Walk	424
13.3	Quantum 1D Walk	425
13.4	Venn Diagram for Shannon Entropy Relationships	431
13.5	Venn Diagram for von Neumann Entropy Relationships	435
13.6	Schematic of the Quantum Information Transmission Process	442

List of Tables

1.1	Key Photon Polarization States	9
1.2	Summary of Photon Polarization States and Observables	23
2.1	Special Spin States	49
2.2	Spin State and Polarization State Correspondence	50
3.1	Examples of Qubit Implementations	61
5.1	Common Single-Qubit Quantum Gates	116
5.2	Transformations by Common Single-Qubit Quantum Gates	118
7.1	Common Multi-Qubit Quantum Gates	177
7.2	Boolean Representation of Common Qubit Gates	188
7.3	Gate Sequence Equivalence Involving X, Z and CNOT	194
7.4	Gate Sequence Equivalence Involving H	196
7.5	Expressing Common Gates Using Toffoli Gate	201
10.1	Summary of Key Applications of Quantum Entanglement	252
10.2	Correction Gates for General Qubit States	253
10.3	Conversion Gates for Bell States	254
10.4	Boolean Representation of Selected Qubit Gates	260
10.5	Gate Sequence Equivalence for Two-Qubit Gate Teleportation	282
10.6	Gate Sequence Equivalence for CNOT Teleportation	283
11.1	Hadamard Transform H_4 and Period Functions	300
11.2	Relationships Between Terms Related to Quantum Optimization	308
12.1	Error Syndrome and Correction for the Three-qubit Bit-flip Code	387
12.2	Error Syndrome and Correction for the Three-Qubit Phase-Flip Code	389
12.3	Error Syndrome for the Nine-Qubit Shor Code	393
13.1	Evolution of Coin & Walker States in a 1D Quantum Walk	426

Index

A

adiabatic quantum computing . 95, 319
anyons 44, 162

B

basis equivalance 22
basis states 12
basis-dependent measurement 75
Bell analyzer 215
Bell basis 209
Bell inequalities 224, 229
    Bell test 236
    derivation 241
    experimental verification 236
    measurement bases 241
Bell measurement 214
Bell states 207
    conversion 215
    creation 211
    definitions 208
    generalization 218
    maximum entanglement 209
    orthonormal basis 209, 217
    summary 252
Bell test
    early experiments 237
    experimental setup 236
    findings and implications 238
    loophole-free experiments 238
Bell-CHSH inequality 229
    classical cases 231
    experimental verification 236
    quantum cases 233
Bloch sphere 50, 63
    bit-flip error 372
    depolarizing error 375
    general qubit states 63
    key states 51
    mixed states 352
    Pauli X, Y, Z gates 51
    phase-flip error 374
    rotation gates 120
Boolean representation 187
    caveat 189
    common gates 188
    interpretation 189
    properties 187

C

change of basis 65
    eigenbasis 70
    general qudits 68
    Hadamard basis 66
classical correlation 224
classical random walk 423
classically controlled-U gate 182
CNOT gate 178
    flipped version 179
    negative version 182
CNOT gate implementation 170
code distance 394
coherent errors 369
    cross-talk error 370
    gate calibration error 369
common single qubit gates 115
compatible observables 34
composite system postulate 157
computation in place 125
conditional entropy 431, 434
conditional probability 413
continuous-time quantum walk . . . 426
CPTP maps 358

D

data processing inequality (DPI) . . 438
decoherence by measurement 356
deferred measurement principle . . . 197
density operator 346
    measurements 356
    partial trace 362
    POVM measurements 358

density operator:Bloch sphere.....352
density operator:time evolution...354
Deutsch-Jozsa algorithm..........297
Dirac notation.....................22

E

E91 QKD protocol
    comparison to BB84.........286
    eavesdropping check.........286
    key sifting...................286
    procedure....................284
empirical expectation value........77
energy eigenstates.................91
energy levels......................91
entanglement.................224, 225
    applications..................252
    misconceptions...............240
    vs. classical correlation.......224
entanglement swapping
    analysis.....................267
    Bell measurement............268
    classical communication......269
    initial entanglement..........268
    procedure....................266
    state reconstruction..........269
EPR paradox.....................227
    locality......................228
    realism......................228
equivalent gate sequence..........191
    analysis approaches..........192
    CNOT gate sequences........194
    H gate sequences.............195
    single-qubit gates.............125
error correction codes.............384
error syndrome measurement.....391
excited states.....................92
expected (expectation) value.......29

F

fault-tolerant quantum computing 344
fermions and bosons...............44

G

gate-based quantum computing....94
general qubit state.................62
general qudit state.................64
generalized DJ algorithm..........300

GHZ state...................160, 367
GHZ states.......................218
graph coloring problem...........337
ground state......................92

H

Hadamard basis....................66
Hadamard gate...................118
hidden variable theory............228
Holevo (HSW) theorem..........440
Holevo's bound...................440
holonomic quantum computation.398

I

incoherent errors..................371
    amplitude damping error.....376
    bit-flip error.................372
    dephasing error..............375
    depolarizing error............375
    Pauli channel error...........376
    phase-flip error..............374
    two-qubit bit-flip error.......376
    two-qubit dephasing error....376
    two-qubit Pauli channel......376
informationally secure vs. computa-
    tionally secure...........330
Ising model..................169, 310

J

joint entropy.................430, 434
joint measurements...............154
joint probability..................413

K

Kraus decomposition.............359

L

Larmor precessing.................96
lattice-based cryptography........333
logical gates.....................390

M

Mach-Zehnder interferometer......31

magnetic moment of spin 45
marginal probability 413
max-cut problem 307
measurement basis 76
measurement errors 377
measurement framework 72
measurement probability 24
measurement state collapse 26
mixed state 347
monogamy of entanglement 367
multi-qubit gates 175
    CNOT (Controlled NOT) 178
    Controlled-U 180
    CZ (Controlled Z) 183
    interaction gates 183
    ZZ gate 183
multi-qubit systems 143, 156
    basis states 157
    entangled states 160
    Hamiltonian 167
    measurements 162
    product states 159
    time evolution 167
mutual information 432, 435

N

nine-qubit Shor code 389
NISQ era 343
NISQ/hybrid algorithms 305
no-cloning theorem 114, 138, 367
no-communication theorem .. 240, 367
Nobel prize in physics, 2022 239
number partitioning problem 336

O

operator diagonalization 70
operator-sum representation 359
orthogonal qubit state 62

P

parallel gates 184
parity measurements 394
partial measurement 163
partial product notation 163
partial trace 362
Pauli gates 117
performance metrics 377
    algorithmic qubits (AQ) 382
    application footprint 383
    circuit depth 381
    CLOPS 382
    coherence time (T_2) 378
    coherence to gate time ratio .. 379
    composite metrics 381
    connectivity 379
    cross-talk 380
    error rate 381
    fidelity 380
    gate time 379
    QECC decoding complexity .. 383
    quantum volume (QV) 381
    qubit lifetime 379
    relaxation time (T_1) 378
    scalability metrics 383
    subthreshold scaling 383
    the Λ factor 383
phase gates 119
phase kickback 304
phase oracle 303
photon circular polarization 15
photon cross-polarizer experiment 9, 28
photon diagonal polarization 13
photon doubel-slit experiment 33
photon elliptical polarization 16
photon polarization 7
photon polarization observables ... 20
photon rectilinear polarization 11
photons 7
post-measurement ensemble 356
post-quantum cryptography 333
POVM measurements 74, 358
principle of entropy increase 438
projective measurement 74
pure state 346

Q

QAOA algorithm 314
QM postulates 5
    composite systems 157
    quantum measurements 24
    quantum observables 17
    quantum states 17
    time evolution 86
quantum algorithm categories 295
quantum annealing 95, 319
quantum bomb test algorithm 322
quantum channel 359

quantum channel capacity 440
quantum circuit diagram 124
quantum communication
    BB84 QKD protocol 128
    channel capacity 440
    cryptography 333
    E91 QKD protocol 283
    entanglement swapping 266
    entropy 426
    entropy increase principle 438
    quantum gate teleportation .. 270
    quantum teleportation 257
    superdense coding 254
quantum error correction 341
quantum gate teleportation
    Bell measurement 273
    classical communication 274
    Clifford group 275
    CNOT gates 279, 281
    from standard teleportation .. 271
    initial state 273
    Pauli gates 275, 276
    single-qubit gate 271
    state reconstruction 274
    two-qubit gates 277, 279
quantum information 405
quantum interference 31
quantum measurement 71
quantum measurement postulate... 24
quantum money 325
    adaptive attack 328
    challenge-and-response prot. .. 330
    learning attack 327
    private key protocol 326
    public key implementation ... 331
    public key protocol 330
    replication attack 327
    Wiesner quantum money 326
quantum observable postulate 17
quantum probability 408, 410
    distinctive features 408
    entangled states 419
    expected values 412
    joint distribution 415
    marginal probability 416
    multi-variable 415
    non-commuting observables .. 416
    observables 410
    probability distribution 411
    random variables 412
quantum random number generator 126
quantum state postulate 17
quantum state tomography 80
quantum stochastic process 421
quantum teleportation
    basis state approach 258
    Bell measurement 263
    Bell measurement approach .. 262
    classical communication 264
    deferred measurement 259
    establish entanglement .. 259, 261
    logic operation approach 260
    procedure 257
    state reconstruction 259, 262, 264
quantum thermodynamics 439
quantum utility era 343
quantum walk 424
qudit framework 59

R

Rabi oscillations 99
readout errors 377
reduced density operator 362
rotation gates 120

S

sampling errors 77
Schrödinger equation 87, 167
Shannon entropy 427
SPDC photon pair 213
spin 44
spin measurement 51
spin precessing 96
spin-1/2 particles 44
stabilizers 394
state preparation errors 377
stationary states 91
statistical average 29
Stern-Gerlach experiment 52
    cascated 54
Stinespring dilation 361
subadditivity 443
superdense coding
    decoding 256
    encoding 255
superluminal communication 240
superposition 12

T

the circuit model 123
three-qubit bit-flip code 385
three-qubit phase-flip code 388
time evolution of qubit systems 85
    operator 87
    postulate 86
Toffoli (CCNOT) gate 200
topological quantum computation . 44, 162, 398
trace preserving operations 358
traveling salesman problem 317
Trotterization 90
two-path interference 31
two-qubit rotation gates 183
two-qubit systems 144
    entangled states 147
    general states 147
    joint measurements 154
    local measurements 148
    measurements 148
    parity measurements 154
    product states 147

U

uncertainty inequality 35
uncertainty principle 33
unified rotation gate 122
unitary evolution properties 87
universal gate set 185
    Clifford group $+\ T$ 186
    Toffoli $+\ H$ 186
universal quantum computing 94

V

von Neumann entropy 429
    strong subadditivity 444
von Neumann equation 355
VQE algorithm 311

W

W state 367

NOTES

NOTES

Journey Forward

A Journey Well Traveled

Congratulations to all of you who have navigated the complexities of this book, with a special nod to those who delved into the exercises. You've successfully laid a solid foundation in quantum computing and quantum information science, equipping yourselves for further endeavors in quantum exploration.

The Quantum Horizon

As with the dawn of classical computing, we stand at a significant juncture in the evolution of quantum computing. However, the path ahead is not without its challenges. Achieving scalability, mitigating noise, and ensuring fault tolerance remain the holy grails of the field. While we have yet to fully realize these goals, the quantum community remains optimistic, knowing that transformative breakthroughs may be just around the corner.

Exploring quantum algorithms opens doors to myriad applications in optimization, machine learning, and quantum simulations. However, their full potential will be unlocked only alongside advancements in quantum hardware, with technologies like superconducting qubits, neutral atoms, and trapped ions leading the charge.

Confronted with the challenges of decoherence and noise, research into quantum error correction becomes indispensable, ensuring the transition from theoretical ideas to practical, fault-tolerant quantum operations. Beyond the immediate realm of computation, disciplines such as quantum information theory, sensing, and post-quantum cryptography enrich the broader quantum narrative, providing layered insights into this ever-evolving domain.

To remain relevant in this dynamic field, staying updated is crucial. From groundbreaking work in quantum cryptography to the emergence of innovative algorithms and their applications, the story of quantum science is continually unfolding. Armed with the foundational knowledge from this book, you are well-positioned to both witness and contribute to the future chapters of this quantum tale.

Looking Forward

As you progress, consider:

- Delving deeper into quantum algorithms, which offer solutions across varied application domains.
- Keeping pace with the latest developments in quantum computing, including hardware innovations, error correction methods, advancements in cryptography, and emerging algorithms.
- Exploring the broader canvas of quantum information science, where disciplines like quantum sensing and post-quantum cryptography add additional depth.

In conclusion, while this book provides a sturdy foundation, the journey is ongoing. With relentless curiosity and continuous learning, traverse the nuances of the quantum world. Wishing you a quantum voyage that is as enlightening as the principles upon which it is built.

Made in United States
Troutdale, OR
06/20/2024